A History of
the Supreme Court

A History of
the Supreme Court

BERNARD SCHWARTZ

New York   Oxford
OXFORD UNIVERSITY PRESS
1993

Oxford University Press

Oxford   New York   Toronto
Delhi   Bombay   Calcutta   Madras   Karachi
Kuala Lumpur   Singapore   Hong Kong   Tokyo
Nairobi   Dar es Salaam   Cape Town
Melbourne   Auckland   Madrid

and associated companies in
Berlin   Ibadan

Copyright © 1993 by Bernard Schwartz

Published by Oxford University Press, Inc.
200 Madison Avenue, New York, New York 10016

Oxford is a registered trademark of Oxford University Press

Library of Congress Cataloging-in-Publication Data
Schwartz, Bernard, 1923–
A history of the Supreme Court / Bernard Schwartz.
p.   cm.   Includes bibliographical references and index.
ISBN 0–19–508099–8
1. United States.   Supreme Court—History.   I. Title.
KF8742.S39   1993   347.73'26'09—dc20   [347.3073509]   92–44097

2 4 6 8 9 7 5 3 1

Printed in the United States of America
on acid-free paper.

Semper uxori suae

Preface

There is no good one-volume history of the United States Supreme Court. The Holmes Devise history is too massive to be usable except for scholars engaged in research, and other histories, such as that by Robert McCloskey, are too short and hence largely superficial. I hope that this book will correct the situation. It tells the story of the Court over the years since its first session in 1790. The emphasis is on the history of the Court in relation to the development of the nation. The theme is that of the Court as both a mirror and a motor—reflecting the development of the society which it serves and helping to move that society in the direction of the dominant jurisprudence of the day.

The organization of the book is chronological. In addition to the chapters on the Court under the different Chief Justices, there are chapters on four watershed cases—landmark cases that bring into sharp focus both how the Court operates and the impact of major decisions upon the nation and on the Court itself. My hope is that the audience for this book will not be limited to specialists in the subject. I have tried to write in a nontechnical manner to make the work less forbidding to the general reader who is normally "turned off" by a book written by a law professor. War, according to the famous aphorism, is too important a matter to be left to the generals. The work of the Supreme Court is similarly too significant in a country such as ours to be left only to the lawyers and law professors. This is particularly true of the historical functioning of the highest tribunal. It is scarcely possible to understand American history fully without an understanding of the part played in that history by the Supreme Court. In so many cases, the decisions of the Court have become a vital part of the story of the nation's development.

I will be more than rewarded for my efforts if this survey of the Supreme Court's history proves useful to the growing number of those who desire to learn more about the institution that plays such a vital part in the polity.

Tulsa, Okla.   B. S.
1993

Contents

A History of
the Supreme Court

Introduction:
"The Very Essence of
Judicial Duty"

"Human history," says H. G. Wells, "is in essence a history of ideas."[1] To an American interested in constitutional history, the great theme in the country's development is the idea of law as a check upon governmental power. The institution that best embodies this idea is the United States Supreme Court. But the Court itself is the beneficiary of a constitutional heritage that starts centuries earlier in England.

Seedtime of Judicial Review

Chief Justice John Marshall tells us that the power to determine constitutionality "is of the very essence of judicial duty."[2] To an American interested in the development of the Marshall concept, as good a starting point as any is the dramatic assertion of the supremacy of law by Sir Edward Coke on November 13, 1608.[3] For it was on that day that James I confronted "all the Judges of England and Barons of the Exchequer" with the claim that, since the judges were but his delegates, he could take any case he chose, remove it from the jurisdiction of the courts, and decide it in his royal person. The judges, as James saw it, were "his shadows and ministers . . . and the King may, if he please, sit and judge in Westminster Hall in any Court there and call their Judgments in question."[4]

"To which it was answered by me," states Chief Justice Coke, "in the presence, and with the clear consent of all the Judges . . . that the King in his own person cannot adjudge any case . . . but that this ought to be determined and adjudged in some Court of Justice, according to the law and custom of England." To this James made the shrewd reply "that he thought the law was founded upon reason, and that he and others had reason as well as the Judges."

Coke then delivered his justly celebrated answer, "that true it was, that God had endowed His Majesty with excellent science, and great endowments of nature; but His Majesty was not learned in the laws of his realm of England, and causes which concern the life, or inheritance, or goods, or fortunes of his subjects, are not to be decided by natural reason but by the artificial reason and judgment of law, which law is an act which requires long study and experience, before that a man can attain to the cognisance of it: that the law was the golden met-wand and measure to try the causes of the subjects."[5]

It is hardly surprising that the King was, in Coke's description "greatly offended." "This means," said James, "that I shall be under the law, which it is treason to affirm." "To which," replied Coke, "I said, that Bracton saith, *quod Rex non debet esse sub homine, sed sub Deo et lege* [that the King should not be under man but under God and law]."

Needless to say, the King's anger only increased. According to one onlooker, in fact, "[H]is Majestie fell into that high indignation as the like was never knowne in him, looking and speaking fiercely with bended fist, offering to strike him, etc."[6]

James's indignation was well justified. Coke's articulation of the supremacy of law was utterly inconsistent with royal pretensions to absolute authority. In the altercation between Coke and the King, indeed, there is personified the basic conflict between power and law which underlies all political history. Nor does it affect the importance of Coke's rejection of James's claim that, with the King's fist raised against him, Coke was led personally to humble himself. That he "fell flatt on all fower"[7] to avoid being sent to the Tower does not alter the basic boldness of his clear assertion that the law was supreme even over the Crown.

Nor did Coke stop with affirming that even the King was not above the law. In *Dr. Bonham's Case*[8]—perhaps the most famous case decided by him—Coke seized the occasion to declare that the law was above the Parliament as well as above the King. Dr. Bonham had practiced physic without a certificate from the Royal College of Physicians. The College Censors committed him to prison, and he sued for false imprisonment. The college set forth in defense its statute of incorporation, which authorized it to regulate all physicians and punish with fine and imprisonment practitioners not admitted by it. The statute in question, however, gave the college one-half of all the fines imposed. This, said Coke, made the college not only judges, but also parties, in cases coming before them, and it is an established maxim of the common law that no man may be judge in his own cause.

But what of the statute, which appeared to give the college the power to judge Dr. Bonham? Coke's answer was that even the Parliament could not confer a power so contrary to common right and reason. In his words, "[I]t appears in our books, that in many cases, the common law will controul Acts of Parliament, and sometimes adjudge them to be utterly void: for when an Act of Parliament is against common right and reason, or repugnant, or impossible to be performed, the common law will controul it, and adjudge such Act to be void."[9]

Modern scholars have debated the exact meaning of these words. To the men of the formative era of American constitutional history, on the other hand, the meaning was clear. Chief Justice Coke was stating as a rule of positive law that there was a fundamental law which limited Crown and Parliament indifferently. Had not my Lord Coke concluded that when an Act of Parliament is contrary to that fundamental law, it must be adjudged void? Did not this mean that when the British government acted toward the Colonies in a manner contrary to common right and reason, its decrees were of no legal force?

The men of the American Revolution were nurtured upon Coke's writings. *"Coke's Institutes,"* wrote John Rutledge of South Carolina, "seem to be almost the foundation of our law."[10] Modern writers may characterize Coke as an obsolete writer whom the British Constitution has outgrown. Americans of the eighteenth century did not have the benefit of such ex post facto criticism. To them, Coke was the contemporary colossus of the law—"our juvenile oracle," John Adams termed him in an 1816 letter[11]—who combined in his own person the positions of highest judge, commentator on the law, and leader of the Parliamentary opposition to royal tyranny. Coke's famous *Commentary upon Littleton,* said Jefferson, "was the universal elementary book of law students and a sounder Whig never wrote nor of profounder learning in the orthodox doctrines of . . . British liberties."[12] When Coke, after affirming the supremacy of the law to royal prerogative, announced, "It is not I, Edward Coke, that speaks it but the records that speak it,"[13] men on the western side of the Atlantic took as the literal truth his assertion that he was only declaring, not making law.

Coke's contribution to constitutionalism was thus a fundamental one. He stated the supremacy of law in terms of positive law. And it was in such terms that the doctrine was of such import to the Founders of the American Republic. When they spoke of a government of laws and not of men, they were not indulging in mere rhetorical flourish.

Otis and Unconstitutionality

The influence of Coke may be seen at all of the key stages in the development of the conflict between the Colonies and the mother country. From Whitehall Palace, to which King James had summoned the judges in 1608, to the Council Chamber of the Boston Town House a century and a half later was not really so far as it seemed. "That council chamber," wrote John

Adams over half a century after the event, "was as respectable an apartment as the House of Commons or the House of Lords in Great Britain. . . . In this chamber, round a great fire, were seated five Judges, with Lieutenant Governor Hutchinson at their head, as Chief Justice, all arrayed in their new, fresh, rich robes of scarlet English broadcloth; in their large cambric bands, and immense judicial wigs."[14] For it was in this chamber that, in 1761, James Otis delivered his landmark attack in *Lechmere's Case*[15] against general writs of assistance.

The Otis argument in *Lechmere's Case* has been characterized as the opening gun of the American Revolution. In it, Otis with "a torrent of impetuous eloquence . . . hurried away everything before him." He argued the cause, Otis declared, "with the greater pleasure . . . as it is in opposition to a kind of power, the exercise of which, in former periods of English history, cost one King of England his head, and another his throne."[16] If Patrick Henry came close to treason in his famous 1765 speech attacking the Stamp Act, he at least had an excellent model in this Otis speech.

To demonstrate the illegality of the writs of assistance, Otis went straight back to Coke. As Horace Gray (later a Justice of the Supreme Court) put it in an 1865 comment, "His main reliance was the well-known statement of Lord Coke in *Dr. Bonham's Case*."[17] This may be seen clearly from John Adams's summary of the Otis argument: "As to acts of Parliament. An act against the Constitution is void: an Act against natural Equity is void; and if an Act of Parliament should be made in the very words of the petition, it would be void. The . . . Courts must pass such Acts into disuse."[18]

The Otis oration, exclaimed Adams, "breathed into this nation the breath of life," and "[t]hen and there the child Independence was born."[19] To which we may add that then and there American constitutional law was born. For Otis, in Justice Gray's words, "denied that [Parliament] was the final arbiter of the justice and constitutionality of its own acts; and . . . contended that the validity of statutes must be judged by the courts of justice; and thus foreshadowed the principle of American constitutional law, that it is the duty of the judiciary to declare unconstitutional statutes void."[20]

Coke's biographer tells us that he would have been astonished at the uses to which *Dr. Bonham's Case* was put.[21] Certain it is that Otis and those who followed in his steps went far beyond anything the great English jurist had expressly intended. Yet had not Coke's own attitude been stated in his picturesque phrase: "Let us now peruse our ancient authors, for out of the old fields must come the new corne."?[22] That is precisely what Americans have done in using Coke as the foundation for the constitutional edifice which, starting with Otis's argument, they have erected. Coke himself would not have been disturbed by the fact that, though the fields were old, the corn was new.

State Precursors

Throughout the Revolutionary period, Americans relied upon their possession of the rights of Englishmen and the claim that infringement upon those rights was unconstitutional and void. That claim could not, however, rest upon a secure legal foundation until the rights of Americans were protected in written organic instruments. Such protection came with the adoption of written constitutions and bills of rights in the states, as soon as independence had severed their ties with the mother country.

How were the rights guaranteed by these new constitutions to be enforced? The American answer to this question was, of course, ultimately, judicial review. That answer was first given during the period between the Revolution and the ratification of the Federal Constitution. By the end of the period, an increasing number of Americans accepted the view that laws might "be so unconstitutional as to justify the Judges in refusing to give them effect."[23] Oliver Ellsworth, later the third Chief Justice of the United States, was stating far from radical doctrine when he asserted in the 1788 Connecticut ratifying convention, "If the United States go beyond their powers, if they make a law which the Constitution does not authorize, it is void and the judicial power . . . will declare it to be void."[24]

Between 1780 and 1787 cases in a number of states saw direct assertions of judicial power to rule on constitutionality. There has been some dispute about whether these cases really involved judicial review. Much of the difficulty in assessing their significance arises from the fact that no meaningful reporting of cases in the modern sense existed at the time these cases were heard and decided. Reported opinions were mainly skimpy or nonexistent. For most of these early cases, recourse has to be had to other materials (such as newspapers and pamphlets) rather than to law reports of the modern type.

The first of the pre-Constitution review cases was the 1780 New Jersey case of *Holmes v. Walton*.[25] A 1778 statute, aimed at traffic with the enemy, permitted trial by a six-man jury and provided for punishment by property seizures. The statute was attacked on the ground that it was "contrary to the constitution of New Jersey." The claim was upheld by the court, though the actual decision has been lost. From other materials, it appears that the decision was based on the unconstitutionality of the six-man jury.[26] Some recent commentators have attacked the conclusion that *Holmes v. Walton* set a precedent for judicial review. It was, however, widely thought of as such at the time the Federal Constitution and Bill of Rights were adopted. Soon after the case was decided "a petition from sixty inhabitants of the county of Monmouth" was presented to the New Jersey Assembly. It complained that "the justices of the Supreme Court have set aside some of the laws as unconstitutional." In 1785, Gouverneur Morris sent a message to the Pennsylvania legislature that mentioned "a law as once passed in New Jersey, which the judges pronounced unconstitutional, and therefore void."[27] In addition, there is an 1802 case that states that in *Holmes v. Walton,* an "act upon solemn argument was adjudged to be unconstitutional and in that case inoperative."[28] At the least, these indicate that comtemporaries did regard *Holmes v. Walton* as a precedent for judicial review.

The second case involving judicial review was *Commonwealth v. Caton,*[29] decided in 1782 by the Virginia Court of Appeals. It has been widely assumed, relying on the report of the case in Call's Virginia Reports, that *Caton* was the strongest early precedent for judicial review. The language in Call is, indeed, unequivocal. "[T]he judges, were of opinion, that the court had power to declare any resolution or act of the legislature, or of either branch of it to be unconstitutional and void; and, that the resolution of the house of delegates, in this case, was inoperative, as the senate had not concurred in it."[30]

Call's report on *Caton* was not published until 1827; it was based upon the reporter's reconstruction of the case from surviving records, notes, and memoranda. There are significant differences between the Call report and the contemporary notes of Edmund Pendleton, who presided over the *Caton* court. According to Pendleton's account, only one of the eight judges ruled that the statute at issue was unconstitutional, though two others did assert judicial power to declare a law void for repugnancy to the Constitution. The two judges in question were Chancellor George Wythe and Pendleton himself. Wythe, perhaps the leading jurist of the day, delivered a ringing affirmation of review authority, declaring that if a statute conflicted with the Constitution, "I shall not hesitate, sitting in this place, to say, to the general court, Fiat justitia, ruat coelum; and, to the usurping branch of the legislature, you attempt worse than a vain thing."[31] Pendleton also stated that the "awful question" of voiding a statute was one from which "I will not shrink, if ever it shall become my duty to decide it."[32]

The *Caton* Court did not exercise the power to hold a law unconstitutional; the majority held "that the Treason Act was not at Variance with the Constitution but a proper exercise of the Power reserved to the Legislature by the latter."[33] Yet three judges did assert power in the courts to void statutes on constitutional grounds, including the two most prestigious members of the court. And it was Wythe's words, in particular, Pendleton's biographer tells us, that were "preserved in the court reports, and they were never forgotten by lawyers and students of government, by whom they were repeated again and again to men who would arrogate to themselves unconstitutional powers or seek to circumvent constitutional limitations."[34]

The most noted of the pre-Constitution review cases[35] was the 1784 New York case of *Rutgers v. Waddington.*[36] It was noted in its day because of Alexander Hamilton's argument for the defendant and because the court's opinion, published at the time, made a considerable stir. It is the best documented of these early cases. Strictly speaking, *Rutgers v. Waddington* did not involve a review of constitutionality but only of judicial power to annul a state statute contrary to a treaty and the law of nations. The statute in question provided for a trespass action against those who had occupied property during the British occupation of New York and barred defendants from any defense based on the following of military orders. Waddington was a British merchant who had occupied Mrs. Rutgers's abandoned property under license of the Commander-in-Chief of the British army of occupation. Hamilton argued that the statutory bar was in conflict both with the law of nations (since defendant had occupied the premises under British authority and thus derived the right of the military occupier over abandoned property sanctioned by the law of war) and the peace treaty with Britain. As stated by the editor of Hamilton's legal papers, he urged that "a court must apply the law that related to a higher authority in derogation of that which related to a lesser when the two came in conflict."[37]

The court agreed that the statute could not override a treaty or international law and refused to apply it to the extent that there was any conflict. Whether or not *Rutgers v. Waddington* may be regarded as a precedent for judicial review, its lesson was not lost on the Framers' Convention; by the Supremacy Clause state judges were directed to set aside state laws that conflicted with treaties. Certainly, Hamilton's assertion of review power in the courts made *Rutgers v. Waddington* "a marker on the long road that led to judicial review."[38]

The next case involving judicial review was the 1786 Rhode Island case of *Trevett v. Weeden*. That case, too, was unreported, but it was widely known through a 1787 pamphlet published by James M. Varnum (better known as one of Washington's generals), who argued the case against the statute.[39] Varnum's argument received wide dissemination and demonstrated the unconstitutionality of a legislative attempt to deprive Weeden of his right to trial by jury. Weeden, a butcher, was prosecuted under a statute making it an offense to refuse to accept paper money of the state in payment for articles offered for sale—in this case, meats. Appearing for the defense Varnum resorted to the modern distinction between the constitution and ordinary statute law,[40] arguing that the principles of the constitution were superior because they "were ordained by the people anterior to and created the powers of the General Assembly." It was the duty of the courts to measure laws of the legislature against the constitution. The judiciary's task was to "reject all acts of the Legislature that are contrary to the trust reposed in them by the people."[41] That the Rhode Island judges agreed with Varnum is shown by the following brief newspaper account:

> The court adjourned to next morning, upon opening of which, Judge Howell, in a firm, sensible, and judicious speech, assigned the reasons which induced him to be of the opinion that the information was not cognizable by the court, declared himself independent as a judge, the penal law to be repugnant and unconstitutional, and therefore gave it as his opinion that the court could not take cognizance of the information! Judge Devoe was of the same opinion. Judge

Tillinghast took notice of the striking repugnancy of the expressions of the act . . . and on that ground gave his judgment the same way. Judge Hazard voted against taking cognizance. The Chief Justice declared the judgment of the court without giving his own opinion.[42]

The clearest pre-Constitution case involving review power was the North Carolina case of *Bayard v. Singleton*,[43] decided in May 1787, just before the Philadelphia Convention. The contemporary account in the North Carolina Reports shows that the judges there realized the implications of what they were doing when they held that a statute contrary to the guaranty of trial by jury in cases involving property in the North Carolina Declaration of Rights "must of course . . . stand as abrogated and without any effect." No "act they could pass, could by any means repeal or alter the constitution" so long as the constitution remains "standing in full force as the fundamental law of the land."[44]

James Iredell, later a Justice of the Supreme Court, had been attorney for the plaintiff in *Bayard v. Singleton*. While attending the Framers' Convention, Richard Dobbs Spaight wrote to Iredell condemning the *Bayard* decision as a "usurpation," which "operated as an absolute negative on the proceedings of the Legislature, which no judiciary ought ever to possess." Iredell replied that "it has ever been my opinion, that an act inconsistent with the Constitution was void; and that the judges, consistently with their duties could not carry it into effect." Far from a "usurpation," the power to declare unconstitutional laws void flowed directly from the judicial duty of applying the law: "[E]ither . . . the *fundamental unrepealable* law must be obeyed, by the rejection of an act unwarranted by and inconsistent with it, or you must obey an act founded on authority not given by the people." The exercise of review power, said Iredell, was unavoidable. "It is not that the judges are appointed arbiters . . . but when an act is necessarily brought in judgment before them, they must, unavoidably, determine one way or another. . . . Must not they say whether they will obey the Constitution or an act inconsistent with it?"[45]

To be sure, these pre-Constitution assertions of review power did not go unchallenged. After *Rutgers v. Waddington,* the New York Assembly passed a resolution attacking the asserted power of the courts.[46] An open letter in a newspaper went even further, asserting, "That there should be a power vested in courts of judicature whereby they might controul the supreme Legislative power we think is absurd in itself. Such power in courts would be destructive of liberty, and remove all security of property."[47]

The reaction to *Trevett v. Weeden* was even stronger. The Rhode Island Legislature ordered the judges to appear before it "to render their reasons for adjudging an act of the General Assembly unconstitutional, and so void."[48] Then, as Madison was to explain it to the Framers' Convention, "In Rhode Island the judges who refused to execute an unconstitutional law were displaced, and others substituted, by the Legislature who would be willing instruments of the wicked and arbitrary plans of their masters."[49]

The important thing, however, is that despite such opposition, the judges did exercise the review power during the pre-Constitution period. The judicial groundwork was thus laid for the assertion of the power that has made the U.S. Supreme Court the fulcrum of our constitutional system.

Constitution and Ratification

The men who came to Philadelphia in the sultry summer of 1787 had the overriding aim of making such alterations in the constitutional structure as would, in the words of the Confederation Congress calling the convention, "render the Federal Constitution adequate to the exigencies of Government."[50] To accomplish that goal, they drafted a new charter providing for a Federal Government endowed with the authority needed to enable it to operate effectively.

The Articles of Confederation had concentrated all the governmental authority provided for under it in a unicameral legislative body. Before the Constitution, "Congress was the general, supreme, and controlling council of the nation, the centre of union, the centre of force, and the sun of the political system."[51] In the Confederation, there was no separate Executive and the only federal courts were those Congress might set up for piracy and felony on the high seas and for appeals in prize cases.

From the beginning of their deliberations, the Framers agreed that the new government they were creating should be based upon the separation of powers. It was, Madison told the Convention, "essential . . . that the Legislative: Executive: and Judicial powers be separate . . . [and] independent of each other."[52] Accordingly, the Virginia Plan drafted by him, which served as the basis for the new Constitution, provided expressly: "That a national government ought to be established consisting of a supreme legislative, judiciary and executive."[53]

In basing their deliberations upon Madison's plan, the Framers decided, at almost the outset of their deliberations, that there should be a federal judiciary and that it should be "supreme." The resolutions introduced to give effect to the plan provided that "a National Judiciary be established," to consist of both a supreme tribunal and inferior tribunals.[54] The federal courts were thus to be modeled upon the colonial and state court systems—consisting, as they did, of both inferior courts and a central high court.

Although attention was paid to the judiciary during the Convention debates, "[t]o one who is especially interested in the judiciary, there is surprisingly little on the subject to be found in the records of the convention."[55] The only serious objection was that to inferior federal courts, which some saw as an encroachment upon the states. The difficulty was resolved by a compromise: inferior courts were not required, but Congress was permitted to create them.[56]

The Framers were, of course, familiar with the preindependence judicial system, under which appeals could be taken to a central appellate tribunal. The proposal for a Federal Supreme Court was adopted with practically no discussion. The Convention considered and rejected a number of motions that would have been inconsistent with the judicial function and might have impaired the independence of the new supreme tribunal—notably one to set up a Council of Revision composed of the Executive and "a convenient number of the National Judiciary" to veto acts passed by Congress, as well as one "that all acts before they become laws should be submitted both to the Executive and Supreme Judiciary Departments."[57] There was debate about who should appoint the federal judges as well as their jurisdiction. In the end, however, the Judiciary Article was adopted essentially as it had been drafted by those who sought a strong national government—especially by its principal draftsman, Oliver Ellsworth.

The Constitution does not, to be sure, specifically empower the federal courts to review the constitutionality of laws. But it does contain provisions upon which judicial review authority can be based. The jurisdiction vested in the new judiciary extends to all "Cases . . . arising under this Constitution, the Laws of the United States, and Treaties." Even more important was the Supremacy Clause of article VI, added on motion of John Rutledge, who was appointed to the first Supreme Court and also served briefly as the second Chief Justice. The supremacy of the Constitution was expressly proclaimed as the foundation of the constitutional structure. And since only "Laws . . . made in pursuance" of the Constitution were given the status of "supreme law," laws repugnant to the Constitution were excluded from the imperative of obedience.

At various times during their debates, the Framers asserted that, in Elbridge Gerry's words, "the Judiciary . . . by their exposition of the laws" would have "a power of deciding on their Constitutionality."[58] Those who did so included Gerry himself, James Wilson,[59] and James Madison,[60] as well as opponents of the Constitution such as Luther Martin[61] and George Mason.[62] Even those who were troubled by such a power in the courts conceded that they saw no workable alternative. As John Dickinson put it, "He thought no such power ought to exist. He was at the same time at a loss what expedient to substitute."[63]

Madison, too, saw dangers in judicial review, which, he said in 1788, "makes the Judiciary Department paramount in fact to the Legislature, which was never intended and can never be proper."[64] Yet Madison also concluded his discussion of the matter by stating, "A law violating a constitution established by the people themselves, would be considered by the Judges as null and void."[65]

During the ratification debates of 1787–1788, both supporters and opponents of the Constitution assumed that judicial review would be an essential feature of the new organic order. "If they were to make a law not warranted by any of the powers enumerated," declared John Marshall in the Virginia ratifying convention, "it would be considered by the judges as an infringement of the Constitution which they are to guard. . . . They would declare it void."[66]

The Anti-Federalist *Brutus* letters, which Hamilton sought to answer in *The Federalist,* agreed with Marshall. If Congress, Brutus wrote, should "pass laws, which, in the judgment of the court, they are not authorized to do by the constitution, the court will not take notice of them . . . they cannot therefore execute a law, which, in their judgment, opposes the constitution."[67]

To opponents of the Constitution, judicial review was one of the new instrument's great defects. Because of it, Brutus wrote, "I question whether the world ever saw . . . a court of justice invested with such immense powers" as the Supreme Court.[68] There would be no power to control their decisions. In such a situation and vested with full judicial independence, the new Justices would "feel themselves independent of Heaven itself."[69]

The most effective defense of the federal judiciary and its review power was, of course, that by Hamilton in *The Federalist*—itself called forth by the challenge presented by the *Brutus* argument. Most important for our purposes was the defense of judicial review in No. 78 of *The Federalist.* Hamilton's essay there stands as the classic pre-Marshall statement on the subject. American constitutional law has never been the same since it was published.

Hamilton's *Federalist* reasoning on review is based upon the very nature of the Constitution as a limitation upon the powers of government. "Limitations of this kind can be preserved in practice in no other way than through the medium of the courts of justice; whose duty it must be to declare all acts contrary to the manifest tenor of the constitution void. Without this, all the reservations of particular rights or privileges would amount to nothing."[70]

The courts, Hamilton urged, were designed to keep the legislature within constitutional limits. "The interpretation of the laws is the proper and peculiar province of the courts. A constitution is, in fact, and must be regarded by the judges as a fundamental law. It must therefore belong to them to ascertain its meaning, as well as the meaning of any particular act proceeding from the legislative body. If there should happen to be an irreconcilable variance between the two, that which has the superior obligation and validity ought, of course, to be preferred: in other words, the constitution ought to be preferred to the statute; the intention of the people to the intention of their agents."[71]

Hamilton's reasoning here, and even his very language, formed the foundation for the *Marbury v. Madison*[72] confirmation of judicial review as the core principle of the constitutional system. The *Marbury* opinion can, indeed, be read as more or less a gloss upon *The Federalist,* No. 78.

Judiciary Act

The Constitution's Judiciary Article was, of course, not self-executing. Before the federal courts, including the Supreme Court, could come into existence, they had to be provided for by statute. The first Congress passed the necessary law when it enacted the Judiciary Act of 1789—the law that both created the Supreme Court and set forth its jurisdiction. The key members of the Senate committee that drafted the statute were Oliver Ellsworth, later the third Chief Justice, and William Paterson, who was to become a Supreme Court Justice. The principal draftsman was Ellsworth; according to a senatorial opponent, "this Vile Bill is a child of his."[73]

The 1789 statute resolved the issue of whether there should be inferior federal courts in favor of their creation. It established the federal judiciary with a Supreme Court, consisting of six Justices, at its apex, and a two-tiered system of inferior courts, with district courts in each state at the base, and three circuit courts grouped into three circuits—the eastern, the middle, and the southern, each composed of two Supreme Court Justices and one district judge. The federal courts were given only limited jurisdiction; as it was put by a member of the Senate drafting committee, "it will not extend to a tenth part of the causes which *might* by the constitution come into the Federal Court."[74]

There was again little discussion on the establishment of the Supreme Court, though some opponents sought to reduce the number of Justices and even to dispense with a Chief Justice. The jurisdiction of the Court was provided for in the form it has retained throughout its history. Crucial was the fact that the Court was given appellate jurisdiction not only over the lower federal courts, but also, under section 25 of the Judiciary Act, over the state courts in cases involving federal questions. "A vital chapter of American history," Justice Frankfurter tells us, "derives from the famous twenty-fifth section of the Judiciary Act."[75] From 1789 to our own day, the Supreme Court's power to review state court decisions has been what the Court historian characterized as "the keystone of the whole arch of federal judicial power."[76] Because of section 25, indeed, William Paterson could state, in his notes on the Judiciary Bill debate, "The powers of the Supreme Court are great—they are to check the excess of Legislation."[77]

With the passage of the first Judiciary Act, the stage was set for the Supreme Court to play its part in the unfolding drama of the new nation's development. The actual scenario would, however, depend upon the personnel of the new tribunal and the manner in which they performed their awesome constitutional role.

1

The First Court, 1790–1801

Chief Justice Warren E. Burger once said "that he himself should be in a wig and gown, and had been cheated out of it by Thomas Jefferson."[1] The question of the Justices' attire was a controversial issue when the United States Supreme Court first met. As was to be expected, Hamilton was for the English wig and gown. Jefferson was against both, but he said that if the gown was to be worn, "For Heaven's sake, discard the monstrous wig which makes the English judges look like rats peeping through benches of oakum!"[2]

Jefferson's opposition (as well as that by other public figures, including Aaron Burr) carried the day. One of the first Justices, William Cushing, came to New York for the first Court session wearing his old-fashioned judicial wig. His appearance with it caused a commotion: "The boys followed him in the street, but he was not conscious of the cause until a sailor, who came suddenly upon him, exclaimed, 'My eye! What a wig!'" Then, we are told, Cushing, "returning to his lodgings, . . . obtained a more fashionable covering for his head. He never again wore the professional wig."[3] Nor did any of the other Supreme Court Justices.

When, on February 2, 1790, the Supreme Court met in its first public session in the Royal Exchange, at the foot of Broad Street in New York City, the Justices did not wear wigs. But they were elegantly attired in black and red robes, "the elegance, gravity and neatness of which were the subject of remark and approbation with every spectator."[4] The elegance of the Justices' attire could, however, scarcely serve to conceal the relative ineffectiveness of the first Supreme Court, at least by comparison with

15

what that tribunal was later to become. To understand the Court's position, it is necessary to look at the new judicial department not through twentieth-century spectacles but through the eyes of men living a decade after the Constitution went into effect. "The judiciary," wrote Hamilton in *The Federalist*, "is beyond comparison the weakest of the three departments of power."[5] This remark was amply justified by the situation of the fledgling Supreme Court.

It is hard for us today to realize that, at the beginning at least, a seat on the supreme bench was anything but the culmination of a legal career that it has since become. John Jay, the first Chief Justice, resigned to become Governor of New York, and Alexander Hamilton declined Jay's post, being "anxious to renew his law practice and political activities in New York." John Rutledge resigned his seat on the first Supreme Court to become Chief Justice of the South Carolina Court of Common Pleas. "[S]ince Marshall's time," as Justice Felix Frankfurter tells us, "only a madman would resign the chief justiceship to become governor"—much less a state judge.[6]

The weakness of the early Supreme Court is forcefully demonstrated by the fact that, in the building of the new capital, that tribunal was completely overlooked and no chamber provided for it. When the seat of government was moved to Washington, the high bench crept into an undignified committee room in the Capitol beneath the House Chamber.

Getting Under Way

With the opening of the Supreme Court, the tripartite governmental structure provided by the Framers was at last fully operative. Two days after the first session, a New York merchant wrote to a British friend, "Our Supreme Court was Opened the 2 Instant . . . we [are] now in Every Respect a Nation."[7]

The Supreme Court itself is established directly by the Constitution, which provides expressly for the existence of "one supreme Court." The Court could not, however, come into operation until the details of its organization and operation were provided by Congress. The Judiciary Act of 1789 set up a Supreme Court consisting of a Chief Justice and five Associate Justices and set forth the jurisdiction vested in it. But the President still had to appoint its members before the Court could come into existence.

President Washington took his responsibility in nominating Supreme Court Justices most seriously. In a famous letter to Edmund Randolph, the first Attorney General, Washington declared, "Impressed with a conviction, that the due administration of justice is the firmest pillar of good government, I have considered the first arrangement of the judicial department as essential to the happiness of our country, and to the stability of its political system." Because of this, the President wrote in his letters to his Supreme Court appointees, "I have thought it my duty to nominate for the high offices in that department, such men as I conceived would give dignity and lustre to our national character."[8] Despite Washington's clear comprehension of the responsibility of making suitable appointments to the first Court, he found it most difficult to get men of stature to accept. The low prestige of the Court led a number of his first choices to prefer other positions.

John Jay was apparently the President's first choice as Chief Justice. A few months before the appointment, Vice President Adams had written, "I am fully convinced that Services, Hazard, Abilities and Popularity, all properly weighed, the Balance, is in favour of Mr. Jay."[9] Jay accepted the position, though he was concerned about the salary. While the appointment was pending, the Secretary of the Senate wrote to a Senator, "The Keeper of the Tower [i.e., Jay] is waiting to see which Salary is best, that of Lord Chief Justice or Secretary of State."[10]

In choosing the five Associate Justices, Washington followed the practice, since followed by many Presidents, of geographic representation. A letter he sent to Hamilton the day after his nominations emphasized the geographic dispersion of Court seats.[11] His choice for Chief Justice was from New York. The other appointees were from South Carolina (John Rutledge, Chief Justice of the state's Chancery Court), Pennsylvania (James Wilson, one of the first American law professors, who had played a leading part in the Constitutional Convention), Massachusetts (William Cushing, Chief Justice of the state's highest court), Maryland (Robert H. Harrison, Chief Judge of the Maryland General Court), and Virginia (John Blair, a judge on the Virginia Supreme Court of Appeals).

Though all the nominees were speedily confirmed by the Senate, Harrison declined the appointment because of poor health. He died two months later. "Poor Col. Harrison," wrote Washington to Lafayette, "who was appointed one of the Judges of the Supreme Court, and declined, is lately dead."[12] In Harrison's place, Washington chose James Iredell of North Carolina, who had served as a state judge and attorney general and had been a leader in the struggle for ratification of the Constitution in his state.

In a letter soon after his appointment, the new Justice described the functioning of the fledgling Court:

> There are to be 2 Sessions of the Supreme Court held at the seat of Gov. in each year & a Circuit Court twice a year in each State. The United States are divided into three Circuits—one, called the [Middle], consisting of New Jersey, Pennsylvania, Delaware, Maryland & Virginia; and the Southern, consisting at present of South Carolina & Georgia, to which I imagine North Carolina will be added. [There was also an Eastern Circuit, consisting of New York and the New England states.] The Circuit Courts are to consist of two Judges of the Supreme Court, and a Judge in each State appointed by the President who has in other respects a separate jurisdiction of a limited kind—Any two of these may constitute a Quorum.[13]

As already seen, the Supreme Court began its first sittings on February 2, 1790. During that session and the next two terms, there were no cases docketed for argument and the Justices had little to do. "There is little business but to organise themselves," wrote Congressman Abraham Baldwin, "and let folks look on and see they are ready to work at them."[14] De Witt Clinton confirmed this assessment. In a letter to his brother, he stated, "The Supreme Court of the U. States is now in session and ha[ve] done no other business than admitting a few Counsellors and making a few rules."[15]

One of the new Court's rules irked the young Clinton:

> One of their orders "that all process shall run in the name of the President" tho' apparently unimportant smells strongly of monarchy—You know that in G. Britain some writs are prefaced with "George the 3d by the Grace of God & c." A federal process beginning with "George Washington by the grace of God & c." will make the American President as important in Law forms as the British King.[16]

Despite the early Court's lack of business, Justice Iredell could still say when he was appointed, "The duty will be severe."[17] That was true because of the arduous duty of serving in the circuit courts. The 1789 Judiciary Act placed on the members of the highest Court the obligation of personally sitting on the circuit courts that had been set up on a territorial basis throughout the country. At a time when travel was so difficult, the imposition upon the Supreme Court Justices of this circuit duty was most burdensome. In February 1792, Justice Cushing complained in a letter to Washington about the hardships involved in his judicial travels: "The travelling is difficult this Season:—I left Boston, the 13th of Jan in a Phaeton, in which I made out to reach Middleton as the Snow of the 18th began, which fell so deep there as to oblige me to take a Slay, & now again wheels seem necessary."[18]

The situation with regard to one Justice was graphically described in a 1798 letter by Samuel Chase, who had been appointed to the Court two years earlier. He would, Chase wrote,

> shew the very great burthen imposed on one of the Judges Mr. Iredell lives at Edenton, in North Carolina. When he is appointed to attend the Middle Circuit, he holds the Circuit Court for New Jersey at Trenton, on 1st of April; and, at Philadelphia, on the 11th of the same month; he then passes through the State of Delaware (by Annapolis) to hold the Court, on 22nd of May, at Richmond, in Virginia (267 miles.); from thence he must return, the same distance to hold Circuit Court on 27th June, at New Castle in Delaware. . . . A permanent system should not impose such hardship on any officer of Government.[19]

"I will venture to say," Iredell himself wrote in 1791, "no Judge can conscientiously undertake to ride the Southern Circuit constantly, and perform the other parts of his duty. . . . I rode upon the last Circuit 1900 miles: the distance from here and back again is 1800."[20]

When Jay resigned as Chief Justice, one of the reasons, according to a letter from a Congressman, was "the system of making the Judges of the Supreme Court ride the Circuits throughout the Union; this has induced Mr. Jay to quit the Bench; he was Seven months in the Year from his family travelling about the Country."[21] In a letter the previous year, Jay had expressed resentment at being "placed in an office . . . which takes me from my Family half the Year, and obliges me to pass too considerable a part of my Time on the Road, in Lodging Houses, & Inns."[22]

Thomas Johnson of Maryland, who had been appointed to Justice Rutledge's seat in 1791, actually resigned a year later because of the burden of circuit duty. "I cannot resolve," Johnson wrote in his resignation letter, "to spend six Months in the Year of the few I may have left from my Family on Roads at Taverns chiefly and often in Situations where the most moderate Desires are disappointed: My Time of Life Temper and other Circumstances forbid it."[23]

Finally, the Justices themselves publicly complained about what they termed, in a 1792 letter to Washington, "the burdens laid upon us so excessive that we cannot forbear representing them in strong and explicit terms."[24] At the same time, they wrote a remonstrance to Congress which declared, "That the task of holding twenty seven circuit Courts a year, in the different States, from New Hampshire to Georgia, besides two Sessions of the Supreme Court at Philadelphia, in the two most severe seasons of the year, is a task which considering the extent of the United States, and the small number of judges, is too burthensome."[25]

In addition, the remonstrance urged the unfairness of a system in which the Justices sat on appeals from decisions which they had made on circuit: "That the distinction made between the Supreme Court and its Judges, and appointing the same men finally to correct in one capacity, the errors which they themselves may have committed in another, is a distinction unfriendly to impartial justice, and to that confidence in the supreme Court, which it is so essential to the public Interest should be reposed in it."[26]

Though the Justices asked "that the system may be so modified as that they may be relieved from their present painful and improper situation."[27] Congress gave them only what has been called "but a half loaf and a meagre one at that."[28] A 1793 statute dispensed with the attendance of more than one Supreme Court Justice at each Circuit Court. This, wrote Justice Cushing, "eases off near half the difficulty. . . . The Justices are now impowered, at each Session of the Supreme Court, to assign a Circuit to a Single Judge, so that a Judge need go but one Circuit in a Year." But that was all Congress was prepared to do. It did not make what Cushing called the "radical alteration of the present Itinerant System for the better," for which the Justices had hoped. In fact, a law which, in Cushing's phrase, "may take off the fatigues of travelling & the inconvenience of so much absence from home"[29] would not take effect for another century.

Early Decisions

One of the reasons why Congress was unwilling, despite the Justices' remonstrance, to relieve them of circuit duties, was the fact that the Supreme Court itself had little work to do. As her son Thomas wrote in a 1799 letter to Abigail Adams, "The Supreme Court of the United States adjourned this day—Little business was done, because there was little to do."[30] During its first decade, the Supreme Court decided relatively few cases. In the first three years of its existence, in fact, the Court had practically no business to transact; it was not until February 1793 that the Justices decided their first case.

A half year earlier, on August 11, 1792, the Justices had delivered their first opinions in *Georgia v. Brailsford*.[31] The Justices adopted the English practice of delivering their opinions seriatim—a practice which they followed until John Marshall's day. Interestingly, in this first case in which opinions were delivered, the first opinion was a dissent by Justice Johnson, thus establishing at the outset the right of Justices to express publicly their disagreement with the result reached by the Court.

Brailsford granted Georgia a temporary injunction. The decision was described by Edmund Randolph in a letter to Madison which contained an unflattering picture of the first Court:

> The State of Georgia applied for an injunction to stop in the Marshal's hands a sum of money which had been recovered in the last circuit court by a British subject, whose estate had been confiscated. It was granted with a demonstration to me of these facts; that the premier [Jay] aimed at the cultivation of Southern popularity; that the professor [Wilson] knows not an iota of equity; that the North Carolinian [Iredell] repented of the first ebullitions of a warm temper; and that it will take a score of years to settle, with such a mixture of judges, a regular course of chancery.[32]

The first important decision of the Supreme Court was that rendered in February 1793 in *Chisholm v. Georgia*.[33] Chisholm was a citizen of South Carolina and his suit was based upon a claim for the delivery of goods to the state for which no payment had been received. Counsel for Georgia appeared and presented a written remonstrance denying the Court's jurisdiction, "but, in consequence of positive instructions, they declined taking any part in arguing the question." The case was then argued by Randolph, who represented Chisholm. (It was common at the time for the Attorney General to represent private clients; indeed, Randolph's official salary was so small that he depended for his livelihood upon such private clients.)

"This great cause,"[34] as the first *Chisholm* opinion (that by Justice Iredell) characterized it, presented the crucial issue of whether a state could be sued in a federal court by citizens of another state. The Court's answer was given unequivocally in favor of its jurisdiction in such a case in opinions by Justices Blair, Wilson, and Cushing and Chief Justice Jay (Justice Iredell alone dissenting). The most important opinion was delivered by

Justice Wilson. Unfortunately, it suffers from the pedantry and exaggerated rhetoric present in all of Wilson's writing—these defects, as much as his delivery of law lectures at the College of Philadelphia, led to his sobriquet of "the professor," as seen in the quoted Randolph letter. With all its faults, however, the Wilson *Chisholm* opinion remains a powerful justification both of the Court's decision and of the United States as a nation, not merely a league of sovereign states.

Wilson's opinion resoundingly rejected the state assertion of immunity from suit. After going into his conception of a state as a "body of free persons united together for their common benefit," Wilson asked, "Is there any part of this description, which intimates in the remotest manner, that a state, any more than the men who compose it, ought not to do justice and fulfil engagements?" Wilson declared that if a free individual is amenable to the courts, the same should be true of the state. "If the dignity of each singly, is undiminished, the dignity of all jointly must be unimpaired." States are subject to the same rules of morality as individuals. If a dishonest state willfully refuses to perform a contract, should it be permitted "to insult . . . justice" by being permitted to declare, "I am a sovereign state"?[35]

In *Chisholm*, Wilson recognized that though the immediate issue of state subjection to suit was important, it was outweighed by one "more important still; and . . . no less radical than this—'do the people of the United States form a nation?'" Wilson's opinion answered this question with a categorical affirmative. In *Chisholm*, he repudiated the concept of state sovereignty in language as strong as that later delivered by Chief Justice Marshall himself. Sovereignty, he asserted, is not to be found in the states, but in the people. The Constitution was made by the "People of the United States," who did not surrender any sovereign power to the states. "As to the purposes of the union, therefore, Georgia is not a sovereign state."[36]

The people, Wilson concluded, intended to set up a nation for national purposes. They never intended to exempt the states from national jurisdiction. Instead, they provided expressly, "The judicial power of the United States shall extend to controversies, between a state and citizens of another state." Wilson asked, "[C]ould this strict and appropriated language, describe, with more precise accuracy, the cause now depending before the tribunal?"[37]

The *Chisholm* decision, we are told, "fell upon the country with a profound shock."[38] Indeed, it led to such a furor in the states that the Eleventh Amendment (prohibiting suits by individuals against states) was at once proposed and adopted. Though the immediate holding in *Chisholm v. Georgia* was thus overruled, the Court's reasoning there remains of basic importance for what it tells us about the nature of the Union. To decide the case, the Court really had to determine the crucial issue of state sovereignty. If Georgia was intended to be a sovereign state under the Constitution, it could not be sued. In deciding that Georgia was subject to suit, the Court was rejecting the claim that the state was vested with the traits of sovereignty. "As to the purposes of the Union," to repeat the declaration of Justice Wilson, "Georgia is not a sovereign state."

Judicial Review

It is, of course, known to even the beginner in constitutional history that the power of the Supreme Court to review the constitutionality of acts of Congress was established by Chief Justice Marshall's landmark opinion in *Marbury v. Madison.*[39] Yet even Marshall—legal colossus though he was— did not write on a blank slate. On the contrary, the law laid down by Marshall in *Marbury v. Madison* was inextricably woven with that ex- pounded by his contemporaries and predecessors. Judicial review, as an essential element of the law, was part of the legal tradition of the time, derived from both the colonial and revolutionary experience. With the appearance during the Revolution of written constitutions, the review power began to be stated in modern terms. Between the Revolution and *Marbury v. Madison,* state courts asserted or exercised the power in at least twenty cases.[40] Soon after the Constitution went into effect, assertions of review authority were made by a number of federal judges.[41]

Even more important, the Supreme Court began to lay the foundation for judicial review soon after it went into operation. Of particular signifi- cance in this respect were three cases decided during the 1790s. The first was *Ware v. Hylton.*[42] A 1777 Virginia law decreed the confiscation of all debts owed to British subjects. Despite it, an action was brought on a debt due before the Revolution from an American to a British subject. Plaintiff relied upon the Treaty of Peace with Britain under which "creditors, on either side, shall meet with no lawful impediment to the recovery of the full value of all *bona fide* debts, heretofore contracted."

A letter from Justice Iredell to his wife called the case "the great Virginia cause."[43] John Marshall argued in favor of the Virginia law and maintained that "the judicial authority can have no right to question the validity of a law; unless such a jurisdiction is expressly given by the consti- tution."[44] As Marshall's biographer notes, "It is an example of the 'irony of fate' that in this historic legal contest Marshall supported the theory which he had opposed throughout his public career thus far, and to demolish which his entire after life was given."[45] Had Marshall's *Ware v. Hylton* assertion prevailed, the American system of constitutional law would have developed along lines altogether different from the course taken.

The Court, however, rejected Marshall's argument and ruled that the Treaty of Peace with Britain overrode conflicting provisions of state law on the debts owed by Americans to British subjects. "A treaty cannot be the supreme law of the land," declared Justice Chase, "if any act of a state Legislature can stand in its way. . . . It is the declared will of the people of the United States that every treaty made, by the authority of the United States, shall be superior to the constitution and laws of any individual state; and their will alone is to decide."[46]

Ware v. Hylton asserted review power over a state law. A similar power was exercised in *Calder v. Bull*,[47] where a Connecticut law that set aside a probate decree disapproving a will and granting a new trial was attacked as a violation of the Ex Post Facto Clause. The Court held that the law was not an ex post facto law, since the Ex Post Facto Clause reaches only laws that are criminal in nature; the constitutional prohibition, in Justice Iredell's words, "extends to criminal, not to civil cases."[48]

The opinions delivered, nevertheless, left no doubt of the Court's power to strike down the state law, if it had been found to violate the Constitution. "I cannot," declared Justice Chase, "subscribe to the omnipotence of a state Legislature or that it is absolute and without control. . . . An act of the Legislature (for I cannot call it a law) contrary to the great first principles of the social compact, cannot be considered a rightful exercise of legislative authority."[49]

In his opinion, Justice Chase stated that he was not "giving an opinion, at this time, whether this court has jurisdiction to decide that any law made by Congress, contrary to the Constitution of the United States, is void."[50] It was not, to be sure, until *Marbury v. Madison* that the Court came down categorically in favor of such a review power. It was, however, in *Hylton v. United States*,[51] almost a decade before Marshall's classic *Marbury* opinion, that the Court first ruled on the constitutionality of a federal law.

Hylton arose under article I, section 9: "No Capitation, or other direct, Tax shall be laid, unless in Proportion to the Census or Enumeration herein before directed to be taken." This means that direct taxes must be apportioned among the states on the basis of their respective populations. Yet this does not tell us what is a "direct tax" within the constitutional provision. During the Framers' Convention, "Mr. King asked what was the precise meaning of *direct* taxation? No one answered."[52]

In *Hylton v. United States,* it was argued that a fixed federal tax on all carriages used for the conveyance of persons was a direct tax and hence invalid, because it was not apportioned among the states according to population. The Court unanimously held that the tax at issue was not a direct tax within the meaning of article I, section 9. According to the opinions rendered, since the Direct-Tax Clause constitutes an exception to the general taxing power of Congress, it should be strictly construed. No tax should be considered "direct" unless it could be conveniently apportioned. "As all direct taxes must be apportioned," said Justice Iredell, "it is evident, that the constitution contemplated none as direct, but such as could be apportioned. If this cannot be apportioned, it is, therefore, not a direct tax in the sense of the constitution. That this tax cannot be apportioned is evident."[53]

Justice Paterson, himself one of the Framers, who had been appointed to the Court after Justice Johnson resigned, stated that the constitutional provision on direct taxes had been intended to allay the fears of the southern states lest their slaves and lands be subjected to special taxes not equally apportioned among the northern states.[54] From this, it was a natural step to the view, expressed by all the Justices, that the "direct taxes contemplated by the constitution, are only two, to wit, a capitation, or poll tax, simply, without regard to property, profession or any other circumstance; and a tax on land."[55]

More important than the *Hylton* holding is the fact that the case was the first in which an act of Congress was reviewed by the Supreme Court. It is true that Justice Chase stated that "it is unnecessary, at this time, for me to determine, whether this court constitutionally possesses the power to declare an act of Congress void, on the ground of its being made contrary to, and in violation of, the constitution."[56] But the mere fact that the Justices considered the claim that the federal statute was "unconstitutional and void"[57] indicates that they believed that the Court did possess the review power. As such, *Hylton* was an important step on the road to *Marbury v. Madison*.

What the Court Does Not Do

Justice Louis D. Brandeis used to say that what the Supreme Court did not do was often more important than what it did do.[58] The fact that the highest tribunal acts as a law court has been more important than any other factor in determining the things that it does not do in our constitutional system. The Framers deliberately withheld from the Supreme Court power that was purely political in form, such as a forthright power to veto or revise legislation. Instead, they delegated to the Court "The judicial Power" alone—a power which, by the express language of article III, extends only to the resolution of "Cases" and "Controversies." This, Justice Robert H. Jackson once noted,[59] is the most significant and the least comprehended limitation upon the way in which the Court can act. Judicial power, the Court pointed out in 1911, "is the right to determine actual controversies arising between adverse litigants, duly instituted in the courts of proper jurisdiction."[60] The result of the constitutional restriction is that the Court's only power is to decide lawsuits between opposing litigants with real interests at stake, and its only method of proceeding is by the conventional judicial process. "The Court from the outset," says Chief Justice Charles Evans Hughes, "has confined itself to its judicial duty of deciding actual cases."[61]

Of course, in a system such as ours, where the highest Court plays so prominent a political role, there might be great advantages in knowing at once the legal powers of the Government. It would certainly be convenient for the parties and the public to know promptly whether a particular statute is valid. The desire to secure these advantages led to strong efforts at the Constitutional Convention to associate the Supreme Court as a Council of Revision in the legislative process; but these attempts failed and, ever since, it has been deemed, both by the Court itself and by most students of its work, that the disadvantages of such a political role by the judiciary were far greater than its advantages.

Similarly, from the beginning, the Court has rejected the notion that it could avoid the difficulties inherent in long-delayed judicial invalidation of legislation by an advisory opinion procedure. The very first Court felt constrained to withhold even from the "Father of his Country" an advisory opinion on questions regarding which Washington was most anxious to have illumination from the highest tribunal.[62] In 1793 President Washington, through a letter sent to the Justices by Secretary of State Jefferson, sought the advice of the Supreme Court on a series of troublesome "abstract questions" in the realm of international law "which have already occurred, or may soon occur." Chief Justice Jay and his associates first postponed their answer until the sitting of the Court and then, three weeks later, replied politely but firmly, declining to give the requested answers.

According to the Justices' letter to Washington, both "the lines of separation drawn by the Constitution between the three departments of the government . . . and our being judges of a court in the last resort, are considerations which afford strong arguments against the propriety of our extra-judicially deciding the questions alluded to."[63] This, says Chief Justice Hughes, was a statement that the Court "considered it improper to declare opinions on questions not growing out of a case before it."[64]

The Justices' refusal has served as a precedent against the giving of advisory opinions by the Court. Ever since that time, it has, in Chief Justice Harlan F. Stone's phrase, been the Court's "considered practice not to decide abstract, hypothetical or contingent questions."[65] A party cannot, in other words, bring an action for what Justice Oliver Wendell Holmes once called a "mere declaration in the air";[66] on the contrary, "A case or controversy in the sense of a litigation ripe and right for constitutional adjudication by this Court implies a real contest—an active clash of views, based upon an adequate formulation of issues, so as to bring a challenge to that which Congress has enacted inescapably before the Court."[67]

A few years earlier, in *Hayburn's Case*,[68] the Justices had decided that they might not, as judges, render decisions that were subject to revision by some other body or officer. They gave effect to this view even though it meant the effective nullification of a federal statute providing for veterans' pensions.

The statute, passed by Congress in 1792, authorized the federal circuit courts to determine the pension claims of invalid veterans of the Revolution and certify their opinions to the Secretary of War, who might then grant or deny the pensions as he saw fit. *Hayburn's Case* was argued in the Supreme Court, but that tribunal never rendered decision, for Congress intervened by providing another procedure for the relief of the pensioners.[69] But the statute at issue was considered in the different circuit courts and their opinions are given in a note to *Hayburn's Case* by the reporter.

All of the circuit courts (with five of the six Justices sitting) concurred in the holding that they could not validly execute the statute as courts set up under article III. The strongest position was taken by Justices Wilson and Blair sitting in the Pennsylvania Circuit Court. In Chief Justice Taney's words, they "refused to execute [the statute] altogether."[70] They entered an order in a case involving Hayburn as the invalid claimant: "[I]t is considered by the Court that the same be not proceeded upon."[71] Then they sent a letter to the President, undoubtedly drafted by Justice Wilson, which gave the reasons for their action. It asserted that "the business directed by this act is not of a judicial nature." For the court to act under the law would mean that it "proceeded *without* constitutional authority."[72]

That was true, "Because, if, upon the business, the court had proceeded, its *judgments* . . . might, under the same act, have been revised and controuled . . . by an officer in the executive department. Such revision and controul we deemed radically inconsistent with the independence of that judicial power which is vested in the courts; and, consequently, with that important principle which is so strictly observed by the Constitution of the United States."[73]

After Hayburn presented a memorial petitioning Congress for relief, Congressman Elias Boudinot explained the action to the House of Representatives:

> It appeared that the Court . . . looked on the law . . . as an unconstitutional one; inasmuch as it directs the Secretary of War to state the mistakes of the Judge to Congress for their revision; they could not, therefore, accede to a regulation tending to render the Judiciary subject to the Legislative and Executive powers, which, from a regard for liberty and the Constitution, ought to be kept carefully distinct, it being a primary principle of the utmost importance that no decision of the Judiciary Department should under any pretext be brought in revision before either the Legislative or Executive Departments of the government, neither of which have, in any instance, a revisionary authority over the judicial proceedings of the Courts of Justice.[74]

According to Chief Justice Taney, *Hayburn's Case* established that the power conferred upon the federal courts by the 1792 statute "was no judicial power within the meaning of the Constitution, and was, therefore, unconstitutional and could not lawfully be exercised by the courts."[75] Since *Hayburn's Case,* it has been settled that the federal judges may not act in cases where their judgments are subject to revision by the executive or legislative department. The alternative is what District Judge Peters, who sat with Justices Wilson and Blair in the circuit court, termed "the danger of Executive control over the judgments of Courts"[76]—something avoided by the judges' strong stand in *Hayburn's Case.*

The judges' action in *Hayburn's Case* was also an important step on the road to *Marbury v. Madison.* This, said Boudinot in his statement to the House, was "the first instance in which a Court of Justice had declared a law of Congress to be unconstitutional."[77] It was widely recognized that the action of the judges was one, in the phrase of a newspaper, "declaring an act of the present session of Congress, unconstitutional."[78] Writing to Henry Lee, Madison referred to the review power and said that the judges' "pronouncing a law providing for Invalid Pensioners unconstitutional and void" was "an evidence of its existence."[79] At any rate, cases such as *Hayburn's Case* as well as those discussed in the last section were early recognitions of the judicial possession of the review power.

Chief Justices Rutledge and Ellsworth

The first national capital was New York and, as seen, the Supreme Court first met in that city early in 1790. A year later, in February 1791, after the seat of government had moved to Philadelphia, the Court held its sessions there, first in the State House and then in the new City Hall just east of Independence Hall, where both the Supreme Court and the state and city courts sat.

There were significant personnel changes while the Court sat in Philadelphia. As already seen, Thomas Johnson of Maryland was appointed in place of John Rutledge, who had resigned in 1791 to become Chief Justice of the South Carolina Court of Common Pleas. Johnson, however, resigned a year later because he found the circuit duties too strenuous, and his seat was filled by William Paterson, one of the Framers, who was then Governor of New Jersey. Samuel Chase of Maryland was elevated to the Court in 1796 in place of Justice Blair, who had resigned because of ill health. Then, when James Wilson died in 1798, President Adams selected Bushrod Washington, the first President's nephew, who had become a leader of the Virginia Bar. Justice Washington was to serve for thirty-two years as one of the pillars of the Marshall Court. Mention should also be made of the appointment in 1799 of Alfred Moore, a North Carolina judge, to fill the vacancy caused by the death of Justice Iredell.

Even more important were the personnel changes that occurred in the Chief Justiceship. John Jay is, of course, one of the leading names in early American history. He was not, however, a success as the head of the Supreme Court. In part this was due to the lack of business in the early Court, as well as the burden of circuit duty. One must conclude, as Edmund Randolph did in a 1792 letter to Madison, that Jay may have been "clear . . . in the expression of his ideas, but . . . they do not abound in legal subjects."[80]

Jay also set a bad example in the early Court by indicating that he did not consider his judicial position all that important. In 1794, Jay accepted an appointment as Special Ambassador to England, where he negotiated the treaty that bears his name. Though his appointment was denounced as a violation of the separation of powers, Jay did not resign as Chief Justice while carrying out his diplomatic assignment. Jay's successor, Oliver Ellsworth, also served as Minister to France without resigning as Chief Justice. These extrajudicial appointments had an inevitable negative effect upon the prestige of the fledgling Court. "That the *Chief Justiceship* is a sinecure," wrote the *Philadelphia Aurora,* "needs no other evidence, than that in one case the duties were *discharged* by one person who resided at the same time in England; and by another during a year's residence in France."[81]

While he was absent in England, Jay was nominated for Governor of New York and elected to that office soon after his return in 1795. He resigned as Chief Justice to accept the Governorship. A striking indication of the relative importance of the two positions at the time is given in the characterization by a New York newspaper of Jay's new office as a "promotion."[82]

President Washington had difficulty in filling the Chief Justiceship. John Rutledge, who had resigned from the Court to become Chief Justice of the South Carolina Court of Common Pleas, wrote to Washington that he was now willing to be Jay's successor. The President gave him a recess appointment, and Rutledge sat as Chief Justice during the August 1795 Term. However, the Senate voted against his confirmation, both because of his vitriolic attack upon the Jay Treaty in a Charleston speech and rumors of what John Adams called his "accellerated and increased . . . Disorder of the Mind."[83]

Washington next nominated Justice Cushing, the Senate confirmed the nomination, and Cushing actually received his commission. But then, as summarized by Adams in a letter to his wife, "Judge Cushing declines the Place of Chief-Justice on Account of his Age and declining Health."[84] Another indication of the contemporary reputation of the highest judicial position is seen in the comment of a Rhode Island official, "It is generally thought that Neighbor Cushing gave a Clear proof of his Understanding when he refused the Chief Justiceship."[85]

The President then chose Oliver Ellsworth of Connecticut, who had been a member of the Continental Congress, a state judge, and a Senator. He had been an important participant in the Framers' Convention, as well as a leader in the ratification struggle. As Senator he had been the principal author of the Judiciary Act of 1789.

The Ellsworth appointment met with general approval. "The appointment of the C.J.," Adams wrote to his wife, "was a wise Measure," even though, by it, "we loose the clearest head and most diligent hand we had [in the Senate]."[86] Senator Jonathan Trumbull agreed in a letter to his brother John, a famous painter, that Ellsworth's appointment was "a great Loss this to the Senate!" At the same time, he wrote, it was "a valuable acquisition to the Court—an acquisition which has been much needed."[87]

The leading history of the early Court states that Ellsworth was the first to make the position of Chief Justice a place of leadership.[88] His tenure was, however, too short for him to establish a true leadership role. About the only sign of the Justices' following Ellsworth's lead was the indication in the reported cases that his predilection for brief opinions was not without effect.[89]

Chief Justice Ellsworth's most important opinion (about the only significant one he delivered on the Supreme Court) laid down a basic rule on the Court's own jurisdiction. The case was *Wiscart v. D'Auchy*,[90] decided in 1796. The question at issue was whether an equity decree was reviewable in the Supreme Court by a writ of error or an appeal. In the course of the case, the Court considered the nature of its appellate jurisdiction. Chief Justice Ellsworth, in an oft-cited passage, declared that the Court's appellate jurisdiction depended entirely upon statute: "If Congress has provided no rule to regulate our proceedings, we cannot exercise an appellate jurisdiction; and if the rule is provided, we cannot depart from it."[91] Therefore, said Ellsworth, the only question in determining whether the Supreme Court has appellate jurisdiction in a given case is whether Congress has established a rule regulating its exercise in such a case.

The Ellsworth view on the matter was rejected by Justice Wilson, who urged, in a dissenting opinion, that the Supreme Court's appellate jurisdiction was derived from the Constitution: "The appellate jurisdiction, therefore, flowed, as a consequence, from this source; nor had the legislature any occasion to do, what the constitution had already done."[92] Even in the absence of congressional provision, therefore, according to Wilson, the appellate jurisdiction of the Supreme Court may be exercised, resting as it does upon the strong ground of the Constitution itself.

Interestingly enough, both Ellsworth and Wilson had been prominent members of the Framers' Convention. Yet less than a decade after the basic document was drafted, they disagreed sharply on the organic nature of the appellate jurisdiction of the nation's highest tribunal. Subsequent cases confirm the correctness of Chief Justice Ellsworth's view. "By the constitution of the United States," declared the Court in 1847, "the Supreme Court possesses no appellate power in any case, unless conferred upon it by act of Congress."[93] Two decades later, the Court was, if anything, even more blunt, asserting, "In order to create such appellate jurisdiction in any case, two things must concur: the Constitution must give the capacity to take it and an act of Congress must supply the requisite authority." The Supreme Court's appellate jurisdiction is thus "wholly the creature of legislation."[94]

His *Wiscart v. D'Auchy* opinion shows both Chief Justice Ellsworth's legal ability and his potential for molding our public law. But he was able to sit in the court's center chair for less than four years, and during much of that time he was absent because of illness. As a 1798 letter from his brother to Oliver Wolcott, Jr., characterized it, "Mr. Ellsworth . . . is considerably unwell, and I understand quite hypocondriac."[95]

During Ellsworth's last year as Chief Justice, he served as special envoy in France. The Court could barely function during that period. At its August 1800 Term, the last in Philadelphia, not only was Ellsworth absent, but also absent were Justice Cushing, who was ill, and Justice Chase, who was in Maryland working for President Adams's reelection. This led to bitter anti-Federalist attacks, such as that in the *Aurora* which condemned "[t]he suspension of the highest court of judicature in the United States, to allow a *Chief Justice* to add NINE THOUSAND DOLLARS a year to his salary, and to permit Chase to make electioneering harangues in favor of Mr. *Adams*."[96] Even Adams's son wrote to his cousin that Chase was "too much engaged in Electioneering."[97]

Not in the best of condition when appointed, Ellsworth's health completely broke down on his journey to France. The Chief Justice described his condition in a letter sent from Le Havre to Wolcott: "Sufferings at sea, and by a winter's journey thro' Spain, gave me an obstinate gravel, which by wounding the kidneys, has drawn & fixed my wandering gout to those parts. My pains are constant, and at times excruciating."[98] On the same day, October 16, 1800, Ellsworth sent to President Adams a letter resigning the office of Chief Justice.

Judiciary Act of 1801

Circuit duty, we have seen, was the great albatross of the early Supreme Court. It is true that the problem of the Supreme Court Justices sitting on circuit was resolved by the Judiciary Act of 1801. That law provided for the creation of six new Circuit Courts to be staffed entirely by newly appointed judges. Unfortunately, however, the new statute was an integral part of the controversy between the Federalists and the Jeffersonians that dominated the political scene at the turn of the century. The desirable reform of relieving Supreme Court members of their circuit duties was less important than the creation by the lame-duck Federalist Congress of a whole new court system, with vacancies in the new tribunals to be filled by deserving members of the defeated party. The bill was enacted into law on February 13, 1801; within two weeks President Adams had filled the new positions with Federalists; and by March 2 (two days before Jefferson took office) the Senate had confirmed the appointments. The new judges, many of whose commissions were actually filled out on the last day of Adams's term of office, were derisively known as the "midnight judges."

The newly elected Jeffersonians greeted the 1801 statute with indignation. They could scarcely concur in the Federalist attempt to entrench themselves in the life-tenure judiciary by the Midnight Judges Bill. Instead, the Jeffersonian Congress did away with what they called the "army of judges" by abolishing the new courts soon after Jefferson took office, without making any provision for the displaced judges. They did so by a simple Act of March 8, 1802, repealing the 1801 Judiciary Act and providing for the revival of the former circuit court system.

Lost in the partisan controversy was the desirable reform effected by the 1801 act in relieving Supreme Court Justices of circuit court duty. Instead, the obligation of sitting on the circuits continued as a burden upon the members of the highest bench. It was only after that burden was finally removed in 1891 that the Supreme Court was able fully to assert its role as guardian of the constitutional system. Though judicial review was established in 1803, it did not really become an important practical factor in the polity until the 1890s.

The Federalists themselves bitterly attacked the 1802 repealing statute as one which, in Gouverneur Morris's characterization, "renders the judicial system manifestly defective and hazards the existence of the Constitution."[99] The Federalist argument was, however, rejected by the Supreme Court in *Stuart v. Laird*,[100] in a laconic opinion which stated only that Congress had constitutional authority to establish, as the members chose, such inferior tribunals as they deemed proper, and to transfer a cause from one such tribunal to another. "In this last particular," said the Court, "there are no words in the constitution to prohibit or restrain the exercise of legislative power."[101]

2

Marshall Court, 1801–1836

On the north and south walls of the Supreme Court Chamber in Washington are carved two marble panels depicting processions of historical lawgivers. Of the eighteen figures on the panels only one is there because of his work as a judge, and he is the one American represented: John Marshall. This is more than mere coincidence, for it sharply illustrates a basic difference between the making of law in the United States and in other countries. The great lawgivers in other systems have been mighty monarchs of the type of Hammurabi and Justinian, divinely inspired prophets like Moses, philosophers such as Confucius, or scholars like Hugo Grotius and Sir William Blackstone. We in the United States have certainly had our share of the last two types of lawgiver—particularly among the men who drew up the organic documents upon which our polity is based. Significantly enough, however, it is not a Jefferson or a Madison who is depicted as *the* American lawgiver, but the great Chief Justice who, more than any one person, has left his imprint upon the development of our constitutional law.

Marshall's Appointment

In the autumn of 1800, not long before Marshall's appointment as Chief Justice, the United States Government moved to Washington, D.C. The new capital was still in the early stages of construction and except for the north wing of the Capitol and the still unfinished White House, there was, as a Congressman wrote, "nothing to admire but the beauties of nature."[1]

32

At least buildings had been erected for the Legislature and Executive. The same was not true of the Judiciary. When the Federal City was planned, the Supreme Court was completely overlooked and no chamber provided for it: "When the seat of government was transferred to Washington, the court crept into an humble apartment"[2] in what had been designed as a House Committee room.

The failure to provide adequate housing for the Supreme Court "provides further evidence that the Court was not regarded as an institution of great importance in the federal system."[3] Indeed, the outstanding aspect of the Court's work during its first decade was its relative unimportance. When Marshall came to the central judicial chair in 1801, the Court was but a shadow of what it has since become. When he died in 1836, it had been transformed into the head of a fully coordinate department, endowed with the ultimate authority of safeguarding the ark of the Constitution.

Marshall it was who gave to the Constitution the impress of his own mind, and the form of our constitutional law is still what it is because he shaped it.[4] "Marshall," declared John Quincy Adams at news of his death, "by the ascendancy of his genius, by the amenity of his deportment, and by the imperturbable command of his temper, has given permanent and systematic character to the decisions of the Court, and settled many great constitutional questions favorably to the continuance of the Union." It was under Marshall's leadership that the Supreme Court transmuted the federal structure created by the Founders into a nation strong enough to withstand even the shock of civil war. To quote Adams's not unbiased view again, "Marshall has cemented the Union which the crafty and quixotic democracy of Jefferson had a perpetual tendency to dissolve."[5]

Marshall's appointment as Chief Justice was one of the happy accidents that change the course of history. In the first place, had Justice Cushing not declined the appointment or had Chief Justice Ellsworth not made the arduous journey to France, there would have been no vacancy in the Chief Justiceship until well after President Adams's term had expired. After Chief Justice Ellsworth's resignation, the President offered his place to John Jay. The Senate confirmed the appointment and, had Jay accepted, there would, of course, still have been no place for Marshall on the Court. Jay, however, also refused the position both because he wanted to retire to his farm in Bedford, New York, and because his acceptance in "a System so defective would give some Countenance to the neglect and Indifference with which the opinions and Remonstrances of the Judges on this important Subject have been treated."[6]

After Jay declined the Chief Justiceship, it was widely expected that the President would nominate Justice Paterson. Marshall later wrote, "On the resignation of Chief Justice Ellsworth I recommended Judge Patteson [*sic*] as his successor."[7] Adams, however, refused to select him. According to Marshall, "The President objected to him, and assigned as his ground of objection that the feelings of Judge Cushing would be wounded by passing him and selecting a junior member of the bench."[8] The real reason for

Adams's refusal, however, was that, as a letter to Hamilton stated, "Either Judge Paterson or General Pinckney ought to have been appointed, but both those worthies are your friends."[9] Adams was unwilling to consider any person in the Hamiltonian faction of his party.

Marshall himself tells us what happened next: "When I waited on the President with Mr. Jays letter declining the appointment he said thoughtfully 'Who shall I nominate now'? I replied that I could not tell, as I supposed that his objection to Judge Paterson remained. He said in a decided tone 'I shall not nominate him.' After a moments hesitation he said, 'I believe I must nominate you'. . . . Next day I was nominated."[10]

The Marshall appointment was both completely unexpected and resented by Adams's own party, which believed that Judge Paterson should have been given the position. "With grief, astonishment & almost indignation," Jonathan Dayton, a Federalist Senator, wrote to Judge Paterson, "I hasten to inform you, that, contrary to the hopes and expectations of us all, the President has this morning nominated Gen. Marshall. . . . The eyes of all parties had been turned upon you, whose pretensions he knew were, in every respect the best, & who, he could not be ignorant, would have been the most acceptable to our country."[11]

The feeling in the Senate against the nomination was so strong, Dayton went on, that "I am convinced . . . that they would do it [i.e., reject the nomination] if they could be assured that thereby *you* would be called to fill it." The Senate suspended the nomination for a week, but finding the President inflexibly opposed to Paterson and fearing, in Dayton's words, "that the rejection of this might induce the nomination of some other character more improper, and more disgusting,"[12] the Senate yielded and unanimously confirmed Marshall's appointment.

Judge Paterson had not wanted to be Chief Justice and went out of his way to praise the Marshall appointment. "Mr. Marshall," Paterson replied to Dayton, "is a man of genius, of strong reasoning powers, and a sound, correct lawyer. His talents have at once the lustre and solidity of gold."[13] Paterson wrote to Marshall to the same effect in congratulating him on his appointment,[14] and, Marshall writes, "I felt truly grateful for the real cordiality towards me"[15] displayed by his new colleague.

Some time before Marshall's appointment, James Kent heard some friends of Hamilton say that Hamilton was in "every way, suited" to be Chief Justice. Kent, writing to Hamilton's wife about the incident, affirmed, "Of all this there could be no doubt." But, Kent concluded, Hamilton's "versatile talents, adapted equally for the bench & the bar, the field, the Senate house & the Executive cabinet, were fortunately called to act in a more complicated, busy & responsible Station."[16]

This estimate by the man who, next to Marshall, was then the nation's preeminent jurist, is still another indication of the low state of the early Supreme Court. All this, however, was to change after Marshall became Chief Justice. It was Marshall who established the role of the Supreme Court as the authoritative expounder of the Constitution, and it was he who exercised this role to lay the legal foundations of a strong nation, endowed with all the authority needed to enable it to govern effectively.

Marshall's Background

On the morning of Jefferson's first inauguration, Marshall wrote to Charles C. Pinckney, "Of the importance of the judiciary at all times but more especially the present I am very fully impressed & I shall endeavor in the new office to which I am called not to disappoint my friends."[17] Certainly, Marshall as Chief Justice was anything but a disappointment to his "friends." Years after John Adams had nominated Marshall to be Chief Justice, he said, "My gift of John Marshall to the people of the United States was the proudest act of my life. . . . I have given to my country . . . a Hale, a Holt, or a Mansfield."[18]

Almost two centuries later, no one doubts Marshall's preeminence in our law. "If American law," says Justice Oliver Wendell Holmes, "were to be represented by a single figure, skeptic and worshipper alike would agree without dispute that the figure could be one alone, and that one, John Marshall."[19] Marshall's was the task of translating the constitutional framework into the reality of decided cases. He was not merely the expounder of our constitutional law; he was its author, its creator. "Marshall found the Constitution paper and he made it power," said James A. Garfield. "He found a skeleton, and he clothed it with flesh and blood."[20] What Justice Story termed "the extraordinary judgments of Mr. Chief Justice Marshall upon constitutional law"[21] laid the foundation of our constitutional edifice. Ever since, that structure has been associated with the Marshall name and has remained the base upon which the American polity functions.

If we look to the background of the man himself, however, he certainly seemed ill equipped for the task to which he was ultimately called. One who reads the modest account of his early life in his famous autobiographical letter to Joseph Story is bound to be amazed at the meagerness of his education and training, both generally and in the law itself. His only formal schooling consisted of a year under the tuition of a clergyman, as well as another under a tutor who resided with his family. For the rest, his learning was under the superintendence of his father, who, Marshall concedes, "had received a very limited education."[22]

His study for the Bar was equally rudimentary. During the winter of 1779–1780, while on leave from the Army, "I availed myself of this inactive interval for attending a course of law lectures given by Mr. Wythe, and of lectures of Natural philosophy given by Mr. Madison then President of William and Mary College."[23] He attended law lectures for less than three months[24]—a time so short, according to his leading biographer, that, in the opinion of the students, "those who finish this study [of law] in a few months, either have strong natural parts or else they know little about it."[25] We may doubt, indeed, whether Marshall was prepared even to take full advantage of so short a law course. He had just fallen in love with his wife-to-be, and his notebook (which is preserved) indicates that his thoughts were at least as much upon his sweetheart as upon the lecturer's wisdom.[26]

Shakespeare, according to Alfred North Whitehead, wrote better poetry for not knowing too much. It may appear paradoxical to make the same assertion with regard to the greatest of American judges, for judicial ability normally depends, in large measure, upon the depth of legal learning. It must, however, be emphasized that Marshall's was not the ordinary judicial role. Great judges are typically not radical innovators. "I venture to suggest," states Justice Felix Frankfurter, "that had they the mind of such originators, the bench is not the place for its employment. Transforming thought implies too great a break with the past, implies too much discontinuity, to be imposed upon society by one who is entrusted with enforcing its law."[27]

Marshall's role, on the other hand, was as much that of legislator as judge. His was the task of translating the constitutional framework into the reality of decided cases. As one commentator puts it, "[H]e hit the Constitution much as the Lord hit the chaos, at a time when everything needed creating."[28] The need was for formative genius—for the transfiguring thought that the judge normally is not called upon to impose on society. Had he been more the trained lawyer, thoroughly steeped in technical learning and entangled in the intricacies of the law, he might not have been so great a judge; for his role called for the talent and the insight of a statesman capable of looking beyond the confines of strict law to the needs of a vigorous nation entered upon the task of occupying a continent.

One aspect of Marshall's education should not be overlooked, though it was far removed from the traditional type of schooling. This was his service as a soldier of the Revolution. It was, his biographer informs us, his military experience—on the march, in camp, and on the battlefield—that taught Marshall the primary lesson of the necessity of strong, efficient government: "Valley Forge was a better training for Marshall's peculiar abilities than Oxford or Cambridge could have been."[29] Above all, his service with Washington confirmed in him the overriding loyalty to an effective Union. Love of the Union and the maxim "United we stand, divided we fall," he once wrote, were "imbibed . . . so thoroughly that they constituted a part of my being. I carried them with me into the army . . . in a common cause believed by all to be most precious, and where I was confirmed in the habit of considering America as my country and Congress as my government."[30] In his most powerful opinions, it has been well said, Marshall appears to us to be talking, not in terms of technical law, but as one of Washington's soldiers who had suffered that the nation might live.

When all is said and done, nevertheless, an element of wonder remains as we contemplate Marshall's work. The magisterial character of his opinions marching with measured cadence to their inevitable logical conclusion has never been equaled, much less surpassed, in judicial history. Clarity, conciseness, eloquence—these are the Marshall hallmarks, which made his opinions irresistible, combined as they were with what Edward S. Corwin termed his "tiger instinct for the jugular vein,"[31] his rigorous pursuit of logical consequences, his power of stating a case, his scorn of qualifying language, the pith and balance of his phrasing, and the developing momentum of his argument. His is the rare legal document whose words can be read and meaning understood by the layman as well as the learned practitioner. And all this from a man almost without formal schooling, either in literature or the law. Were we not historically certain of the fact, we might have as much doubt that such an individual, possessed as he was only of raw genius and the courage to use it, really wrote the masterful opinions that served as the doctrinal foundation of a great nation as some have expressed with regard to the authorship by an unschooled Elizabethan actor of the supreme literary products of the English language.

The Chief Justice

The Supreme Court met for the first time in Washington on Februrary 2, 1801. Since only Justice Cushing was present, the Court *Minutes* state, "A sufficient number of Justices [was] not . . . convened to constitute a quorum"[32] and the Court was adjourned. A quorum was present on the rainy winter morning of February 4 and the Court proceeded to the business of the day: the swearing in of the new Chief Justice, who then took his seat upon the bench.

Aside from Marshall's induction, the February 1801 Term saw little done by the Court. Nor did the press take much notice either of the new Chief Justice's appointment or of the sessions at which he first presided. The February 5, 1801, issue of the *National Intelligencer,* then the leading Washington newspaper, noted only "The Justices of the Supreme Court have made a court, Marshall, Cushing, Chase, and Washington."[33] The lack of press interest illustrates both the Court's low prestige at the time and the fact that it had not yet begun to play its important role in the constitutional structure.

The Court's lack of prestige was strikingly shown by the fact that, as seen, no chamber was provided for it when the new capital was being built. Instead, soon after it convened, the House resolved "[t]hat leave be given to the Commissioners of the City of Washington to use one of the rooms on the first floor of the Capitol for holding the present session of the Supreme Court of the United States."[34] A similar resolution was passed by the Senate.

The room assigned to the Court pursuant to these resolutions was one of the first-floor committee rooms, under the south end of the hall assigned to the House of Representatives. The room measured thirty to thirty-five feet; it had two windows. The chamber was heated by a fireplace set in the wall. Where the bench was located has not been ascertained.[35] We do know, however, that the bench was not raised—a feature present in most courtrooms. The room itself, Benjamin Latrobe, the architect of the Capitol, wrote to Madison, was only half finished and "meanly furnished, very inconvenient."[36]

Here, in the bare quarters of Committee Room 2, the Court sat when Marshall took his seat. The new Chief Justice was a commanding figure in the courtroom—tall, erect, though slightly ungainly if not awkward. His black hair was tied in a queue, after the fashion of the day, but his clothes were frequently disheveled, his appearance anything but that of a man impressed with his high station. William Wirt tells us that, "in his whole appearance, and demeanor; dress, attitudes, gesture; sitting, standing or walking; he is as far removed from the idolized graces of lord Chesterfield, as any other gentleman on earth."[37]

In the courtroom, according to Joseph Story, Marshall's outstanding characteristic was the "quiet, easy dignity" with which he presided. "You heard him," Story wrote, "pronounce the opinion of the Court in a low, but modulated voice, unfolding in luminous order every topic of argument, trying its strength, and measuring its value, until you felt yourself in the presence of the very oracle of the law."[38]

Marshall's accomplishments as Chief Justice were directly related to his overriding conception of law. To Marshall, the law was essentially a social instrument—with the Constitution itself to be shaped to special and particular ends. The Constitution was not to be applied formalistically; it must be applied in light of what it is for. Marshall never doubted that the overriding purpose behind the organic instrument was to establish a nation that was endowed with all the necessary governmental powers. Marshall, wrote John Quincy Adams in his diary at the Chief Justice's death, "settled many great constitutional questions favorably to the continuance of the Union. Marshall has cemented the Union."[39]

The key to the Marshall conception is his seminal dictum: "[W]e must never forget that it is a *constitution* that we are expounding."[40] Justice Frankfurter once termed this the "most important, single sentence in American Constitutional Law." It set the theme for constitutional construction—that the Constitution is not to be read as "an insurance clause in small type, but a scheme of government . . . intended for the undefined and unlimited future."[41]

Just after Marshall assumed his Court seat, Charles Cotesworth Pinckney complained in a letter that "attempts are making to construe away the Energy of our Constitution, to unnerve our Government, & to overthrow that system by which we have risen to our present prosperity."[42] Marshall's Court tenure was devoted to defeating these attempts. It is customary to designate a particular Supreme Court by the name of its Chief Justice. Such designation was more than formalism when Marshall presided over the Court. From the time when he first took his judicial place to his death thirty-four years later, it was emphatically *the Marshall Court* that stood at the head of the judiciary. Throughout his judicial career, Marshall's consistent aim was to use the Supreme Court to lay the constitutional foundation of an effective nation. Before this aim could be realized, the prestige and power of the Court itself had to be increased, for the bench to which Marshall was first appointed could hardly hope to play the positive role in welding the new nation that the great Chief Justice conceived.

As soon as Marshall began to discharge his duties as head of the highest Court, Beveridge's classic biography informs us, "he quietly began to strengthen the Supreme Court."[43] Before Marshall, the Court followed the English practice of having opinions pronounced by each of the Justices. "For the first time," says Beveridge, "the Chief Justice disregarded the custom of the delivery of opinions by the Justices seriatim, and, instead, calmly assumed the function of announcing, himself, the views of that tribunal."[44] Marshall did so in the very first case decided by his Court. "Opinions of the Court" were made the primary vehicle for announcing decisions, with the opinion in virtually all important cases delivered by the Chief Justice himself.[45]

The change from a number of individual opinions to the Court opinion was admirably suited to strengthen the prestige of the fledgling Court. Marshall saw that the needed authority and dignity of the Court could be attained only if the principles it proclaimed were pronounced by a united tribunal. To win conclusiveness and fixity for its decisions, he strove for a Court with a single voice. How well he succeeded is shown by the reception accorded Justice William Johnson, who sought to express his own views in dissent. "During the rest of the Session," he plaintively affirmed in a letter to Jefferson, "I heard nothing but lectures on the indecency of judges cutting at each other, and the loss of reputation which the Virginia appellate court had sustained by pursuing such a course."[46]

Yet, though American constitutional decisions have thus, since Marshall's innovation, been the offspring of the Supreme Court as a whole, it is important to bear in mind that their expression is individual. As Justice Frankfurter said, "The voice of the Court cannot avoid imparting to its opinions the distinction of its own accent. Marshall spoke for the Court. But *he* spoke."[47] And this enabled him to formulate in his own way the landmarks of American constitutional law.

Judicial Review

The first such landmark was, of course, *Marbury v. Madison* (1803),[48] where Marshall asserted for the judicial department the power needed to enable it to forge the constitutional bonds of a strong nation. In 1974, Chief Justice Burger circulated a draft opinion in which he referred to "the power of judicial review first announced by this Court under the authority of Article III in *Marbury.*"[49] In a July 18, 1974, letter to the Chief Justice, Justice Byron R. White objected to the draft's implication that *Marbury v. Madison* had created judicial review. "Because I am one of those who thinks that the Constitution on its face provides for judicial review, especially if construed in the light of what those who drafted it said at the time or later, I always wince when it is inferred that the Court created the power

or even when it is said that the 'power of judicial review [was] first announced in *Marbury v. Madison.*' . . . But perhaps this is only personal
idiosyncrasy."[50]

Despite the White disclaimer, there is no doubt that it was *Marbury v.
Madison* that made judicial review positive constitutional doctrine. It was
the decision in that case that established the power of the Supreme Court
to rule on the constitutionality of a congressional act. *Marbury v. Madison*
arose out of the passage of the Judiciary Act of 1801 during the last days of
John Adams's Federalist Administration. As seen in the last chapter, that
law provided for the appointment of a large number of new federal circuit
judges. A second law, passed a few days later, established courts in the
District of Columbia, with the judges there also to be appointed by the
outgoing President. The bills angered the Jeffersonians, who were about to
assume office. In the words of the pro-Jefferson *Philadelphia Aurora,* they
would allow Adams, in his last days in the Presidency, to provide "sinecure
places and pensions for thorough going Federal partisans." That, of
course, was the intention of Adams, who wished to fill the judiciary with
men who would carry on Federalist principles. As Henry Adams put it,
"the Federalists felt bound to exclude Republicans from the bench, to
prevent the overthrow of those legal principles in which, as they believed,
national safety dwelt."[51]

Acting under the new laws, President Adams appointed over fifty
men—all right-thinking Federalists. Among them, selected as a justice of
the peace for the District of Columbia was a member of a prominent
Maryland family, a banker and large landowner named William Marbury.

The Senate confirmed the appointees, and the outgoing President
signed their commissions of office. It was now the job of the Secretary of
State, John Marshall—who had just been appointed and sworn in as Chief
Justice—to place the Great Seal of the United States on the commissions
and deliver them. Because of the pressure of last-minute duties, Marshall
overlooked delivering the commissions of the justices of the peace and the
new President, Thomas Jefferson, ordered his Secretary of State, James
Madison, not to deliver them.

Marbury then brought an action for mandamus against Madison, asking the court to order Madison to deliver his commission. Marbury
brought the action directly in the Supreme Court under section 13 of the
Judiciary Act of 1789, which gave the Court original jurisdiction in mandamus cases against federal officials.

The case appeared only to raise the question of whether the Court
could issue a mandamus against the Secretary of State. Seemingly, the
Court could either disavow having such power over the Executive branch
and dismiss Marbury's application, or it could order Madison to deliver the
commission to him. The first course would have meant abdicating the
essentials of "The judicial Power" conferred on the Court by the Constitution. But the second course would have been no better. While it would
have declared the Court's authority to hold the Executive to the law, it

would have remained only a "paper declaration," for the Court would have no power to enforce its mandate.

That Marshall was able to choose neither course was a tribute to his judicial statesmanship. He escaped from the dilemma by ruling that section 13 of the Judiciary Act was unconstitutional on the ground that, since the original jurisdiction conferred upon the Supreme Court by the Constitution was exclusive, it could not be enlarged by congressional law. Thus the Court could deny Marbury's application, not because the Executive branch was above the law (Marshall's opinion, on the contrary, contained a strong repudiation of that claim), but because the Court itself did not possess the original jurisdiction to issue the writ that Marbury requested.

To reach that decision and rule that the congressional law was invalid, the Court was asserting that it had power to review the constitutionality of acts of Congress. From a strategic point of view, the Court could not have chosen a better case through which to declare that power. Since its decision did not rule in favor of Marbury, there was nothing that would bring on a direct conflict with the Jefferson Administration.

More than that, the assertion of the greatest of all judicial powers—review of laws—was made in a case that ruled against authority that had been granted to the Court. The Jeffersonians found it hard to attack the decision in which the Court declined—even from Congress—a jurisdiction to which it was not entitled by the Constitution. "To the public of 1803," Charles Warren tells us, "the case represented the determination of Marshall and his Associates to interfere with the authority of the Executive, and it derived its chief importance then from that aspect." On the other hand, "To the lawyers of today, the significance of Marshall's opinion lies in its establishment of the power of the Court to adjudicate the validity of an Act of Congress—the fundamental decision in the American system of constitutional law."[52]

Marbury v. Madison is *the* great case in American constitutional law because it was the first case to establish the Supreme Court's power to review constitutionality. Indeed, had Marshall not confirmed review power at the outset in his magisterial manner, it is entirely possible it would never have been insisted upon, for it was not until 1857 that the authority to invalidate a federal statute was next exercised by the Court.[53] Had the Marshall Court not taken its stand, more than sixty years would have passed without any question arising as to the omnipotence of Congress. After so long a period of judicial acquiescence in congressional supremacy, it is probable that the opposition then would have been futile.

To be sure, Marshall in *Marbury* merely confirmed a doctrine that was part of the American legal tradition of the time, derived from both the colonial and Revolutionary experience. One may go further. Judicial review was the inarticulate major premise upon which the movement to draft constitutions and bills of rights was ultimately based. The doctrine of unconstitutionality had been asserted by Americans even before the first written constitutions, notably by James Otis in his 1761 attack on general writs of assistance[54] and by Patrick Henry in 1763 when he challenged the right of the Privy Council to disallow the Virginia Two-penny Act.[55] The Otis-Henry doctrine was a necessary foundation, both for the legal theory underlying the American Revolution and the constitutions and bills of rights that it produced.

The doctrine could, however, become a principle of positive law only after independence, when written constitutions were adopted that contained binding limitations, beyond the reach of governmental power. Judicial review started to become a part of the living law during the decade before the adoption of the Federal Constitution. Cases in at least five states between 1780 and 1787 involved direct assertions of the power of judicial review.[56] Marshall himself could affirm, in his *Marbury* opinion, not that the Constitution establishes judicial review, but only that it "confirms and strengthens the principle."[57] Soon after the Constitution went into effect, further assertions of review authority were made by a number of federal judges, including members of the Supreme Court sitting on circuit. In addition, the first Supreme Court itself, as we saw in the last chapter, decided cases such as *Hylton v. United States,*[58] which were based upon its possession of review power. Hence, when Madison introduced the proposed amendments that became the Federal Bill of Rights, he could recognize expressly that the new guaranties would be enforced by the courts. For Madison, as for his compatriots generally, judicial review was an implicit aspect of the constitutional structure.

That Marshall's opinion in *Marbury v. Madison* was not radical innovation does not at all detract from its importance. The great Chief Justice, like Jefferson in writing the Declaration of Independence, may have merely set down in clear form what had already been previously declared. Yet, as Marshall's biographer observes, Thomas Jefferson and John Marshall as private citizens in Charlottesville and Richmond might have written declarations and opinions all their lives, and today none but the curious student would know that such men had ever lived.[59] It is the authoritative positions which those two Americans happened to occupy that have given immortality to their enunciations. If Marshall's achievement in *Marbury v. Madison* was not transformation but only articulation, what has made it momentous is the fact that it was magisterial articulation as positive law by the highest judicial officer of the land.

Marshall's *Marbury* reasoning is a more elaborate version of that used by Hamilton in *The Federalist*, No. 78. But it was Marshall, not Hamilton, who elevated that reasoning to the constitutional plane, and he did so in terms so firm and clear that the review power has never since been legally doubted. As the encomium of a leading constitutional scholar puts it, "There is not a false step in Marshall's argument." Instead, his "presentation of the case . . . marches to its conclusion with all the precision of a demonstration from Euclid."[60]

In Marshall's *Marbury* opinion, the authority to declare constitutionality flows inexorably from the judicial duty to determine the law: "It is emphatically the province and duty of the judicial department to say what the law is. . . . If two laws conflict with each other, the courts must decide that case comformably to the law, disregarding the constitution; or conformably to the constitution, disregarding the law; the court must determine which of these conflicting rules governs the case. This is of the very essence of the judicial duty."[61] One may go further and say that judicial review, as declared in *Marbury v. Madison,* has become the sine qua non of the American constitutional machinery: draw out this particular bolt, and the machinery falls to pieces.

Addressing the court in the 1627 *Five Knights' Case,* the Attorney General, arguing for the Crown, asked, "Shall any say, The King cannot do this? No, we may only say, He will not do this."[62] It was precisely to ensure that in the American system one would be able to say "The State *cannot* do this" that the people enacted a written Constitution containing basic limitations upon the powers of government. Of what avail would such limitations be, however, if there were no legal machinery to enforce them? Even a Constitution is naught but empty words if it cannot be enforced by the courts. It is judicial review that makes constitutional provisions more than mere maxims of political morality.

Review Power over States

To hold as Marshall did in *Marbury v. Madison* that the Supreme Court can review the constitutionality of acts of Congress is, however, to lay down only half of the doctrine of judicial review. According to a noted statement by Justice Holmes, indeed, it is the less important half. "I do not think," he asserted, "the United States would come to an end if we lost our power to declare an Act of Congress void. I do think the Union would be imperilled if we could not make that declaration as to the laws of the several states."[63] The power to pass on the validity of state legislation is a necessary part of the review power if the Constitution is truly to be maintained as supreme law throughout the country.

In was in the 1810 case of *Fletcher v. Peck*[64] that the Supreme Court first exercised the power to hold a state law unconstitutional. In ruling that a Georgia statute violated the Contract Clause of the Constitution, Marshall, who delivered the opinion, declared categorically that the state could not be viewed as a single, unconnected sovereign power, on whom no other restrictions are imposed than those found in its own constitution. On the contrary, it is a member of the Union, "and that Union has a constitution the supremacy of which all acknowledge, and which imposes limits to the legislatures of the several states, which none claim a right to pass."[65]

In *Fletcher v. Peck,* Marshall, in Beveridge's description, laid the second stone in the structure of American constitutional law. Yet even this was still not enough to enable the Supreme Court to maintain the Constitution as the supreme law of the land. In addition to the power to review the validity of legislative acts of both the nation and the states, review power over the judgments of the state courts is also necessary. In a system in which state judicatures coexist with those of the nation, vested with equal competence to pronounce judgment on constitutional issues, it is essential that their judgments be subjected to the overriding control of the highest tribunal. The review power over state courts is necessary if the Supreme Court is to uphold national supremacy when it conflicts with state law or is challenged by state authority.

The appellate power of the Supreme Court over state court decisions, in order to harmonize them with the Constitution, laws, and treaties of the United States, was established in two memorable decisions by the Marshall Court. The first was rendered in 1816 in *Martin v. Hunter's Lessee.*[66] That case arose out of the refusal of the highest court of Virginia to obey the mandate issued by the Supreme Court in an earlier case in which the Virginia court's decision had been reversed on the ground that it was contrary to a treaty of the United States. The Virginia judges had asserted that they were not subject to the highest bench's appellate power "under a sound construction of the constitution of the United States" and ruled that the provision of the first Judiciary Act which "extends the appellate juris-diction of the Supreme Court to this court, is not in pursuance of the constitution of the United States."

The Supreme Court, in an opinion by Justice Story, categorically rejected the holding that it could not be vested with appellate jurisdiction over state court decisions. Marshall did not deliver the opinion because a personal interest in the case led him to decline to participate. There is no doubt, however, that Story's opinion was strongly influenced by the Marshall view on judicial power. Beveridge tells us, indeed, that it was commonly supposed that Marshall "practically dictated" Story's opinion.[67] Be that as it may, the opinion was certainly one that, save for some turgidity of language, the great Chief Justice could have written.

Five years later, in the case of *Cohens v. Virginia,*[68] Marshall was given the opportunity to demonstrate that such was the case. Defendants there had been convicted in a Virginia court of violating that state's law prohibit-ing the sale of lottery tickets. They sought a writ of error from the Supreme Court on the ground that, since the lottery in question had been autho-rized by an act of Congress, the state prohibitory law was invalid since it conflicted with federal law. Again it was claimed that the highest Court had no appellate power over the state courts. With typical force, Marshall declared that such an argument was contrary to the Constitution. The states, he affirmed, are not independent sovereignties; they are members of one great nation—a nation endowed by the basic document with a govern-ment competent to attain all national objects. "The exercise of the appellate power over those judgments of the state tribunals which may contravene the constitution or laws of the United States, is, we believe, essential to the attainment of those objects."[69] Let the nature and objects of the Union be considered, let the great principles on which the constitutional framework rests be examined, and the result must be that the Court of the nation must be given the power of revising the decisions of local tribunals on questions which affect the nation.

According to Marshall's biographer, the opinion in *Cohens v. Virginia* is "one of the strongest and most enduring strands of that mighty cable woven by him to hold the American people together as a united and imperishable nation."[70] Certain it is that Marshall's masterful opinion conclusively settled the competence of the high bench to review the decisions of state courts. Since *Cohens v. Virginia,* state attempts to make themselves the final arbiters in cases involving the Constitution, laws, and treaties of the United States have been foredoomed to defeat before the bar of the highest Court.

With the decision in *Cohens v. Virginia,* the structure of judicial power erected by the Marshall Court was completed. The authority of the judicial department to enforce the Constitution against both the national and state governments became an accepted part of American constitutional law. All governmental acts, whether of the nation or the states, now had to run the gantlet of review by the Supreme Court to determine whether they were constitutional. And that Court itself was now the veritable supreme tribunal of the land, for it was vested with the last word over the state, as well as the federal, judiciaries.

National Power

For Marshall, judicial review, like law itself, was a means not an end. The end was the attainment of the goal intended by "the framers of the Constitution, who were his compatriots"[71]—an effective national government endowed with vital substantive powers, the lack of which had rendered the Articles of Confederation sterile. Judicial review was the tool that enabled Marshall to translate this goal into legal reality.

To Marshall, the overriding end to be served by our public law was nationalism in the broad sense of that term. The law was to be employed to lay down the doctrinal foundations of an effective nation. That end was attained through a series of now-classic decisions that had two principal aims: to ensure that the nation possessed the powers needed to enable it to govern effectively; and to ensure federal supremacy vis-à-vis state powers.

The key case in this respect was *McCulloch v. Maryland,*[72] decided in 1819. It established the doctrine of implied powers in our constitutional law, resolving in the process the controversy between those who favored a strict and those who favored a broad construction of the Necessary-and-Proper Clause of the Constitution. That clause, after enumerating the specific powers conferred on Congress, authorizes it "to make all laws which shall be necessary and proper for carrying into execution the foregoing powers, and all other powers vested by this Constitution in the government of the United States." Conflicting approaches had been taken to the clause by Jefferson and Hamilton. Jefferson had adopted a strict view, emphasizing the word *necessary* in the clause: it endowed the Federal Government only with those powers indispensable for the exercise of its enumerated powers. The broader Hamilton view maintained that to take the word in its rigorous sense would be to deprive the clause of real practical effect: "It is essential to the being of the National Government, that so erroneous a conception of the meaning of the word *necessary* should be exploded."[73]

In *McCulloch v. Maryland,* Marshall adopted the broad Hamiltonian approach. The case itself presented the same issue on which Jefferson and Hamilton had differed—the constitutionality of the Bank of the United States, established by Congress to serve as a depository for federal funds and to print bank notes. Under pressure from its state banks, Maryland imposed a tax upon the federal bank's Baltimore branch and then brought suit in a state court against James William McCulloch, the branch's cashier, when he refused to pay the Maryland tax. The state won its suit; but the Federal Government, facing similar taxes in other states, appealed to the Supreme Court.

To decide whether the Maryland tax law was constitutional the Court had to decide whether Congress had the power to charter the bank. In relying upon the Necessary-and-Proper Clause for an affirmative answer, Marshall relied directly upon the reasoning of Hamilton's *Bank Opinion.* If the establishment of a national bank would aid the government in the exercise of its granted powers, the authority to set one up would be implied. "Let the end be legitimate," reads the key sentence of the Marshall opinion, "let it be within the scope of the constitution; and all means which are appropriate, which are plainly adapted to that end, which are not prohibited, but consist with the letter and spirit of the constitution, are constitutional."[74]

This passage is essentially similar to the "criterion of what is constitutional" contained in Hamilton's *Bank Opinion.* Once again, however, it was Marshall, not Hamilton, who made the implied powers doctrine an accepted element of our constitutional law and finally put to rest the view that the Necessary-and-Proper Clause extended only to laws that were indispensably necessary.

Marshall himself, writing extrajudicially, stressed that if the rejected view "would not absolutely arrest the progress of the government, it would certainly deny to those who administer it the means of executing its acknowledged powers." Indeed, Marshall asserted, "[T]he principles maintained by the counsel for the state of Maryland . . . would essentially damage the constitution, render the government of the Union incompetent to the objects for which it was instituted, and place all powers under the control of the state legislatures."[75] Or, as it was put in a Marshall letter to Justice Story, "If the principles which have been advanced on this occasion were to prevail, the Constitution would be converted into the old confederation."[76]

Marshall used *McCulloch v. Maryland* not only to ensure that the nation had the powers needed to govern effectively; he used it also to cement the federal supremacy declared in article VI. Having decided, as just seen, that Congress had the power to charter the Bank of the United States, the Court then had to determine whether Maryland might tax the bank. The Marshall opinion answered the question of state power with a categorical negative. Since the bank was validly established by Congress, it followed logically that it could not be subjected to state taxation.

The national government, declared Marshall, "is supreme within its sphere of action. This would seem to result necessarily from its nature." National supremacy is utterly inconsistent with any state authority to tax a federal agency. "The question is, in truth, a question of supremacy; and if the right of the states to tax the means employed by the general government be conceded the declaration that the constitution, and the laws made in pursuance thereof, shall be the supreme law of the land, is empty and unmeaning declamation."[77]

Federal supremacy, to Marshall, meant "that the states have no power, by taxation or otherwise, to retard impede, burden, or in any manner control, the operations of the Federal Government, or its agencies and instrumentalities.[78] It also meant that federal action, if itself constitutional, must prevail over inconsistent state action. This second meaning was developed in *Gibbons v. Ogden* (1824).[79] The decision there held that New York statutes which had granted an exclusive license to use steam navigation on the waters of the state were invalid so far as they applied to vessels licensed under a federal statute to engage in coastwise trade. According to Marshall's opinion, "[T]he laws of New York . . . come into collision with an act of Congress, and deprived a citizen of a right to which that act entitles him." In such a case, "[T]he acts of New York must yield to the law of Congress; and the decision [below] sustaining the privilege they confer, against a right given by a law of the Union, must be erroneous. In every such case, the act of Congress . . . is supreme; and the law of the state, though enacted in the exercise of powers not controverted, must yield to it."[80]

Commerce Power

The same expansive approach to federal authority can be seen in the *Gibbons v. Ogden* opinion on the most important substantive power vested in the Federal Government in time of peace: the power "[t]o regulate commerce with foreign nations, and among the several States."[81] The need to federalize regulation of commerce was one of the principal needs that motivated the Constitutional Convention of 1787. Yet the delegates there were interested mainly in the negative aspects of such regulation, concerned as they were with curbing state restrictions that had oppressed and degraded the commerce of the nation. It was Marshall, in *Gibbons v. Ogden,* who first construed the commerce power in a positive manner, enabling it to be fashioned into a formidable federal regulatory tool.

Gibbons v. Ogden itself arose out of the invention of the steamboat by Robert Fulton. Fulton and Robert Livingston, American Minister in Paris when the inventor had demonstrated his steamboat in France in 1803, secured from the New York legislature a monopoly of steam navigation on the waters of that state. Under the monopoly, the partners licensed Aaron Ogden to operate ferryboats between New York and New Jersey. When Thomas Gibbons began to run steamboats in competition with Ogden and without New York permission (though he had a coasting license from the Federal Government), Ogden sued to stop Gibbons.

The case became a sensational battle between the two men and almost wrecked them both. However, the Supreme Court decision nullifying Ogden's monopoly was more than the settling of a quarrel between two combative men. Marshall seized the opportunity to deliver an opinion on the breadth of Congress's authority under the Commerce Clause. The clause vests in Congress the power "to regulate commerce." The noun "commerce" determines the subjects to which congressional power extends. The verb "regulate" determines the types of authority that Congress can exert. Both the noun and the verb were defined most broadly in Marshall's opinion.

"Commerce," in Marshall's view, covered all economic intercourse—a conception comprehensive enough to include within its scope all business dealings: "It describes the commercial intercourse between nations, and parts of nations, in all branches."[82]

Having given such a broad construction to the noun "commerce," Marshall proceeded to take an equally liberal view of the meaning of the verb "regulate." "What is this power?" he asked. "It is the power to regulate; that is, to prescribe the rule by which commerce is to be governed. This power, like all others vested in congress is complete in itself, may be exercised to its utmost extent."[83]

According to the most recent history of the Marshall Court, however, Marshall's *Gibbons* opinion was "highly inconclusive. . . . *Gibbons,* for all the fanfare with which it was received, settled very little and that in an awkward fashion."[84] This surely understates the seminal role of the *Gibbons* decision in the expansion of federal power. Justice William O. Douglas once stated that the Commerce Clause "is the fount and origin of vast power."[85] But that is true only because, in *Gibbons,* "Marshall described the federal commerce power with a breadth never yet exceeded."[86] So interpreted, the Commerce Clause was to become the source of the most important powers the Federal Government exercises in time of peace. If in recent years it has become trite to point out how regulation from Washington controls Americans from the cradle to the grave, that is true only because of the Marshall Court's emphasis at the outset on the embracing and penetrating nature of the federal commerce power.

Looking back, it is easy to conclude that the Marshall conception was demanded by the needs of the developing nation. To Americans today, the broad construction of the federal commerce power was plainly essential to the period of growth upon which the United States was entering. To the men of Marshall's day, the need was not nearly so obvious. To appreciate the very real contribution to national power made by *Gibbons v. Ogden,* we must contrast the opinion there with the restricted scope which President James Monroe had just given to the commerce power in his 1822 veto of the Cumberland Road Act (which provided for the building of a federal road to the West). According to Monroe, "A power . . . to impose . . . duties and imposts in regard to foreign nations and to prevent any on the trade between the States, was the only power granted."[87] Marshall's sweeping opinion ruthlessly brushes aside this narrow theory. That Marshall was able to mold his convictions on effective national power into positive law at the outset made a profound difference to the development of the American nation.

Contracts, Corporations, and Property

According to Vernon L. Parrington, "The two fixed conceptions which dominated Marshall during his long career on the bench were the sovereignty of the federal state and the sanctity of private property."[88] In the landmark cases just discussed, Marshall used constitutional law to establish federal sovereignty as the foundation of the polity. But he also employed the law to further the protection of property rights, which, to him as to other jurists of the day, was a primary end of law. Marshall subscribed completely to the then-prevailing conception of a natural right to acquire property and to use it as one saw fit.[89] He referred to "the right which every man retains to acquire property, to dispose of that property according to his own judgement, and to pledge himself for a future act. These rights are not given by society, but are brought into it."[90]

To Marshall, the natural right of property was one to be exercised free from governmental interference. "I consider," he once wrote, "the interference of the legislature in the management of our private affairs, whether those affairs are committed to a company or remain under individual direction, as equally dangerous and unwise. I have always thought so and still think so."[91]

However, the Marshall conception of property rights, like that of other early American jurists, differed essentially from that in England. American property law was expansive rather than defensive in character. We can see this in the changes already made in this country in the land law, which had freed real property from most of the restrictions that still prevailed in English law. Speaking of the English system, a New York court asserted, "This ancient, complicated and barbarous system . . . is entirely abrogated."[92] Real property in this country became allodial (i.e., unrestricted) with the freehold established as the normal type of land title.[93]

If land could now be dealt with according to the will of the owner, the same was soon to be true of other forms of property. But individual property rights alone, even liberated from common law archaisms, were scarcely enough for the needs of the expanding American economy. The industrial growth which so strikingly altered the nature of the society during the nineteenth century could scarcely have been possible had it depended solely upon the initiative and resources of the individual entrepreneur. It was the corporate device that enabled men to establish the pools of wealth and talent needed for the economic conquest of the continent.

The ability of American courts to adapt the common law to the nation's requirements is nowhere better seen than in the development of corporation law. The American law of corporations, more than most branches of our judge-made law, was an indigenous product.[94] The English courts had for centuries been deciding cases relating to corporate problems, but the law developed by them dealt almost entirely with nonprofit corporations and was of limited value in solving the problems confronting business enterprises in the United States. The development of the business corporation (formed to carry on business for profit) and resolution of the legal issues connected with it were almost entirely the handiwork of American law.

Corporation law itself well illustrates the developing conception of American law. From the beginning, American judges looked with favor upon the corporate device as a method of doing business. It was during the Marshall era that the first steps were taken in "[t]he constant tendency of judicial decisions in modern times . . . in the direction of putting corporations upon the same footing as natural persons."[95] Corporations may, in Coke's famous phrase, "have no souls,"[96] but they gained the essentials of legal personality by the decisions of the Marshall Court.

It was Marshall who laid down the first essential prerequisite to corporate expansion in the *Dartmouth College* case (1819).[97] The decision there vested the corporation with constitutionally protected contract rights by holding that a corporate charter was a contract within the protection of the Contract Clause of the Constitution. The corporate creature of the law— "invisible, intangible, and existing only in contemplation of law"[98]—was endowed with basic legal rights even against its creator.

Sir Henry Maine, writing in 1885, characterized the *Dartmouth College* decision as "the basis of the credit of many of the great American Railway Incorporations." It is, he went on, its principle "which has in reality secured full play to the economical forces by which the achievement of cultivating the soil of the North American Continent has been performed."[99] At a time when no other constitutional provision would serve the purpose, corporate property rights were brought under the fostering guardianship of the Contract Clause. Those who were called upon to pool their wealth and talents in the vast corporate enterprises needed for the nation's development were thus ensured that their contributions would not remain at the mercy of what Justice Story termed "the passions of the popular doctrines of the day."[100] Before *Dartmouth College,* there were still relatively few manufacturing corporations in the country. Under the confidence created by Marshall's decision, those corporations proliferated to such an extent that they soon transformed the face of the nation.[101]

A historian of the Marshall Court concludes that its decisions affecting property and business "facilitated commerce, shaped the law to conform to the dictates and practices of the market, and developed doctrinal rules that were consistent with dynamic and expansive uses of property and mechanisms of commercial exchange."[102] More than that, the justices were well aware of the social and economic ramifications of their decisions. That was particularly true of the man who sat in the Court's center chair. Here, too, Marshall looked at the law as a tool to enable the needs of the society to be served. In cases like *Dartmouth College,* he helped mold legal doctrine both to protect property rights and to further economic expansion.

Of course, Marshall made mistakes in using the law as a means to accomplish the ends he favored. His *Dartmouth College* decision may have fostered the corporate proliferation that soon distinguished American economic development. In Bank of the *United States v. Dandridge,*[103] however, Marshall refused to loosen the bonds of corporate formalism that had so restricted the use of the corporation in English law. At issue was what the Supreme Court historian termed "a vital question of corporation law—whether approval of acts of its agents by a corporation may be shown by presumptive testimony or only by written record and vote."[104] Marshall held on circuit that the record and vote were necessary. As he wrote to a colleague at the time, "I thought the assent of the Bank Directors indispensable . . . and that consent I thought could be given only at the board and could be proved only by the minutes of their proceedings."[105]

An "affirmance of this view by the Court would have retarded the commercial development of this country immeasurably,"[106] for it would have required corporate contracts to be cast in the elaborate forms of English law, with a record and vote required for each contract. Marshall himself recognized that his decision was contrary to corporate practice. "The case," he wrote to Justice Story, "goes to the Supreme Court & will probably be reversed. I suppose so, because I conjecture that the practice of banks has not conformed to my construction of the law." Marshall did, however, indicate, "I shall retain the opinion I have expressed."[107] When the Court, in an opinion by Story, did reverse his circuit decision, Marshall delivered a rare lengthy dissent.

Dandridge was the exception in the Marshall jurisprudence. Normally, Marshall spoke for the Court and his opinions used the law to lay the legal foundations of the political and economic order that he favored. Marshall is considered great because his conception coincides with what we now deem were the dominant needs of the new nation. This is all but self-evident to anyone familiar with our legal history, so far as Marshall's public-law opinions are concerned. They confirmed the expansive interpretation of the Constitution and helped to weave the legal fabric of Union in such a way that it was to prove strong enough to withstand even the shock of civil war.

But the same approach was generally used by Marshall in his opinions dealing with property and business. It should not be forgotten that, even in the Marshall era of burgeoning constitutional law, nonconstitutional cases still made up the bulk of the Court's docket.[108] The developing American common law was also shaped by Marshall and his colleagues to serve the needs of the expanding society and economy. Many of the cases involved adjudication of real property disputes, and the decisions furthered the change in the legal conception of land from a static locus to a commodity that could be bought and sold in a market economy.[109] Tocqueville contrasts the European expectation of passing on to sons land held in a long family line with his observation of the American practice, where the farmer "brings lands into tillage in order to sell it again, . . . on the speculation that, as the state of the country will soon be changed by the increase of population, a good price may be obtained for it."[110] The Marshall Court, like other courts of the day, fostered the change from the European to the American practice.

The same was true of the Court's commercial cases. The legal status of contracts and secured transactions produced a jurisprudence that would facilitate the operation of the developing commercial market.[111] Corporate growth was advanced by the treatment of corporations as property owners having protected constitutional rights. Commercial dealings were furthered by emphasis on the security of transactions freely entered into; the basic legal principle became that promises be kept and undertakings be carried out in good faith. When the Court held in *Coolidge v. Payson* (1817)[112] that a promise to accept a bill of exchange had the same effect as an acceptance, where a person had taken it on the credit of the promise, the Marshall opinion explained that the decision would promote commercial transactions based upon the bill: "The great motive for considering a promise to accept, as an acceptance, is, that it gives credit to the bill, and may induce a third person to take it." Marshall stressed that "[i]t is of much importance to merchants that the question should be at rest" and, according to a note by the reporter, the Marshall "decision may be considered as settling the law of the country on the subject."[113]

However, like Lord Mansfield, his great counterpart on the other side of the Atlantic, Marshall recognized that commercial law could not be based upon common law alone. "Bills of exchange," he affirmed on circuit, "are transferable . . . by the custom of merchants. Their transfer is regulated by usage and that usage is founded in convenience."[114] In the *Coolidge* case, Marshall refused to follow English decisions disapproving of "the doctrine of implied acceptance"[115] because he deemed them to be "anticommercial"[116] and unsuited to an expanding market economy. Instead, under the Marshall decision, a promise to accept a bill of exchange converted the bill into an instrument which could be negotiated by other, even unknown, parties. The original instrument had been turned into a negotiable instrument that could be circulated to third persons.[117]

To the Marshall Court, the key consideration was the need to further negotiability so that commercial paper could properly serve the needs of the developing economy. As Justice Story was later to put it, "It is for the benefit and convenience of the commercial world to give as wide an extent as practicable to the order and circulation of negotiable paper."[118] The result, a history of the Marshall Court concludes, was that "[t]he sanctioning of negotiability meant that commercial paper would be a commonplace of American business life."[119] Promissory notes and other negotiable instruments became an essential medium of commercial exchange.

The legal rules developed by Marshall and his colleagues reflected changing business practices and facilitated their spread. The Court "shaped the law to conform to the dictates and practices of the market, and developed doctrinal rules that were consistent with dynamic and expansive uses of property and mechanisms of commercial exchange."[120] The changes in the economy were paralleled by the rules affecting contracts, property, and negotiable instruments laid down by the Marshall Court.

Marshall's Own Defense

Though, as seen, public opinion at the time largely ignored the *Marbury v. Madison* holding on judicial review, Thomas Jefferson was well aware of its potential. In a letter to Judge Spencer Roane, Jefferson characterized Marshall's *Marbury* opinion as follows: the Constitution "has given, according to this opinion, to one of them alone the right to prescribe rules for the government of the others; and to that one too which is unelected by, and, independent of, the nation, for experience has already shewn that the impeachment it has provided is not even a scare crow. . . . The constitution, on this hypothesis is a mere thing of wax in the hands of the judiciary, which they may twist and shape into any form they please."[121]

Spencer Roane, to whom Jefferson thus wrote, was the President of the Virginia Court of Appeals, and head of the Republican party organization in the state. Jefferson had planned to appoint Roane as Chief Justice upon the death of Chief Justice Ellsworth; his plan had been thwarted when Ellsworth resigned in time to permit Adams to nominate Marshall as his successor. From then on, there was a bitter enmity between Marshall and Roane. In fact, according to Marshall's biographer, Roane was one of only two men who hated the Chief Justice personally (the other was Jefferson).[122]

But the antagonism between Marshall and Roane was based upon more than personal ill will. Their hostility was fueled by fundamental differences in constitutional philosophy and principle. Marshall was, of course, the great exponent of the Federalist conception of the Constitution, elevating that view to the supreme law of the land. Roane had a diametrically opposed view of the legitimate sphere of federal authority. He has been called "the most energetic states' rights ideologue of all."[123]

In 1819 Roane published four essays in the *Richmond Enquirer,* signed "Hampden," in which he bitterly attacked the nationalistic decisions of the

Marshall Court. The Chief Justice was greatly disturbed by the "Hampden" essays, which he likened to a "most furious hurricane" that had burst on the judges' heads. "I find myself," he stated, "more stimulated on this subject than on any other because I believe the design to be to injure the Judges & impair the constitution. I have therefore thoughts of answering these essays."[124]

Marshall published nine essays in the *Alexandria Gazette* in answer to Roane. They were signed "Friend of the Constitution," for the Chief Justice sought to keep his authorship secret and it was not until recently that it was made known that the essays were written by Marshall himself.

His extrajudicial excursus—so untypical of the judge, who otherwise let his opinions speak for themselves—showed how strongly Marshall felt about the principles laid down in his leading opinions. In addition, his "Friend of the Constitution" essays dramatically demonstrate how Marshall used our public law to give effect to the type of polity he deemed appropriate to the developing nation.

Roane's "Hampden" essays had attacked the Marshall decisions as "well calculated to aggrandize the general government, at the expense of the states; to work a consolidation of the confederacy."[125] Their result, Roane asserted, was "to give a Carte Blanche to our federal rulers, and to obliterate the state governments, forever, from our political system." In particular, the *McCulloch v. Maryland* interpretation of the Necessary-and-Proper Clause had been the object of the Roane censure. Under the clause, Hampden asserted, "the only enquiry is whether the power is properly an incident to an express power and necessary to its execution, and if it is not, congress cannot exercise it." Only "such means were implied, and such only, as were *essential* to effectuate the power."

Marshall's essays replied "that this charge of 'in effect expunging those words from the constitution,' exists only in the imagination of Hampden. It is the creature of his own mind." The Constitution as "expounded by its enemies" would "become totally inoperative." They "may pluck from it power after power in detail, or may sweep off the whole at once by declaring that it shall execute its acknowledged powers by those scanty and inconvenient means only which the states shall prescribe." The national government would then "become an inanimate corpse, incapable of effecting" the objects for which it was created.

Marshall strongly objected to Hampden's claim that the grant of implied powers "is limited to things strictly necessary, or without which the obligation could not be fulfilled." Instead, said Marshall, the grant to Congress "carried with it such additional powers as were *fairly incidental* to them."

Hampden had relied upon common-law authorities, particularly Coke. Marshall countered by expanding on his seminal *McCulloch* dictum.[126] The difference, he declared, between "the examples taken . . . from the books of the common law; and the constitution of a nation" are apparent. The Constitution is not a contract: "It is the act of a people,

creating a government, without which they cannot exist as a people
. . . it is impossible to construe such an instrument rightly, without ad-
verting to its nature, and marking the points of difference which distinguish it from ordinary contracts." Such an instrument gives only the "great
outlines" of governmental power and is not to be construed like an instrument with "a single [object] which can be minutely described."

As in his opinions, Marshall's essays stressed the Hamiltonian principle of liberal construction which must govern constitutional doctrine. Under it, the "means for the execution of powers should be proportioned to
the powers themselves."

Roane had repudiated Marshall's famous dictum, that "it is a *constitution* we are expounding,[127] saying, "If it is a constitution, it is also a *compact*
and a limited and defined compact." To this Marshall sarcastically noted,
"[H]e is so very reasonable as not to deny that it is a constitution." All
Marshall meant, he wrote, was "only that, in ascertaining the true extent of
those powers, the constitution should be fairly construed." Under this
approach, "the choice of these means devolve on the legislature, whose
right, and whose duty it is, to adopt those which are most advantageous to
the people, provided they be within the limits of the constitution."

In particular, Marshall rejected the notion that the Necessary-and-
Proper Clause limited Congress "to such [laws] as are indispensable, and
without which the power would be nugatory." Such a principle would be
disastrous: "[T]his principle, if recognized, would prove many of those
acts, the constitutionality of which, are universally acknowledged . . . to
be usurpations."

Toward the end of his essays, Hampden had attacked the Court's
jurisdiction to decide cases involving conflicts between federal and state
power. Marshall realized that the core issue here between him and his
adversary was that of the nature of the Constitution. "[T]he point to which
all his arguments tend," Marshall stated, is "his idea that the ligament
which binds the states together, is an alliance, or a league." To support this
principle, "an unatural [*sic*] or restricted construction of the constitution is
pressed upon us . . . which would reduce the constitution to a dead
letter." The Roane attack upon the Court's jurisdiction was based upon
what Marshall termed the "unaccountable delusion . . . that our constitution is . . . a compact, between the several state governments, and the
general government."

As in his opinions, Marshall's essays completely rejected the compact
theory upon which the position of states'-rights advocates such as Roane
was based. "Our constitution," Marshall affirmed in his essays, "is not a
compact. It is the act of a single party. It is the act of the people of the
United States, assembling in their respective states, and adopting a government for the whole nation."

Marshall next went into the Court's constitutional role. "For what
purpose," he asked, "was [the judicial] department created?" The answer
was apparent to "any reasonable man . . . must it not have been the

desire of having a tribunal for the decision of all national questions?" The Constitution clearly answered in the affirmative when it provided for federal judicial power over "cases . . . arising under the constitution, and under the laws and treaties of the U. States."

Roane had, however, urged that the Supreme Court could not decide such a case "without treading under foot the principle that forbids a man to decide his own cause." Marshall countered by asking, "To whom more safely than to the judges are judicial questions to be referred? . . . It is not then the party sitting in his own cause. It is the application to individuals by one department of the acts of another department of the government. The people are the authors of all; the departments are their agents; and if the judge be personally disinterested, he is as exempt from any political interest that might influence his opinion, as imperfect human institutions can make him."

Marshall asked what alternative there was to Supreme Court jurisdiction in cases arising under the Constitution. Roane had said, "[T]hey must of course be decided in the state courts." But see where that would leave us. "It follows then that great national questions are to be decided, not by the tribunal created for their decision by the people of the United States, but by the tribunal created by the state which contests the validity of the act of congress, or asserts the validity of its own act." The result was summed up in Marshall's concluding sentences: "Let Hampden succeed, and that instrument will be radically changed. The government of the whole will be prostrated at the feet of its members; and that grand effort of wisdom, virtue, and patriotism, which produced it, will be totally defeated."

The Roane-Marshall essays show how the opposing schools of constitutional construction used their jurisprudence to further the type of polity and society that they favored. Leader of the Republican Junto in Virginia, Roane shared the Jeffersonian vision of national development—emphasizing an idealistic agrarianism instead of the centralizing capitalism that was emerging. Jefferson had urged the danger of a consolidated government, which invariably tended toward self-aggrandizement—with the inevitable result a political Leviathan. The danger could be avoided by circumscribing central power and emphasizing states' rights.

Marshall's vision was an entirely different one. For him, what was required was a truly national government that would neet the needs of an expanding people and promote the physical and economic conquest of the continent. The sovereignty of the federal state was the fixed conception that dominated Marshall throughout his judicial tenure. His constitutional jurisprudence was consistently molded to give effect to this overriding conception.

If Roane and his fellow states' righters had obtained the palm, they would have used their victory to secure a polity and society very different from that which did evolve. The legal doctrines on which Roane rested his "Hampden" critique were employed to reach the constitutional results conducive to attainment of his political and social ends.

But the same was true of Marshall. His "Friend of the Constitution" essays, like his judicial opinions, were intended to supply the legal support for the values in which he believed. At the core of the Marshall conception was the supremacy of federal power, exercised by a government endowed with the means necessary to give effect to his vision of a strong nation. In the emerging struggle between the commercial and agrarian interests, Marshall's jurisprudence emphatically supported the former. If the compact theory meant keeping the Federal Government secondary in all but "indispensably requisite" powers to state sovereignties, it had to be repudiated. Instead of a rigid approach to national authority, constitutional doctrine must be the plastic one exemplified by *McCulloch*'s construction of the Necessary-and-Proper Clause.

Chase Impeachment

Marshall's work in strengthening the judicial department was strongly resisted by the Jeffersonian party, which was dominant in the other two departments. Jefferson was, indeed, Marshall's principal antagonist throughout his life. To the great democrat, control of the validity of governmental acts by nonelected judges "would place us under the despotism of an oligarchy."[128] He never really appreciated the need for judicial review as the true safeguard of constitutional rights against the power of government.

The Jeffersonians did not confine their opposition to the judges to verbal criticism such as that in Roane's "Hampden" essays. Instead, they sought to use the weapon of impeachment to bend the judicial department to their will. Their efforts in that direction culminated in the 1805 attempt to secure the removal by impeachment of Justice Samuel Chase, then a member of the Marshall Court. The charges against Chase were based on his acts while on the bench and were far removed from the "high Crimes and Misdemeanors" required by the Constitution. Rather, it was generally recognized that the impeachment was political in purpose. As Senator William Branch Giles, the Jeffersonian leader in the upper House, candidly expressed it to John Quincy Adams while the Chase trial was pending, "We want your offices, for the purpose of giving them to men who will fill them better."[129] It was widely believed that the Chase impeachment was to be but the first step in the Jeffersonian plan. "The assault upon Judge Chase," wrote John Quincy Adams to his father, "was unquestionably intended to pave the way for another prosecution, which would have swept the Supreme Judicial Bench clean at a stroke."[130]

The arrangements for the Chase trial were as dramatic as the event itself. The pomp of the Warren Hastings impeachment, when, says Macaulay, "The grey old walls were hung with scarlet," was still vivid in the minds of all, and perhaps in imitation, the Senate Chamber was, in the words of one Senator, "fitted up in a stile beyond anything which has ever appeared, in the Country." The Senate Chamber too was "aglow with theatrical color . . . [the] benches . . . covered with crimson cloth." As Henry Adams characterized it, "The arrangement was a mimic reproduction of the famous scene in Westminster Hall; and the little society of Washington went to the spectacle with the same interest and passion which had brought the larger society of London to hear the orations of Sheridan and Burke."[131]

The Chase trial itself resulted in an acquittal, for enough Senators of the Jeffersonian party were convinced by the argument of the defense— "Our property, our liberty, our lives, can only be protected and secured by [independent] judges"[132]—to make the vote for conviction fall short of the constitutional majority. "The significance of the outcome of the Chase trial," says Chief Justice William H. Rehnquist, "cannot be overstated."[133] Had Justice Chase been removed, it would have made impossible the independence of the judiciary upon which the constitutional structure rests. The Chase acquittal, as a matter of history, put an end to the danger of judicial removal on political grounds. Since 1805, though impeachment proceedings have been brought against other federal judges, in none of these cases was the effort to secure removal based upon political reasons.

The Chase acquittal meant that the Marshall Court could exercise the constitutional authority asserted by it without fear or favor. To Marshall, as has been stressed, judicial power was not an end in itself, but only a means to attain the end of a sound national structure. Once he had obtained for the Court the authority to enforce the Constitution, and once the independence had been secured that is the essential prerequisite for the fearless exercise of such authority, Marshall could turn the judicial instrument to the forging of the legal bonds of a strong Union. The nationalism nurtured at Valley Forge was to flower in the great decisions by which were hewn the high road of the nation's destiny.

Marshall's Colleagues

Describing her 1835 visit to the Supreme Court, the British writer Harriet Martineau told how she heard Chief Justice Marshall deliver an opinion in his "mild voice" and gave her impression of his colleagues on the bench: "[T]he three Judges on either hand gazing at him more like learners than associates."[134] Virtually every estimate of the Marshall Court agrees with the Martineau characterization. Though the Brethren on his bench were men of intellectual stature, they were overshadowed by the Chief Justice, who dominated his Court as few judges have ever done.

To be sure, during his first years on the bench, the Court was composed of Federalists who were expected to share Marshall's nationalistic views. However, in 1811, after Gabriel Duval and Joseph Story replaced Justices Chase and Cushing, the Court had a Republican majority. Yet, as his leading biographer tells us, "Marshall continued to dominate it as fully as when its members were of his own political faith and views of the government. In the whole history of courts there is no parallel to such supremacy."[135]

What gave Marshall his commanding influence? Certainly it was not his intellect and learning. Joseph Story was his equal in the first and by far his superior in the second and William Johnson had both a strong intellect and a far better education. Bushrod Washington would also have been a prominent judge on any other Court. All three Justices, however, stood in the Chief Justice's shadow throughout their judicial tenure.

In the end, Marshall's dominance over his associates rests upon two things. The first is the elusive quality we call leadership—to which we can apply Justice Potter Stewart's celebrated aphorism: "I could never succeed in [defining it]. But I know it when I see it."[136] It may be impossible to say what makes a great leader. But we know leadership when we see it; and we *know* that Marshall was *the* leader par excellence of the highest Court. Whatever the qualities of judicial leadership may be, Marshall plainly possessed them to the ultimate degree.

Just as important, however, was that the principal Marshall decisions corresponded to the "felt necessities"[137] of the developing nation. The Republican appointees came to see this as clearly as the Chief Justice himself. In a day when, to most Americans, one's state was still one's country, all the Justices understood the need to assert national power and the need for a powerful Union.

Even a strong Jeffersonian such as Justice William Johnson worked as constantly as Marshall for principles that would strengthen the bonds of union.[138] In *Gibbons v. Ogden*,[139] for example, Johnson delivered a concurrence which, according to Beveridge, was more nationalist than Marshall's opinion of the Court itself.[140] The adherence of the entire Court to the principles articulated by Marshall bears out the assertion once made by Johnson that the "pure" men of both parties were in basic agreement on fundamentals.[141]

Joseph Story

Of Marshall's associates, the most important was Justice Story. Joseph Story is, of course, one of the great names in American law; he was perhaps the outstanding Justice during the nineteenth century and stands high on every list of the greatest American judges. Story was the most learned scholar ever to sit on the Supreme Court and also the youngest person ever named to the Court. He was only thirty-two when President Madison appointed him in 1811. Yet he had already been a Congressman, Speaker of the Massachusetts House, a leader at the Bar, and author of two volumes on pleading as well as three American editions of standard English works. More than that, he enjoyed a reputation as a minor poet. While studying law, he composed a lengthy poem, *The Power of Solitude*, referring to it in a letter as "the sweet employment of my leisure hours."[142] Story rewrote the poem, with additions and alterations, and published it with other poems in 1804. One who reads the extracts contained in his son's biography quickly realizes that it was no great loss to literature when Story decided to devote his life to the law. Story himself apparently recognized this, for he later bought up and burned all copies of the work he could find.

When the Supreme Court vacancy filled by Story occurred in 1810, Jefferson gloated "old Cushing is dead. At length then, we have a chance of getting a Republican majority in the Supreme Judiciary." Yet the former President also predicted "it will be difficult to find a character of firmness enough to preserve his independence on the same Bench with Marshall."[143] Story was soon the Court's leading supporter of Marshall's nationalistic views and became a vital second to the Chief Justice in constitutional doctrine.

In his opinions supporting Marshall, Story supplied the one thing the Chief Justice lacked—legal scholarship. "Brother Story here . . . can give us the cases from the Twelve Tables down to the latest reports," Marshall is reputed to have said.[144] If Marshall disliked the labor of investigating legal authorities to support his decisions, Story reveled in legal research. His opinions were usually long and learned and relied heavily on prior cases and writers.

Story may have been appointed as a Republican. But it soon became apparent that he fully shared Marshall's nationalistic views. His opinion in *Martin v. Hunter's Lessee,*[145] discussed earlier in this chapter, contributed as much as any the Chief Justice delivered to laying the jurisprudential foundation of a strong nation. And its impact was as great as any Marshall opinion. After Story had delivered his opinion, there was no disputing the appellate power of the Supreme Court over state court decisions. Story had demonstrated conclusively that the Union could not continue if state courts were able to defy it.

But it was not as a junior Marshall that Story left his main imprint on American law. If Marshall was the prime molder of early American public law, Story was his Supreme Court counterpart so far as private law was concerned. Early American commercial and admiralty law were largely the creation of Story's decisions. Story's opinions strikingly exemplified the common law's capacity to reshape itself to meet changing needs and develop the legal framework of the new industrial order. Important Story decisions blended the law of trusts with the rudimentary law of corporations that had developed in England to produce the modern business corporation and enable it to conduct its affairs on the same basis as natural persons.

The key case here was the *Dandridge* case.[146] Our previous discussion of the case showed Marshall taking an unusual narrow view and refusing to allow approval of acts of its agents by a corporation to be shown by presumptive testimony, but only by written record and vote. The Marshall approach, we saw, would have required corporate contracts to be cast in the elaborate forms of English law, with a record and vote required for each contract. That would have substantially hindered the growth of the business corporation in this country.

In *Dandridge,* however, it was Story, not Marshall, who spoke for the Court. Story's opinion held that the fact of approval could be shown by presumptive evidence. It stated specifically that the common law governing corporations was irrelevant to an American corporation created by statute. Instead, the corporation was to be treated like a natural person: "[T]he acts of artificial persons afford the same presumptions as the acts of natural persons."[147]

The *Dandridge* decision played a vital part in the development of the business corporation, which (as Kent's *Commentaries* then noted) was beginning to "increase in a rapid manner and to a most astonishing extent."[148] By permitting corporations to operate as freely as individuals, Story played a crucial part in accommodating the corporate form to the demands of the expanding American economy. The ability of the American courts to adapt the common law to the new nation's requirements is nowhere better seen than in cases such as *Dandridge.*

In his classic constitutional commentaries Story pointed out that the American colonists did not adopt the whole body of the common law, but only those portions which their different circumstances did not require them to reject.[149] "The common law of England," he wrote in another opinion, "is not to be taken in all respects to be that of America. Our ancestors brought with them its general principles, and claimed it as their birthright; but they brought with them and adopted only that portion which was applicable to their situation."[150] Indeed, the principal contribution of judges such as Story was to remold the common law to fit the different situation that existed on the western side of the Atlantic.

The quotation in the preceding paragraph is from Story's 1829 opinion in *Van Ness v. Pacard,*[151] where the traditional land law was adapted to meet the needs of the new mobile business economy rather than a static agricultural one. The Story decision modified the rigid conception of property that underlay the common-law rule on ownership of fixtures, rejecting the rule that the landlord owned all fixtures which the tenant had annexed to the land. Story's opinion "broke new ground in treating fixtures in terms of an increasingly mobile, business-oriented economy rather than a static agricultural one."[152] In an 1833 article, the common-law rule on fixtures was denounced as one that "must operate as a restraint upon the improvement of real property; and have a general tendency to lessen the amount of its productions and profits."[153] The Story decision helped avoid that consequence in American law. Under it, as Story's opinion put it, the law provided "every motive to encourage the tenant to devote himself to agriculture, and to favor any erection which should aid this result." Otherwise, "what tenant could afford to erect fixtures of much expense or value, if he was to lose his whole interest therein by the very act of erection?"[154]

Story's contribution in helping lay the American legal foundation was not limited to his work as a judge. In 1829, while he was still on the Supreme Court, Story became the first Dane Professor of Law at Harvard. His appointment signaled the reorganization of Harvard Law School and its emergence as the first modern school of law in the country. Despite his heavy judicial duties, he taught two of the three yearly terms at the school. He also found time to publish an amazing number of significant works that constituted the first great specialized treatises on American law. "I have now published seven volumes," he wrote in 1836, "and, in five or six more, I can accomplish all I propose."[155] By the end of his career he had published nine treatises (in thirteen volumes) on subjects ranging from constitutional to commercial law. They confirmed the victory of the common law in the United States and presented judges with authoritative guides. As a judge, Story may have been overshadowed by Marshall; as a law teacher and writer, he had no peer.

Other Associates

The Associate Justices on the Court during Marshall's early years as Chief Justice were succinctly characterized in a letter from Justice William Johnson to Jefferson: "Cushing was incompetent. Chase could not be got to think or write—Paterson was a slow man and willingly declined the trouble, and the other two judges you know are commonly estimated as one judge."[156]

The last two Justices referred to by Johnson were John Marshall and Bushrod Washington. The latter would have been an outstanding judge on any other court. Story wrote about him, "[H]e is highly esteemed as a profound lawyer, and I believe not without reason. His written opinions are composed with ability."[157] On the Marshall Court, however, he was, as the Johnson estimate indicates, virtually considered the Chief Justice's alter ego; the two differed in opinion only three times in their twenty-nine years together on the Court.[158]

Yet, when he worked outside of Marshall's shadow, Justice Washington did make important contributions of his own. It was Washington's opinion on circuit in *Corfield v. Coryell* (1823)[159] that gave the definitive answer to the question of what specific privileges and immunities are protected by the Comity Clause of article IV, which entitles "[t]he Citizens of each State . . . to all Privileges and Immunities of Citizens in the several States." Under Washington's opinion, the constitutional provision does not create any rights; it only ensures that such rights, so far as they do exist within a state, will be afforded equally to citizens of the state and other states. The Washington approach has been uniformly followed in later cases; it also set the pattern for the restricted interpretation of the Fourteenth Amendment's Privileges and Immunities Clause that has prevailed.[160]

The Court described by Justice Johnson began to change when Johnson was appointed in 1804 in place of Justice Moore, who had resigned.

Moore was said to possess a mind of uncommon strength and great power in analysis. On the Court, nevertheless, he was a virtual nonentity, having delivered an opinion in only one reported case.[161] William Johnson was an entirely different judge. His leading biography is titled *Justice William Johnson: The First Dissenter*.[162] Technically the subtitle is inaccurate. The first dissenter in the Supreme Court was Justice Johnson; but it was Justice Thomas Johnson who, we saw in the last chapter, delivered a dissent in the very first opinion reported in the Court.

The first Justice Johnson is, however, remembered, if at all, only for that first opinion; he did nothing more of consequence on the Court. Justice William Johnson, on the other hand, stands as Marshall's most competent colleague, aside from Justice Story himself. More than that, the second Justice Johnson was the member of the Marshall Court who displayed the most independence from the Chief Justice. His, says Justice Frankfurter, was one of the strongest minds in the Court's history.[163]

Johnson was appointed at the age of thirty-three (the second youngest Justice after Story). He was a member of South Carolina's highest court and was one of the leading Republicans of his state, having served in the legislature and been speaker of its lower House. He was Jefferson's first appointee, and the President desired most of all a loyal Republican who would serve as a counterweight to the Chief Justice and his jurisprudence.

His nominee was, in a Senator's characterization, "a zealous democrat," but he did not wholly fulfill Jefferson's expectations even though he was, as stated, the most independent member of the Marshall Court—the Justice least under the Chief Justice's sway. He was the first to demonstrate the potential of the dissenting opinion; he delivered his first dissent in his third term and delivered many more separate opinions than any other Marshall colleague. There were thirty-five concurring opinions and seventy-four dissents while Johnson was on the Court; of these he wrote twenty-one concurrences and thirty-four disssents.[164]

Yet Johnson himself virtually acquiesced in Marshall's unanimity rule after his first decade and a half on the Court. As Johnson explained it in a noted 1822 letter to Jefferson, "I . . . was not a little surprised to find our Chief Justice in the Supreme Court delivering all the opinions in cases in which he sat, even in some instances when contrary to his own judgment and vote. But I remonstrated in vain; the answer was he is willing to take the trouble and it is a mark of respect to him." After trying to resist the Marshall practice, Johnson went on, "At length I found that I must either submit to circumstances or become such a cipher in our consultations as to effect no good at all. I therefore bent to the current."[165]

Jefferson, however, persisted in his opposition to what he termed, in reply to Johnson, "cooking up opinions in conclave, [which] begets suspicions that something passes which fears the public ear." Jefferson stressed "the importance and the duty of [the judges] giving their country the only evidence they can give of fidelity to its constitution and integrity in the administration of its laws; that is to say, by every one's giving his opinion seriatim and publicly on the cases he decides."[166]

Jefferson's letters led Justice Johnson to renew his practice of speaking out—alone if need be. During his last ten terms on the Court, Johnson issued nine concurring and eighteen dissenting opinions.[167] In the end, however, if Jefferson had hoped that Johnson would lead a Republican opposition on the Court that would curb the Marshall jurisprudence, it proved a false hope. Jefferson had warned Johnson to resist Marshall's centralization of power in Washington.[168] "[I]n truth," he wrote to Johnson in 1823, "there is no danger I apprehend so much as the consolidation of our government by the noiseless, and therefore unalarming, instrumentality of the Supreme Court. This is the form in which federalism now arrays itself, and consolidation is the present principle of distinction between Republicans and these pseudo-republicans but real federalists."[169]

Johnson on the bench, however, revealed himself as nationalistic as any of his associates. He remained silent in the face of the *McCulloch v. Maryland* opinion,[170] thus acquiescing in what Jeffersonians deemed the heresy of implied powers.[171] Two years later, in 1821, the Justice wrote an opinion strongly supporting the implied powers doctrine in language that could have been written by Marshall himself.[172] In addition, Johnson raised no objection to the Chief Justice's resounding affirmation of the Court's jurisdiction to hear appeals from state courts.

Though Justice Johnson was silent in *McCulloch,* he indicated soon thereafter that he fully agreed with Marshall's broad construction of federal power there. In 1822, he answered President Monroe's reliance upon a narrow theory of federal power in his veto of the Cumberland Road Act. Johnson sent a letter to the President in which he asserted the *McCulloch* principle "that the grant of the principal power carries with it the grant of all adequate and appropriate means of executing it. That the selection of those means must rest with the general government and as to that power and those means the Constitution makes the government of the U.S. supreme." Indeed, Johnson wrote, he considered the *McCulloch* opinion so sound that the President should have it "printed and dispersed through the Union."[173]

These views were enough to make Johnson a virtual traitor in the eyes of the Jeffersonians.[174] Johnson and the other apostate Republican Justices were denounced in another series of newspaper articles by Spencer Roane, under the pen name "Algernon Sidney." "How else is it," Roane asked, referring to the Republican Justices, "that they also go to all lengths with the ultra-federal leader who is at the head of their court? That leader is honorably distinguished from you messieurs judges. He is true to his former politics. . . . He must be equally delighted and surprised to find his Republican brothers going with him."[175]

President Jefferson and his successors may have expected the Justices they appointed to change the course of Supreme Court jurisprudence. But it did not work out that way. Roane, to be sure, exaggerated in his

"Sidney" articles when he asserted that the Chief Justice's sway over his Court "is the blind and absolute despotism which exists in an army, or is exercised by a tyrant over his slaves."[176] Yet it cannot be doubted that Marshall remained, in Beveridge's characterization, as much the master of the Supreme Court after it acquired a Republican majority as he had been before that time.

The other Justices appointed by Jefferson and his successors appear in the Court's history even more as mere appendages to the Chief Justice. Certainly none of them has attained anything like the stature associated with Marshall's most important colleagues, Justices Story and Johnson, and, to a lesser extent, Justice Washington. Three of them have been called "silent Justices" in the Holmes Devise study of the Marshall Court.[177] They were Brockholst Livingston (appointed from New York in 1806 to succeed Justice Paterson), Thomas Todd (a Kentuckian chosen in 1807 to fill an additional seat created by Congress), and Gabriel Duval (appointed from Maryland in 1811 to succeed Justice Chase). Todd is now best known for the large number of terms (five out of nineteen) that he missed, Duval for his deafness, and Livingston for the duels he fought, in one of which he killed his adversary. All three could be counted on, in Story's phrase, to "steadfastly support . . . the constitutional doctrines which Mr. Chief Justice Marshall promulgated, in the name of the court."[178]

Of the other Marshall Court Justices, the only one who served for some time was Smith Thompson, a New Yorker appointed in 1823 to Justice Livingston's seat, in which he sat for twenty years. He is largely known for his active participation in politics even after his appointment; he ran for Governor of New York in 1828, thus breaking the practice established by Marshall and his colleagues of withdrawal from political activities (except for Justice Chase, of course, whose blatant electioneering for Adams had been one of the factors that led to his impeachment). Next to Justice Johnson, Thompson was the Justice who wrote the most concurrences and dissents. But his views were not consistent and he tended to support Marshall in the Court's opposition to the Jeffersonian Virginia School, though he did recognize a greater concurrent state power to regulate commerce than did the Chief Justice.[179]

Robert Trimble, another Kentuckian selected in 1826 in place of Justice Todd, gave signs of a potentially distinguished Court career. However, the potential was never realized, since he died two years after his appointment. The Marshall Court in its last years also had three Justices appointed by Andrew Jackson: John McLean (1829), Henry Baldwin (1830), and James Wayne (1835). There are indications that they would have supplied the opposition that might have ended Marshall's monolithic tenure. But they served too briefly with the great Chief Justice for that to happen. Instead, they made their principal contributions under Marshall's successor, Roger B. Taney.

Marshall's Jurisprudence

More than any other jurist, Marshall employed the law as a means to attain the political and economic ends that he favored. In this sense, he was the very paradigm, during our law's formative era, of the result-oriented judge. More than that, the law which he thus used was, in major part, molded as well as utilized by him. That is all but self-evident as far as Marshall's public-law decisions are concerned. The constitutional principles Marshall proclaimed in his cathedral tones were, in large part, principles of his own creation.

Marshall was undoubtedly one of the greatest legal reasoners. But his ability in that respect only masked the fact that he was the author as well as the expounder of his legal doctrines. His public-law opinions were based on supposedly timeless first principles which, once accepted, led, by unassailable logic, to the conclusions that he favored. "The movement from premises to conclusion is put before the observer as something more impersonal than the working of the individual mind. It is the inevitable progress of an inexorable force."[180]

Even Marshall's strongest critics were affected by the illusion. "All wrong, all wrong," we are told was the despairing comment of one critic, "but no man in the United States can tell why or wherein."[181]

It is not generally realized that Marshall also followed his instrumentalist approach in nonconstitutional cases. The common assumption has been that the Marshall Court was not a great common-law court and that its head was anything but a master of the common law. Thus it has often been pointed out that, in the Marshall–Story correspondence, there are requests from the Chief Justice for advice on various nonconstitutional issues. Now it is certainly true that Marshall never approached such issues with the mastery he displayed in his public-law opinions. Yet it unduly denigrates Marshall's legal ability to assume that he was a mere slouch as far as private law was concerned.

Such denigration of Marshall's legal ability has unfortunately not been uncommon. But "the fact is that he was a brilliant attorney whose expertise extended to both domestic and international law."[182] Few men have come to the Supreme Court with Marshall's legal experience. His legal education may, we saw, have been limited, but his years of practice had made him a leader of the Virginia bar and his cases covered the whole gamut of private law, particularly property and commercial law. Marshall's conception of private law was, of course, based primarily upon the common law. The young Marshall had been given a copy of Blackstone by his father, one of the original subscribers to the American edition, and the common law became the basis of his jurisprudence. During his brief study at William and Mary under George Wythe, Marshall wrote almost two hundred pages of manuscript notes, which were intended as a summary of the common law as it was then practiced in Virginia.[183]

Marshall had no doubt that American law had a common-law foundation. "My own opinion," he wrote in 1800, "is that our ancestors brought with them the laws of England both statute and common law, so far as they were applicable to our situation." With the Revolution and adoption of the Federal Constitution, "the common and statute law of each state remained as before and . . . the principles of the common law of the state would apply themselves to magistrates."[184]

To Marshall as to other common lawyers, the law was essentially judge-made law. That was why he placed such stress upon the role of the judge and the need for judicial independence and prestige. "That in a free country," he declared to Story, "any intelligent man should wish a dependent judiciary, or should think that the constitution is not a law for the court . . . would astonish me." Much of his hostility toward Jefferson was based upon Marshall's belief that Jefferson "looks of course with ill will at an independent judiciary."[185] During the Chase impeachment proceedings, Marshall asserted to the beleaguered Justice, "The present doctrine seems to be that a Judge giving a legal opinion contrary to the opinion of the legislature is liable to impeachment."[186]

But the judicial role was to Marshall plainly a means, not an end. The judge was to use his independence and prestige to mold the law in accordance with the needs of American society. In public law, the goal was to lay down constitutional principles that would give effect to the Marshall nationalistic vision. In private law, the end was both the protection of property rights and their expansion to permit them to be used to foster the growing entrepreneurial economy. Common-law principles were to be adapted, not transported wholesale in their English form. Both property and commercial law began to receive their modern cast, as a common law appropriate to the new nation's situation was being developed.

Oliver Wolcott once described Marshall as "too much disposed to govern the world according to rules of logic."[187] Marshall the logician is, of course, best seen in his magisterial opinions, which, to an age still under the sway of the syllogism, built up in broad strokes a body so logical that it baffled criticism from contemporaries.[188] To Marshall, however, logic, like law, was only a tool. Indeed, the great Chief Justice's opinions may be taken as an early judicial example of the famous Holmes aphorism: "The life of the law has not been logic: it has been experience."[189] Marshall, more than any early judge, molded his decisions to accord with "the felt necessities of the time." If this was often intuitive rather than conscious on his part, the intuitions were those that best furthered his notion of the public interest.[190]

For Marshall, then, the Constitution was a tool, and the same was true of the common law. Both public law and private law were to be employed to lay down the doctrinal foundations of the polity and economy that served his nationalistic vision of the new nation. Compared to Jefferson, the Marshall vision may have been a conservative one. Yet though their visions may not have been the same, both the great Chief Justice and his lifelong antagonist looked at the law as an instrument to serve the needs of the new nation.

Jefferson could write of "the rancorous hatred which Marshall bears to the government of his country." The truth is that each man had a different conception of what the American polity should become. Marshall saw all too acutely that the Jeffersonian theme was sweeping all before it. "In democracies," he noted in 1815, "which all the world confirms to be the most perfect work of political wisdom, equality is the pivot on which the grand machine turns." As he grew older, Marshall fought the spread of the equality principle, notably in the Virginia Convention of 1829–1830. For, as he saw it, "equality demands that he who has a surplus of anything in general demand should parcel it out among his needy fellow citizens."[191]

Yet if Marshall's last effort—against the triumph of Jacksonian democracy—was doomed to failure, his broader battle for his conception of law was triumphantly vindicated. One the least conversant with our public law knows that it is the Marshall conception of the Constitution that has dominated Supreme Court jurisprudence, particularly during the present century. But the same has also been true of the Marshall concept of private law. The Marshall Court decisions adapting the common law to the needs of the expanding market economy led the way to the remaking of private law in the entrepreneurial image. Free individual action and decision became the ultimate end of law, as it became that of the society itself. The law became a prime instrument for the conquest of a continent and the opening of the economy to men of all social strata. Paradoxically perhaps, it was Marshall, opponent of Jefferson–Jackson democracy though he may have been, whose conception of law furthered opportunity and equality in the marketplace to an extent never before seen.

3

Taney Court, 1837–1864

"Imagine a Court . . . conducting its business not in a massive building but in a basement room of an unfurnished Capitol."[1] One familiar with the marble temple in which the highest bench now sits will find it hard to picture the dingy quarters which that tribunal occupied during its early years. We saw in Chapter 2 that when the new capital was built, no chamber had been built for the Supreme Court. Instead, when the Government moved to Washington, the Court had to hold its sessions in a House committee room on the first floor of the Capitol. When the Capitol north wing was rebuilt in 1809, the Court moved to a new courtroom underneath the reconstructed Senate Chamber. After the British burned the Capitol in 1814, the Court moved to temporary quarters for five years. They were back in their basement courtroom for the 1819 Term.

That courtroom was almost as unimpressive as the committee room in which the Court first sat. In the first place, as a newspaper described it, "The apartment is not in a style which comports with the dignity of that body, or which wears a comparison with the other Halls of the Capitol . . . it is like going down cellar to reach it." Second, the paper complained, it was "a room which is hardly capacious enough for a ward justice." Hence it lacked the dignity appropriate for a puisne court, much less for the highest Court of the land. "Owing to the smallness of the room," the same article tells us, "the Judges are compelled to put on their robes in the presence of the spectators, which is an awkward ceremony, and destroys the effect intended to be produced by assuming the gown."[2]

Even the elevated bench that adds to a court's dignity was lacking.

"The part where the judges sit, is divided from the bar, by a neat railing; . . . beyond the railing, are the judges' seats, upon pretty nearly a level with the floor of the room, not elevated as are our judges' seats."[3]

Charles Sumner described the Court's basement chamber as "a dark room almost down cellar."[4] "A stranger," wrote a newspaper, "might traverse the dark avenues of the Capitol for a week, without finding the remote corner in which Justice is administered to the American Republic."[5] According to the sardonic comment of a contemporary, housing the Court underneath the Senate was "an arrangement wholly unjustifiable unless perhaps by the idea that Justice should underlie legislation."[6]

When Benjamin F. Butler visited a Supreme Court session with his son, the latter recalled that his "boyish attention was fastened upon the seven judges as they entered the room—seven being the number then composing the Court. It was a procession of old men—for so they seemed to me—who halted on their way to the bench, each of them taking from a peg hanging on the side of the wall near the entrance a black robe and donning it in full view of the assembled lawyers and other spectators."[7]

Marshall's Successor

It was just such a session over which Roger Brooke Taney first presided as Chief Justice at the beginning of 1837. The new Chief Justice was tall and impressive. He was, said an 1838 magazine article, "full six feet high: spare, but yet so dignified in deportment that you are at once impressed with an instinctive reverence and awe."[8] As he led the Supreme Court into its chamber on January 9, 1837, he looked the picture of a model judge. On that day Chief Justice Taney first took his seat in the Court's central chair[9] and began a new era in the constitutional history of the nation. Taney, of course, started with the handicap of having to fill the place of the greatest judge in our history.

The void created by Chief Justice Marshall's death can scarcely be overestimated. Wrote Joseph Story (who, as senior Associate Justice, had to "act as *locum tenens* of the Chief Justiceship"), "I miss the Chief Justice at every turn . . . the room which he was accustomed to occupy . . . wears an aspect of desolation."[10] The dejection of Marshall's admirers was compounded by apprehension with regard to his potential successor. "It is much to be feared," gloomily wrote John Quincy Adams, "that a successor will be appointed of a very different character. The President of the United States now in office . . . has not yet made one good appointment. His Chief Justice will be no better than the rest."[11]

Marshall's adherents had hoped against hope that Joseph Story would become the new Chief Justice. "The Supreme Court," Harvard President Josiah Quincy toasted, "may it be raised one Story higher."[12] President Andrew Jackson, however, could scarcely appoint one so opposed to his views. Instead, Story became a vigorous dissenter in the Taney Court, delivering caustic dissents like his opinion in the soon-to-be-discussed

Charles River Bridge case. Several weeks after it was delivered, he declared, "I am the last of the old race of Judges."[13]

Marshall himself toward the end of his life had expressed misgivings about the future of the Supreme Court. The advent of Jacksonian democracy, with its "enormous pretentions of the Executive," appeared but a portent of the fate that awaited both his constitutional labors and the strong national government which he sought to construct through them. "To men who think as you and I do," he wrote Story during the last year of his life, "the present is gloomy enough; and the future presents no cheery prospect."[14] Toward the end of his career, the great Chief Justice saw the Supreme Court defied, both by the State of Georgia and by the President himself. "Georgia," complained Adams, "has planted the Standard against the Supreme Court of the United States—and I hear the twenty-fifth Section of the Judiciary Law is to be repealed. To these proceedings there is an apparent acquiescence of the People in all Quarters."[15]

President Jackson, too, was the author of a vehement attack upon the very basis of the Supreme Court's review power in his famous 1832 message vetoing a bill to extend the charter of the Bank of the United States.[16] Well might Marshall feel that his long effort to construct judicial power as the cornerstone of an effective and enduring Union had been all but in vain.

Now it was Jackson who, on Marshall's death, was given the occasion to remold the Court in the feared radical image. He had an opportunity denied to most Presidents. He was able to place more men on the Supreme Court than any President before him except Washington or after him except Franklin D. Roosevelt. Not unnaturally, Jackson chose his six appointees from men of his own party, whom he felt he could trust. Not unnaturally, the Whigs attacked the packing of the Court with Democrats and Southerners (by 1837, soon after Jackson went out of office, a majority of the high tribunal, newly enlarged to nine, came from below the Mason-Dixon line). Such a bench, its opponents were convinced, would be all too ready to write the principles of Jacksonian democracy into the law. Above all, the opposition was apprehensive about Jackson's choice of a new Chief Justice. To Marshall's admirers, it must be admitted, no selection Jackson might have made, except perhaps Joseph Story, would have been satisfactory. Jackson's choice, asserted Story, "will follow a man who cannot be equalled and all the public will see . . . the difference."[17]

The apprehension of men like Story appeared justified when, on December 28, 1835, Jackson nominated Roger B. Taney as Chief Justice of the United States. It was Taney who had drafted the key portions of the 1832 Veto Message and who had been the instrument for carrying out the President's plan for the removal of Government deposits from the Bank of the United States. The veto had questioned the very review power of the Supreme Court, asserting that "the opinion of the judges has no more authority over Congress than the opinion of Congress has over the judges, and on that point the President is independent of both. The authority of the Supreme Court must not, therefore, be permitted to control the Congress or the Executive."[18] The choice of the author of these words to head a bench dominated by Jacksonian Democrats appeared to presage the virtual undoing of all that the Marshall Court had accomplished. Wrote Daniel Webster soon after the Taney nomination, "Judge Story . . . thinks the Supreme Court is *gone* and I think so too."[19]

Despite the bitterness of the opposition—"[T]he pure ermine of the Supreme Court," acidly affirmed one Whig newspaper, "is sullied by the appointment of that political hack"[20]—and the fact that Taney had, during the preceding two years, been turned down by the Senate as Secretary of the Treasury and as a Justice of the Supreme Court, this time his nomination was confirmed by the upper House, on March 15, 1836, by a nearly two-to-one majority. The majority in Taney's favor is somewhat surprising since the Jackson and anti-Jackson forces were equally divided in the forty-eight-member Senate. When John Tyler received his "walking papers from the [Virginia] Legislature"[21] and, the day before the vote, was replaced by a Democrat, the Jackson forces still had only the slimmest of majorities.

Apparently, what happened was that most of the Senate opposition abstained from the proceedings, largely because they felt that the Administration now had the votes necessary for the nomination. As Francis Scott Key tells it in a letter to his sister (Taney's wife), "Taney, Kendall & Barbour have all passed . . . those who did not choose to vote for them went off, knowing it was of no use to stay." Therefore, concluded Key, "I must greet you as Mrs. *Chief Justice Taney*."[22]

Taney and Jacksonian Democracy

Chief Justice Taney first sat with his Brethren when he was just under sixty years old. He was to serve until he was eighty-seven, a tenure as Chief Justice second only to that of Marshall. He was the first Chief Justice to wear trousers; his predecessors had always given judgment in knee breeches.[23] There was something of portent in his wearing democratic garb beneath the judicial robe,[24] for under Taney and the new majority appointed by Jackson, the Supreme Court for the first time mirrored the Jacksonian emphasis upon public power as a counterweight to the property rights stressed by the Federalists and then the Whigs.

Taney had been one of the foremost exponents of Jacksonian democracy; it has been asserted that the Jacksonian political theory is more completely developed and more logically stated in Taney's writings and speeches than anywhere else.[25] This assertion may be extreme, but it cannot be denied that Taney's years on the Court marked growing judicial concern for safeguarding of the rights of the community as opposed to property rights—of the public, as opposed to private welfare. "We believe property should be held subordinate to man, and not man to property," declared a leading Jacksonian editor, "and therefore that it is always lawful to make such modifications of its constitution as the good of Humanity requires."[26] The Taney Court was to elevate this concept to the constitutional plane.

Taney was well aware that his relationship to Jackson played a crucial part in his career. When the Senate approved his nomination, Taney wrote the President expressing warm gratitude, saying he would rather owe the honor to Jackson than to any other man in the world. It was a particular gratification, he declared, that "it will be the lot of one of the rejected of the panic Senate, as the highest judicial officer of the country to administer in your presence and in the view of the whole nation, the oath of office to another rejected of the same Senate, when he enters into the first office in the world."[27]

It is hard to understand Taney and his judicial work without awareness of his constant concern with the rejections that occurred during different stages of his career, from his early defeats in elections for Congress and the Maryland legislature[28] to the refusals of the Senate in 1834 and 1835 to confirm his nomination by Jackson as Secretary of the Treasury and then as Associate Justice of the Supreme Court. In this respect, Taney is one of the most difficult judicial historical subjects, much more so than Marshall, for his character was far more complex than that of his relatively straightforward predecessor. Like Marshall, he left an autobiographical sketch, but it is longer, rambling and abstruse, and unfinished.[29] It does clearly reveal his more complicated character—a constant emphasis on what he himself termed "morbid sensibility."[30] This sensibility, exaggerated perhaps by his delicate health and the fact that he was a Catholic in the Know-Nothing era, was to remain an essential part of the true Taney, beneath the stern facade shown to contemporaries. "I do not exactly understand why *Friday* has become the fashionable day for dinners here," he plaintively complained in an 1845 letter to his son-in-law, indicating his acute susceptibility to supposed slights at his religion.[31]

Taney was not only overly sensitive but had an exaggerated conception of himself as the very paragon of rectitude, an attitude that was to lead directly to the judicial fall that followed the *Dred Scott* decision. When he was first appointed to Jackson's Cabinet, he served briefly as both Attorney General and Acting Secretary of War. During that period he drew the salaries of both positions—something which strikes the observer as unethical. Yet he stoutly defended his action in an 1841 letter, declaring that "as I performed the duties of both offices, I received the salaries of both. I thought then and still think that it was right."[32]

Even at Marshall's death, it should have been evident that the doctrines of national power the great Chief Justice had espoused were bound to prevail. Contemporary admirers of the Marshall constitutional edifice might look upon Taney as the instrument chosen for its destruction, but Taney was not the man to preside at the liquidation of the tribunal he was called upon to head. On the contrary, the Supreme Court under him continued the essential thrust of constitutional development begun by Marshall and his colleagues.

In fact, if we look at Taney's constitutional work, avoiding the tendency to compare his accomplishments with the colossal structure erected by his predecessor, we find it far from a mean contribution. The shadow of the *Dred Scott* decision, it is now generally recognized, for too long cast an unfair pall over his judicial stature. To be sure, there was an inevitable reaction after Marshall's death, but it was not as great as has often been supposed. Chief Justice Taney may not have been as nationalistic in his beliefs as his predecessor, but his greater emphasis on states' rights should not obscure the continuing theme of his Court: that of formulating the principles needed to ensure effective operation of the Constitution.

In addition, it should be borne in mind that, however far-reaching Jacksonian democracy might have seemed to its contemporary opponents, it was, by present-day conceptions, quite limited. The Jacksonians did, it is true, go further than the Founders in the direction of both political and economic equality, but their notion of the democratic ideal as providing both liberty and equality for all must be sharply distinguished from the twentieth-century conception of the meaning of the word "all." The Jacksonians, like the Framers before them, did not understand the ideal of liberty and equality *for all men* to require the abolition of slavery, the emancipation of women from legal and political subjection, or the eradication of all constitutional discriminations based on wealth, race, or previous condition of servitude.[33] Yet, though the Jacksonian conception was limited, one should not underestimate its significance: it made substantial contributions to both the theory and practice of equality.

When the occasion demanded it, indeed, the Jacksonians could eloquently articulate the concept of equality and the premises upon which it was based. "In the full enjoyment of the gifts of Heaven and the fruits of superior industry, economy, and virtue," declared Jackson in his 1832 veto of the bill rechartering the Bank of the United States,

> every man is equally entitled to protection by law; but when the laws undertake to add to these natural and just advantages artificial distinctions, to grant titles, gratuities, and exclusive privileges, to make the rich richer and the potent more powerful, the humble members of society—the farmers, mechanics, the laborers—who have neither the time nor the means of securing like favors to themselves, have a right to complain of the injustice of their Government. . . . If it would confine itself to equal protection, and, as Heaven does its rains, shower its favors alike on the high and the low, the rich and the poor, it would be an unqualified blessing.[34]

The language quoted (which may have been written by Taney) is a statement, in positive terms, of the equal right of all persons to the equal protection of equal laws, comparable to the negative version in the Fourteenth Amendment, which was adopted thirty-six years later.[35]

Charles River Bridge Case

We need not, in Justice Frankfurter's phrase, subscribe to the hero theory of history to recognize that great men do make a difference, even in the law.[36] Certainly it made a difference that the 1837 Term of the Supreme Court was presided over by Taney instead of Marshall. In all likelihood Marshall would have decided differently in the three key decisions rendered in 1837.[37] The three cases had been argued while Marshall was still Chief Justice, but the Court had been unable to reach a workable decision. The cases were inherited by the Taney Court, and the new Chief Justice galvanized the Court into speedy action; the cases were all reargued and decided within less than a month after Taney first sat with his Brethren.

The Marshall Court had been concerned with strengthening the power of the fledgling nation so that it might realize its political and economic destiny. Like the Framers themselves, it stressed the need to protect property rights as the prerequisite to such realization. To Jacksonians like Taney, private property, no matter how important, was not the be-all and end-all of social existence. "While the rights of property are sacredly guarded," declared the new Chief Justice in his first important opinion, "we must not forget that the community also have rights, and that the happiness and well being of every citizen depends on their faithful preservation."[38] The opinion was delivered in *Charles River Bridge v. Warren Bridge*,[39] a case which was a cause célèbre in its day, both because it brought the Federalist and Jacksonian views on the place of property into sharp conflict and because stock in the corporation involved was held by Boston's leading citizens and Harvard College.

The Charles River Bridge had been operated as a toll bridge by a corporation set up under a charter obtained by John Hancock and others in 1785. Each year two hundred pounds was paid from its profits to Harvard. The bridge, opened on a day celebrated by Boston as a "day of rejoicing," proved so profitable that the value of its shares increased tenfold and its profits led to public outcry. The bridge became a popular symbol of monopoly, and, in 1828, the legislature incorporated the Warren Bridge Company to build and operate another bridge near the Charles River Bridge. The second charter provided that the new bridge would become a free bridge after a short period of time. This would, of course, destroy the business of the first bridge, and its corporate owner sued to enjoin construction, alleging that the contractual obligation contained in its charter had been impaired.

The case was elaborately argued on January 24, 1837, with Daniel Webster appearing for the Charles River Bridge, "and at an early hour all the seats within and without the bar . . . filled with ladies, whose beauty and splendid attire and waving plumes gave to the Court-room an animated and brilliant appearance such as it seldom wears."[40] Less than three weeks later, the Court was ready for decision. On February 11, Justice Story wrote to his Harvard colleague Professor Simon Greenleaf (who had argued in opposition to Webster), "[T]omorrow . . . the opinion of the Court will be delivered in the Bridge case. You have triumphed."[41]

The new Chief Justice delivered the opinion of the Court. Taney's opinion refused to hold that there had been an invalid infringement upon the first bridge company's charter rights. There was no express provision in the charter making the franchise granted exclusive or barring the construction of a competing bridge, and the basic principle is "that in grants by the public, nothing passes by implication."[42] Since there is no express obligation not to permit a competing bridge nearby, none may be read in.

In deciding as it did, the Taney Court laid down what has since become a legal truism: the rights of property must, where necessary, be subordinated to the needs of the community. The Taney opinion declined to rule that the charter to operate a toll bridge granted a monopoly in the area. Instead, the charter should be construed narrowly to preserve the rights of the community: where the rights of private property conflict with those of the community, the latter must be paramount. "The object and end of all government," Taney declared in words virtually setting forth the theme of Jacksonian democracy in the economic area, "is to promote the happiness and prosperity of the community by which it is established, and it can never be assumed that the government intended to diminish its power of accomplishing the end for which it was created." Governmental power in this respect may not be transferred, by mere implication, "to the hands of privileged corporations."[43]

Justice Story, who delivered a characteristically learned thirty-five-thousand-word dissent, bitterly attacked the majority decision, asserting in a letter to his wife that "a case of grosser injustice, or more oppressive legislation, never existed."[44] In his dissent, he declared that the Court, by impairing the sanctity of property rights, was acting "to alarm every stockholder in every public enterprise of this sort, throughout the whole country."[45] Yet, paradoxical though it may seem, it was actually the Taney decision, not the Story dissent, which ultimately was the more favorable to the owners of property, particularly those who invested in corporate enterprise.

Though in form the *Charles River Bridge* decision was a blow to economic rights, it actually facilitated economic development by providing the legal basis for public policy choices favoring technological innovation and economic change, even at the expense of some vested interests. The case arose when the corporate form was coming into widespread use as an instrument of capitalist expansion. In the famous *Dartmouth College* case,[46] the Marshall Court had ruled the grants of privileges in corporate charters to be contracts and, as such, beyond impairment by government. The Marshall approach here would have meant the upholding of the first bridge company's monopoly. Such a result would have had most undesirable consequences, for it would have meant that every bridge or turnpike company was given an exclusive franchise which might not be impaired by the newer forms of transportation being developed.

"Let it once be understood," declared Taney's *Charles River Bridge* opinion, "that such charters carry with them these implied contracts, and give this unknown and undefined property in a line of travelling; and you will soon find the old turnpike corporations awakening from their sleep, and calling upon this court to put down the improvements which have taken their place. The millions of property which have been invested in railroads and canals, upon lines of travel which had been before occupied by turnpike corporations, will be put in jeopardy."[47] To read monopoly rights into existing charters would be to place modern improvements at the mercy of existing corporations and defeat the right of the community to avail itself of the benefits of scientific progress.

"Taney's decision in the *Charles River Bridge* case," as Justice Frankfurter sums it up, "shows the statesman, Story's dissent proves that even vast erudition is no substitute for creative imagination."[48] Those who believed as Justice Story did, however, refused to see the beneficial implications of Chief Justice Taney's decision. Like all those wedded to the old order, they knew only that a change had been made in the status quo, and that was sufficient for their condemnation. "I stand upon the old law," plaintively affirmed the Story dissent, "upon law established more than three centuries ago . . . not . . . any speculative niceties or novelties."[49] In a letter written several weeks later, he dolefully declared, "I am the last of the old race of Judges."[50] To men like Story, Taney and the majority had virtually "overturned . . . one great provision of the Constitution."[51]

The truth, of course, is that the Taney Court had only interpreted the Contract Clause in a manner that coincided with the felt needs of the era of economic expansion upon which the nation was entering. Because of the Taney decision, that expansion could proceed unencumbered by inappropriate legal excrescences. By 1854, a member of the highest Court could confidently assert, with regard to the *Charles River Bridge* decision, "No opinion of the court more fully satisfied the legal judgment of the country, and consequently none has exerted more influence upon its legislation."[52]

State Power

Two other cases decided by the Taney Court during its first term received much public attention. They too had been inherited from the Marshall Court and were decided differently than they would have been before Taney's accession. Both *Briscoe v. Bank of Kentucky* and *New York v. Miln*[53] dealt with the question of reserved state power—an issue crucial in a federal system such as ours in which national and state governments coexist, each endowed with the complete accoutrements of government, including the full apparatus of law enforcement, both executive and judicial. Marshall, with his expansive nationalistic tenets, had perhaps tilted the scale unduly in favor of federal power. The Taney Court sought to redress the balance by shifting the emphasis to the reserved powers possessed by the states.

In the *Briscoe* case (1837), the Court held that the issuance by a state-owned bank of small-denomination notes which circulated as currency did not violate the constitutional prohibition against issuance by the states of bills of credit. According to the majority opinion of Justice John McLean, the notes were not bills of credit put out by the state since they were issued by the bank, not the state, even though the state owned the bank. Again Justice Story dissented. He tells us that Marshall himself—"a name never to be pronounced without reverence"—would not have been with the majority: "Had he been living, he would have spoken in the joint names of both of us."[54] As it was, Story was alone in his effort to preserve the reign of the dead hand.[55]

The *Briscoe* decision is today of purely historical significance since the problem of state power at issue there has long been academic. Yet the decision does show the willingness of the Taney Court to uphold state action if at all possible, even though it involved a refusal to look behind the form to the substance of the challenged action. Perhaps the best explanation of *Briscoe* is to be found in the Jacksonian fear of the growing power of finance, particularly as exemplified by "the power which the moneyed interest derives from a paper currency which they are able to control."[56] To avoid the evils of financial monopoly, the state was ruled able to regulate its circulating medium through the issue of notes by its own bank.[57]

More interesting to the present-day observer is *New York v. Miln* (1837), for it dealt with an aspect of state power that is still most pertinent, namely, the police power and its impact upon commerce. At issue in *Miln* was a New York law which required masters of vessels to report the names, places of birth, ages, health, occupations, and last legal residence of all passengers landing in New York City and to give security to the city against their becoming public charges. The city was seeking to collect the statutory penalty against the ship *Emily* because of its master's failure to file the report required by the statute. Defendant contended that the statute involved an invalid state regulation of foreign commerce since the power over such commerce was vested exclusively in Congress by the Constitution.

The opinion of the Court in *Miln* was written by Justice Philip P. Barbour (characterized by John Quincy Adams as that "shallow-pated wildcat . . . fit for nothing but to tear the Union to rags and tatters"),[58] who had been appointed to the Court at the same time as Taney. The Barbour opinion avoided the Commerce Clause issue, holding that the statute was valid as a matter of "internal police." The state's powers with regard to such "internal police" were not surrendered or restrained by the Constitution; on the contrary, "in relation to these, the authority of a state is complete, unqualified, and exclusive." The law at issue was passed by the state "to prevent her citizens from being oppressed by the support of multitudes of poor persons, who come from foreign countries, without possessing the means of supporting themselves. There can be no mode in which the power to regulate internal police could be more appropriately exercised."[59]

Perhaps the most significant aspect of *New York v. Miln* is its role in the development of the police power concept. That subject is so important that it deserves separate treatment. Here let us note another phase of the decision, which plainly appears anomalous today yet is useful to illustrate the restricted scope of the Jacksonian notions of freedom and equality. The *Miln* opinion asserts a general power in the states to exclude undesirables. "We think it," declared the opinion, "as competent and as necessary for a state to provide precautionary measures against the moral pestilence of paupers, vagabonds, and possibly convicts; as it is to guard against the physical pestilence, which may arise from unsound and infectious articles imported."[60] Similar language, it should be noted, was repeated in other decisions down to the turn of the century.[61]

Relying upon these Supreme Court dicta, many states enacted laws restricting the movement of indigent persons. Those laws were consistent with the historical common-law tradition of restricting the liberty of the pauper. To the judges of a century ago, to strike down such a restriction would be to include in citizenship "a right of the indigent person to live where he will although the crowding into one State may be a menace to society. No such right exists."[62] But this justification for a statutory restriction upon the poor person's freedom of movement is now considered wholly inconsistent with our concepts of personal liberty.

In 1941 the Supreme Court ruled that the power to restrict freedom of movement may never be based upon the economic status of those restricted. The Court then stated that it did not consider itself bound by the contrary language in the *Miln* case, emphasizing that *Miln* was decided over a century ago: "Whatever may have been the notion then prevailing, we do not think that it will now be seriously contended that because a person is without employment and without funds he constitutes a 'moral pestilence.'"[63]

Police Power

Taney's leading biographer asserts that more credit has been given Taney in recent years for the development of the police power than he is entitled to or than he himself would have been willing to accept.[64] This assertion unduly denigrates the contribution of Taney and his Brethren in the development of what has become so seminal a concept in our public law. It was the Taney Court which first gave to the notion of police power something like its modern connotation. In his opinion in the *Charles River Bridge* case, Chief Justice Taney affirmed the existence of police power in the states: "We cannot . . . by legal intendments and mere technical reasoning, take away from them any portion of that power over their own internal police and improvement, which is so necessary to their well-being and prosperity."[65] And, as noted previously, the decision in *New York v. Miln* turned expressly upon the police power concept.

In the 1847 *License Cases* Taney himself gave to the police power the broad connotation that has been of such influence in molding the development of constitutional law:

> But what are the police powers of a State? They are nothing more or less than the powers of government inherent in every sovereignty to the extent of its dominions. And whether a State passes a quarantine law, or a law to punish offenses, or to establish courts of justice, or requiring certain instruments to be recorded, or to regulate commerce within its own limits, in every case it exercises the same power; that is to say, the power of sovereignty, the power to govern men and things within the limits of its dominion.[66]

In the Taney conception, police powers and sovereign powers are the same.[67] In this sense, the states retain all powers necessary to their internal government which are not prohibited to them by the Federal Constitution. Of course, such a broad conception of state power over internal government may be inconsistent with the fullest exertion of individual rights. Taney saw the inconsistency as inherent in the very nature of the police power. Indeed, it was his chief contribution to recognize and articulate the superior claim, in appropriate cases, of public over private rights. The Taney Court developed the police power as the basic instrument through which property might be controlled in the public interest. Community rights were thus ruled "paramount to all private rights . . . , and these last are, by necessary implication, held in subordination to this power, and must yield in every instance to its proper exercise."[68]

It was the Taney opinion in the *License Cases* which gave currency to the phrase "police power." In 1851 was decided the first case in the state courts to speak of the police power, the now classic Massachusetts case of *Commonwealth v. Alger,*[69] with its oft-cited definition of the term by Chief Justice Lemuel Shaw. Only four years later, a Missouri court could say of this power that it was "known familiarly as the police power."[70] By the time of the Civil War, certainly, the term was in common use throughout the land.

The Taney Court's articulation of the police power concept was a necessary complement to the expansion of governmental power that was an outstanding feature of the Jacksonian period. During that period "the demand went forth for a large governmental programme: for the public construction of canals and railroads, for free schools, for laws regulating the professions, for anti-liquor legislation."[71] In the police power concept, the law developed the constitutional theory needed to enable the states to meet the public demand. The Taney Court could thus clothe the states with the authority to enact social legislation for the welfare of their citizens. Government was given the "power of accomplishing the end for which it was created."[72] Through the police power a state might, "for the safety or convenience of trade, or for the protection of the health of its citizens,"[73] regulate the rights of property and person. Thenceforth, a principal task of the Supreme Court was to be determination of the proper balance between individual rights and the police power.

Commerce Regulation

One of the most difficult tasks of the Taney Court was that of determining the reach of the Commerce Clause and the proper scope of concurrent state power over commerce. In *New York v. Miln,*[74] it will be recalled, the Justices had avoided direct resolution of the question of whether the commerce power was exclusively vested in the Federal Government. The question came before the Supreme Court with increasing frequency because of the growing resort by the states to regulatory legislation.

Taney and his colleagues vacillated on the commerce issue, confirming, in the 1847 *License Cases,*[75] the power of the states to regulate the sale of liquor which had been imported from abroad, and then, in the 1849 *Passenger Cases,*[76] striking down state laws imposing a tax on foreign passengers arriving in state ports. The confusion in the Court was shown by the plethora of judicial pronouncements to which this issue gave rise. Nine opinions were written in the first case and eight in the second; in neither was there an opinion of the Court in which a majority was willing to concur.

To understand the problem presented in these cases involving commerce regulation, we should bear in mind that the Commerce clause itself is, as Justice Wiley Rutledge tells us, a two-edged sword.[77] One edge is the positive affirmation of congressional authority; the other, not nearly so smooth or keen, cuts down state power by negative implication. By its very inferential character, the limitation is lacking in precise definition. The clause may be a two-edged blade, but the question really posed is the swath of the negative cutting edge.[78] To put it more specifically, did the Commerce Clause, of its own force, take from the states any and all authority over interstate and foreign commerce, so that state laws on the subject automatically dropped lifeless from the statute books for want of the sustaining power that had been wholly relinquished to Congress?[79] Or was the effect of the clause less sweeping, so that the states still retained at least a portion of their residual powers over commerce?

According to Justice Story's dissent in *New York v. Miln,* the Marshall Court had rejected the notion that the congressional power was only concurrent with that of the states. Marshall, said Story, held that the power given to Congress was full and exclusive: "Full power to regulate a particular subject implies the whole power, and leaves no residuum; and a grant of the whole to one, is incompatible with the grant to another of a part."[80]

In actuality, despite the Story statement to the contrary, the Marshall view of the commerce power was not that of unequivocal federal exclusiveness, as shown by his opinion in *Willson v. Black Bird Creek Marsh Co.*

(1829).[81] In that case, a state law had authorized the construction of a dam across a small navigable creek for the purpose of draining surrounding marshland. It was claimed that the law was repugnant to the federal commerce power. Chief Justice Marshall rejected this contention, emphasizing in his opinion the benefits to be derived from draining the marsh in enhanced land values and improved health. "Measures calculated to produce these objects," he said, "provided they do not come into collision with the powers of the general government, are undoubtedly within those which are reserved to the states."[82]

The *Willson* opinion indicates that Marshall's interpretation of the negative aspect of the Commerce Clause was not as far from that of Taney as is generally believed. "It appears to me to be very clear," declared Taney in an 1847 opinion, "that the mere grant of power to the general government cannot, upon any just principles of construction, be construed to be an absolute prohibition to the exercise of any power over the same subject by the States."[83] Yet the *Willson* case indicates that Marshall shared this view as far as the commerce power was concerned. Both Marshall and Taney, then, refused to follow the notion of complete exclusiveness of federal power under which the Commerce Clause, of its own force, removed from the states any and all power over interstate and foreign commerce.

Where Marshall and Taney differed was in their conception of just how much power over commerce remained in the states. Taney followed his rejection of the complete exclusiveness theory to the opposite extreme and asserted in the states a concurrent power over commerce limited only by the Supremacy Clause of the Constitution. The states, in his view, might make any regulations of commerce within their territory, subject only to the power of Congress to displace any state law by conflicting federal legislation.[84] His concurrent power theory (under which the states possess, concurrently with Congress, the full power to regulate commerce) is, however, incompatible with the basic purpose which underlies the Commerce Clause—that of promoting a system of free trade among the states protected from state legislation inimical to that free flow. For that goal to be achieved, the proper approach to the commerce power lies somewhere between the antagonistic poles of extreme exclusiveness and coextensive concurrent power.

Such an approach had been urged by William Wirt, the Attorney General, as cocounsel with Daniel Webster in *Gibbons v. Ogden*.[85] Webster, whose view of national power was similar to Marshall's, argued in the case for exclusive federal power, which would have barred all state economic regulation. Wirt was a Virginian whose views were closer to those of Jefferson and his followers. His argument for concurrent state power left an area open for state commercial regulation, consistent with the Jeffersonian view that the Constitution did not provide for a federal monopoly of the enumerated powers.

The potential in the Wirt approach may not have been immediately apparent. But it furnished the basis for state economic regulation during the years before Washington began to intervene actively in economic affairs. Wirt also set forth the theory upon which state regulation of commerce was to be sustained. In his *Gibbons* argument Wirt put forth the proposition, "not that all the commercial powers are exclusive, but that those powers being separated, there are some which are exclusive in their nature,"[86] while others might be left open to the states.

In addition, he set out a criterion by which to judge when the federal commerce power was exclusive—that of *uniformity*. In his argument, he said, "[L]et us suppose that the additional term, uniform, had been introduced into the constitution, so as to provide that Congress should have power to make uniform regulations of commerce throughout the United States." In his view, the express insertion was not necessary. Federal power under the Commerce Clause "necessarily implies uniformity, and the same result, therefore, follows as if the word had been inserted."[87] The implication was that where uniformity of regulation was required, only Congress might regulate; where it was not, the states had concurrent regulatory power—subject, of course, to federal supremacy where there were conflicting regulations.

What Wirt was urging in his *Gibbons v. Ogden* argument was that some, but not all areas of commercial regulation are absolutely foreclosed to the states by the Commerce Clause. Here was a doctrine of what might be termed "selective exclusiveness," with the Supreme Court determining, in specific cases, the areas in which Congress possessed exclusive authority over commerce. Its great advantage was that of flexibility. Since it neither permitted nor foreclosed state power in every instance in advance, it might serve as a supple instrument to meet the needs of the future.

Unfortunately, history has not given Wirt credit for this major creative accomplishment. Thus Justice Frankfurter refers to "Webster's doctrine of selective 'exclusiveness'" and states that in *Cooley,* "Webster's analysis became Supreme Court doctrine."[88] As already indicated, Webster's *Gibbons* argument was devoted entirely to the proposition that the federal power over commerce was exclusive. It was Wirt's alternative argument that ultimately became accepted doctrine and served as the basis for state regulatory power.

Cooley Case

It was not until *Cooley v. Board of Port Wardens* (1852)[89] that the Supreme Court was to adopt the middle approach urged by Wirt to the question of state power to regulate commerce. Before that case, as already indicated, the Taney Court had vacillated in its answer to that question. Taney had urged the existence of a commerce power in the states coextensive with that of Congress, to yield only where state regulation was in conflict with federal law, but he could not induce a majority to acquiesce in his view. A compromise was necessary if the question was to be resolved in a way that rejected the opposite extreme of exclusive congressional authority, urged by the "high-toned Federalists on the bench."[90]

The *Cooley* opinion was delivered by Justice Benjamin R. Curtis, who is remembered today almost entirely because of his dissent in the *Dred Scott* case.[91] Yet Curtis's contribution to American law was more significant than mere authorship of his now-classic dissent from the most discredited judicial decision in our history. Indeed, a 1972 evaluation by professors of law, history, and political science rates Curtis as a Supreme Court Justice higher than all the other members of the Taney Court except the Chief Justice.[92] "No one," says Justice Frankfurter, "can have seriously studied the United States Reports and not have felt the impact of Curtis' qualities—short as was the term of his office."[93]

Curtis had attended Harvard Law School, where he was one of Justice Story's outstanding students,[94] and then built up a reputation as leader of the Boston Bar, particularly in commerical law cases. His most noted argument while in practice defended the right of a slaveholder visiting Massachusetts to hold the slave and take her back to the owner's home in Louisiana[95]—arguing for the very principle that his *Dred Scott* dissent was to dispute so vigorously. Aside from brief service in the state legislature, Curtis's career was entirely in practice when he was appointed in 1851, largely through Webster's influence, to the New England seat on the Supreme Court.

As the Frankfurter quote indicates, Curtis's Court tenure was brief; he resigned after only six years, soon after the *Dred Scott* decision. During his short term, however, Curtis delivered opinions that indicated his judicial potential. Had he, as a laudatory article on his appointment put it, "consented to devote the rest of his days to dispensing justice on the highest tribunal in the world,"[96] he undoubtedly would have become one of the outstanding Supreme Court Justices.

The major Curtis contribution to our law came in the *Cooley* case, where his opinion resolved the issue of the reach of the Commerce Clause and the proper scope of concurrent state power over commerce. Before *Cooley,* we saw, the Justices had avoided direct resolution of that issue, with the Court divided between the view that the commerce power was exclusively vested in the Federal Government and the Taney view that the states possessed, concurrently with Congress, the full power to regulate commerce.

In *Cooley,* the Curtis opinion of the Court adopted the "selective exclusiveness" compromise that had been urged by William Wirt in his *Gibbons v. Ogden* argument. It is not known whether Justice Curtis was familiar with Wirt's argument. But he adopted the Wirt approach in his *Cooley* opinion as a necessary compromise to resolve the issue on which the Court had until then vacillated. Curtis had taken his seat only two months before *Cooley* was argued and was thus an ideal judge to write a compromise opinion between the extremes of exclusive congressional power advocated by Justices McLean and Wayne (who dissented in *Cooley*) and the Taney view of coextensive concurrent power.

Chief Justice Taney concurred silently in the Curtis *Cooley* opinion. Why he did so has always been a matter for speculation. As Chief Justice, he could, if he chose, make himself spokesman for the Court. That he did not do so shows that he could not carry a majority for his own approach.[97] If he did not accept the *Cooley* compromise, it would have meant the same fragmented resolution of the commerce issue that had occurred in the prior cases. Taney's concurrence in the *Cooley* compromise made it possible for the law at last to be settled with some certainty on the matter (it was only after *Cooley,* asserts the Court's historian, "that a lawyer could advise a client with any degree of safety as to the validity of a State law having any connection with commerce between the states").[98]

Taney's biographer asserts that the author of the *Cooley* opinion "brought to the Court no new ideas on the subject of the interpretation of the commerce power."[99] The assertion is unfair. Of course, Justice Curtis followed the time-honored judicial technique of pouring new wine into old bottles. He based his opinion on Wirt's "selective exclusiveness" doctrine, but he went beyond Wirt's argument to make a truly original contribution which has since controlled the law on the matter. Well could Curtis write, just before the *Cooley* decision was announced, "I expect my opinion will excite surprise. . . . But it rests on grounds perfectly satisfactory to myself . . . although for twenty years no majority has ever rested their decision on either view of this question, nor was it ever directly decided before."[100]

The *Cooley* case itself arose out of a Pennsylvania law requiring vessels using the port of Philadelphia to engage local pilots or pay a fine, amounting to half the pilotage fee, to go to the Society for the Relief of Distressed and Decayed Pilots. Since there was no federal statute on the subject, the question for the Supreme Court was that of the extent of state regulatory power over commerce where Congress was silent on the matter. It was contended that the pilotage law was repugnant to the Consitution because the Commerce Clause had vested the authority to enact such a commercial regulation exclusively in Congress. To the question whether the power of Congress was exclusive, Justice Curtis answered, "Yes and No"—or, to put it more accurately, "Sometimes Yes and sometimes No." There remained the further inquiry: "When and why, Yes? When and why, No?"[101]

In his *Gibbons v. Ogden* argument, Wirt's cocounsel, Daniel Webster, had said that "the power should be considered as exclusively vested in Congress, so far, and so far only, as the nature of the power requires."[102] *Cooley* followed the same basic approach. If the states are excluded from power over commerce, Curtis said, "it must be because the nature of the power, thus granted to Congress, requires that a similar authority should not exist in the states."[103] If that be true, the states must be excluded only to the extent that the nature of the commerce power requires. When, Curtis asked, does the nature of the commerce power require that it be considered exclusively vested in Congress? This depends not upon the abstract "nature" of the commerce power itself but upon the nature of the "subjects" over which the power is exercised, for "when the nature of a power like this is spoken of, when it is said that the nature of the power requires that it should be exercised exclusively by Congress, it must be intended to refer to the subjects of that power, and to say they are of such a nature as to require exclusive legislation by Congress."[104]

Having thus transferred the focus of inquiry from the commerce power in the abstract to the subjects of regulation in the concrete, Curtis then examined them pragmatically. If we look at the subjects of commercial regulation, he pointed out, we find that they are exceedingly various and quite unlike in their operation. Some imperatively demand a single uniform rule, operating equally on commerce throughout the United States; others as imperatively demand that diversity which alone can meet local necessities. "Either absolutely to affirm, or deny," said Curtis, "that the nature of this power requires exclusive legislation by Congress, is to lose sight of the nature of the subjects of this power, and to assert concerning all of them, what is really applicable but to a part."[105]

Whether the states may regulate depends upon whether it is imperative that the subjects of the regulation be governed by a uniform national system. As the *Cooley* opinion put it, "Whatever subjects of this power are in their nature national, or admit only of one uniform system, or plan of regulation, may justly be said to be of such a nature as to require exclusive legislation by Congress."[106] On the other hand, where national uniformity of regulation is not necessary, the subject concerned may be reached by state law. That is the case with a law for the regulation of pilots like that at issue in *Cooley*.

Almost two decades after the *Cooley* decision, Justice Samuel F. Miller, speaking for the Supreme Court, stated, "Perhaps no more satisfactory solution has ever been given of this vexed question than the one furnished by the court in that case."[107] Over a century later, much the same comment can be made, despite the attempts by the Court since *Cooley* to formulate other tests. Those tests have proved unsatisfactory, and the Court has basically continued to follow the *Cooley* approach in cases involving the validity of state regulations of commerce.

The Curtis approach in *Cooley* was a necessary modification of the developing conception of law. Judges like Marshall and Story had developed legal principles to accord with their vision of the emerging society and economy. In doing so, they had rejected the opposing vision of men like Jefferson and the legal doctrines of jurists like Spencer Roane, who sought to give effect to that vision. What was to happen, however, when neither vision was able to command majority support? That had become the situation with regard to the *Cooley* issue. The "high-toned Federalists on the bench"[108] refused to yield on the Marshall vision of a nation vested with exclusive power over commerce. Chief Justice Taney was equally unyielding on the opposite Jacksonian posture.

Ultimately, the difference on the matter came down to a difference in the protection of property rights. In particular, the principle of federal exclusiveness meant the virtual immunity of property from public power, since congressional power over commerce was to remain in repose during most of the century. What regulation of business there was occurred at the state level. Hence the Taney conception of concurrent state power gave effect to the Jacksonian emphasis upon public power as a counterweight to the property rights stressed by the Federalists and then the Whigs.

As it turned out, neither the Federalist nor the Jacksonian view could command the needed juristic support. More important, neither was appropriate to the nation's commercial needs. Federal exclusiveness would have led to a complete absence of control over business abuses for almost a century. Taney's opposite approach would have resulted in the crazy quilt of commercial regulations that had led to the Constitution and the Commerce Clause themselves.

What was needed was the *Cooley* compromise, which could both secure the necessary votes and further the needs of commerce in a federal system. The *Cooley* test was one that could be adjusted to the differing demands of a polity characterized at first by absense of federal regulation and later by one dominated by control from Washington. The *Cooley* approach neither permitted nor prohibited state power in advance. As such, it could be molded by future jurists to meet the "felt necessities" of their times. The same would not have been true if the simple universality of the rules[109] rejected by Curtis had been elevated to the plane of accepted jurisprudence.

Curtis did more in *Cooley* than resolve the commerce issue by compromise. He stated a balancing test which makes the validity of a state regulation depend upon a weighing of the national and local interests involved. "More accurately," the Court more recently informed us, "the question is whether the State interest is outweighed by a national interest in the unhampered operation of interstate commerce."[110] An affirmative answer must be given only when a case falls within an area of commerce thought to demand a uniform national rule. But in the absence of conflicting legislation by Congress, there is a residuum of power in the states to make laws governing matters of local concern which nevertheless affect, or even regulate, interstate commerce.[111]

In marking out the areas of permissible state regulation, *Cooley* makes the primary test not the mechanical one of whether the particular activity regulated is part of interstate commerce, but rather whether, in each case, the competing demands of the state and national interests involved can be accommodated.[112] State regulations are to be upheld where it appears that the matter involved is one which may appropriately be regulated in the interest of the safety, health, and well-being of local communities. *Cooley* recognizes that there are matters of local concern which may properly be subject to state regulation—matters which, because of their local character and the practical difficulties involved, may in fact never really be adequately dealt with by the Congress.[113]

Under *Cooley*, then, regulation depends upon balancing the circumstances of the locality which may tilt in favor of local regulation, on the one hand, and the national need for uniformity, on the other. As one commentator summarizes it,

> In his recognition of the complexity of commercial activity, his desire to strike a balance between upholding federal regulatory power while safeguarding local freedom of action, his indication to leave to the future and to the courts the job of drawing lines of responsibility in the gray areas of jurisdiction, his requirement that judges look hard at the specific facts on which a particular case turns and avoid Federalist or Jeffersonian dogmatizing, and his insistence upon results—the arrangement that works best—Curtis grafted onto the Constitution a flexible approach with a pragmatic method of analysis.[114]

In *Cooley*, Justice Curtis stated a new balancing approach to law that foreshadowed modern constitutional jurisprudence.

Corporate Expansion

Justice Frankfurter once said that the history of our constitutional law in no small measure is the history of the impact of the modern corporation upon the American scene.[115] While the Taney Court sat, new economic forces were bringing new issues for judicial resolution; the corporate device and the concentrations of economic power made possible by it began to come before the Justices with increasing frequency.

If there was one tenet common to advocates of Jacksonian democracy, it was that of opposition to what Jackson's Bank Veto Message termed "the rich and powerful [who] too often bend the acts of government to their selfish purposes."[116] They deeply distrusted corporations as aggregations of wealth and power—the "would-be lordlings of the Paper Dynasty"[117]—which posed a direct danger to the democratic system. In his 1837 Farewell address, Jackson had warned of the perils posed by "the great moneyed corporations": "[U]nless you become more watchful . . . and check this spirit of monopoly and thirst for exclusive privileges you will in the end find that the most important powers of Government have been given or bartered away, and the control over your dearest interests has passed into the hands of these corporations."[118]

The Jacksonian view of corporate power was shared by Chief Justice Taney and most of his colleagues on the bench. Taney himself had been a leader in the war against the Bank of the United States—the corporate monster, "citadel of the moneyed power,"[119] which the Jacksonians had finally overthrown. An 1834 Taney speech affirmed that "in every period of the world . . . history is full of examples of combinations among a few individuals, to grasp all power in their own hands, and wrest it from the hands of the many."[120] Certainly Taney subscribed to the view that "the extent of the wealth and power of corporations among us, demands that plain and clear laws should be declared for their regulation and restraint."[121]

In theory at least, the Jacksonians on the bench shared the agrarian persuasion of their most extreme member, Justice Peter V. Daniel, who wrote in 1841 that, though he perceived the spread of banks and corporations to every hamlet, he still hoped that they might be weeded completely out of society.[122] In practice, however, even the Jacksonian Justices had to recognize that the corporation had a proper place in the legal and economic systems. The result was that, as Jacksonians, men like Taney might fear its abuses, but as practical men of affairs they had to recognize its utility in an expanding nation. To the United States of the first half of the nineteenth century, the corporate device was an indispensable adjunct of the nation's growth. The corporation enabled men to establish the pools of wealth and talent needed for the economic conquest of a continent. Even these Jacksonian judges realized the relationship between the corporation and economic development and made decisions favorable to the corporate personality—notably the 1839 decision in *Bank of Augusta v. Earle*.[123]

The question presented in that case has been characterized by the Supreme Court's historian as "of immense consequence to the commercial development of the country—the power of a corporation to make a contract outside of the State in which it was chartered."[124] It should be borne in mind that the corporation is entirely a creation of law; its very existence and legal personality have their origin in some act of the law. Corporations, of course, appear at an early stage of American history, for chartered companies first settled the Colonies of Virginia and Massachussets Bay. Yet, though the corporation as a legal person was developed under English law and recognized from the beginning in American law (especially in the classic 1819 *Dartmouth College* case), it was not until the decision in *Bank of Augusta v. Earle* that it could really be made to serve the needs of the burgeoning American economy.

In the *Bank of Augusta* case two banks and a railroad, incorporated respectively in Georgia, Pennsylvania, and Louisiana, brought an action in the federal court in Alabama on bills of exchange purchased by them in that state; the makers had refused to pay on the ground that the corporations had no power to do business in Alabama, or, indeed, outside their own states. Their contention was upheld by Justice John McKinley, sitting in the circuit court. As explained in an oft-quoted letter of Justice Story to Charles Sumner, "He has held that a corporation created in one State has no power to contract (or, it would seem, even to act) in any other State, directly or by an agent."[125]

The McKinley ruling was characterized by Story as "a most sweeping decision . . . which has frightened half the lawyers and all the corporations of the country out of their proprieties."[126] Its practical effect was to limit corporate business to the states in which the corporations were chartered, which would have rendered all but impossible the growth of interstate enterprises of any consequence. Well might Webster, in his argument, characterize McKinley's decision as "anti-commercial and anti-social . . . and calculated to break up the harmony which has so long prevailed among the States and people of this Union."[127]

The Supreme Court opinion in the *Bank of Augusta* case was delivered by the Chief Justice. Taney rejected the notion that a corporation could have no existence beyond the limits of the state in which it was chartered. He held that a corporation, like a natural person, might act in states where it did not reside. Comity among the states provided a warrant for the operation throughout the Union of corporations chartered in any of the states: "We think it is well settled that by the law of comity among nations, a corporation created by one sovereignty is permitted to make contracts in another, and to sue in its courts; and that the same law of comity prevails among the several sovereignties of this Union."[128]

Chief Justice Taney did not go as far as Webster had urged in his argument. Though he upheld the power of corporations to act outside their domiciliary states, he also recognized the power of a state to legislate against the entrance of outside corporations. Corporations could operate nationwide, but each state was given the authority to regulate corporate activities within its own borders.[129]

In this respect, the *Bank of Augusta* opinion is a clear reflection of the mixed attitude of judges like Taney toward the corporation. Taney gave legal recognition to the fact that a corporation has the same practical capacity for doing business outside its home state as within its borders. But he refused to go further and adopt the Webster theory of citizenship for corporations within the protection of the Privileges and Immunities Clause of the Constitution. Instead, he carefully circumscribed the basis of their constitutional rights.[130]

Historically speaking, the most important aspect of the decision is the stimulus it provided to economic expansion. The view of the rising capitalist class was expressed by Story, when he wrote to Taney, "Your opinion in the corporation cases has given very general satisfaction to the public; . . . it does great honor to yourself as well as to the Court."[131] Because of the Taney decision, said Webster, "we breathe freer and deeper."[132] *Bank of Augusta v. Earle* was the first step in what the Supreme Court in 1898 was to term "the constant tendency of judicial decisions in modern times . . . in the direction of putting corporations upon the same footing as natural persons."[133] This tendency has been the essential jurisprudential counterpart of the economic unfolding of the nation. Looked at this way, the *Bank of Augusta* decision was as nationalistic as those rendered by Marshall himself.

Taney and Judicial Power

In 1842 John J. Crittenden, recently resigned as President Tyler's Attorney General, was commenting about a case he had just lost before the highest bench. "If it was not a decision of the Supreme Court," he declared, "I should say it was Supremely erronious [/sic]—It is thoroughly against us on all the questions of law and evidence. . . . Sic Transit &c."[134] The Crittenden complaint is one which has been directed against the Supreme Court throughout its history, and not only on behalf of disappointed litigants or their counsel. At times the propensity of the high tribunal toward error has been animadverted upon by Justices themselves. "There is no doubt," caustically commented Justice Robert H. Jackson over a century after the Crittenden reproof, "that if there were a super Supreme Court, a substantial proportion of our reversals . . . would also be reversed. We are not final because we are infallible, but we are infallible only because we are final."[135]

That the Taney Court was far from infallible is apparent to even a casual student of its work. Yet it is amazing to note how, once Taney had established his imprint upon the Court, the opposition that had greeted his appointment was quickly stilled. Even his bitterest enemies soon saw, from his work on the bench, that their partisan censures were unjustified. Until the *Dred Scott* case, the stature of the Supreme Court compared favorably with what it had been under Marshall, and, if anything, its decisions were more generally accepted. Criticisms by disappointed litigants and political opponents, of course, continued, but its prestige as an institution never stood higher than in Chief Justice Taney's first twenty years.

It was doubtless of Taney, who had been a Federalist before he became a supporter of Andrew Jackson,[136] that James K. Polk was thinking when, in 1845, he wrote in his diary, "I have never known an instance of a Federalist who had after arriving at the age of 30 professed to change his opinions, who was to be relied on in his constitutional opinions. All of them who have been appointed to the Supreme Court Bench, after having secured a place for life became very soon broadly Federal and latitudinarian in all their decisions involving questions of Constitutional power."[137]

To accuse Taney of a "relapse into the Broad Federal doctrines of Judge Marshall and Judge Story"[138] was unfair. The Taney Court did make important doctrinal changes, particularly in shifting the judicial emphasis from private to community rights and stressing the existence of power to deal with internal problems. That the Court's reaction after Marshall's death was not as great as has often been supposed does not alter the fact that there was a real change.

Even with the changes in constitutional law which it continuously made, however, the Taney Court did not (Justice Story to the contrary notwithstanding) seek to destroy the constitutional structure built by Marshall and his colleagues. Instead, it used that structure as the base for its own jurisprudence, making only such modifications as it deemed necessary to meet the needs of the day. In no respect was this more apparent than in its decisions on the place of judicial power in the governmental system.

In the 1832 Veto Message on the bill to recharter the Bank of the United States, President Jackson had denied the authority of the Supreme

Court to make decisions binding upon the political branches. The implication was that the Court's review power could not control the President. Jackson's opponents charged (as Taney put it years later) that he was, in effect, asserting "that he, as an executive officer, had a right to judge for himself whether an Act of Congress was constitutional or not, and was not bound to carry it into execution, even if the Supreme Court had decided otherwise."[139] Writing to Martin Van Buren in 1860, Taney denied that the Veto message meant any such thing: "[N]o intelligent man who reads the message can misunderstand the meaning of the President. He was speaking of his rights and duty when acting as part of the legislative power, and not his right or duty as an executive officer."[140] If all Jackson meant was that the President could veto bills on constitutional grounds despite Supreme Court decisions going the other way, the Veto Message was far from heretical doctrine, despite the contrary view of Jackson's Whig opponents.

Whether Taney's later justification of the words he wrote for Jackson in the heat of political battle was valid is not as relevant as the fact that almost three decades as Chief Justice gave ample proof of his full adherence to the notion of judicial power expounded by the Marshall Court. In the very first case in which he sat, even before he presided over a Supreme Court session, he declared (in an 1836 charge to a circuit court grand jury), "In a country like ours, blessed with free institutions, the safety of the community depends upon the vigilant and firm execution of the law; every one must be made to understand, and constantly to feel, that its supremacy will be steadily enforced by the constituted tribunals, and that liberty cannot exist under a feeble, relaxed or indolent administration of its power."[141]

The Taney Court was just as insistent as the tribunal headed by Marshall in vindicating the position of the Supreme Court as guardian of the Constitution and ultimate interpreter of its provisions. The judiciary, in the Taney view, was plainly the sine qua non of the constitutional machinery—draw out this particular bolt and the machinery falls to pieces: "For the articles which limit the powers of the Legislative and Executive branches of the Government, and those which provide safeguards for the protection of the citizen in his person and property, would be of little value without a Judiciary to uphold and maintain them which was free from every influence, direct or indirect, that might by possibility, in times of political excitement, warp their judgments."[142]

The point is illustrated most clearly by *Ableman v. Booth*[143]—a case which, during the 1850s, excited an interest comparable to that aroused by *Dred Scott* itself. The *Booth* case arose out of the prosecution of an abolitionist newspaper editor in Milwaukee for his part in rescuing a fugitive slave from federal custody. After his conviction in a federal court early in 1855 for violating the Fugitive Slave Act, Booth secured a writ of habeas corpus in the Wisconsin courts on the ground that the act was unconstitutional. A writ of error was taken to the United States Supreme Court, but the highest state court directed its clerk to make no return, declaring that its judgment in the matter was final and conclusive.

In effect, the Wisconsin judges were asserting a power to nullify action taken by the federal courts. In Taney's characterization, "[T]he supremacy of the State courts over the courts of the United States, in cases arising under the Constitution and laws of the United States, is now for the first time asserted and acted upon in the Supreme Court of a State." To uphold the power thus asserted would, he said, "subvert the very foundations of this Government."[144] If the state courts could suspend the operation of federal judicial power, "no one will suppose that a Government which has now lasted nearly seventy years enforcing its laws by its own tribunals, and preserving the union of States, could have lasted a single year, or fulfilled the high trusts committed to it." The Constitution, in its very terms, refutes the claimed state power; its language, in this respect, "is too plain to admit of doubt or to need comment."[145] The federal supremacy "so carefully provided in . . . the Constitution . . . could not possibly be maintained peacefully, unless it was associated with this paramount judicial authority." In affirming its authority to set federal judicial action at naught, Wisconsin really "has reversed and annulled the provisions of the Constitution itself . . . and made the superior and appellate tribunal the inferior and subordinate one."[146]

The Court's decision (correct though it was in law) was the subject of bitter political attack. To the public, the legal issues were inextricably interwined with the slavery controversy. More than a century later, we are scarcely concerned with the partisan censures of Taney's day, and *Ableman v. Booth* stands as a ringing affirmation of federal judicial power, as strong as any made by Marshall himself.

Judicial Self-Restraint

Just after Andrew Jackson took office Secretary of War J. H. Eaton informed him that "the Cherokees have filed here a protest against the laws of Georgia being extended over them. As it is a delicate matter will you think as to the course of the reply, to be given." Jackson took note of the problem thus presented and wrote on the address leaf of Eaton's letter, "[T]he answer to be well considered on constitutional grounds."[147] Despite his realization that Georgia was acting in violation of federal treaties, the President declined to support the Indians. Instead, he upheld Georgia and helped induce the "voluntary" removal of most of the Indians across the Mississippi.[148] The Cherokees then sought a judicial remedy, and the Supreme Court held that the Constitution barred Georgia from extending its laws over Indian lands and ruled invalid the arrest and imprisonment by the state of two missionaries working with the Cherokees.[149]

Georgia defied the Court's mandate and refused to release the imprisoned missionaries. "The Constitution, the laws and treaties of the United States," declared John Quincy Adams, "are prostrate in the State of

Georgia. Is there any remedy for this state of things? None. Because the Executive of the United States is in league with the State of Georgia."[150] Jackson is reported to have said that John Marshall had made his decision, now let him enforce it, but this may be only apocryphal.[151] Even so, it accurately describes Jackson's actions.[152] He did not seek in any way to enforce the judgment. Instead, he stated, "The decision of the supreme court has fell still born, and they find that it cannot coerce Georgia to yield to its mandate."[153]

This example of judicial impotence in the face of refusal by the political departments and the state to carry out the Supreme Court judgment is one that inevitably had great influence upon Chief Justice Taney, at least during most of his tenure on the bench. He developed a strong tendency to restrict the area of judicial discretion in constitutional decision.[154] Judicial self-restraint became for the first time an essential element of Supreme Court doctrine. With the *Cherokee Nation* example before him, Taney strove to steer the Court away from unduly political issues.

Speaking of Chief Justice Taney, Dean Acheson said that "judicial self-restraint . . . was his great contribution to the law and custom of the Constitution. . . . [T]he giant stature which Taney assumes in the history of the Supreme Court is due chiefly to his insistence that the judge, in applying constitutional limitations, must restrain himself and leave the maximum of freedom to those agencies of government whose actions he is called upon to weigh."[155] The concept of self-restraint cuts across the work of the Taney Court and distinguishes it most sharply from its predecessor. Where Marshall and his colleagues did not hesitate to involve themselves in issues that were essentially political in character, the Taney Court was more cautious. Until the *Dred Scott* case, Taney was largely successful in keeping the Court out of the "political thicket"[156] of party controversies. The basic Taney philosophy was to leave every opportunity for the solving of political problems elsewhere than in the courtroom. "In taking jurisdiction as the law now stands," he asserted in one case, "we must exercise a broad and indefinable discretion, without any certain and safe rule to guide us. . . . [S]uch a discretion appears to me much more appropriately to belong to the Legislature than the Judiciary."[157]

This statement was made in *Pennsylvania v. Wheeling & B. Bridge Co.* (1852),[158] a case that shows the Taney approach and the danger of departing from it. "The stupendous structure that spans the Ohio at Wheeling," wrote a contemporary of the bridge involved in the case, "strikes the eye of the traveller passing beneath it, as it looms above him in the darkness, as one of the great architectural wonders of the age."[159] At its dedication, Henry Clay had declaimed, "You might as well try to take down the rainbow."[160] But Pennsylvania sought to do just that by a suit in the Supreme Court for an injunction directing the removal of the bridge on the ground that it blocked river traffic.[161]

The majority of the Court was willing to order that the rainbow at least be raised to meet minimum ship clearances.[162] Chief Justice Taney dissented, urging (as the quote from his opinion already given indicates) that the matter was one for Congress, not the Court, to regulate under the commerce power. The Court's acceptance of jurisdiction was, however, short-lived. Congress passed a statute declaring the bridge to be a lawful structure and not an obstruction to navigation.[163] The Court then upheld the congressional power to enact such a law.[164] The ultimate result was thus precisely what Taney had urged in his original dissent—though only at the cost of congressional intervention to, in effect, reverse a Supreme Court decision.

Most of the time, Chief Justice Taney was able to carry the Court with him in adherence to the self-restraint doctrine, particularly in cases involving judicial attempts to dictate action by the other branches. "The interference of the courts with the performance of the ordinary duties of the executive departments of the government," he once affirmed, "would be productive of nothing but mischief."[165] In line with this view, he refused to order a state governor to extradite a fugitive from another state: "[I]f the governor of Ohio refuses to discharge this duty, there is no power delegated to . . . the judicial department . . . to use any coercive means to compel him."[166] At the back of Taney's mind must have been the need to avoid a clash such as that involved in the *Cherokee Nation* case.

The most famous case in which the Taney Court applied the self-restraint doctrine was *Luther v. Borden*.[167] It arose out of the only revolution that occurred in a state of the Union after the Revolutionary War itself—the so-called Dorr Rebellion in Rhode Island in 1841. That state was then still operating under the royal charter granted in 1663. It provided for a very limited suffrage and, worse still from the point of view of those who considered it completely out of date, no procedure by which amendments might be made. Popular dissatisfaction led to mass meetings in 1841, which resulted in the election of a convention to draft a new constitution. It was drawn up and provided for universal suffrage. Elections were held under it, and Thomas Wilson Dorr was elected Governor. All these acts were completely unauthorized by the existing charter government, which declared martial law and called out the militia to repel the threatened attack. In addition, it appealed to the Federal Government for aid, and President Tyler, expressly recognizing the charter government as the rightful government of the state, took steps to extend the necessary help, declaring that he would use armed force if that should prove necessary. The announcement of the President's determination caused Dorr's Rebellion to die out. *Luther v. Borden,* decided several years later, was left as its constitutional legacy.

The actual case arose out of the efforts of the charter government to suppress the Dorr Rebellion. When one of its agents broke into the house of a strong Dorr supporter and arrested him, the latter brought an action of trespass. Defendant justified his action by the plea that he was acting under the authority of the legal government of the state. Plaintiff countered with the contention that the charter government was not republican in form, as required by the Constitution. Therefore, he asserted, that government had no valid legal existence and the acts of its agents were not justified in law. Essentially, he was claiming that the action of the charter government violated his constitutional right to live under a republican government and that that claim was cognizable in a court.

The Supreme Court rejected the claim, denying that it was within judicial competence to apply the consitutional guaranty. On the contrary, the enforcement of the guaranty is solely for Congress. Under article IV, section 4, declares the opinion of Chief Justice Taney, "[I]t rests with Congress to decide what government is the established one in a State . . . as well as its republican character." Moreover, the congressional decision in the matter is not subject to any judicial scrutiny: "[I]ts decision is binding on every other department of the government, and could not be questioned in a judicial tribunal."[168]

Likewise, it is up to Congress "to determine upon the means proper to be adopted to fulfill this guaranty."[169] Under an Act of Congress that body had delegated to the President the responsibility of determining when the Federal Government should interfere to effectuate the constitutional guaranty,[170] and, in this case, as we saw, the President acted to support the charter government. After such action by the President, asked Taney, "[I]s a circuit court of the United States authorized to inquire whether his decision was right? Could the court, while the parties were actually contending in arms for the possession of the government, call witnesses before it and inquire which party represented a majority of the people? . . . If the judicial power extends so far, the guarantee contained in the Constitution of the United States is a guarantee of anarchy, and not of order."[171]

In *Luther v. Borden,* Taney refused to go into the issue of the legal authority of the government actually in power, holding that the questions involved were political and beyond the sphere of judicial compentence. The overriding consideration was to steer clear of political involvement; the question of governmental legitimacy was left exclusively to the political departments. The wisdom and authority of Taney's restraint in this respect has not been generally questioned.[172]

The judicial reluctance to approach too close to the founts of sovereignty was a dominant characteristic of the Taney Court. The soundness of such an attitude was amply demonstrated when the Court refused to follow the rule of abnegation and sought instead to resolve in the judicial forum the basic controversy over slavery which had come to tear the nation apart. Even masterful judges are not always restrained by the wisdom of self-denial: the *Dred Scott* case was the one occasion when Taney yielded to the temptation, always disastrous, to save the country, and put aside the judicial self-restraint which was one of his chief contributions to our constitutional law.

Taney's Associates

An 1859 article in the *New York Daily Tribune* described a visit to the Supreme Court:

> [Y]ou are ushered into a queer room of small dimensions, shaped over head like a quarter section of a pumpkin shell, the upper and broader rim crowning three windows, and the lower and narrower coming down garret-like to the floor; the windows being of ground glass, and the light trickling through them into the apartment. That which most arrests your attention is a long pew, just in front of these windows, slightly elevated above the floor, along which are ranged in a straight line nine ancient persons, clad in black-silk gowns.[173]

The "ancient persons" during the Taney tenure were, of course, different from those who had sat on the bench in Marshall's day. In 1836, just before Chief Justice Taney took his seat, Webster showed a visitor around the Capitol. At the Supreme Court, Webster referred to the great changes since he had first appeared there, saying, "No one of the Judges who were here then, remains."[174] When Story died in 1845, none of the Justices who had contributed to the work of the Marshall Court was on the bench.[175]

The supreme bench was completely remade by President Jackson, who, as already stated, appointed six Justices. Of the Justices who had sat sometime with Marshall, only Justices Story and Thompson served on the Taney Court and they realized that they were leftovers from another era—in Story's phrase, "in the predicament of the last survivor."[176] Justice Thompson had indicated a state's-rights leaning in Marshall's day, having differed from the great Chief Justice in cases on state power to enact bankruptcy laws, tax imports, and issue bank notes which circulated as money.[177] Now, however, he was as dismayed as Story at the tilt away from Marshall's nationalistic emphasis; John Quincy Adams in 1831 noted, "He is alarmed for the fate of the Judiciary," since "the leading system of the present Administration is to resolve the Government of the Union into the national imbecility of the old Confederation."[178]

It was that Jackson Administration that sought to remake the Court in its own image. Its first opportunity came when Justice Trimble died in 1828. In his place, President Jackson chose John McLean of Ohio, who had been on the highest court of his state and John Quincy Adams's Postmaster General. McLean served on the Court thirty-two years. Despite the many opinions he wrote, he is remembered today for his dissent in the *Dred Scott* case and, even more so, because he was the first Justice who, in Lincoln's phrase about Chief Justice Salmon P. Chase, "had the Presidential maggot in his brain."[179] In practically every election after his appointment, McLean was, actively or passively, a candidate for the Presidency.[180]

McLean's role as what a newspaper termed "a judicial politician"[181] could not help but have a harmful effect upon the Court's reputation—he was "dragging the ermine in the mire of politics," declaimed the *National Intelligencer*.[182] To others, McLean may, as Webster once said, have had "his head turned too much by politics."[183] McLean himself, however, indicated that he had no doubt about the propriety of a Justice being a candidate for the Presidency. "I did not suppose," he wrote in answer to a critic, "that you or any other person who had reflected upon the subject could entertain the least apprehension of any improper influence being used by a Judge who comes before the people in a popular election, and especially that it could lend [sic] to corrupt the Bench."[184]

McLean also did not hesitate to speak out publicly on controversial political issues. He published a letter strongly attacking the Mexican war and publicly expressed his views on slavery—most notably in an 1848 letter on the power of Congress over slavery in the territories. He thus undertook, in the words of another critic, "to adjudicate a question" that could come before the Court,[185] as it ultimately did in the *Dred Scott* case. Though we may have serious doubts about the propriety of such conduct by a Justice, here, too, McLean insisted, "As a citizen, I claim the right and shall exercise it, of forming and expressing my opinion on public measures."[186]

As a Justice, McLean disappointed the expectations of the President who had appointed him. During the years he served, McLean was one of the most nationalistic of the Taney Court Justices. In his brief tenure under Marshall, he was characterized in a newspaper "as sound as Marshall himself" on issues of federalism and property rights.[187] He continued to follow the same approach under Marshall's successor. Indeed, in an 1852 letter, Justice Curtis referred to "McLean & Wayne, who are the most high-toned Federalists on the bench."[188]

James M. Wayne, the third Justice appointed by Jackson, also was closer to Marshall's views than those of the President who chose him. Wayne was appointed in 1835 to succeed Justice Johnson, who had died the previous year. He had been active in Georgia politics and had supported Jackson as a Congressman in both the nullification controversy and the struggle over the Bank of the United States. These, more than Wayne's legal ability or his prior service as a judge in his state, led to his selection. Wayne, like McLean, continued to be a strong Unionist on the bench, as is shown by his statement of dissent in the *Cooley* case, which led to the quoted Curtis comment. Despite his role in the *Dred Scott* case, he remained an adherent of the Union to the end; during the Civil War, he abandoned his Savannah home for residence in Washington. Because of this, a Confederate court branded him an enemy alien and confiscated his Georgia property.[189]

When Wayne was appointed, James Buchanan wrote that he would "[n]ever make [an] able judge."[190] According to one commentator, "Buchanan's prophecy proved accurate. Wayne served on the Court for thirty-two years without making any conspicuous contribution."[191] On the other hand, it is said that Wayne was influential in the Court's internal deliberative process[192]—a conclusion apparently borne out by the part he played in steering the Court to its *Dred Scott* decision.

Between the appointments of McLean and Wayne, President Jackson was given the opportunity to select another Justice by the death in 1829 of Bushrod Washington. The President chose Henry Baldwin, who had been Jackson's outspoken defender while a Pennsylvania Congressman. Justice Baldwin was unhappy on the Marshall Court. Within a year, Martin Van Buren could write in his *Autobiography,* "Judge Baldwin is dissatisfied with his situation for reasons which . . . grow out of opposition to what he regards as an unwarrantable extension of its powers by the Court, and has given the President notice of his intention to resign."[193]

Jackson, however, prevailed on Baldwin not to leave the Court, where he remained a disturbing influence in part because, as Story wrote, "he is partially deranged."[194] His mental problems continued during most of his tenure, causing him to be erratic in his jurisprudence. Quick to concur or dissent, he followed no consistent constitutional approach. He was more of a jarring influence than a real contributor to the Taney Court.

President Jackson was, of course, also able to appoint Roger B. Taney as Marshall's successor, as well as to appoint two other Justices. The first was Philip P. Barbour, confirmed the same day as the new Chief Justice, who took the seat of Justice Duval, who had resigned in 1835. According to Senator Thomas Hart Benton, "Judge Barbour was a Virginia country gentleman, after the most perfect model of that class."[195] He had been a Congressman and a federal district judge. On his appointment, a Richmond newspaper wrote that he was "eminently fitted to adorn the Bench . . . and enlighten it with his inflexible and uncompromising State-Rights principles."[196] To Marshall's adherents, however, Barbour's appointment was an unhappy portent. John Quincy Adams referred to him as a "shallow-pated wild-cat, . . . fit for nothing but to tear the Union to rags and tatters."[197] Barbour did not, however, serve long enough (he died at the beginning of 1841) to tilt the Court toward the doctrines of the Virginia school.

President Jackson's last Court appointment filled one of two new seats created in 1837. "The Supreme Court," Justice Story wrote in that year, "now consists of nine Judges, two having been lately added by an act of Congress."[198] Jackson now had the opportunity to select two new Justices. However, one of his appointees, his lifelong friend William Smith, declined and Jackson's successor made the appointment. For the other new seat, Jackson chose John Catron of Tennessee, who met the key litmus requirements of Jacksonian loyalty, opposition to the Bank of the United States, and support of the President's stand against nullification.

Catron was not deterred from expressing fervent public support for Jackson's position by the fact that he was then Chief Justice of his state's highest court. In fact, Catron was another early Justice who never followed the traditional judicial ethic of divorcing himself from politics. He was a close adviser of Presidents Polk and Buchanan. His activities in this respect leave more recent Justices who have advised the White House in the shade. Catron's contacts with President Buchanan in the *Dred Scott* case (to be discussed in the next chapter) violated all notions of judicial propriety, even by the more relaxed standards of his day.

When we think of the Presidents between Jackson and Lincoln we must ask, with James Bryce, "who now knows or cares to know anything about the personality of James K. Polk or Franklin Pierce? The only thing remarkable about them is that being so commonplace they should have climbed so high."[199] The same is true of most of the Justices chosen by our pre-Civil War Presidents. Almost all of them, too, have been relegated to the obscurity reserved for the Tylers, Buchanans, and other lesser lights who failed to measure up to what those in their high positions should be.

The Presidents who succeeded Jackson added the following Justices to the Taney Court:

President Van Buren: John McKinley, a former Senator and loyal Jacksonian, appointed from Kentucky to fill the new ninth seat; Peter V. Daniel of Virginia, another ardent Jackson man, to replace the deceased Justice Barbour.

President Tyler: Samuel Nelson, a judge from New York, who was to serve for twenty-seven years, replacing Justice Thompson, who had died.

President Polk: Levi Woodbury of New Hampshire, a former judge, Governor, Senator, and Cabinet member, who could, of course, scarcely fill the seat left vacant by Justices Story's death in 1845; Robert C. Grier, a Pennsylvania judge appointed after Justice Baldwin's death.

President Fillmore: Benjamin R. Curtis, already discussed, who despite the shortness of his tenure, was the outstanding Associate Justice on the Taney Court.

President Pierce: John A. Campbell, a preeminent Alabama lawyer, whose appointment, Campbell himself informs us, was "one recommended by the Justices—Justices Catron and CURTIS bearing their recommendation to the President,"[200] who succeeded Justice McKinley on the latter's death.

President Buchanan: Nathan Clifford from Maine, a former Attorney General, to replace Justice Curtis, who resigned after the *Dred Scott* decision.

Writing about one of these Justices, Taney's biographer tells us that he was a "man of moderate ability who achieved neither distinction nor notoriety."[201] Except for Justice Curtis, the same can be said of the other Justices just listed. Attempts by revisionist biographers to change this conclusion can scarcely alter history's estimate. The only place in legal history attained by some of these Justices stems from their role in the *Dred Scott* case—which, of course, finally gave them the "notoriety" that they did not otherwise achieve.

The appointments by President Jackson and his successors did, however, have one important consequence for the Court. After Justice McKinley of Kentucky was appointed, a majority of the Justices were from below the Mason-Dixon Line. This was to prove of great significance, since it led directly to the Court's proslavery position, which reached its disastrous climax in the *Dred Scott* case. There is irony in the fact that the Southern Court majority was brought about by the appointment of Justice McKinley by President Van Buren, who was himself an opponent of slavery and ran as the candidate of the antislavery Free Soil Party in 1848.

Just before the end of the Taney tenure, at the beginning of the December 1860 Term, the Court moved into the old Senate Chamber in the Capitol. The Court was to occupy this as its chamber until the construction of the present Supreme Court building. In addition, twelve other rooms were provided for the use of the Court, its officers, and records—particularly a separate robing room, which enabled the Justices to make the dignified formal entry that had been impossible when they had donned their robes from pegs on the wall in the presence of the audience.

The new Court Chamber was certainly an improvement over the one previously used in the Capitol basement. The new courtroom was much larger and more elegant. Though it was still semicircular in shape, like the previous basement chamber, it was forty-five feet long and the same distance wide at its widest point. The ceiling was a low half-dome, with a suspended chandelier. As in the lower chamber, the bench was arranged with judicial backs to the wall, and it was now substantially elevated. Ionic marble columns formed a colonnade along one side of the room, while pilasters of marble decorated the circular wall. Marble busts of Chief Justices were arranged around the walls. The added dignity of the chamber was enhanced by soft brown carpeting and red velvet cushioning of the benches available to the public.[202]

Certainly this was an improvement over what a newspaper called the old "potato hole of a place"[203] in the Capitol basement. It was, however, still a far cry from the Court's present marble palace—the first setting that housed the court in a manner befitting its august constitutional role and its position as the highest court in the land.

Taney and His Court

It is customary to point to the drastic change that occurred in constitutional jurisprudence when Tanley succeeded Marshall. The traditional historical view was summarized over a generation ago by Justice Frankfurter: "[E]ven the most sober historians have conveyed Taney as the leader of a band of militant 'agrarian,' 'localist,' 'pro-slavery' judges, in a strategy of reaction against Marshall's doctrines. They stage a dramatic conflict between Darkness and Light: Marshall, the architect of a nation; Taney, the bigoted provincial and protector of slavery."[204]

Such an approach is based upon ignorance of the manner in which a tribunal like the Supreme Court functions. It is incorrect to suppose that Taney accomplished a wholesale reversal of Marshall's doctrines. He did not and could not do so: the institutional traditions of the Supreme Court have always exercised an overpowering influence. Even the Jacksonian neophytes on the bench were molded, more than is generally realized, into the Court's institutional pattern. To be sure, with Taney's accession, the supreme bench was now safely in the hands of the Democrats. That fact alone implies much.[205] The Justices appointed by Jackson and Van Buren inevitably had a different outlook than their predecessors, products of an earlier day. As already emphasized, Taney and his colleagues shared the Jacksonian belief that property rights must be subject to control by the community. Acting on that belief, they sought to redress the balance of constitutional protection which they felt the Marshall Court had thrown unfairly against the public interest in favor of property.

Yet, as indicated previously, it is an error to assume that the Taney Court translated wholesale the principles of Jacksonian democracy into constitutional law. The performance of the Jacksonian Justices shows, as well as anything, the peril of predicting in advance how new appointees to the Supreme Court will behave after they don the robe. "One of the things that laymen, even lawyers, do not always understand," Justice Frankfurter once stated, "is indicated by the question you hear so often: 'Does a man become any different when he puts on a gown?' I say, 'If he is any good, he does.'"[206] Certainly, Taney and his Brethren must have seemed in many cases altogether different men as judges than they had been off the bench. Paradoxically, perhaps, the erstwhile Jacksonian politicians did as much as Marshall and his colleagues to promote economic development and the concentrations of wealth and financial power that were its inevitable concomitants.

Chief Justice Taney may have had the strong Jacksonian bias against what Jackson called "the multitude of corporations with exclusive privileges which [the moneyed interest] have succeeded in obtaining in the different States,"[207] but it was the Taney opinions in cases like *Charles River Bridge v. Warren Bridge* and *Bank of Augusta v. Earle* which opened the door to the greatest period of corporate expansion in our history. The corporation first became common in the 1820s and 1830s[208]—stimulated both by the *Dartmouth College* case and by the decisions favorable to corporate personality rendered during the early years of the Taney Court. The statistics underline the stimulus given to economic expansion by the decisions of the high tribunal. In the 1830s and 1840s there was a sharp increase in the number of corporations, particularly those engaged in manufacturing.[209] Before Taney, only $50 million was invested in manufacturing; that figure had grown to $1 billion by 1860.[210]

Perhaps the major change in the jurisprudence of the Taney Court arose from its tendency, in doubtful cases, to give the benefit of the doubt to the existence of state power far more than had been the case in Marshall's day; but this is far from saying that Taney and his confreres were ready to overturn the edifice of effective national authority constructed so carefully by their predecessors. On the contrary, like Jackson himself, they were firm believers in national supremacy where there was a clear conflict between federal and state power. When state authorities acted to interfere with federal power, Chief Justice Taney and his colleagues were firm in upholding federal supremacy. Hence, despite its greater willingness to sustain state authority, it is unfair to characterize the Taney Court as concerned only with states' rights.

When the occasion demanded, Taney could assert federal power in terms characterized by Chief Justice Huges as "even more 'national' than Marshall himself."[211] This is shown dramatically by the 1852 case of *The Genesee Chief v. Fitzhugh*,[212] which arose out of a collision between two ships on Lake Ontario. A damage suit was brought in a federal court under an 1845 statute extending federal admiralty jurisdiction to the Great Lakes and connecting navigable waters. The constitutionality of this law was upheld in *The Genesee Chief*. In an earlier case,[213] the Supreme Court had confined the territorial extent of federal admiralty jurisdiction substantially to that followed under English doctrine, namely, to the high seas and to rivers only as far as the ebb and flow of the tide extended. In a small island like Britain, where practically all streams are tidal, such a limitation might be adequate, but it hardly proved so in a country of continental extent.

The Taney opinion in the *Genesee Chief* well illustrates the manner in which the law changes to meet changed external conditions. When the Constitution went into operation, the English "tidal flow" test of admiralty jurisdiction may well have sufficed. In the original thirteen states, as in England, almost all navigable waters were tidewaters. With the movement of the nation to the west and the consequent growth of commerce on the inland waterways, the English test became inadequate. "It is evident," says the Taney opinion, "that a definition that would at this day limit public rivers in this country to tide water rivers is utterly inadmissible. We have thousands of miles of public navigable waters, including lakes and rivers in which there is no tide. And certainly there can be no reason for admiralty power over a public tide water, which does not apply with equal force to any other public water used for commercial purposes."[214]

An inexorable advocate of states' rights would scarcely have written the *Genesee Chief* opinion. In fact, the extreme Jacksonian on the Court, Justice Daniel, flatly refused to countenance the revolutionary[215] enlargement of federal jurisdiction approved by the decision and delivered a stinging dissent. But Daniel's opinion was (as he himself conceded) "contracted and antiquated, unsuited to the day in which we live."[216] The Taney opinion was dictated by sound common sense; it was a legitimate nationalizing decision brought on by the changed conditions resulting from the geographic growth of the nation. As Ralph Waldo Emerson put it, in commenting on the case, "The commerce of rivers, the commerce of railroads, and who knows but the commerce of air balloons, must add an American extension to the pondhole of admiralty."[217]

A decision like the *Genesee Chief* shows how difficult it is to pigeonhole judges like Taney. His states'-rights heritage did not blind him to the need for effective governmental power. His distrust of corporations did not make him disregard the practical possibilities of the corporate device and

its utility in an expanding economy. Indeed, it was Taney, Justice Frankfurter tells us, "who adapted the Constitution to the emerging forces of
modern economic society."[218] Jacksonianism was at bottom only an ethical
conception of the social responsibilities of private property.[219] To translate
that conception into decisions like that in the *Charles River Bridge* case was
the great constitutional contribution of the Taney Court.

Henry Clay, who had led the fight against Taney's confirmation, was
later to tell the new Chief Justice that "no man in the United States could
have been selected, more abundantly able to wear the ermine which Chief
Justice Marshall honored."[220] The judgment of history has confirmed the
Clay estimate. The pendulum has shifted from the post–*Dred Scott* censures
by men like Charles Sumner to the more sober estimate of those who sat
with Taney on the bench or argued before him at the Bar. According to a
vituperative denunciation published at his death, Taney "was, next to Pontius Pilate, perhaps the worst that ever occupied the seat of judgment."[221]
Today we reject such partisan bias and agree with the estimate of Justice
Frankfurter: "The devastation of the Civil War for a long time obliterated
the truth about Taney. And the blaze of Marshall's glory will permanently
overshadow him. But the intellectual power of his opinions and their
enduring contribution to a workable adjustment of the theoretical distribution of authority between two governments for a single people, place
Taney second only to Marshall in the constitutional history of our country."[222]

4

Watershed Cases:
Dred Scott v. Sandford, 1857

Supreme Court history is marked by landmark cases which have drastically affected both the country and the Court itself. These are watershed cases, which, in Holmes's words, "exercise a kind of hydraulic pressure which makes what previously was clear seem doubtful, and before which even well settled principles of law will bend."[1]

If any case deserves to be treated as such a case, it is the *Dred Scott* case.[2] Before the decision there, the prestige of the Supreme Court had never been greater. Taney was universally acclaimed worthy of his predecessor, destined to rank almost with Marshall himself in the judicial pantheon. After the *Dred Scott* decision all was changed. "The name of Taney," declared Charles Sumner early in 1865, "is to be hooted down the page of history. . . . The Senator says that he for twenty-five years administered justice. He administered justice, at last, wickedly, and degraded the Judiciary of the country and degraded the age."[3]

Soon after Taney died, an anonymous pamphlet was published entitled *The Unjust Judge.* In it, Taney, dead less than a year, was excoriated "with hatred so malignant that it seems obscene." Its vilification culuminated in the assertion that "as a jurist, or more strictly speaking as a Judge, . . . he was, next to Pontius Pilate, perhaps the worst that ever occupied the seat of judgment among men."[4]

To so many of his contemporaries, *Dred Scott* made Taney the very prototype of "the unjust judge." Conceding that the deceased Chief Justice may have had "good qualities and . . . ability," Gideon Welles wrote in his famous diary, "But the course pursued in the Dred Scott case . . . forfeited respect for him as a man or a judge."[5] For more than a century the case has stood as a monument of judicial indiscretion: as Justice Robert Jackson acidly commented: "One such precedent is enough!"[6]

Almost a century and a half later, however, we can say that *Dred Scott* was not so much a judicial crime as a judicial blunder—a blunder that resulted from the Taney Court's failure to follow the doctrine of judicial self-restraint that was one of Taney's great contributions to our law. In it, in Justice Frankfurter's phrase, "[T]he Court disregarded its settled tradition against needlessly pronouncing on constitutional issues."[7] The *Dred Scott* Court fell victim to its own success as a governmental institution. The power and prestige that had been built up under Marshall and continued under Taney had led men to expect too much of judicial power. The Justices themselves too readily accepted the notion that judicial power could succeed where political power had failed. From this point of view, Taney may be characterized not as an "unjust judge" but as an "unwise judge." His essential mistake was to imagine that a flaming political issue could be quenched by calling it a "legal" question and deciding it judicially.[8]

Slavery in the Territories

To understand the issues in the *Dred Scott* case, it is necessary to have some knowledge of what had by the 1850s become the thorniest aspect of the slavery controversy—the question of slavery in the territories. From the founding of the Republic, the question had been dealt with by Congress. When the Constitution went into effect, the United States possessed vast territories which had been ceded by Virginia and other states. The Confederation Congress had provided for the government of the territory northwest of the Ohio River by the famous Northwest Ordinance of 1787, which flatly prohibited slavery in the territory governed by it.

One of the earliest measures enacted by the first Congress that convened under the Constitution was a law providing that the Northwest Ordinance should "continue to have full effect."[9] In 1790, Congress passed an act accepting a deed of cession by North Carolina of the territory that later became the State of Tennessee. That statute declared that no regulations were to be made in the territory which "tend to emancipate slaves."[10] These early assertions of congressional power to govern slavery in the territories were reinforced in scores of later statutes, some of which contained the express prohibition of the Northwest Ordinance.

Then, in 1820, came the Missouri Compromise, by which, it was hoped, the question of slavery in the territories had finally been settled. It prohibited slavery in the remainder of the territory included in the Louisiana Purchase north of a prescribed line, 36° 30' of north latitude.[11] As it turned out, this provision did not really resolve the issue of the extension of slavery. All it did was establish a temporary armistice in the growing conflict between the pro- and antislavery forces. In this sense, John Quincy Adams was correct when he wrote in his diary at the time "that the present question is a mere preamble—a title-page to a great tragic volume."[12]

The question of slavery in the territories arose again at the end of the Mexican war, when large areas were acquired. Conflict over the extension of slavery into the new regions grew in intensity in the decade that followed and towered over other political issues; the crisis that sounded in Jefferson's ears "like a fire bell in the night"[13] in 1819 had become a primary cause of sectional animosity.

The slavery issue was brought to the fore by the so-called Wilmot Proviso. In the summer of 1846, President Polk asked for an appropriation to enable him to negotiate a cession of Mexican territory. On August 8, he wrote in his diary, "I learned that after an excited debate in the House a bill passed that body, but with a mischievous & foolish amendment to the effect that no territory which might be acquired by treaty from Mexico should ever be a slave-holding country. What connection slavery had with making peace with Mexico it is difficult to conceive."[14]

If the President, eager to vindicate his Mexican policy by expansion of the Union, could not see the intimate relationship between adding "to the U.S. an immense empire"[15] and the question of whether the new territories would be slave or free, others in the political arena did. The war on the battlefield was minor compared to the one that now arose. The Wilmot Proviso never became part of Polk's appropriation bill, but the issue it raised overshadowed all others. "The United States," Emerson foresaw in 1846, "will conquer Mexico, but it will be as the man who swallows the arsenic which brings him down in turn. Mexico will poison us."[16] The struggle over slavery in the new territories soon showed how valid this prophecy was.

Constitutional Theories

The controversy catalyzed by the Wilmot Proviso brought to the fore sharply opposed constitutional theories on the issue of slavery in the territories. The antislavery men relied upon the express congressional power to "make all needful Rules and Regulations respecting the Territory . . . belonging to the United States" and urged that it included the authority to deal with slavery in the territories. Such authority had been exercised by the national legislature from the beginning, and its constitutionality had not been questioned. From the Wilmot Proviso debates a new version of this theory emerged: Congress had the moral duty to prohibit slavery wherever its jurisdiction extended; freedom must be national, slavery only sectional.[17] This version, soon to be adopted by the Free Soil and Republican parties, also rested upon the constitutional power of Congress to regulate slavery in the territories.

The Southerners put forward an opposing constitutional theory denying that congress had any legitimate authority to exclude slavery from the territories. This rejection of congressional power represented a shift in the Southern position. At the time of the Missouri Compromise, John Quincy Adams could say that only some "zealots . . . on the slave side" argued "that Congress have not power by the Constitution to prohibit slavery . . . in any territory."[18] Responsible Southern leaders did not take any such extreme position. When President Monroe put the question to his Cabinet in March 1820, "it was unanimously agreed that Congress have the power to prohibit slavery in the Territories."[19] Among those who strongly argued in support of that authority was Secretary of War Calhoun.

However, Southerners came to believe that the Union itself depended upon an equal division between slave and free states. "Sir," Calhoun declared to the Senate in 1847, "the day that the balance between the two sections of the country—the slaveholding States and the nonslaveholding States—is destroyed, is a day that will not be far removed from political revolution, anarchy, civil war, and wide-spread disaster."[20] The controversy over the Wilmot Proviso made it plain that the delicate free–slave equilibrium would soon be upset. Calhoun sought to preserve the balance by having the Missouri Compromise line extended to the Pacific. His efforts proved futile: all proposals to extend the Missouri line were voted down.

The rejection of the Compromise approach led Calhoun to reexamine his constitutional position. Now he saw that the westward march of the nation meant the inevitable end of equality for the slave states. Senate protection had to be replaced by some other instrument to defend slaveholding interests. That need was met by a change in constitutional theory. The only hope now, declared Calhoun in an 1847 speech, lay in the basic document itself: "The constitution . . . is a rock. . . . Let us be done with compromises. Let us go back and stand upon the Constitution."[21]

It is usually said that Chief Justice Taney, in his *Dred Scott* opinion, was simply elevating to the constitutional plane the new Calhoun theory on slavery in the territories. This view, as we shall see, is an oversimplification. While Calhoun, like Taney in *Dred Scott,* denied the constitutional power of Congress to prohibit slavery in the territories, the Carolinian's approach was far more extreme in its rejection of federal power than that later adopted by the Supreme Court. The Calhoun theory was based upon the doctrine of state sovereignty pushed almost to absurdity. The territories, he argued, were "the common property of the States of this Union. They are called 'the territories of the United States,' and what are the 'United States' but the States united? Sir, these territories are the property of the States united; held jointly for their common use."[22] The Federal Government, as the agent of the sovereign states, held the territories in trust for their common benefit; consequently it could not prevent a citizen of any one state from carrying with him into the territories property whose legal status was recognized by his home state.[23]

In February 1847, Calhoun introduced resolutions before the Senate stating the essentials of his new position: the territories were the joint property of the states; Congress, as the states' agent, could not make any discriminations between states depriving anyone of them of its equal right in any territory; a law depriving citizens of any state of the right to emigrate into any territory with their property would violate the Constitution and the rights of the states.[24] These resolutions, adopted by many Southern legislatures, became the virtual platform of the South. Under the Constitution, wrote Jefferson Davis, there was an "obligation of the U.S. Govt. to recognize property in slaves, as denominated in the compact, . . . to enforce the rights of its citizens to equal enjoyment of the territorial property which had been acquired and held as a common possession."[25]

This constitutional issue gave rise to most of the increasingly bitter political dialogue after the Mexican war. To the South particularly, defense of the Calhoun theory was seen as a matter of life and death. Only by denying congressional authority to prohibit slavery in the territories could the South prevent itself from being swamped by a vast new free-soil area that would reduce the slave states to an ever-smaller minority. If the balance of power were altered, the very ability of the South to defend itself would be at an end. "The surrender of life," Calhoun warned in a famous 1847 speech, "is nothing to sinking down into acknowledged inferiority."[26]

Need for Judicial Resolution

Although the primary error of the Supreme Court in the *Dred Scott* case was its assumption that the issue of slavery in the territories could be resolved judicially, it is a mistake to picture the Justices as blithely rushing into the political arena, officiously seeking to save the nation. Perhaps Taney and his colleagues should never have tried to settle the slavery issue, particularly since the case before them could have been disposed of without consideration of the slavery question; yet it is fair to say that their action was a response to a widespread popular desire to have the issue decided by the highest Court.

The opposing constitutional theories on congressional power could scarcely be resolved through normal political processes. Upon the issue joined by those theories, Congress itself was largely helpless:[27] "[N]o Bill to establish a Territorial Government could be passed through the Ho. Repts. without having the Wilmot Proviso attached to it as a condition . . . with this provision the Bill would probably be rejected by the Senate, . . . and . . . the people of California would be left without a Government."[28] Settlers in the Far West had to do without government because Congress could not decide whether they should have slaves.[29]

In this situation, it was not unnatural to turn to the tribunal vested with the primary function of resolving disputed constitutional issues. The impasse between the Northern and Southern views led a Senate select committee to propose the so-called Clayton Compromise. Under it, Congress was to provide for governments in California and New Mexico, and "they should be restrained by Congress from Legislating on the subject of slavery, leaving that question, if it should arise, to be decided by the judiciary."[30] In this way, the right to introduce or prohibit slavery was to rest "on the Constitution, as the same should be expounded by the judges, with a right to appeal to the Supreme Court."[31]

With the support of Calhoun,[32] the Clayton attempt to have Congress "avoid the decision of this distracting question, leaving it to be settled by the silent operation of the Constitution itself"[33] passed the Senate, but it was defeated in the House. In his last Annual Message, Polk restated the essence of the Clayton proposal, "to leave the subject to the decision of the Judiciary,"[34] as a possible solution: "If the whole subject be submitted to the judiciary, all parts of the Union should cheerfully acquiesce in the final decision of the tribunal created by the Constitution for the settlement of all questions which may arise under the Constitution."[35]

Buchanan used strikingly similar language in his Inaugural Address, referring to the then-pending *Dred Scott* case,[36] and called down upon himself the vitriolic abuse of the antislavery press. By then, opponents of slavery feared an adverse decision, but almost a decade earlier the movement to have the Supreme Court resolve the issue was supported by political leaders on both sides. It was widely recognized, Jefferson Davis wrote later, that "it was necessary to settle finally the asserted right of the Southern people to migrate with their slaves to the territories and there to have for that property the protection which was given to other property of citizens by the U.S. Govt."[37]

The Compromise of 1850 itself was essentially based upon the need to have the key constitutional question settled judicially. On that point, Senator Davis could agree with Senator E. J. Phelps of Vermont: "The Constitution has provided its remedy . . . that tribunal which sits in the chamber below us, Mr. President . . . we are entitled to a decision of the Supreme Court."[38] The Compromise established territorial governments for Utah and New Mexico and provided that "the legislative power of the Territory shall extend to all rightful subjects of legislation consistent with the Constitution of the United States."[39] Provision was made for judicial settlement of the constitutional question by special provisions liberalizing federal court jurisdiction in slavery litigation. In "all cases involving title to slaves," appeals to the Supreme Court were to be allowed without regard to the jurisdictional amount normally required for such appeals.[40]

It has been questioned whether the framers of the Compromise really intended to turn over the constitutional issue to the highest tribunal.[41] During the debates, however, Henry Clay (the prime author of the 1850 measure) did indicate that this was his intent. What, he asked the Senate, could be "more satisfactory to both sides" than to have Congress keep its hands off the issue "and to leave the question of slavery or no slavery to be decided by the only competent authority that can definitely settle it forever, the authority of the Supreme Court?"[42] And the judicial review provisions of the Compromise are a virtual copy of the parts of the Clayton Compromise which proposed leaving the slavery question to the highest Court.[43]

The Kansas–Nebraska Act of 1854 followed the Compromise approach, containing, like the 1850 measure, provisions indicating congressional intent not to deal with slavery in the territory (this time in terms of Stephen A. Douglas's theory of popular sovereignty) and authorizing liberalized appeals to the Supreme Court in slavery cases. The purpose was to leave "the question where . . . it should be left—to the ultimate decision of the courts. It is purely a judicial question."[44] Douglas himself affirmed in 1856 that "I stated [in the Kansas–Nebraska debate] I would not discuss this legal question, for by the bill we referred it to the Courts."[45]

This widespread sentiment is plainly relevant to the charge that the Court's decision in the *Dred Scott* case amounted to mere judicial usurpation. It acted in response to congressional invitation[46] and did no more than yield to the prevalent public demand for judicial pronouncement on the matter. Even Lincoln, severe critic of the decision though he later showed himself, welcomed Supreme Court action in 1856. Noting the Democratic view that restrictions of slavery in the territories would be unconstitutional, he declared that he was not bound by such political construction of the Constitution: "The Supreme Court of the United States is the tribunal to decide such questions, and we will submit to its decisions."[47]

Facts and Issues

Dred Scott was originally called Sam and was so listed in the inventory of his first owner's estate. The name made so famous by the Supreme Court decision was acquired in Illinois or the Wisconsin Territory, where Sam was taken by his new owner, Dr. Emerson, an army surgeon. The case made the short, stubby slave "the hero of the day, if not of the age. He has thrown Anthony Burns, Bully Bowlegs, Uncle Tom and Fred Douglas into . . . oblivion."[48]

In 1846 Scott brought suit in a Missouri court for his freedom against Mrs. Emerson, who had acquired title to him on her husband's death. Scott's counsel argued that his service for Dr. Emerson in Illinois and in territory from which slavery had been excluded by the Missouri Compromise made him a free man. The jury returned a verdict in Scott's favor, but the Missouri Supreme Court reversed on the ground that Missouri law governed, and under it Scott was still a slave.

Scott's attorneys next maneuvered the case into the federal courts. Mrs. Emerson had remarried, and Scott found himself the purported property of her brother, John Sanford, of New York. Scott, claiming Missouri citizenship, could now sue in a federal court on the ground of diversity of citizenship. In 1853 an action was instituted in the United States Circuit Court for Missouri. Scott, as a citizen of Missouri, brought an action for damages, alleging that Sanford,[49] a citizen of New York, had assaulted him. Defendant filed a plea in abatement, alleging that plaintiff was not a citizen of Missouri "because he is a Negro of African descent; his ancestors . . . were brought into this country and sold as negro slaves." The court sustained a demurrer to this plea, and defendant then pleaded that Scott was his slave and that, therefore, no assault could have occurred. After a jury verdict, judgment was given for defendant on the ground that Scott was still Sanford's property. A writ of error was taken by Scott to the Supreme Court.

Until the high bench appeal, the *Dred Scott* case was like many others heard in the courts on behalf of slaves, scarcely noted except by the participants. But from the beginning it was really "enclosed in a tumultuous privacy of storm,"[50] for inherent in it was "the much vexed [question] whether the removal by the master of his slave to Illinois or Wisconsin marks an absolute emancipation."[51] And that, in turn, involved consideration of the effect of the provisions prohibiting slavery found in the Illinois Constitution and the Missouri Compromise. Necessarily included in that issue was the question of power over slavery in the territories.

When *Dred Scott* first instituted his suit, debate over the crucial constitutional issue had been relatively low-keyed. Between that time and the date of the Supreme Court decision, however, it intensified, and just before the case was appealed to the highest Court the whole question was brought to the boiling point by the Kansas–Nebraska Act and its repeal of the Missouri Compromise. The potential of the case for resolution of the issue of congressional power over slavery in the territories was now widely grasped. "This is a question of more importance, perhaps," Scott's attorney could say in his Supreme Court argument, "than any which was ever submitted to this court; and the decision of the court is looked for with a degree of interest by the country which seldom attends its proceedings. It is, indeed, the great question of our day."[52]

But there was more in the case than this: defendant's plea in abatement had posed the question of whether even a free black could be a citizen. In some ways, that question was more fundamental than that of congressional authority over slavery. Legislative power to eliminate slavery would be empty form if those freed could not attain citizenship. If even the free black would have to remain "like some dishonoured stranger"[53] in the community, the Northern majority who hoped that slavery would gradually disappear throughout the country was doomed to disappointment. Extralegal means would be needed to end the degraded status of the enslaved race. What had come to the Supreme Court as a question of law now became a matter of morality.

Maneuverings Toward Decision

The day after the first *Dred Scott* conference Justice Curtis wrote his uncle, "The court will not decide the question of the Missouri Compromise line—a majority of the judges being of the opinion that it is not necessary to do so."[54] At the conference, a majority were of the opinion that the case should be decided without consideration of the two crucial issues. They felt that the issue of citizenship was not properly before them and also took the position that they need not consider the Missouri Compromise because Scott's status was a matter for Missouri law and had already been determined against him by the state's highest court. Justice Nelson was selected to write an opinion disposing of the case in this manner.

Had the Nelson opinion (limiting itself to Scott's status under Missouri law after his return to that state) prevailed as the opinion of the Court, the *Dred Scott* case would scarcely be known today except to the curious student of high bench miscellany. Pressures were, however, building up which soon led the Justices to abandon their original intent.

Soon after his election to the Presidency, James Buchanan wrote to Justice Grier that "the great object of my administration will be, if possible to destroy the dangerous slavery agitation and thus restore peace to our distracted country."[55] The pending decision of the Supreme Court gave him hope that a major part of the problem could be solved at a single stroke.[56] On February 3, the President-elect wrote to Justice Catron, a close friend, asking him whether the case would be decided before March 4, the date of the inauguration. Catron replied that "it rests entirely with the Chief Justice to move in the matter" and that he had said nothing about it. Then, on February 10, Catron wrote Buchanan that the case would be decided four days later but that no opinion would be announced before the end of the month. The decision would not help Buchanan in his Inaugural Address, he said, since the question of congressional power over the territories would probably not be touched on in it.[57]

In the meantime, the Justices had been shaken in their initial resolve to decide the case without considering the issues of citizenship or slavery in the territories. Justice Curtis later said that the change was brought about by Justice Wayne, a Georgian who, while serving as a judge in Savannah, had sentenced an offender for "keeping a school for Negroes." Two years before *Dred Scott*, he had declared that there was no possibility that even free blacks "can be made partakers of the political and civil institutions of the states, or those of the United States."[58] As Curtis recalled it, "it was urged upon the court, by Judge Wayne, how very important it was to get rid of the question of slavery in the Territories, by a decision of the Supreme Court, and that this was a good opportunity of doing so."[59]

Wayne moved in conference that the decision deal with the two vital issues Justice Nelson was omitting. "My own and decided opinion," he said, "is that the Chief Justice should prepare the opinion on behalf of the Court upon all of the questions in the record."[60] The five who voted in favor of Wayne's motion were from slave states. Wayne himself told a Southern Senator that he had "gained a triumph for the Southern section of the country, by persuading the chief-justice that the court could put an end to all further agitation on the subject of slavery in the territories."[61]

The Chief Justice himself apparently did not play a major part in the conference that adopted Wayne's motion, though he clearly was in favor of it and undoubtedly spoke to that effect. On the other hand, according to Curtis's brother, the Justice "in the conferences of the court, explained in the strongest terms that such a result, instead of putting an end to the agitation in the North, would only increase it." In addition, Curtis stressed that it was "most unadvisable to have it understood that the decision of these very grave and serious constitutional questions had been influenced by considerations of expediency."[62] The fact that the five votes for the new decision were by Southerners would lead to anything but Wayne's conference prediction that "the settlement . . . by judicial decision" would result in "the peace and harmony of the country."[63] Instead, as Horace Greeley noted in the *New York Tribune,* settlement of the slavery issue by the Court meant submitting it to five slaveholders and "I would rather trust a dog with my dinner."[64]

After Wayne's motion had been adopted, two of the Justices—Catron and Grier—had written President-elect Buchanan that, in Catron's words, "[A] majority of my brethren will be forced up to this point [i.e., to rule on the constitutional issues of citizenship and slavery in the territories]."[65] These letters to Buchanan were intended to inform the President-elect of the Court's plan to have the Chief Justice write a broad majority opinion. Justice Catron urged Buchanan to write Justice Grier (a fellow Pennsylvanian), who hesitated to join the new majority, telling him "how necessary it is—and how good the opportunity is to settle the agitation by an affirmative decision of the Supreme Court." Buchanan did write to Grier, who showed the letter to Chief Justice Taney and Wayne and then wrote Buchanan that he fully concurred with the need for decision on "this troublesome question." He was afraid that the case would be decided on sectional lines; so "that it should not appear that the line of latitude should mark the line of division in the Court," he would concur with Taney. Both Justices wrote Buchanan that the Court's decision would not be announced until just after the inauguration because of the Chief Justice's poor health.[66]

By present-day standards, the correspondence between Buchanan and two members of the Court was improper. Even more so was Buchanan's pressure on Justice Grier, at the invitation of another Justice, to join the majority. To say, as the Supreme Court historian does,[67] that it was not infrequent at the time for Justices to tell a friend or relative the probable outcome of a pending case scarcely excuses Buchanan and the Justices concerned. Buchanan was not just a friend; he was the new President and was hoping to use the information and his influence over Grier for political purposes. Even given the more permissive standards of an earlier day, the propriety of their conduct cannot be defended.

Yet this is not to say, as was widely charged at the time, that the *Dred Scott* decision itself was the result of a conspiracy between Buchanan and the Southern members of the Court. Though Lincoln asserted, in his famous "House Divided" speech, that the decision was the product of an understanding between "Stephen, Franklin, Roger, and James" in which "all worked upon a common plan . . . drawn up before the first lick was struck,"[68] all we know about the case indicates that it did not happen that way.

Decision

Chief Justice Taney may not have played a key role in the Court's changed posture. However, once the conference voted to decide the merits of the two crucial issues, he became the principal protagonist of the majority view. It was Taney who wrote the opinion of the Court, which stated the polar view against Scott's case, just as Justice Curtis was to write the dissenting opinion that best set forth the opposite position.

On March 6, 1857, the nine Justices filed into their basement courtroom, led by the now-feeble Chief Justice, exhausted by age and illness—a mere shadow, save in intellect, of the man who first presided over the Court two decades earlier. Taney began the reading of the *Dred Scott* opinions in a voice so weak that, during much of the two hours in which he spoke, it sank to a whisper. Each of the majority Justices read his own opinion, and Justices McLean and Curtis read lengthy dissents.[69] The reading of the opinions took two days.

"No wonder," declaimed Greeley's *Tribune,* soon after the decision was announced, "that the Chief Justice should have sunk his voice to a whisper . . . knowing that he was engaged in a pitiful attempt to impose upon the public."[70] To Greeley and other abolitionist editors, the decision was a patent triumph for slavery—a view that was accepted by the South as well.

Chief Justice Taney's opinion for the Court contained three main points: (1) Negroes, even those who were free, were not and could not become citizens of the United States within the meaning of the Constitution; (2) Scott had not become a free man by virtue of residence in a territory from which slavery had been excluded by the Missouri Compromise because the Compromise provision excluding slavery was itself beyond the constitutional power of Congress; (3) Scott was not free by reason of his stay in Illinois because the law of Missouri alone governed his status once he returned to that state.

Only on the third point (which was the sole ground upon which the majority had originally agreed to decide)[71] was the Taney opinion relatively uncontroversial.[72] The seven majority Justices concurred on this point, and the Court's opinion was but a reaffirmation of the law laid down in earlier decisions.[73] What burst with such dramatic impact upon the nation was the fact that the highest Court in the land had denied both the right of blacks to be citizens and the power of Congress to interfere with slaveholding in the territories.

At any rate, it is clear that the effort to have the Court settle the troublesome constitutional issues once and for all failed dismally. In his concurring opinion, Justice Wayne stated that the issues involved had become so controversial "that the peace and harmony of the country required the settlement of them by judicial decision."[74] Seldom has wishful thinking been so spectacularly wrong.[75] Whatever *Dred Scott* brought about, it was anything but peace and harmony—either for the Court or for the country. Instead, as Justice Frankfurter tells us, *"Dred Scott . . . probably helped to promote the Civil War, as it certainly required the Civil War to bury its dicta."*[76]

Congressional Power

To the contemporary observer, the most important part of the opinion of the Court was its holding that the Missouri Compromise was unconstitutional. Five justices[77] concurred with Chief Justice Taney in this holding. This aspect of the opinion was also most strongly censured by critics. Speaking on the Court's view "on the question of the power of the Constitution to carry slavery" into the territories, a leading Northern Senator declared, "beyond all question, to any fair and unprejudiced mind, that the decision has nothing to stand upon except assumption, and bad logic from the assumptions made."[78]

The Taney holding on congressional power was deceptively simple. It began by recognizing congressional authority to acquire new territory and to determine what rules and regulations to make for any territory. That authority was, however, subject to the limitations imposed by the Constitution upon governmental power, including those designed to safeguard private property rights. In particular, property rights are protected by the Due Process Clause of the Fifth Amendment: "And an act of Congress which deprives a citizen of the United States of his liberty or property, merely because he came himself or brought his property into a particular Territory of the United States, and who had committed no offence against the laws, could hardly be dignified with the name of due process of law."[79] Hence the Missouri Compromise prohibition against the holding of property in slaves is unconstitutional and void.

Two aspects of the reasoning just summarized are of special significance. In the first place, Taney depends upon the assumption that congressional power over slavery in the territories is limited by the Constitution itself. This basic holding was sharply censured by former Senator Thomas H. Benton, who, though near death from cancer, published a lengthy attack on the decision soon after it was handed down: "The Court sets out with a fundamental mistake, which pervades its entire opinion, and is the parent of its portentous error. That mistake is the assumption, that the Constitution extends to Territories as well as to States."[80]

Though the law on the matter may not have been settled over a century ago, today we can see that Taney was correct in his approach to the question of the applicability of constitutional limitations in the territories.

The alternative is to hold that Congress may, "when it enters a Territory . . . , put off its character, and assume discretionary or despotic powers which the Constitution has denied to it."[81] Americans migrating to a territory would, if that were true, be mere colonists, dependent upon the will of the Federal Government.

Later decisions of the Supreme Court confirm the Taney reasoning. The landmark *Insular Cases,*[82] decided at the turn of the last century, held that a fundamental provision such as the Due Process Clause is definitely binding in all American territory, including conquered territory subject to military government. More recently, Justice Black used language recalling that of Taney himself; according to Black, whenever our government acts, regardless of locale, it can act only "in accordance with all the limitations imposed by the Constitution."[83]

Having held Congress bound by the requirements of due process in legislating for the territories, the Chief Justice next proceeded to hold that the prohibition of slavery violated due process. Here we come to the second significant aspect of his holding on congressional power. In his ruling that the Missouri Compromise was unconstitutional, Taney was, for the first time in Supreme Court jurisprudence, holding that the Due Process Clause has a substantive as well as a procedural aspect. It was as a violation of substantive due process that the congressional prohibition of slavery was stricken down; what Taney was saying was that a law which deprives a citizen of his property in slaves simply because he brings such property into a territory is arbitrary and unreasonable and hence violative of due process.

Although *Dred Scott* was the first case in which the Supreme Court used the Due Process Clause as a substantive restriction upon governmental power, the Taney approach was not something made up out of legal whole cloth. On the contrary, the development of substantive due process was one of the outstanding judicial achievements of the last century. It began in several state courts during the 1830s and 1840s and culminated in 1856 in *Wynehamer v. People,*[84] decided by the highest court of New York in the period between the first and second arguments in *Dred Scott.*[85] That decision (recognized as epoch-making almost as soon as it was rendered) may well have been the immediate source of Taney's opinion on due process.

Nor was the notion that congressional prohibition of slavery violated due process original with Taney. In the debates preceding the Missouri Compromise, several members of Congress expressed the view that prohibiting slavery in Missouri would violate the Due Process Clause.[86] In 1841 a Northern member of the Supreme Court declared, with regard to slaves, "Being property . . . , the owners are protected from any violations of the rights of property by Congress, under the fifth amendment."[87] Plainly, this was getting very close to Taney's approach in *Dred Scott.*

Even more relevant to the Taney treatment of due process as a substantive restraint was the fact that Justice Curtis, while on circuit in 1852, had stricken down a state liquor law on substantive due process grounds.[88] Particularly suggestive is the fact that Curtis discussed his approach in this case with Taney, who approved the reasoning employed.[89] Taney had only to change the word "liquor" in the Curtis opinion to "slave" and he had the substance of the reasoning by which he invalidated the Missouri Compromise.[90] In view of this, it is surprising that the Curtis dissent in *Dred Scott* emphasized the novel nature of the Taney approach on due process.

Today the Taney application of substantive due process may be considered unduly simplistic, if not naive. The mere fact that a law destroys property rights, we now know, does not necessarily mean that it violates due process. Governmental power does, in appropriate circumstances, include the power to prohibit as well as the power to regulate. It should, however, be borne in mind that Taney was speaking at the very infancy of substantive due process. If his approach was relatively unsophisticated, the same was true of the other early opinions that developed it.

In the era after the Civil War the Taney approach became established in the law. Toward the end of the century, substantive due process was to be used as the fundamental restriction upon govermental action interfering with property rights. The discrediting of the *Dred Scott* decision did not really affect the seminal nature of the concept invoked by Taney. Though the particular property interest which he sought to protect was soon to become anachronistic, the doctrine he articulated opened a new chapter in our constitutional law.

To opponents of slavery, however, the Taney denial of congressional power to interfere with slavery in the territories was a disaster. Acquiescence in the ruling was fatal to the Republicans and the advocates of popular sovereignty alike. It frustrated the hopes of those who sought to confine slavery to an area that would become an ever-smaller portion of an expanding nation. It meant instead that slavery was a national institution; there was now no legal way in which it could be excluded from any territory.

Black Citizenship

In 1834 the status of the free colored population in Pennsylvania was elaborately set forth in a pamphlet published by a member of the Bar of that state. It arrived at the conclusion that the free Negro was neither a citizen of the United States nor a citizen of Pennsylvania. A copy of the pamphlet was sent to Chief Justice Marshall, and he sent the author a letter expressly endorsing his conclusion on Negro citizenship.[91] Thus Marshall, not long before his death, came to the same conclusion as Taney with regard to Negro citizenship. This fact that (so far as the present writer could determine) has been unknown to previous commentators indicates that the Taney ruling may not have been as contrary to law as most of its critics have contended.

Despite this, without a doubt, the Chief Justice's categorical denial of black citizenship, even for free blacks, is the aspect of Taney's opinion that is most difficult to grasp. It seems completely out of line with constitutional conceptions to doom the members of a particular race to live in permanent limbo, forever barred from the dignity of citizenship. Yet that was exactly the result under Taney's holding. As *Harper's Weekly* summed it up, "[T]he Court has decided that free negroes are not citizens of the United States."[92]

Taney's answer to the question of black citizenship resulted from the manner in which he framed the question: "The question is simply this: can a negro, whose ancestors were imported into this country and sold as slaves, become a member of the political community formed and brought into existence by the Constitution of the United States, and as such become entitled to all the rights, and privileges, and immunities, guarantied by that instrument to the citizen. One of these rights is the privilege of suing in a court of the United States in the cases specified in the Constitution."[93]

Taney's answer was based upon a distinction between "the rights of citizenship which a state may confer within its own limits, and the rights of citizenship as a member of the Union." Since adoption of the Constitution, national citizenship has been federal, not state, in origin. National citizenship was created by the Constitution and, under it, "every person, and every class and description of persons, who were at the time of the adoption of the constitution recognized as citizens in the several States, became also citizens of this new political body; but none other."[94]

Taney went into a lengthy analysis of the situation in this respect and concluded "that neither the class of persons who had been imported as slaves, nor their descendants, whether they had become free or not, were then acknowledged as a part of the people, nor intended to be included in the general words used in [the Declaration of Independence]."[95]

The Taney conclusion rested ultimately upon the concept of Negro inferiority, which was also the basis of the Southern slavery jurisprudence: "They had for more than a century before been regarded as beings of an inferior order; and altogether unfit to associate with the white race, either in social or political relations; and so far inferior, that they had no rights which the white man was bound to respect; and that the negro might justly and lawfully be reduced to slavery for his benefit. He was bought and sold, and treated as an ordinary article of merchandise and traffic."[96]

Legislation, as well as practice, Taney asserted, "shows, in a manner not to be mistaken, the inferior and subject condition of that race at the time the Constitution was adopted, and long afterwards." It can hardly "be supposed that they intended to secure to them rights, and privileges, and rank, in the new political body [when] they had deemed it just and necessary thus to stigmatize, and upon whom they had impressed such deep and enduring marks of inferiority and degradation." In consequence, "Dred Scott was not a citizen of Missouri within the meaning of the Constitution of the United States, and not entitled as such to sue in its courts."[97]

Curtis Dissent

Taney's *Dred Scott* opinion takes up fifty-fives pages of small print in Howard's *Reports,* with over four-fifths devoted to the issues of citizenship and congressional power to prohibit slavery in the territories. There were five concurring and two dissenting opinions. The principal dissent was by Justice Curtis, and his response to Taney is generally considered the most effective statement of the law the other way. The Curtis opinion was the longest of all in *Dred Scott,* covering seventy pages in the *Reports.*

The bulk of the Curtis dissent was devoted to answering the Taney opinion on its two principal rulings. Curtis argued that United States citizenship depended upon state actions. Citizenship within the Constitution for persons born within the United States was through the states and did not depend upon national authority: "[T]he citizens of the several States were citizens of the United States under the Confederation" and the same was true under the Constitution. Curtis cites both statutes and decisions to "show, in a manner which no argument can obscure, that in some of the original thirteen States, free colored persons, before and at the time of the formation of the Constitution, were citizens of those states."[98] That, in turn, under the Curtis approach, made them citizens of the United States.

Curtis dealt with the issue of the Missouri Compromise slavery prohibition by strongly reaffirming the congressional power over the territories. Taney's opinion had recognized congressional authority to acquire new territory and to determine what rules and regulations to make for any territory. But it did so in what we should now consider a peculiar way— recognizing it as an implied power rather than one expressly provided in article IV, section 3, giving Congress power to make rules and regulations for the territories. Curtis gave full effect to the Territories Clause, saying that the power "to make all needful rules and regulations respecting the Territory, is a power to pass all needful laws respecting it."[99]

To be sure, the power "finds limits in the express prohibitions on Congress not to do certain things." Thus it cannot pass an ex post facto law or bill of attainder for a territory any more than it can for any other part of the country. There is, however, no such prohibition for laws relating to slavery and none can be implied. "An enactment that slavery may or may not exist there, is a regulation respecting the Territory."[100] Hence it is within the power of Congress under the Territories Clause. Curtis here referred with particular effect to the Northwest Ordinance prohibition against slavery in the Northwest Territory and the enactment in the First Congress that it should "continue to have full effect."[101]

To Curtis, then, the Territories Clause plainly gave Congress the power to prohibit slavery. There is "no other clause of the Constitution . . . which requires the insertion of an exception respecting slavery" and nothing to indicate that such an exception was intended by the Framers. And "where the Constitution has said all needful rules and regula-

tions, I must find something more than theoretical reasoning to induce me
to say it did not mean all."[102]

To us today, the most important part of the Curtis dissent was its
refutation of Taney on the citizenship issue. Commentators today assume
that the materials relied on by Curtis showed conclusively that free blacks
were citizens of at least some of the original thirteen states and hence were
citizens of the United States for purposes of the case. And, as the issue was
raised by the plea of abatement filed by defendant, "[I]t is only necessary to
know whether any such persons were citizens of either of the States under
the Confederation at the time of the adoption of the Constitution."[103]

Curtis relied upon constitutional and statutory provisions and decisions in five states[104] to show that "free persons of color"[105] had the right
to vote and were consequently citizens of those states. In reality, the evidence in support of Curtis's position was stronger than he indicated. Taney
had relied upon the Militia Law of 1792, which limited military service to
white males, to show that blacks could not perform one of the essential
duties of citizenship.[106] Yet, as pointed out by Judge John Appleton of
Maine's highest court, who had once been Curtis's teacher, in an opinion
issued only a few months after *Dred Scott,* "[T]here are no historic facts
more completely established, than that during the revolution they were
enlisted, and served as soldiers; that they were tendered and received as
substitutes; that they were required to take, and took the oath of allegiance."[107]

"If these things be so," Appleton went on, "and that they are so cannot
be denied or even doubted, and if they had been known to the learned
Chief Justice, his conclusions would have been different, for he says, 'every
person and every class and description of persons, who *were at the time of
the adoption of the constitution recognized as citizens of the several states, became
also citizens of this new political body.*" Appleton concluded that Taney's
"published opinion, therefore, rests upon a remarkable and most unfortunate misapprehension of facts, and his real opinion upon the actual facts
must be considered as in entire and cordial concurrence with that of his
learned dissenting associates."[108]

The Appleton conclusion here was ingenuous. It is, to say the least,
unlikely that Taney would have concurred with Curtis had the facts with
regard to military service been pointed out to him. Moreover, it should be
recognized that the case for black citizenship in the 1850s was not as
conclusive as the Curtis presentation indicates. Before the Civil War the
question of black citizenship was by no means settled clearly. Indeed, there
was substantial authority that tended to support Taney rather than Curtis
on the matter. Several attorneys general (including Taney himself in
1832)[109] and a number of state courts had concluded that free Negroes
were not citizens.[110] Their decisions were based upon the many disabilities
from which blacks suffered, which made it plain that they did not enjoy the
full rights of citizens.

It is true that there were the state decisions cited by the Curtis dissent holding that free blacks were citizens. But those decisions use the notion of citizenship in a manner which now seems most peculiar. Thus *State v. Manuel*[111]—the case most relied on in the Curtis dissent—involved a North Carolina law providing that, where a free Negro had been convicted of a misdemeanor and could not pay the fine, his services could be sold for up to five years to the highest bidder. This statute (which applied only to blacks, not whites) was upheld by the state court, which stated in its opinion that what "citizenship" the Negro had was of a most restricted sort—on a lower level, as it were, than that possessed by other citizens. Taney could easily have used the *Manuel* case to support his basic thesis of inequality in the treatment of the races.

Curtis also used *State v. Newsom,*[112] a later North Carolina case, to support the *Manuel* citation, but *Newsom* sustained a law making it a crime for "any free Negro" to carry a gun or knife. Here again we have a disability which seems inconsistent with the rights of citizenship. The law was attacked on the ground that free blacks, as citizens, were entitled to the same rights as other citizens. The court stated that *Manuel* was a "controlling influence." Yet, in upholding the law, it went on to say, "[T]he free people of color cannot be considered as citizens, in the largest sense of the term, or if they are, they occupy such a position in society, as justifies the Legislature in adopting a course of policy in its acts peculiar to them."[113]

The truth seems to be that implied in the *Newsom* case: Negro citizenship was a legal euphemism. The current of judicial decisions was relegating the free black to a subordinate status, regardless of whether he was clothed with the formal title of citizen. Actually, a third class of free residents in this country was being created in the law: there were now citizens, free Negroes, and aliens.[114] In this sense, the *Dred Scott* decision was, despite the Curtis dissent, only confirming one line of pre–Civil War jurisprudence.

Yet if the Taney conclusion on citizenship had stronger support in the law than most commentators have recognized, it must still be conceded that the decision was little short of disastrous. It meant that, without constitutional amendment, the Negro was consigned to a permanent second-class status which could not be changed even if all the slaves were ultimately freed. More fundamentally, it gave the lie to the very basis of the American heritage: the notion of equality that was the central theme of the Declaration of Independence—in Lincoln's words, "the electric cord in that Declaration that links the hearts of patriotic and liberty-loving men together."[115] The ruling would have aborted the effort to give effect to the "progressive improvement in the condition of all men"[116] that had been a dominant force since the founding of the Republic.

Result-oriented Jurisprudence

The Taney and Curtis opinions in *Dred Scott* are prime illustrations of the result-oriented judge that is assumed by so many to be a unique characteristic of our own day. Both Justices were employing the legal materials used by them to reach the results they favored. Each was treating the law as an instrument to foster his own societal vision.

If the *Dred Scott* decision brought about anything but the peace and harmony intended by the majority Justices, it did result from the desire of Taney and his Southern colleagues to settle the slavery issue by authoritative judicial decision. The Taney opinion was written to support the Southern position, which in turn was deemed a sine qua non for the preservation of Southern society and its way of life. Only by denying congressional authority to prohibit slavery in the territories could the South prevent itself from being swamped by a vast new free-soil area that would reduce the slave states to an ever-smaller minority. If the balance of power were altered, the very ability of the South to defend itself would be at an end. "The surrender of life," Calhoun warned in an already-quoted speech, "is nothing to sinking down into acknowledged inferiority."[117]

To support his decision, Chief Justice Taney was ready not only to use but to make legal doctrine in a transforming manner that Marshall himself might have envied. To demonstrate that even free blacks were not citizens, Taney made a most selective use of the available materials. Despite Judge Appleton, Taney must have known about the black military experience during the Revolution. But his opinion refers only to the 1792 law which provides for militia service by every "free . . . white male citizen."[118] Yet Curtis also gave a partisan cast to the statutes and cases cited by him. This has been shown, for example, of the North Carolina cases relied upon by Curtis. He quoted the general language on citizenship in the opinions, but he ignored the facts and statements indicating that free blacks still had a subordinate status inconsistent with true citizenship.

With regard to congressional power to prohibit slavery in the territories, Taney went even further by using legal doctrine that may not have been made up out of whole cloth but was certainly new in Supreme Court jurisprudence. And he did so with the mere ipse dixit that a law which deprives a person of his property because he brought it into a particular territory violated due process.

The Supreme Court later pointed out that the Taney conclusion here depends upon the proposition that a slave is property just the way "an ordinary article of merchandise"[119] is property: "If the assumption be true, that slaves are indistinguishable from other property, the inference from the *Dred Scott Case* is irresistible that Congress had no power to prohibit their introduction into a territory."[120] But the crucial weakness of slavery law was that the slave was not ordinary property, the way a house or a horse was. "The difficulty with the *Dred Scott Case* was that the court refused to make a distinction between property in general, and a wholly exceptional class of property."[121]

Taney stretched the law to protect this exceptional property because, like other Southern judges, he was most concerned with preserving what he considered the indispensable foundation of his society and its economy.

Indeed, as the leading modern student of *Dred Scott* puts it, by the time of the case, Taney "had become as resolute in his determination to protect [slavery] as Garrison was in his determination to destroy it."[122]

Ultimately, however, *Dred Scott* stands as a monument of judicial hubris—with both Taney and Curtis assuming that they could resolve in the judicial forum the basic controversy that was tearing the country apart. By trying to act as the deus ex machina on the slavery issue, the Court was stretching judicial power to the breaking point. The case could have been disposed of without consideration of the slavery question and was thus one where the Justices should have adhered to the doctrine of judicial self-restraint that, we saw in Chapter 3, was one of Taney's great contributions to public-law jurisprudence.

On June 12, 1857, Stephen A. Douglas, the chief political victim of the *Dred Scott* decision, addressed a grand jury at Springfield, vigorously defending the Supreme Court and rejecting the charge that the Justices had gone out of their way to decide the crucial constitutional issues. According to Douglas, if the Court had relied on a technicality to avoid the main issues, the outcry against it would have been even worse: "If the case had been disposed of in that way, who can doubt . . . the character of the denunciations which would have been hurled upon the devoted heads of those illustrious judges, with much more plausibility and show of fairness than they are now denounced for having decided the case . . . upon its merits?"[123]

The Court might, as Douglas claimed, have disappointed some, but it could scarcely have tarnished its reputation to the extent that the actual decision did. As a general proposition, it may be said that the Supreme Court as an institution has never been harmed by abstention from political issues. On the contrary, most of the controversies in which it has been embroiled have been caused by failure to follow the doctrine of judicial self-restraint.

Regardless of legal logic, the opponents of slavery could not accept the Court's decision as final, particularly the Republican party, whose very raison d'être was undercut by it. This explains (though it may not justify) the vituperation which Republican orators directed against both the decision and the Court. Lincoln's repeated claim that there was a master conspiratorial plan which sought to use the Supreme Court to make the country "an entire slave nation"[124] by a decision "ere long . . . declaring that the Constitution . . . does not permit a *state* to exclude slavery"[125] must be laid to a lack of understanding of constitutional doctrine. *Dred Scott* could be based upon the Fifth Amendment, for it dealt with congressional authority; before the Fourteenth Amendment, there was no organic provision upon which the Court could base a comparable limitation upon state power.[126]

After the Civil War, Jefferson Davis prepared some notes for Major W. T. Walthall, who was helping him prepare his well-known history of the Confederacy, "on the assigned causes for the invasion of the South." In these notes he made the curious assertion that "the unjust and offensive denial of an equal right to occupy the territories with any species of property recognized by the laws of their states was one of the causes which provoked the Southern people to withdraw from an association in which the terms of the partnership were disregarded."[127]

Davis was confused in his recollection of what had happened. It was not until 1862, well after the Southern states had seceded, that Congress passed a law expressly prohibiting slavery in the territories[128] (thus legislatively denying the right in slave property in the territories which *Dred Scott* had recognized). Davis's recollection does, however, demonstrate the crucial importance of *Dred Scott* in the events leading to the Civil War. In his notes for Walthall, he stated flatly that "the territorial question . . . is another . . . pretext for the war waged against the Southern states."[129]

The *Dred Scott* decision was thus a major factor in precipitating the political polarization of the nation. It was actually the catalyst for the civil conflict that soon followed. With it collapsed the practical possibility of resolving by political and legal means the issues which divided the nation. Thenceforth, extremists dominated the scene. Bloodshed alone could settle the issue of slavery—and of the very nature of the Union, which that issue had placed in the balance.

5

War and Reconstruction, 1861–1877

In many ways the Civil War was the test of fire of the American constitutional system. In an 1862 article, Lord Acton referred to it as the Second American Revolution[1]—a characterization that has since been made often. Like the Revolution, the Civil War represented an extralegal appeal to force to settle the ultimate legal issue of the nature of the polity. And the issue itself was decided, not by the tribunal to which the resolution of such questions was confided by the Constitution, but by the victorious Union armies. When the Supreme Court in 1869 decided that secession was illegal, since "[t]he Consititution in all its provisions, looks to an indestructible Union,"[2] it was only confirming a decision already made at Appomattox Courthouse.

Yet if our law broke down in the face of the nation's most serious crisis, that was not so much the fault of the law itself. The men of the day expected too much of both the law and the courts. In particular, the power and prestige that had been built up under Marshall, and continued under Taney, had led to these too great expectations. And the Justices themselves had succumbed to the lure of seeking to save the country from the bench, losing sight of the limitations inherent in judicial power.

If the Civil War represented an appeal from law to the sword, that was true because the opposing extremes no longer accepted the underlying premises of the legal order. Americans too often forget that the rule of law draws only limited strength from judicial guaranties; it must have roots far deeper than a formal fundamental document and decisions of the judges enforcing it. Our public law depends for its efficacy on popular acceptance of its basic presuppositions. Acceptance, rather than formal legal machinery, is the decisive force in the law's implementation. With Learned Hand in a famous passage, we may "wonder whether we do not rest our hopes too much upon constitutions, upon laws and upon courts. These are false hopes, believe me, these are false hopes."[3]

Merryman and Military Power

"Determining the proper role to be assigned to the military in a democratic society," declared Chief Justice Earl Warren in his 1962 James Madison lecture, "has been a troublesome problem for every nation that has aspired to a free political life."[4] The claims of military power were first asserted in extreme form in the American constitutional system during the Civil War. It cannot be said that a proper balance between military power and law was achieved during that conflict. On the contrary, in the midst of civil strife most of all, as Burke pointed out in his *Reflections on the French Revolution,* "laws are commanded to hold their tongues amongst arms; and tribunals fall to the ground with the peace they are no longer able to uphold."

At the outset of the Civil War the extreme claims of both war and law were presented. The former was personified by President Lincoln, the latter by Chief Justice Taney. To deal with the life-and-death crisis facing the Government after Sumter, Lincoln assumed unprecedented powers. On his own authority he suspended the writ of habeas corpus and ordered wholesale arrests without warrants, detentions without trials, and imprisonments without judicial convictions. Newspapers were seized and their publication suppressed;[5] persons were arrested and held incommunicado by military officers acting under presidential authority. As Taney put it, the military had "thrust aside the judicial authorities and officers to whom the constitution has confided the power and duty of interpreting and administering the laws, and substituted a military government in its place, to be administered and executed by military officers."[6]

The passage quoted is from an opinion Taney delivered in May 1861, in the celebrated *Merryman* case.[7] On April 27 Lincoln authorized the Commanding General of the Army to suspend the right of habeas corpus along any military line between Philadelphia and Washington. A month later Taney, sitting in the federal circuit court in Baltimore, was petitioned for habeas corpus by John Merryman, who had been arrested by the Army and confined in Fort McHenry for his secessionist activities, particularly his participation in the attack upon the Sixth Massachusetts Militia while it was en route to Washington and the destruction of railroad bridges to prevent the passage of troops. Sitting in chambers, Taney granted a writ of habeas corpus directed to the general commanding the fort.

On the return date, an aide-de-camp (in full military uniform and wearing, appropriately, a sword and bright red sash) appeared in the court-room and declined obedience to the writ on the ground that it had been suspended by the Commanding General pursuant to the April 27 order of the President. Taney issued a writ of attachment for contempt against the general, but the marshal seeking to serve it was refused entry to the fort. Taney then delivered his *Merryman* opinion, in which he sharply con-demned as illegal the suspension of habeas corpus by the President and the arrest, without warrant and hearing, of a civilian by military order. But his attempt to uphold the letter of the law against military claims of emergency was fruitless. As he himself plaintively put it, "I have exercised all the power which the constitution and laws confer upon me, but that power has been resisted by a force too strong for me to overcome."[8]

Taney filed his opinion with the clerk of the circuit court, with the direction that a copy be sent to the President: "It will then remain for that high officer, in fulfillment of his constitutional obligation, to 'take care that the laws be faithfully executed,' to determine what measures he will take to cause the civil process of the United States to be respected and enforced."[9] We do not know what Lincoln did with his copy of the Taney opinion (or whether he ever received it), but we do know that he went right on exercising the power that Taney had branded unconstitutional. In addition to other limited suspensions, such as that in the *Merryman* case, he issued an order on September 24, 1862, suspending the writ throughout the country for all persons confined by military authority.[10] Despite the refusal of the military to obey Taney's writ, however, Merryman was released from military custody shortly afterward, and a subsequent indictment against him was never prosecuted.[11]

Over a century later, looking back at the conflict, we can see that neither the Lincoln nor the Taney philosophy alone is adequate. What is needed at such times is a reconciliation of the extreme demands of war and law, not exclusion of the one or the other. It was with keen perception that Justice Robert H. Jackson wrote, "Had Mr. Lincoln scrupulously observed the Taney policy, I do not know whether we would have had any liberty, and had the Chief Justice adopted Mr. Lincoln's philosophy as the philoso-phy of the law, I again do not know whether we would have had any liberty."[12]

When all is said and done, however, there remains something admi-rable in Taney's action in the *Merryman* case:

> There is no sublimer picture in our history than this of the aged Chief Justice—the fires of Civil War kindling around him . . . serene and unafraid, . . . interposing the shield of law in the defense of the liberty of the citizen. Chief Justice Coke when the question was put to him by the King as to what he would do in a case where the King believed his prerogative concerned, made the answer which has become immortal. "When the case happens, I shall do that which shall be fit for a judge to do." Chief Justice Taney when presented with a case of presidential prerogative did that which was fit for a judge to do.[13]

The Court and Civil Liberties

"In the interval between April 12 and July 4, 1861," says W. A. Dunning, "a new principle thus appeared in the constitutional system of the United States, namely, that of a temporary dictatorship. All the powers of government were virtually concentrated in a single department, and that the department whose energies were directed by the will of a single man."[14] According to Lincoln's critics, this situation did not really change even after Congress met, at the President's call, on July 4, 1861. Wendell Phillips continued to denounce Lincoln's government as a "fearful peril to democratic institutions,"[15] and a law professor at Harvard characterized Lincoln as a government in himself—"an absolute, . . . uncontrollable government; a perfect military despotism."[16]

In Lincoln's expansive view of presidential power in wartime, even constitutional doctrine might have to give way if it conflicted with the national necessity. "By general law," he asserted, at the height of what must still be considered our greatest national emergency, "life and limb must be protected, yet often a limb must be amputated to save a life; but a life is never wisely given to save a limb."[17] In assessing this philosophy, we should recognize the difficult choices which confronted the President when strong measures seemed the only alternative to disintegration and defeat. In a famous statement he posed the "grave question whether any government, not *too* strong for the liberties of its people, can be strong *enough* to maintain its own existence, in great emergencies."[18] If the war were lost, government, country, and Constitution itself would all fall together: "I felt that measures, otherwise unconstitutional, might become lawful, by becoming indispensable to the preservation of the constitution, through the preservation of the nation."[19]

It is true that throughout his Presidency Lincoln expressed his distaste for the extraconstitutional measures which he had taken. At the very outset of the conflict he told General Winfield Scott "how disagreeable it is to me to do a thing arbitrarily."[20] This theme he developed in both public and private utterances. When he was informed of military arrests of civilians in the District of Columbia, he wrote, "Unless the necessity for these arbitrary arrests is manifest and urgent, I prefer they should cease,"[21] and to Benjamin Butler, at the height of that officer's conflict with the "restored" Pierpoint government of Virginia, he declared, "Nothing justifies the suspending of the civil by the military authority, but military necessity."[22]

Perhaps the best statement of Lincoln's inner conflict is found in an 1862 letter: "I am a patient man—always willing to forgive on the Christian terms of repentance; and also to give ample *time* for repentance. Still I must save this government if possible."[23] If measures of dubious constitutionality were necessary to accomplish that end, that was, to paraphrase the just-quoted letter, a card that had to be played to prevent losing the game.[24]

It was a card that, despite the distaste expressed by him, Lincoln constantly played. In particular, arrests and suspensions of habeas corpus, such as that in the *Merryman* case, were carried out on a broad scale throughout the war. Nor did the President authorize these measures with the belief that they were unconstitutional. On the contrary, he sought to justify his "supposed unconstitutional action such as the making of military arrests" in a noted 1863 letter to Erastus Corning. Far from conceding that he had abused the power to make military arrests, the time would come, he asserted, "when I shall be blamed for having made too few arrests rather than too many."[25]

Lincoln went on in the Corning letter to note that the arrests which his critics attacked were preventive, rather than vindictive. Such preventive detention could scarcely be accomplished by the traditional processes of the ordinary law: "Nothing is better known to history than that courts of justice are utterly incompetent to try such cases."[26] At any rate, it is clear that, during the Civil War, the courts were not able to decide the legality of Lincoln's measures restricting personal liberty, particularly his suspensions of habeas corpus. It is true, as we saw, that Chief Justice Taney ruled against the presidential suspension in the *Merryman* case; true also that the consensus of learned opinion has agreed that Taney was right and Lincoln wrong on the question of presidential power to suspend the writ. But Taney's decision, as seen, had no practical effect because of the refusal of the military authorities to obey the *Merryman* writ.

Nor was the issue decided by the tribunal set up by the Constitution to resolve such matters. The legality of the habeas suspensions, as well as the other restrictions imposed by military authorities during the war, did not come to decision before the Supreme Court at all. The Court, to be sure, dealt with important aspects of the civil liberties issue in the soon-to-be-discussed *Milligan* case.[27] But *Milligan* was decided after the war and the restrictions laid down by the Court there had no practical effect upon the military violations committed during the conflict.

What was true of the wartime restrictions upon civil liberties was also true of other measures taken during the conflict. The Supreme Court did little more than passively confirm the measures taken by the government to cope with the Southern rebellion. During the whole period, the Court remained in the state of recession which its *Dred Scott* decision had induced. Unconstitutional though many of the wartime measures may have been (at least by present-day standards), the Court itself was unable to rule upon almost all of them. In the one case to reach it during the war involving a violation of civil liberties—*Ex parte Vallandigham*,[28] where the notorious Copperhead had been arrested and tried by the military—the Court avoided the issue by holding that, under the Judiciary Act, it had no appellate jurisdiction over a military commission.

Prize Cases

The Supreme Court did hand down one decision of consequence during the Civil War—that in the 1863 *Prize Cases*.[29] It arose out of one of the emergency measures taken by President Lincoln just after the fall of Fort Sumter, his April 1861 proclamation of a blockade of Southern ports. Four ships had been captured by Union naval vessels enforcing the blockade and had been brought into ports to be sold as prizes. Their owners contended that they had not been lawfully seized because a blockade was a belligerent act which could not be proclaimed in the absence of a state of war declared by Congress.

The formal proclamation of a blockade at the outset of the Civil War has been criticized as a tactical error.[30] In international law, a blockade implies a state of belligerency. In its neutrality proclamation of May 1861, Great Britain took note of such belligerency, and the British Foreign Secretary was able to state, in reply to the claim that the proclamation was "precipitate," "It was, on the contrary, your own Government which, in assuming the belligerent right of blockade, recognized the Southern States as belligerent."[31]

Legally speaking, the proclamation of a blockade, with its recognition of belligerency, constitutes an act of war. To the Supreme Court, indeed, it was Lincoln's blockade proclamation that constituted the beginning of the Civil War.[32] But could the President thus begin a war without violating the claimed "inexorable rule" that the country could be involved in war legally only by declaration of Congress?

The Court in the *Prize Cases* avoided a direct answer to this question by stating that the President did not, by his blockade proclamation or any other act, initiate the conflict. In his argument on behalf of the Government, William Evarts urged that "war is, emphatically, a question of actualities."[33] Whenever a situation assumes the proportions and pursues the methods of war, he said, the peace is driven out, and the President may assert the warlike strength of the nation. Evarts's approach was essentially that followed by the Court. The actuality of the situation confronting the President after Sumter was one of war; "However long may have been its previous conception, it nevertheless sprung forth from the parent brain, a Minerva in the full panoply of War." The President did not initiate such war, but he was bound to accept the challenge "in the shape it presented itself, without waiting for Congress to baptize it with a name; and no name given to it by him or them could change the fact."[34]

A commentary on the *Prize Cases* asserts that that case was as important as a case can be in shaping the contours of presidential power for future occasions when Presidents would wage war without congressional authorization.[35] The decision itself holds only that the President could deal with the situation presented after Sumter as a war and employ what belligerent measures he deemed necessary without waiting for Congress to declare war.[36] It would be absurd for the President to be required, simply because Congress had not declared its existence, "to affect a technical ignorance of the existence of a war, which all the world acknowledges to be the greatest civil war known in the history of the human race, and thus cripple the arms of the Government and paralyze its power by subtle definitions and ingenious sophisms."[37]

It is when we look beyond the bare holding to the language of the Court and its import that we can understand the wide implications of the *Prize Cases*. War, said the opinion of Justice Grier, is "that state in which a nation prosecutes its right by force."[38] Under such a definition, the President can in fact (if not in the technical contemplation of the Constitution) initiate a war. The *Prize Cases* constitute a rejection of the doctrine that only Congress can stamp a hostile situation with the character of war and thereby authorize the legal consequences which ensue from a state of war.[39] Rejection of the rule that only Congress can initiate a war has been of tremendous practical significance. If the President can initiate belligerent measures to cope with civil rebellion, why may he not do so to deal with hostile invasion? In such a case also he would be empowered to meet the challenge of war without waiting for Congress to act. It would be no less a war for the fact that it was begun not by formal declaration but by unilateral act.[40] Must the President take belligerent measures only after the first blow has been struck? Such a limitation on his powers could mean national annihilation in an age in which a nation has the absolute power to destroy an enemy. A constitution which did not permit the Commander in Chief to order belligerent acts whenever necessary to defend his country's interests would be little more than a suicide pact.[41]

On the other hand, as Lincoln himself noted when he was a Congressman during the Mexican war, "Allow the President to invade a neighboring nation, whenever *he* shall deem it necessary to repel an invasion, and you allow him to do so, whenever he may choose to say he deems it necessary for such purpose—and you allow him to make war at pleasure."[42] If we give the President the power to order the commission of belligerent acts whenever he deems necessary, we invest him with the power to make war without legal check. His authority is to act to defend the interests of the United States, but history shows that this is no real restraint, since even aggressive action can be framed in ostensibly defensive terms.

Nature of the Union

In an 1839 letter to Gerrit Smith, an abolitionist, John Quincy Adams predicted that the slavery conflict would ultimately lead to both secession and civil war:

> I believe that long before [emancipation] . . . the slave holding representation would secede in a mass, and that the States represented by them would secede from the Union. I know that among the abolitionists there are some leading and able men, who consider this a desirable event. I myself believe that it would

naturally and infallibly lead to the total abolition of Slavery but it would be through the ultimate operation of a War, more terrible than the thirty years war . . . and I shrink from it with horror."[43]

The conflict which Adams so graphically foresaw raised the overriding issue of the legal nature of the Union itself. That issue had been at the core of most of the pre–Civil War constitutional controversy. "That the Slave holders of the South," Adams went on in his letter to Smith, "should flatter themselves that by seceding from this Union they could establish their peculiar institutions in perpetuity, is in my judgment one of those absurd self-delusions which would be surprising if they did not compose the first chapter in the history of human nature." Yet, he accurately guessed, "the Slaveholders do so flatter themselves, and will act accordingly."[44]

This "delusion" was a natural consequence of the Southern conception of the Constitution. Secession was but the logical culmination of the doctrine of states' rights and state sovereignty which dominated thinking below the Mason-Dixon Line. That doctrine was so ingrained in the Southern mind that it enabled the movers of secession to (in Lincoln's phrase) "sugar-coat" their rebellion "by an insidious debauching of the public mind. They invented an ingenious sophism, which, if conceded, was followed by perfectly logical steps, through all the incidents, to the complete destruction of the Union. The sophism itself is, that any state of the Union may, consistently with the national Constitution, and therefore lawfully, and peacefully, withdraw from the Union, without the consent of the Union, or of any other state."[45]

The truth is that secession and union are constitutionally incompatible. If the Civil War accomplished anything in the constitutional sphere, it was to reject categorically "the position that secession is consistent with the Constitution—is lawful and peaceful."[46] That position was relegated to the realm of constitutional heresy along with the Calhoun doctrine of the states as separate sovereignties upon which it was based.

From a Southern point of view, the Civil War may be looked upon as an attempt to overrule the nationalistic conception of the Constitution which had prevailed since Marshall became Chief Justice. Virginian though he was, Marshall's dominant aim was to establish a strong nation, with the powers needed to govern a continent. The decisions of Marshall and his colleagues constructed federal supremacy upon so strong a base that it has never since been subjected to successful *legal* attack. The adherents of state sovereignty could hope to prevail only by resorting to methods outside the judicial arena. To render the Constitution workable, it had to incorporate "a coercive principle"—the question being, as one of the Framers put it, whether it should be "a coercion of law, or a coercion of arms."[47] With national supremacy so firmly established by the coercion of law, its opponents deemed themselves relegated to the coercion of arms if their view of the nature of the nation was to prevail.

The defeat of the South placed the imprimatur of arms upon both the intent of the Framers and the Marshall interpretation of federal power. The conduct of the war itself furnished ample proof of the soundness of the Marshall conception. "The Federal Government," Winston Churchill tells us, "gaining power steadily at the expense of the states, rapidly won unquestioned control over all the forces of the Union. The Southern 'Sovereign States,' on the other hand, were unable even under the stress of war to abandon the principle of decentralization for which they had been contending."[48] The kind of government Southerners wanted was not the type that could win a lengthy war; states' rights was a hopeless base for total conflict.

The war not only confirmed but accelerated the trend toward strong national government that has been the underlying theme of our constitutional development. "The South," says a Southern writer, "with whatever justification, tried in '61 to break the Union. She succeeded only in strengthening what she fought against."[49] Until the war the advocates of state sovereignty could, despite the uniform case law in opposition to their view, continue to assert the temporary contractual nature of the Union. The defeat of the South meant the final repudiation of such an assertion. In the law itself this repudiation was marked by *Texas v. White*,[50] decided by the Supreme Court shortly after the war ended.

In *Texas v. White*, the State of Texas brought an original action to enjoin the payment of certain United States bonds owned by the state before the war and negotiated by the Confederate state government to the defendants. The key issue presented was whether Texas was then a state of the Union and, as such, capable of bringing suit. Defendants contended that it was not—that having seceded and not yet being represented in Congress, it was still out of the Union. According to the Court's opinion, the ordinance of secession by Texas was a legal nullity. Texas consequently always remained a state within the purview of the Constitution: "When, therefore, Texas became one of the United States, she entered into an indissoluble relation. . . . The act which consummated her admission into the Union was something more than a compact; it was the incorporation of a new member into the political body. And it was final. . . . There was no place for reconsideration, or revocation."[51]

It is all too easy to dismiss the case as only the judicial ratification of the real decision on the validity of secession made at Appomattox Courthouse. To be sure, if the actual outcome of the conflict had been different, the Supreme Court decision could never have been made, but that is true because the constitutional nature of the Union would have been completely altered by military power. As a purely legal decision, under the Constitution as it is written, *Texas v. White* is sound. It is "self-evident that the Union could scarcely have had a valuable existence had it been judicially determined that powers of sovereignty were exclusively in the States":[52] the very language of the Constitution refutes the notion that the states have a sovereign right to secede at will. The Articles of Confederation declare the Union's character to be "perpetual." Says the Court,

"[W]hen these Articles were found to be inadequate to the exigencies of the country, the Constitution was ordained 'to form a more perfect Union.' It is difficult to convey the idea of indissoluble unity more clearly then by these words. What can be indissoluble if a perpetual Union made more perfect, is not?"[53]

The Constitution is thus a bond of national unity, not a mere league which may be dissolved at the pleasure of any party. "The Constitution of the United States," said a member of the highest Court in 1871, "established a government, and not a league, compact or partnership. . . . The doctrine so long contended for, that the Federal Union was a mere compact of States, and that the States, if they chose, might annul or disregard the acts of the National legislature, or might secede from the Union at their pleasure, and that the General Government had no power to coerce them into submission to the Constitution, should be regarded as definitely and forever overthrown."[54]

Reconstruction and the Constitution

When the Civil War ended, the "peace" that followed was anything but a return to the status quo ante—at least so far as the defeated South was concerned. In 1849 John C. Calhoun had evoked what must have seemed to his supporters an apocalyptic vision of the consequences of forcible emancipation. If emancipation ever should be effected, he asserted, "it will be through the agency of the Federal government, controlled by the dominant power of the Northern States." Emancipation itself would come "under the color of an amendment of the Constitution," forced through by the North. It "would lead to consequences unparalleled in history."

Nor, according to Calhoun, would the North stop at emancipation of the slaves: "Another step would be taken—to raise them to a political and social equality with their former owners, by giving them the right of voting and holding public offices under the Federal Government." The ex-slaves would become "the fast political associates of the North," cementing the political union by acting and voting with them. "The blacks, and the profligate whites that might unite with them, would become the principal recipients of federal offices and patronage, and would, in consequence, be raised above the whites of the South in the political and social scale."[55]

The Carolinian's forecast bears a striking resemblance to what used to be the accepted view of the constitutional history of the postbellum South. If, in the end, it did not turn out as Calhoun had predicted—Negro supremacy, he said in his 1849 address, would force the whites to flee the very homes of their ancestors and leave the South "to become the permanent abode of disorder, anarchy, poverty, misery and wretchedness"[56]—that was because of the extreme nature of the Radical Republican program. The measures taken during Reconstruction, in the traditional view, were bound to result in a reaction in the opposite direction once Southerners regained control over their own destiny.

Reconstruction was traditionally seen as an aberration in American constitutional history. It was, wrote Sir Henry Maine soon after it ended, "a Revolutionary period of several years, during which not only the institutions of the Southern States, but the greater part of the Federal institutions were more or less violently distorted to objects not contemplated by the framers of the Constitution."[57] This language has been echoed by more recent writers, even those who have contributed to the changed climate of opinion on Reconstruction. "Frankly revolutionary in mood," concedes C. Vann Woodward, "Thaddeus Stevens and his followers overrode constitutional restraints right and left."[58]

Reconstruction can be considered a patent violation of the Constitution only if we ignore the fact that it posed issues which the Framers had neither foreseen nor provided for in the document they drafted. To be sure, throughout our history "the Court has viewed the separation and subordination of the military establishment as a compelling principle."[59] Yet that is true only in states which are not in an actual theater of military operations and in whch the civil courts are open and functioning.[60] Ever since the Mexican war, a different rule had been applied to territory occupied by American forces; there the prevailing principle has been that of military conquest, with government set up by the occupying forces, though subject to ultimate congressional control.[61]

What happens, however, if the occupied territory consists of states which have sought to secede and have been prevented from doing so by federal force? Such a situation plainly was not considered by the Framers, who intended the Union to be perpetual and indissoluble. "The Constitution," declared the Supreme Court in *Texas v. White*, "in all its provisions, looks to an indestructible Union."[62] Nor were there constitutional provisions for the legal status of the defeated states after their attempt at secession had been suppressed.

How to deal with these unprecedented constitutional problems? The one thing clear to the Republican leaders was that the constitutional clock could not simply be turned back to 1860. That had, of course, become the "Southern" theory of Reconstruction—that with the end of the war, all affairs should revert to their previous condition. By 1865, however, the North had come to recognize the truth of Elizur Wright's statement that "the general facts . . . make a restoration of the state [of affairs] before the war equivalent to defeat."[63]

This meant that the congressional leaders could not agree with President Andrew Johnson's theory of Reconstruction, for it was essentially similar to the Southern position—based as it was upon the conception that the Southern states had only had "their life-breath . . . suspended:[64] and that it could be speedily restored by presidential action.

The congressional leaders could not accept the Johnson conception of the postwar task as that of bringing about a prompt restoration of the South to the Union. The President, said Thaddeus Stevens, "preferred 'restoration' to 'reconstruction.' He chooses that the slave States should remain as nearly as possible in their ancient position."[65] Though there were differences of detail and degree between the various theories advanced by different Radical spokesmen—from the extreme "conquered provinces" theory of Stevens to the somewhat more moderate ones of "state suicide" espoused by Charles Sumner or "forfeited rights" of Samuel Shellabarger—they all refused to accept the view that the Southern states needed only the restoration of "loyal" governments.

Instead, however they were phrased, the congressional Reconstruction theories meant in practice that the rebel states would be treated (in George W. Julian's phrase) "as outside of their constitutional relations to the Union, and as incapable of restoring themselves to it except in conditions to be prescribed by congress."[66] The Southern states could no longer be considered in the same relationship to the Union as other states. On the contrary, "[S]uch States and their people ceased to have any of the rights or powers of government as States of this Union."[67] Consequently, it was for Congress to decide the conditions upon which the Southern states were to be reconstructed as full states of the Union. "The Southern states have ceased to be states of the Union—their soil has become National territory"[68]—subject, as such, to congressional power.

Supreme Court Reconstruction Theory

In the already-discussed case of *Texas v. White*[69] the Supreme Court stated its position on the constitutional basis of Reconstruction. In that case, it will be recalled, the Court upheld, in strong language, the indissoluble nature of the Union. From a constitutional point of view, the attempted secession of a state like Texas was ruled "absolutely null . . . utterly without operation in law."[70] Texas remained a state, with its obligations as a member of the Union unimpaired.

The fact that secession was legally void did not mean that it had never happened, however, and that "the governmental relations of Texas to the Union remained unaltered."[71] There was a difference between the constitutional existence of a state itself and the existence within it of a government competent to represent it in its relations with the nation. When the war ended there was no such government in Texas, and it became the duty of the United States to provide for the reestablishment of such a government. The President had initial authority in the matter, since as Commander-in-Chief he could institute governments in areas occupied by federal forces. However, "the action of the President must . . . be considered as provisional";[72] presidential power was ruled subject to the overriding authority of Congress to provide for lawful governments in the Southern states.

The answer given to the question of Reconstruction in *Texas v. White* is one that has been repeated in more recent cases involving occupied territory, particularly those growing out of the war with Spain and World War II.[73] Those cases confirm the overriding congressional power to establish governments in areas occupied by American forces. Under them, too, the military power of the President to govern such areas continues only until it has been terminated by congressional action.[74]

Texas v. White supports both the claim of overriding Reconstruction authority in Congress and the constitutionality of the congressional Reconstruction measures. The Supreme Court recognized that it was for Congress to provide for governments in the defeated states to fill the vacuum that existed after the Southern defeat. Just as important was the source of the congressional power: "[A]uthority was derived from the obligation of the United States to guarantee to every state in the Union a republican form of government."[75] *Texas v. White* followed Chief Justice Taney in *Luther v. Borden* (1849)[76] in holding that the power to carry out the republican guaranty clause "resides in Congress."[77] Under *Luther v. Borden* and the later cases, legislative action to enforce the Guaranty Clause presents a political, rather than a judicial, question.

What this means is shown by *Georgia v. Stanton*,[78] where an action was brought to enjoin enforcement of the Reconstruction Acts. As more recently explained by Justice Brennan, "It seemed to the Court that the only constitutional claim that could be presented was under the Guaranty Clause, and Congress having determined that the effects of the recent hostilities required extraordinary measures to restore governments of a republican form, this Court refused to interfere with Congress' action at the behest of a claimant relying on that very guaranty."[79]

In substance, the Supreme Court adopted the constitutional position upon which congressional Reconstruction was based. The framers of the Wade-Davis bill had expressly relied on *Luther v. Borden*. According to Congressman H. W. Davis, that case meant that "it is the exclusive prerogative of Congress—of Congress, and not of the President—to determine what is and what is not the established government of the State."[80] In *Texas v. White* the Supreme Court placed its imprimatur upon this view. That one may disapprove of some or all of the Reconstruction measures shoud not obscure this fact. To characterize Reconstruction as a constitutional aberration is an exercise in rhetoric, not in law—certainly not in the law laid down by the Supreme Court.

Congressional Reconstruction

That the general theory of congressional Reconstruction was accepted by the Supreme Court does not mean that the details of Reconstruction in operation were approved by the Justices. The 1866 election had given the Republicans an overwhelming majority in both houses. This majority the congressional leaders used to remake the Reconstruction process in the legislative image. The governments instituted under the presidential plan of Reconstruction were swept aside, and Congress assumed control over the reestablishment of Southern governments. Most members of Congress were prepared to do so because they had come to feel that drastic measures were needed to deal with the results of President Johnson's laissez-faire Reconstruction program. Rather than permit an unreconstructed South "to substitute a degrading peonage for slavery and make a mockery of the moral fruits of northern victory,"[81] the congressional majority decided upon the drastic measure of military rule.

The basic statute in the congressional Reconstruction program was the First Reconstruction Act of March 2, 1867.[82] Its two principal features were the imposition of military rule and the complete reorganization of government. It declared that "no legal State governments or adequate protection for life or property" existed in the ten "rebel States." Those states were then divided into five military districts, each under the command of an army general. These commanders were given broad powers "to protect all persons in their rights of person and property, to suppress insurrection, disorder, and violence, and to punish . . . all disturbers of the public peace." To make such powers effective, they were authorized to make arrests, conduct trials in military courts,[83] and use federal troops to preserve order.

Writing of the Reconstruction Act in 1902, John W. Burgess asserted, "There was hardly a line in the entire bill which would stand the test of the Constitution." This assertion is wide of the legal mark. Burgess himself concedes, "There can be no question in the mind of any sound political scientist and constitutional lawyer that Congress was in the right, logically, morally, and legally, in insisting upon brushing aside the results of executive Reconstruction in the winter of 1867, and beginning the work itself from the bottom up." If Congress had the power to brush aside the governments set up during presidential Reconstruction, it surely had the power to set up governments in their place. Perhaps, as Burgess contends, Congress should have set up "regular Territorial civil governments."[84] The choice, however, was within the discretion of Congress, and that body could reasonably conclude that a probationary period under military rule was necessary. That an immediate transition to civil government might have been a wiser choice does not affect the legality of the congressional action.

The decision of the Supreme Court in *Ex parte Milligan*[85] does not alter the constitutional picture. The decision there was referred to by Chief Justice Warren as a "landmark," which "established firmly the principle that when civil courts are open and operating, resort to military tribunals for the prosecution of civilians is impermissible."[86] Yet vital though the *Milligan* case has been as the foundation of the wall of separation between the military and civil classes in the community, it had little immediate practical effect, since during the war the Supreme Court had refused to rule on the legality of military arrests, notably in the case of Clement Vallandigham. The holding of illegality in the similar *Milligan* fact pattern came over a year after the war was over. Milligan may have had the satisfaction of being immortalized in the *Supreme Court Reports,* but that hardly was an adequate substitute for the imprisonments suffered by Copperheads and others while the Court declined to come to grips with the constitutionality of military arrests and trials.

In *Milligan,* the Court went out of its way to lay down limitations upon resort to martial law and use of military tribunals to punish civilians. This led Thaddeus Stevens to assert that the *Milligan* decision, "although in terms perhaps not as infamous as the Dred Scott decision, is yet far more dangerous in its operation."[87] But it was the exercise of military jurisdiction in Indiana—a loyal state not in an actual theater of war, in which the civil courts were functioning—that called forth restrictive language from the Court. Military jurisdiction, it said, "can never by applied to citizens in states which have upheld the authority of the government, and where the courts are open."[88]

This *Milligan* reasoning did not necessarily apply to states whose attempt to secede had been overcome by force and which Congress had not yet provided with lawful governments, entitling them to resume their full place in the Union. Justice Davis, the author of the Court's opinion, indicated that *Milligan* did not necessarily imply the unconstitutionality of Reconstruction. In a letter early in 1867, he noted that there was "not a word said in the opinion about reconstruction, & the power is conceded in insurrectionary States."[89]

McCardle and Judicial Power

Despite what has just been said, it was widely assumed at the time that *Milligan,* in deciding against the military trial of civilians, had indicated that the Supreme Court would invalidate the military governments set up under the Reconstruction statutes. The occasion for doing so, it was believed, was furnished by *Ex parte McCardle,*[90] then on its way to the highest bench. McCardle was a Mississippi newspaper editor who had been arrested in 1867 and held for trial by a military commission. The charges were based upon editorials published by McCardle, particularly one which asserted that General Sheridan and the other generals commanding Southern military districts "are each and all infamous, cowardly, and abandoned villains, who, instead of wearing shoulder straps and ruling millions of people, should have their heads shaved, their ears cropped, their foreheads branded, and their precious persons lodged in a penitentiary."[91]

McCardle petitioned for a writ of habeas corpus in the federal circuit court, challenging the validity of the First Reconstruction Act's provision authorizing the military detention and trial of civilians. The writ was denied by the circuit court, and an appeal was taken to the Supreme Court under an 1867 statute authorizing appeals from circuit court decisions in all cases involving detentions in violation of the Constitution or federal laws.[92] The Supreme Court unaminously decided that it had jurisdiction to hear the appeal.[93] The case was thoroughly argued on the merits and taken under advisement by the Justices.

Congressional leaders feared that the Court would seize the opportunity presented by *McCardle* to invalidate the military governments authorized by the Reconstruction Act. "Should the Court in that case, as it is supposed they will," wrote Gideon Welles in his famous diary, "pronounce the Reconstruction laws unconstitutional, the military governments will fall and the whole Radical fabric will tumble with it."[94]

To avoid this danger, the congressional leaders considered various maneuvers. The most extreme was embodied in a bill passed by the House early in 1868,[95] which required a two-thirds vote of the Justices before any act of Congress could be ruled unconstitutional. That bill, said Gideon Welles, "is a scheme to change the character of the Supreme Court."[96] It ultimately died in the Senate, as did a bill introduced by Senator Lyman Trumbull forbidding the Court to take jurisdiction in any case arising out of the Reconstruction acts.[97] Congress then passed a law repealing the 1867 statute authorizing an appeal to the Supreme Court from circuit court judgments in habeas corpus cases and prohibiting the Court's exercise of any jurisdiction on appeals which had been or which might be taken.[98] The *McCardle* case was then reargued on the question of the authority of Congress to withdraw jurisdiction from the Supreme Court over a case which had already been argued on the merits.

The Court in its *McCardle* decision unanimously answered the question of congressional power over its appellate jurisdiction in the affirmative even in such a case. The effect of the repealing act, said the Court, upon the case before it was plain: to withdraw jurisdiction over the appeal. It is quite clear, therefore, the Court decided, "that this court cannot proceed to pronounce judgment in this case, for it has no longer jurisdiction of the appeal."[99]

If the Court had been able to decide the merits in *McCardle*, the decision might well have been in petitioner's favor—that, at any rate, was what Chief Justice Chase and Justice Stephen J. Field told two of their contemporaries.[100] But the point of *McCardle* is that Congress was able to prevent a decision on the merits. "The Judges of the Supreme Court," Gideon Welles plaintively wrote, "have caved in, fallen through, failed, in the McCardle case."[101] The *McCardle* law is the only instance in American history in which Congress rushed to withdraw the appellate jurisdiction of the Supreme Court for the purpose of preventing a decision on the constitutionality of a particular statute. That law, in the pithy phrase of a newspaper, "put a knife to the throat of the *McCardle* Case."[102] And the *McCardle* decision permitted Congress to do just that: "Congress," wrote former Justice Curtis, "with the acquiescence of the country, has subdued the Supreme Court."[103]

Commentators, both on and off the bench, have construed *McCardle* as the ultimate illustration of the unlimited legislative power over the appellate jurisdiction of the Supreme Court. The result of *McCardle*, we are told,[104] is to vest an unrestrained discretion in Congress to curtail and even abolish the appellate jurisdiction of the highest tribunal. After referring to *McCardle*, Justice Owen J. Roberts asked: "What is there to prevent taking away, bit by bit, all the appellate jurisdiction of the Supreme Court?"[105]

One seeking to understand the *McCardle* case should certainly not unduly minimize its impact. The repealing act there had as its aim the prevention of a decision by the Court on the constitutionality of the Reconstruction Acts. That was its sole purpose and end[106]—clearly understood as such by the Congress and the country, especially after President Johnson's veto message, which directly attacked such purpose as contrary to the spirit of the basic document.

Ex parte McCardle is, however, a case more celebrated than understood. Far-reaching though the *McCardle* decision may be, it cannot be taken as a judicial confirmation of congressional omnipotence with regard to the appellate jurisdiction of the highest Court. The organic position of the Supreme Court is specifically provided for in article III and its appellate jurisdiction is also given directly by that article. The appellate powers of the court flow, as a consequence, from the constitutional source.

The power given to Congress to prescribe exceptions and regulations to the appellate jurisdiction of the Supreme Court cannot be taken to include the authority to do away with such jurisdiction. Instead, the purpose of the Exceptions-and-Regulations Clause was to authorize exceptions and regulations by Congress not incompatible with the essential function of the Court as ultimate arbiter of the constitutional system.

The congressional power to prescribe exceptions to the Supreme Court's appellate jurisdiction may thus not be treated as authorizing exceptions which engulf the rule—even to the point of eliminating the appellate jurisdiction altogether. The Court affirmed that the congressional authority in this respect is limited by the fundamental purposes of the Constitution: "What such exceptions and regulations should be it is for Congress, in its wisdom, to establish, having, of course, due regard to all the provisions of the Constitution."[107] The exceptions and regulations laid down by Congress must not be such as will destroy the basic role of the Supreme Court in the constitutional scheme. Reasonably interpreted, the organic clause means: "With such exceptions and under such regulations as the Congress may prescribe, not inconsistent with the essential functions of the Supreme Court under the Constitution."[108]

But is not such an interpretation wholly contrary to *Ex parte McCardle*, where, as already seen, Congress was sustained in ousting the Court of competence over a case already at the bar, which was of such drastic import to the liberty of the individual concerned?

Those who would unqualifiedly assert that an affirmative response to this query is the only correct one overlook the real meaning of the *McCardle* case. Far-reaching though its decision there was, the Court in *McCardle* did not hold that the Congress could validly oust it of all appellate jurisdiction in habeas corpus cases. The repealing act at issue there did not have that extreme result. In the words of the *McCardle* opinion: "Counsel seem to have supposed, if effect be given to the repealing Act in question, that the whole appellate power of the court, in cases of habeas corpus, is denied. But this is an error. The Act of 1868 does not except from that jurisdiction any cases but appeals from circuit courts under the Act of 1867. It does not affect the jurisdiction which was previously exercised."[109]

Prior to the *McCardle* statute, the Supreme Court could review denials of habeas corpus by lower courts either on appeals[110] or on petitions to it for habeas corpus.[111] The *McCardle* statute did no more than eliminate the first of these methods for obtaining review of decisions denying the Great Writ. But, as the Supreme Court held in *Ex parte Yerger*,[112] only half a year after its *McCardle* decision, it left unimpaired the other method of invoking the Court's appellate jurisdiction in habeas corpus cases.

The *McCardle* case consequently did not really present the question of the congressional power to destroy the Supreme Court's appellate jurisdiction over denials of the Great Writ. The *McCardle* statute withdrawing jurisdiction, though successful to frustrate decision of the appeal in that case, left intact the power to review denials of the writ through a habeas petition in the Court itself. Indeed, under *Ex parte Yerger*, McCardle could presumably have petitioned the Supreme Court for a writ of habeas corpus to test the constitutionality of his military detention.

In *Yerger,* the same Court that decided *McCardle* strongly intimated that Congress lacked the power to deprive the high bench of all habeas corpus jurisdiction. "It would have been, indeed, a remarkable anomaly," declares the *Yerger* opinion, "if this court, ordained by the Constitution for the exercise, in the United States, of the most important powers in civil cases of all the highest courts of England, had been denied, under a constitution which absolutely prohibits the suspension of the writ, except under extraordinary exigencies, that power in cases of alleged unlawful restraint, which the Habeas Corpus Act of Charles II expressly declares those courts to possess."[113]

The implication is that, by virtue of the Habeas Corpus Clause, the jurisdiction of the Supreme Court to issue the writ, once conferred, may not be withdrawn by the legislature, except in cases of rebellion or invasion, despite its power to prescribe exceptions and regulations with regard to the high Court's appellate powers. As the *Yerger* Court put it: "[I]t is too plain for argument that the denial to this court of appellate jurisdiction in this class of cases must greatly weaken the efficacy of the writ, deprive the citizen in many cases of its benefits, and seriously hinder the establishment of that uniformity in deciding upon questions of personal rights which can only be attained through appellate jurisdiction."[114] To permit Congress to push its power over appellate jurisdiction to the extreme of abolishing the Supreme Court's competence in habeas corpus cases is to empower it, at will, to abrogate the guaranty of the Habeas Corpus Clause. For, in the language of the *Yerger* opinion, "[I]t is evident that the imprisoned citizen, however unlawful his imprisonment may be in fact, is wholly without remedy unless it be found in the appellate jurisdiction of this court."[115]

Despite *McCardle,* then, the Exceptions-and-Regulations Clause of article III does not vest Congress with unlimited power over the appellate jurisdiction of the Supreme Court. The clause does not permit Congress to negate the essential functions of a coordinate department. The power to make exceptions may not be pushed so far that those prescribed eliminate the appellate jurisdiction altogether or leave only a trifling residuum of jurisdiction—as by an exclusion of everything but patent cases.[116] The power to regulate is not the power to destroy where the subject of regulation is so essential to the existence of the Constitution itself. The congressional power is that to modify—not that to abolish.

Era of the Oath

To one who remembers all too vividly the "cold war" era of emphasis on individual loyalty, a particularly interesting aspect of the Civil War and postbellum period was the pervasive use of loyalty tests. Before Sumter, Americans took it for granted that public servants were loyal to the United States. As soon as hostilities began, that assumption gave way. Two weeks after Sumter fell, Lincoln's Attorney General recommended that "all the employees of the Departments—from the head secretary to the lowest messenger, be required to take anew the oath of allegiance."[117]

But the simple oath to defend the Constitution (which was contained in the first statute enacted by the first Congress)[118] was soon deemed inadequate to ensure loyalty. In August 1861, Congress passed a law prescribing a new and more elaborate oath for government employees. They were required to swear to "support, protect, and defend" the Constitution and the Federal Government and to declare their "loyalty to the same, any ordinance, resolution, or law of any State Convention or Legislature to the contrary notwithstanding."[119] The swearing process was duly repeated a second time in all federal offices.[120] This new loyalty oath was soon believed to be insufficient, and on July 2, 1862, Congress enacted into law the so-called ironclad oath of loyalty.[121]

That oath became the very backbone of the congressional system of disenfranchisement and disqualification for office during the war and in the Reconstruction period. All persons (except the President) "elected or appointed to any office . . . under the government of the United States, either in the civil, military, or naval departments" were required to take the ironclad oath. The oath was in two parts. The affiant had to swear that he had never voluntarily borne arms against the United States or given any voluntary aid to those engaged in armed hostilities against the Union, and that he had not held office under or yielded voluntary support to any government hostile to the United States. The second part of the oath was a pledge to support and defend the Constitution and bear true faith and allegiance to it in the future.

During Reconstruction, the congressional leaders insisted upon the ironclad oath as the fundamental test. "I have ceased to hope anything that justice or humanity demands," wrote Judah P. Benjamin in October 1866, "from the men who now seem to have uncontrolled power over public affairs in the U.S."[122] The inronclad oath had become the fulcrum upon which Reconstruction turned. The Supplemental Reconstruction Act (March 23, 1867)[123] prescribed an oath of past loyalty comparable to the ironclad oath for all who sought to register as voters; all registration officials were required to take the ironclad oath itself. The Third Reconstruction Act (July 19, 1867) required that "all persons hereafter elected or appointed to office in said military district" should take the ironclad oath.[124] The final step was a law of 1869 requiring the military commanders in Virginia, Texas, and Mississippi (which were still unreconstructed) to remove all officials who could not take the ironclad oath and to replace them with persons who could.[125]

The first people usually required to take loyalty oaths are public employees; concern with loyalty, however, soon extends to other areas. By mid-1862, oaths of loyalty to the Union were required of government contractors,[126] shipmasters,[127] claimants before federal agencies,[128] pensioners, telegraphers, and passport applicants.[129] Congress imposed the ironclad oath upon attorneys seeking to practice in the federal courts, [130] as well as a comparable oath upon federal jurors.[131] In addition, as we have seen, the Reconstruction laws required Southern voters to take such an oath.

"It is unfortunate," wrote Georges Clemenceau in 1867, "that the Republicans have not in all cases shown good judgment in their Reconstruction measures. One of the principal tests of loyalty . . . is the oath. But the Anglo-Saxons have always abused the oath. . . . It does not in the least hamper a rogue who becomes as accustomed to taking an oath as a dealer in church furniture to handling a pyx."[132] Even more important is the impact of the requirement upon honorable men. As Lincoln put it with regard to the Tennessee loyalty oath, "I have found that men who have not even been suspected of disloyalty, are very averse to taking an oath of any sort as a condition to exercising an ordinary right of citizenship."[133]

More fundamental is the question whether the oath technique can possibly attain its purpose of fostering loyalty. The observer who has lived through a second "era of the oath" has more than a modest doubt about what the oath technique accomplishes apart from indicating the malaise of the society in which it is employed. While the Decii are rushing with devoted bodies on the enemies of Rome, what need is there of preaching patriotism? When loyalty is made a principal object of the state's concern, it has already become less than all-transcendent.

The ironclad oath was put to the constitutional test and found wanting in *Ex parte Garland*[134] and *Cummings v. Missouri*[135]—two companion cases decided in 1867. In *Garland* the Court held invalid the ironclad oath which Congress had required for admission to practice in the Supreme Court. In *Cummings* the Missouri oath required to practice any profession (in this case, that of a Catholic priest) was ruled unconstitutional. In both cases, as the Court more recently explained, the oaths were ruled unconstitutional "as bills of attainder on the ground that they were legislative acts inflicting punishment on a specific group: clergymen and lawyers who had taken part in the rebellion and therefore could not truthfully take the oath."[136] In addition, since they imposed a penalty for an act not so punishable at the time it was committed, they also violated the prohibition against ex post facto laws.

Cummings and *Garland* may be leading cases on loyalty oaths. However, the post–*Dred Scott* shadow in which the Court still labored ensured that the decisions would have little immediate practical effect. Under *Cummings* and *Garland,* it is clear that much of the Reconstuction loyalty oath program was unconstitutional. The imposition of disqualifications upon those who could not swear to their past loyalty to the Union comes directly within the reach of the Supreme Court decisions. Yet only two monts after *Cummings* and *Garland,* Congress (as already seen) prescribed an oath of past loyalty for all voters in the reconstructed South and, a few months later, required the ironclad oath of all officeholders in the South. Despite the Supreme Court's categorical condemnation, these oath requirements were continued during the entire Reconstruction period. Nor did the Supreme Court's censure affect the use of the ironclad oath in the Federal Government, both for civil servants and in other cases. Four years after *Cummings* and *Garland,* Benjamin F. Butler declared, "I hope the iron-clad oath will never be repealed—ay, even after every disability is removed from every rebel. . . . I roll it as a sweet morsel under my tongue."[137] It was not, indeed, until 1884 that it was finally repealed.[138]

6

Chase and Waite Courts, 1864–1888

On June 26, 1857, Abraham Lincoln replied to the contention of Stephen A. Douglas that the Declaration of Independence, in declaring "that all men are created equal," was only "speaking of British subjects on this continent being equal to British subjects born and residing in Great Britain." Lincoln emphatically rejected this interpretation. The authors of the Declaration, he said, "intended to include all men. . . . They meant to set up a standard maxim for free society, which should be familiar to all, and revered by all; constantly . . . spreading and deepening its influence, and augmenting the happiness and value of life to all people of all colors everywhere."[1]

In emphasizing the concept of equality as a central theme of the Declaration of Independence Lincoln echoed what has always been a driving force in American history, despite the fact that the Framers of the Constitution did not repeat the unqualified assertion of the Declaration of Independence. Nowhere in the basic document is there any guaranty of equality or even any mention of that concept. Yet whatever may have been the Framers' intent, their work disseminated the ideals of Liberty and Equality throughout the world. "What Archimedes said of the mechanical powers," wrote Tom Paine in *Rights of Man,* "may be applied to Reason and Liberty. 'Had we,' said he, 'a place to stand upon, we might raise the world.' The revolution of America presented in politics what was only theory in mechanics."[2]

The concept of equality, however, could scarcely complete its triumphant march while slavery not only existed but was protected by the Constitution. "Liberty and Slavery," declared Frederick Douglass, "—opposite as Heaven and Hell—are both in the Constitution," and the Constitution itself was "a compromise with Slavery—a bargain between the North and the South."[3] While that bargain persisted, an express guaranty of equality would have only been hypocritical. With the Civil War, the situation completely changed. "The bond of Union being dissolved," Jefferson Davis conceded, "the obligation of the U.S. Govt. to recognize property in slaves, as denominated in the compact, might be recognized as thereby no longer binding."[4] William Lloyd Garrison, who had earlier committed the Constitution to the fire, could now fervently support the Union. When charged with inconsistency, he replied: "Well, ladies and gentlemen, when I said I would not sustain the Constitution, because it was a 'covenant with death and an agreement with hell,' I had no idea that I would live to see death and hell secede."[5]

When slavery was abolished and the American system repudiated the heresy that "all men are created equal, except Negroes,"[6] it was no longer inconsistent with reality for the Constitution to contain an express guaranty of equality. It came with ratification of the Fourteenth Amendment in 1868. That amendment and the other postbellum additions to the Constitution made equality regardless of race a fundamental constitutional principle.

The postbellum amendments constituted the first changes in the organic text in over sixty years. From a legal point of view, the changes were fundamental, for they made for a nationalization of individual rights that was completely to transform the constitutional system. The protection of life, liberty, and property now became a national responsibility—federalizing, as it were, the vindication of individual rights throughout the land.

One must, however, note a shift in constitutional emphasis as the post–Civil War period developed. The early Reconstruction concern with vindication of civil rights was soon replaced by primarily economic concerns. The headlong industrialization of the period inevitably raised new problems for the law, and the constitutional history of the nation after Appomattox must largely be written in terms of the reaction of the legal order to the new economy. If, before the Civil War, the major constitutional theme was the nation–state problem, in the period that followed the dominant concern became the government–business relationship.

The key constitutional provision of the new industrial era was the Fourteenth Amendment, which was the most significant of the legal changes imposed as part of the price of Southern defeat. Its Due Process Clause was to serve as the Great Charter for the protection of the private enterprise that was so transforming society. Due process as the great bulwark of private property did not, however, develop fully until the last decade of the century. During the period to which this chapter is devoted, it had not yet been converted into the cornerstone of American constitutionalism. The first decisions under the Fourteenth Amendment manifested a restrictive attitude toward its effect, but they were only the first steps in the interpretation of the new amendment. The dissents delivered in them ultimately served as the foundation upon which due process was to be elevated to the foremost place in the organic pantheon. Before dealing with the constitutional protection of property rights under the post–Civil War Constitution, however, we should take note of the personnel changes that took place as the Taney Court gave way to its successor.

Taney's Successor

During the Civil War and Reconstruction, the Supreme Court was anything but the master of the Constitution. In the main, control of the constitutional machine was concentrated in the legislative department, and the Court could play only a minor part. When Chief Justice Taney died in 1864, the age of giants on the bench ended. Now it was the turn of the political jobbers and manipulators. Not for more than half a century would a man of true stature again sit in the Court's central chair.

Lincoln was frank about the political considerations that governed his choice of Taney's successor. "We wish," he said, "for a Chief Justice who will sustain what has been done in regard to emancipation and the legal tenders. . . . Therefore, we must take a man whose opinions are known."[7] Such a man Lincoln thought he had found in Salmon P. Chase, his former Secretary of the Treasury and a leader of the Republican Radicals. Chase was appointed to Taney's vacant seat on December 6, 1864.

The Supreme Court historian declares, of Chase's appointment, that "it was of inestimable value to the country to have at the head of the Court not only a great lawyer, but a great statesman."[8] One familiar with Chase's legal career can only be amazed at this characterization. Regardless of what history may think of Chase as a statesman, he was anything but a "great lawyer." A leading biography concedes "his modest qualifications for the position of Chief Justice,"[9] and Chase himself tells us that when he applied in 1829 to Judge Cranch of the District of Columbia Circuit Court for admission to the Bar, Cranch was so skeptical of his professional attainments that he agreed to admit him only after Chase explained that he did not intend to practice in Washington but expected to go to the "western country."[10] He was able, after some years, to build up a practice at the Cincinnati Bar. A recent account terms his practice "a most distinguished and lucrative" one,[11] but an advertising circular sent out by Chase in 1839 indicates that it was largely that of a glorified collection agent,[12] and his political ambitions soon interfered even with that practice. Spending more and more time on politics, he all but gave up his legal work in the decade and a half before his appointment.

Noting Chase's probable appointment, Gideon Welles wrote, "The President sometimes does strange things, but this would be a singular mistake."[13] The Welles assessment proved accurate, and not primarily because of Chase's lack of learning. More important was the overriding fact that (as Lincoln is said to have put it) Chase "had the Presidential maggot in his brain, and he never knew anybody who once had it to get rid of it."[14] Certainly, Chase did not get rid of his ambition merely because of his appointment to the highest judicial position. As Chief Justice, says Henry Adams, "He loved power as though he were still a Senator"[15] and throughout his judicial career still nourished the hope that the presidential mantle would at last descend upon him. "In my judgment," wrote Morrison R. Waite (Chase's successor) in 1875, "my predecessor detracted from his fame by permitting himself to think he wanted the Presidency. Whether true or not it was said that he permitted his ambitions in that direction to influence his judicial opinions."[16] His confreres on the bench, like the rest of the country, felt that Chase's judicial actions were governed primarily by political considerations. The inevitable result was a decline in the leadership role of the Chief Justice.

Chase's Associates

In the Marshall and Taney Courts, the Chief Justice towered over his colleagues, both in leadership and in legal ability. That was clearly not the case after Chief Justice Chase took his place in the Court's center chair. The new Chief Justice was not the intellectual leader of the Court. In fact, in both leadership and intellect, he was inferior to several of his associates— particularly to Justices Miller and Field, two of the outstanding Justices in Supreme Court history.

Justice Miller described how Chase had to adapt to the new role of primus inter pares which was thenceforth to characterize the position of Chief Justice. "He liked to have his own way," Miller said about Chase, "but when he came upon the bench it was admirable to see how quietly and courteously the Court resisted his imperious will, never coming to direct conflict, and he finally had to take the position which he held, that he was the Moderator and presiding officer over the Supreme Court, and not possessed of any more authority than the rest of the Bench chose to give him."[17]

When Chief Justice Chase took his seat he joined a Court composed of four Justices appointed by Lincoln's Democratic predecessors (Justices Wayne, Catron, Nelson, and Grier) and four appointed by Lincoln himself. The first of the latter was Noah H. Swayne, a former Attorney General and leader of the Ohio Bar, who was chosen in January 1862 to succeed Justice McLean, who had died. On Taney's death, a majority of the Justices proposed Swayne as his successor. He was a learned and amiable colleague and presumably the Justices, particularly the Democratic half, preferred his appointment to the Republican Radical Lincoln might otherwise select.

Samuel Freeman Miller was Lincoln's second Court appointee, having been chosen in 1862 to replace Justice Daniel, who had died two years earlier. Miller had been the leader of the Iowa Bar and a leader among the state's Republicans. As already indicated, Miller is one of the great Justices in the Court's history. During Chase's tenure, he was the Court's intellectual leader. His constitutional posture became the Chase Court's doctrine—especially in the *Slaughter-House Cases*,[18] where that Court handed down its most important decision. Miller was the first Justice appointed from the new Northwest and his selection signaled the shift that was occurring in the national center of gravity.

The shift was also reflected in Lincoln's next two appointees, Justices Davis and Field. Justice Campbell had resigned to follow his State of Alabama out of the Union. To replace him, Lincoln chose his friend David Davis, an Illinois judge before whom the President had practiced and who had been one of those present on those fabulous nights on the Eighth Illinois Circuit when the elect gathered to hear Lincoln talk.

In 1863, Congess added a tenth seat to the Court and Lincoln appointed Stephen J. Field, then Chief Justice of the California Supreme Court. Field was one of the most colorful men appointed to the highest bench. He was the brother of David Dudley Field, the leader of the nineteenth-century codification movement, and Cyrus W. Field, who laid the Atlantic Cable. Justice Field lived a more flamboyant life than his brothers. In 1849 he joined the gold rush to California, becoming a frontier lawyer and carrying a pistol and bowie knife. He became involved in a quarrel with a judge, during which he was disbarred, sent to jail, fined, and embroiled in a duel. His lengthy feud with another judge, David Terry (Chief Justice of the California Supreme Court when Field was elected to that body in 1857), led to a threat to shoot Field. Years later, in 1889, when Field had long been Justice of the U.S. Supreme Court, Terry assaulted him in a restaurant and was shot by a federal marshal assigned to guard Field. The marshal was indicted for murder, but the Supreme Court held the killing justified.[19]

During Chief Justice Chase's tenure, Justice Miller was the intellectual leader of the Court, particularly in its narrow conception of due process. Then, under Chief Justices Morrison R. Waite and Melville W. Fuller, the leadership gradually shifted to Justice Field and his broad notion of substantive due process. Before his retirement, Field was to see the elevation of his earlier dissents on the matter into the law of the land. Ultimately, as Justice Frankfurter put it, the Justices "wrote Mr. Justice Field's dissents into the opinions of the Court."[20]

Field served on the Court for thirty-four years, eight months, and twenty days—the longest tenure save that of Justice William O. Douglas. Toward the end, Field's mind began to falter. In 1896, Justice John Marshall Harlan was deputized to suggest that Field resign. He reminded the aged Justice that Field had done the same years earlier in suggesting that another Justice step down. "Yes!" replied Field. "And a dirtier day's work I never did in my life!" In April 1897, however, he sent a letter of resignation to take effect December 1—the postponement enabling him to stretch the length of his tenure beyond that of John Marshall, the longest up to that time.

The Court in Operation

Toward the end of the Taney tenure, as we saw, the Supreme Court moved into a chamber more befitting its exalted constitutional status. When new wings were added to the Capitol for the Senate and the House, the former Senate Chamber was adapted for the high bench's use as its courtroom. At last, the Justices had a chamber comparable to those possessed by other courts—with a raised bench, marble pillars, and a separate robing room.

The Justices could thenceforth begin their sessions with the now-traditional formal entry, entering the chamber in their robes through parted curtains as the Marshal announced the Court. The order of seating was that which has since prevailed—the senior Associate Justice at the Chief Justice's right, the next at his left, and so on, in order of seniority. When the Justices were at their places, the Crier would proclaim the Court open: "God save the United States and this Honorable Court." The sessions lasted four hours and the Justices also met in conference on Saturdays to consider pending cases.[21]

By the end of Taney's Chief Justiceship, the Court had developed the internal process of decision that is still current. In particular, the opinion-writing process had evolved from the practice in John Marshall's day to that which has since been followed. When Chase became Taney's successor, the most important function of the Chief Justice was already that of assigning opinions. The power of the Chief Justice to assign the opinions probably goes back to Marshall's tenure. During Marshall's early years, it is probable that he delivered the opinion of the Court even in cases where he dissented. Apparently the practice then was to reserve delivery of the opinion of the Court to the Chief Justice or the senior Associate Justice present on the bench and participating in the decision.[22] But as time went on, other Justices also began to deliver opinions. By Taney's day, the Chief Justice assigned each opinion.

In the early years of opinion assignment by the Chief Justice, he may well have assigned all opinions. It was not very long, however, before the Chief Justice's assigning power was limited to cases where he had voted with the majority. It is probable that this practice developed under Chief Justice Taney. In his history of the Supreme Court under Chief Justice Chase, Charles Fairman describes the procedure at the beginning of Chase's tenure: "The writing of opinions was assigned by the Chief Justice—save that if he were dissenting, the Senior Justice in the majority would select the one to write."[23]

Writing in his diary, Chief Justice Chase noted, "Field intimated that Miller was displeased with my assignment of cases and after we adjourned I took occasion to speak to M. frankly on the subject."[24] Apparently the Justices were already starting to complain of the way in which the Chief

Justice assigned opinions—a complaint that has been a common theme in the Court since then.

There was, however, still one major difference between the work of the Justices in Chase's day and at the present time—the duty the Justices then had to make the rounds of their respective circuits. It was the duty of every Justice to sit in the circuit court once each year for each district within his circuit. In one sense, the burden had eased with the spread of the railroad—a definite improvement over having to ride thousands of miles "on the circuit" on horseback or even in carriages or coaches. As William Herndon, Lincoln's last law partner, put it with regard to riding the much smaller circuit in Illinois, "No human being would now endure what we used to do on the circuit . . . and oh—such victuals."[25]

An 1839 Senate report gave the following mileage summary for the Justices' circuit duties at the time:

Taney	458 Miles
Baldwin	2,000
Wayne	2,370
Barbour	1,498
Story	1,896
Thompson	2,590
McLean	2,500
Catron	3,464
McKinley	10,000[26]

Justices Catron and McKinley, assigned to the two Southern circuits, had the greatest distances to cover and the most backward travel conditions to endure.

A few years later, Justice Daniel, who had succeeded to a Southern circuit, wrote to Van Buren, who had appointed him, from Jackson, Mississippi, "I am here two thousand miles from home (calculating by the travelling route,) on the pilgrimage by an exposure to which, it was the calculation of federal malignity that I would be driven from the Bench. Justice to my friends, and a determination to defeat the machinations of mine and their enemies, have decided me to undergo the experiment, and I have done so at no small hazard, through yellow fever at Vicksburg and congestive and autumnal fevers in this place and vicinity."[27]

The circuit burden was particularly severe upon Justice Field during his first years on the Court. He had to follow the route he had used when he first went to California—by sail via the railroad at Panama—until the transcontinental railroad was opened in 1869.[28] Even with that railroad, travel to the Far West remained burdensome, particularly as Field grew older.

Throughout the tenure of Chief Justice Chase and his successor, complaints were constantly voiced by the Justices about their circuit obligation. One of the problems, as Charles Fairman pointed out, was that Congress had not appropriated funds for this travel. In their stead, the Justices apparently accepted passes from the railroads, which they used for both judicial and personal travel. The practice was basically improper; it made the Justices beholden to the railroads even though they were constantly deciding cases involving them.[29]

Despite the burden and the complaints, Congress did nothing to relieve the Justices of their circuit duties until almost the end of the century. During the period covered by this chapter, the circuit obligation remained the Justices' great albatross.

Judicial Nadir?

The traditional view of the Supreme Court during the Civil War and Reconstruction has been that it played a more subdued role than at any other time in its history—that it had been weakened, if not impotent, ever since the *Dred Scott* decision. "Never," states a leading history of the period, "has the Supreme Court been treated with such ineffable contempt, and never has that tribunal so often cringed before the clamor of the mob."[30] This view was challenged a quarter century ago by Stanley I. Kutler, who asserts that "the Court in this period was characterized by forcefulness and not timidity, . . . by boldness and defiance instead of cowardice and impotence, and by a creative and determinative role with no abdication of its rightful powers." Kutler's study is scholarly and full of suggestive insights. One may wonder, nevertheless, whether his attempt to change the accepted picture completely is justified. The postbellum Court may have played a more important role than the traditional view admitted; all the same, the historical evidence does not support the Kutler conclusion that "the Supreme Court under Salmon P. Chase was of only 'little less importance' than that under John Marshall."[31]

The Kutler case for judicial activism during Reconstruction is based upon the increasing use by the Chase Court of the power to hold laws of Congress unconstitutional. From 1865 to 1873 ten congressional acts were voided, a statistic which, Kutler points out, must be compared with two judicial vetoes in the previous seventy-six years.[32]

Two things should be noted before accepting his conclusion, however. First, the Chase Court was exercising a review power that had already been confirmed in both law and practice. By the postbellum period, there was no doubt of the legal power of the courts to review constitutionality. The Supreme Court may have exercised the power over congressional acts rarely before the Civil War, but it was exercised in many cases by state courts and was accepted without question by the leading text writers and other legal commentators. Even the critics of *Dred Scott* did not dispute the power of judicial review; their strictures were directed to the merits of the Court's decision.

In the second place, it should be recognized that, of the ten cases cited by Kutler, seven were of little practical importance and received scant notice either at the time or from constitutional historians since then.[33] This cannot be said of *Hepburn v. Griswold*,[34] where the Chase Court made its boldest assertion of review authority; yet even there the decision was limited in its effect. It applied only to contracts entered into before 1862, and, even more important, as we shall see, it was overruled by the Supreme Court the next year.[35] Reargument of the issue was granted only three months after the case had been decided. The promptness of the Court in allowing reargument "had the effect of apprising the country that the decision was not fully acquiesced in, and of obviating any injurious consequences to the business of the country by its reversal."[36] Certainly, *Hepburn v. Griswold* and the *Legal Tender Cases* focused attention upon the Court's review function, but their main immediate impact was in their demonstration of the manner in which the Court could be "packed" by new appointments to secure a desired decision.

The remaining two cases arose out of the congressional Reconstruction program. The significance of *United States v. Klein*[37] as a limitation upon congressional control over Supreme Court jurisdiction has been demonstrated by me in another book.[38] Its impact at the time was nevertheless highly limited; it invalidated a statute that had been scarcely noted when it was passed by Congress. In addition, its practical effect was largely nullified by a later decision.[39] The other case dealing with an aspect of Reconstruction was *Ex parte Garland*.[40] The decision there, as we saw in Chapter 5, held invalid the ironclad oath which Congress had required for admission to practice in the Supreme Court.

Under *Garland* and the companion decision in *Cummings v. Missouri*[41] it is clear that much of the loyalty oath program was unconstitutional. Yet even after those decisions, as seen in the last chapter, similar loyalty oath requirements were enacted and enforced. Here, too, the Supreme Court decisions, however important they may have been in constitutional jurisprudence, had no practical effect during the Reconstruction period.

Hepburn v. Griswold and Original Intention

Nothing illustrates the position of the Supreme Court during Chief Justice Chase's tenure better than its two legal tender decisions. In addition, nothing shows the inadequacy of "original intention" as the be-all-and-end-all of constitutional interpretation as well as those decisions.

During the early history of the United States, federal paper currency, with notes issued as legal tender, did not exist. Instead, as John Kenneth Galbraith points out, "[T]he money of the United States was precious metal. . . . The only paper currency was the notes of banks."[42] During the Civil War, however, Congesss was forced to make subsantial changes in the currency system. In three Legal Tender Acts, it provided for the issuance of $450 million in United States notes not backed in specie (the so-called greenbacks) and provided that those notes were to be legal tender at face value in all transactions. A constitutional controversy soon arose over Congress's power to make its paper money legal tender.

During the Civil War the Supreme Court astutely avoided deciding a case challenging the validity of the greenback laws. After the war, the issue could not be evaded. In *Hepburn v. Griswold,*[43] a bare majority ruled the Legal Tender Acts invalid. One of the main reasons Lincoln had appointed Chase as Chief Justice was to ensure a favorable decision on the constitutionality of the legal tender laws, for Chase, as Secretary of the Treasury, had been their chief architect. But the new Chief Justice disappointed the presidential expectation. Writing of Chase's attitude toward legal tender, Henry Adams comments, "As Secretary of the Treasury he had been its author; as Chief Justice he became its enemy."[44] It was Chase who delivered the majority opinion in *Hepburn v. Griswold.*

What in another judge might have been considered high moral courage was in Chase condemned as but another example of political jobbery. His act was interpreted not as an indication of judicial independence but as a bid for the nomination for President.

The young Holmes pointed out, in a contemporary comment, that *Hepburn v. Griswold* "presented the curious spectacle of the Supreme Court reversing the determination of Congress on a point of political economy."[45] At the same time, it cannot be denied that the *Hepburn* decision was in exact accord with the original intention of the Framers of the Constitution. If there was one point on which the men of 1787 were agreed, it was the need to prevent a repetition of the paper money fiasco of the American Revolution, when the expression "not worth a Continental" was born. The Framers "had seen in the experience of the Revolutionary period the demoralizing tendency, the cruel injustice, and the intolerable oppression of a paper currency not convertible on demand into money, and forced into circulation by legal tender provisions and penal enactments."[46] They therefore determined to give the government they were establishing the power to issue only a metallic currency.

This can be seen from both the constitutional text and the Framers' debates. They gave Congress the power "to coin Money," which clearly indicates "their determination to sanction only a metallic currency."[47] As Chief Justice Chase put it, "The power conferred is the power to coin money, and these words must be understood as they were used at the time the Constitution was adopted. And we have been referrred to no authority which at that time defined coining otherwise than as minting or stamping metals for money; or money otherwise than as metal coined for the purposes of commerce."[48]

The accuracy of the Chase statement is confirmed by the only dictionary available to the Framers—the one compiled by Samuel Johnson. It defines the verb "coin" as "to mint or stamp metals for money" and the noun "money" as "metal coined for the purposes of commerce."[49]

The available records of the Philadelphia Convention also bear out Chase's view of the Framers' intent. The original constitutional draft gave Congress power to "emit Bills on the Credit of the United States."[50] Gouverneur Morris moved to strike out these words. Except for one delegate, who said that he "was a friend to paper money," those who spoke on the matter supported the motion, which carried, nine to two. Madison appended a note to the debate, explaining his affirmative vote by stating that he "became satisfied that striking out the words . . . would only cut off the pretext for a paper currency and particularly for making the bills a tender either for public or private debts."[51]

The Framers' intent with respect to paper money and making it legal tender is as clear as anything that we know about the Philadelphia convention. As Luther Martin, a delegate at Philadelphia, explained it in November 1787, a "majority of the convention, being wise beyond every event, and being willing to risk any political evil, rather than admit the idea of a paper emission, in any possible event, refused to trust this authority to [the federal] government."[52]

Nor can it be doubted that the decision in *Hepburn v. Griswold* was completely in accord with the original intention of the Framers on the matter. Yet if the *Hepburn* decision was thus categorically correct in terms of original intention, it was plainly wrong so far as the needs of the nation were concerned.

It is all but impossible to conceive of a functioning modern economy without paper money, in which the only currency is specie. Yet that is exactly what would have been required under *Hepburn v. Griswold*. Well might the Supreme Court later say that its decision on the matter would

> affect the entire business of the country, and take hold of the possible continued existence of the government. If it be held by this court that Congress has not constitutional power, under any circumstances, or in any emergency, to make Treasury notes a legal tender for the payment of all debts (a power confessedly possessed by every independent sovereignty other than the United States), the government is without those means of self-preservation which, all must admit, may, in certain contingencies, become indispensable.[53]

Hence, the *New York Herald* could assert, if *Hepburn v. Griswold* meant what it said, it "involved the whole country in financial chaos and the Government perhaps in bankruptcy and repudiation."[54]

Legal Tender Cases

Hepburn v. Griswold, however consistent with the Framers' original intention, was not destined to achieve this disastrous result. When the case was decided, the Supreme Court consisted of only seven members, who divided four to three on the ruling. To deprive President Andrew Johnson of the opportunity of filling expected vacancies, Congress had passed a law providing that no vacancy on the Court was to be filled until it was reduced to fewer than seven members.[55] With President Grant's election the situation was changed, and an 1869 statute raised the number of Justices to nine and authorized the President to make the necessary appointments.[56]

On the very day when the decision adverse to the government was announced in *Hepburn v. Griswold,* Grant appointed two new Justices (Strong and Bradley), who were known to support the constitutionality of the Legal Tender Acts. After they took their seats, the Court permitted argument again on the validity of the greenback laws. This time, in the *Legal Tender Cases*[57]—decided only a year after *Hepburn v. Griswold*— Justices Strong and Bradley, plus the Hepburn dissenters, made up a new majority. Finally putting to rest the controversy over congressional authority, the Court ruled that the nation's fiscal powers included the authority to issue paper money vested with the quality of legal tender.

Historians today reject the charge that Grant "packed the Court" for the deliberate purpose of obtaining a reversal of *Hepburn v. Griswold.*[58] At the same time, it is clear that the President chose the new Justices not only because he was convinced of their fitness but because he believed they would sustain the Legal Tender Acts. For years after the cases were decided there was strong criticism because of the coincidence of the change in constitutional interpretation with the change in Court personnel. The Court's action, a contemporary newspaper commented, "will greatly aggravate the growing contempt for what has long been the most respected . . . department of our government, its Judiciary."[59] From this point of view, Chief Justice Hughes writes, "the reopening of the case was a serious mistake and the overruling in such a short time, and by one vote, of the previous decision shook popular respect for the Court."[60]

The legal tender decisions demonstrate the congressional predominance over the Court in the post–Civil War period. Even if there was no specific intent to "pack" the Court to secure a favorable legal resolution of the greenback controversy, the eventual outcome was the same. Yet even here, the picture is not entirely one-sided. The very fact that the weighty issue of legal tender was accepted as a judicial issue to be resolved by the Supreme Court is ultimately more important than the political injury inflicted on the Court. As it turned out, the vote of one Justice (however the new majority was really secured) decided a matter crucial to the economic life of the nation.

Slaughter-House Cases

The history of the Supreme Court in the post-Reconstruction period can be written largely in terms of the Court's interpretation of the Fourteenth Amendment and its Due Process Clause. That clause was ultimately to serve as the legal foundation for the great era of economic expansion upon which the nation was entering. During the Chase tenure the amendment was construed most narrowly. The key Chase Court decision here was that rendered in the 1873 *Slaughter-House Cases.*[61]

The fundamental role of the Supreme Court in the constitutional system, even in a period when the judicial power is essentially in repose, was underscored by the *Slaughter-House* decision. Like so many landmark decisions rendered by the Court, its effect was scarcely noted at the time.

"The decision," wrote a newspaper reporter the day after it was announced, "was given to an almost empty Courtroom . . . and has as yet attracted little attention outside of legal circles, although the Judges of the Court regarded the case as the most important which has been before them since the *Dred Scott* decision."[62]

Section 1 of the Fourteenth Amendment defined United States citizenship so as to include the newly freed blacks, and it prohibited states from making laws abridging the "privileges or immunities" of that citizenship or denying "due process of law" or the equal protection of the laws. Yet though the *Slaughter-House Cases* were the first cases involving the interpretation of the Fourteenth Amendment, they had nothing to do with the rights of the freedmen. They arose out of an 1869 statute passed by the "carpetbag" legislature of Louisiana. The law, secured by widespread bribery (the governor, legislators, various state officials, and two newspapers had all been paid for their support), had incorporated the Crescent City Live Stock Landing and Slaughter House Company and had given it the exclusive right to slaughter livestock in New Orleans. It had driven from business all the other butchers in the city, and the Butchers' Benevolent Association had brought an action challenging "the Monopoly," as the new corporation was called, for operating in violation of the Fourteenth Amendment.

The case was argued by legal giants of the day: John A. Campbell (leader of the Southern Bar, who had resigned from the Supreme Court when his state had seceded) for the butchers, and former Senator Matthew H. Carpenter (who had helped draft the Fourteenth Amendment) for the Monopoly. The Court ruled for the Monopoly, adopting the view that the provisions of the Fourteenth Amendment were intended only to protect the Negro in his newly acquired freedom, and that the Due Process Clause of the amendment was irrelevant to the case.

"The banded butchers are busted," Carpenter announced exultingly after the decision.[63] But it was not only the plaintiff butchers who were "busted" by the decision. The *Slaughter-House* opinion virtually emasculated section 1 of the Fourteenth Amendment itself. Had the Court's restrictive interpretation not ultimately been relaxed, the amendment could scarcely have come to serve as the legal instrument for the protection of property rights, particularly those of corporations.

The congressional debates on the Fourteenth Amendment indicate that its framers (particularly Representative John Bingham, who drafted most of section 1, and Senator Jacob Howard, who opened the Senate debate) placed particular stress upon the clause prohibiting the states from abridging "the privileges and immunities of citizens of the United States." It is possible, indeed, that the privileges and immunities to be protected included all the rights covered in the first eight amendments, with the Privileges and Immunities Clause intended to make the Bill of Rights binding upon the states.

If that was the intent of the draftsmen, it was soon frustrated. The clause was all but read out of the amendment by the *Slaughter-House Cases,* where the Court found crucial decisional significance in the difference in language between its Citizenship Clause and its Privileges and Immunities Clause. The opinion of Justice Miller stressed the fact that, while the first sentence of the amendment makes all persons born or naturalized in this country both "citizens of the United States and of the State wherein they reside," the next sentence protects only "the privileges or immunities of citizens of the United States" from state abridgement. The distinction was intended to leave the fundamental rights of life and property untouched by the amendment; they remained, as always, with the states.

Under *Slaughter-House,* the Privileges and Immunities Clause did not transform the rights of citizens of each state into rights of national citizenship enforceable as such in the federal courts. It protected against state encroachment only those rights "which owe their existence to the federal Government, its national character, its Constitution, or its laws."[64] Rights which antedate and thus do not owe their existence to that government are privileges and immunities of state citizenship alone. Earning a living is such a right. Hence the Louisiana law in *Slaughter-House* was not violative of the clause.

If the *Slaughter-House* decision rendered the Privileges and Immunities Clause "a practical nullity"[65] within five years after it became part of the Constitution, what of the amendment's Due Process Clause, upon which the *Slaughter-House* butchers had also relied? The opinion adopted the limited view that the Fourteenth Amendment was intended only to protect blacks in their newly acquired freedom. That being the case, the Due Process Clause was all but irrelevant in considering the constitutionality of the law which conferred upon the Monopoly the exclusive right to slaughter livestock. Referring to the Due Process Clause, the Court declared that "under no construction of that provision that we have ever seen, or any that we deem admissible, can the restraint imposed by the State of Louisiana upon the exercise of their trade by the butchers of New Orleans be held to be a deprivation of property within the meaning of that provision."[66] With the Due Process Clause inapplicable, the states were left almost as free to regulate the rights of property as they had been before the Civil War.

The *Slaughter-House* Court was, however, sharply divided on this restrictive interpretation of due process. Four Justices strongly disputed the Court's casual dismissal of the Due Process Clause. Foremost among them were Justices Field and Bradley, who delivered vigorous dissents. In their view, the Fourteenth Amendment "was intended to give practical effect to the declaration of 1776 of inalienable rights, rights which are the gift of the Creator, which the law does not confer, but only recognizes."[67] From the rights guaranteed in the Declaration of Independence to due process was a natural transition in the Field-Bradley approach: "Rights to life, liberty, and the pursuit of happiness are equivalent to the rights of life, liberty and property. These are the fundamental rights which can only be taken away by due process of the law."[68] A law like that in *Slaughter-House,* in the dissenting view, did violate due process: "In my view a law which prohibits a large class of citizens from adopting a lawful employment, or following a lawful employment previously adopted, does deprive them of liberty as well as property, without due process of law."[69]

What the Field-Bradley dissents were doing was to urge adoption of a substantive due process approach similar to that used in Chief Justice Taney's ill-fated *Dred Scott* opinion.[70] Like Taney, the *Slaughter-House* dissenters rejected the limitation of due process to a procedural guaranty. They urged that it also contemplated judicial review of the substance of challenged state action. In their view a monopoly law which deprived the New Orleans butchers of their right to earn their living was an arbitrary violation of due process. Much of the substance of constitutional history in the quarter century following *Slaughter-House* involved the writing of the Field-Bradley dissents into the opinions of the Supreme Court;[71] but for over a decade after it was decided, *Slaughter-House* sharply restricted the reach of the Fourteenth Amendment and its Due Process Clause. "When this generation of mine opened the reports," says a federal judge who came to the bar at that time, "the chill of the Slaughter House decision was on the bar . . . the still continuing dissents of Judge Field seemed most unorthodox. The remark in another judgment,[72] that due process was usually what the state ordained, seemed to clinch the matter."[73]

Waite and His Associates

Three weeks after the *Slaughter-House* decision, Chief Justice Chase suddenly died. President Grant's attempts to find a successor were so ludicrous that they might be relegated to the realm of the comic but for the baneful effect they had on the Supreme Court's reputation. Charles Sumner is reported to have said, "We stand at an epoch in the country's life, in the midst of revolution in its constitutional progress . . . and I long for a Chief Justice like John Marshall, who shall pilot the country through the rocks and rapids in which we are."[74] Instead, Grant used the office as a political plum—a gift to be bestowed on those who had won his personal gratitude. Only after he had failed in his stumbling efforts to appoint various associates of his did Grant choose Morrison R. Waite, a little-known Ohio lawyer, who was accepted by the Senate and the country with a collective sigh of relief. "The President," declared *The Nation,* "has, with remarkable skill, avoided choosing any first-rate man. . . . [But], considering what the President might have done, and tried to do, we ought to be very thankful."[75]

Waite was a competent legal craftsman, though scarcely endowed with the personality or prestige usually associated with the highest judicial office. "The touch of the common-place about him was, indeed, the key to his appointment. . . . Grant doubtless felt confident that the relative obscurity of Waite was the best assurance for his confirmation."[76] Certainly,

Waite had nothing of the grand manner—the spark that made Marshall and Taney what they were. A humdrum, pedestrian lawyer, he remains a dim figure in our constitutional history. "I can't make a silk purse out of a sow's ear," wrote Justice Miller a year after Waite's appointment. "I can't make a great Chief Justice out of a small man."[77]

If Waite was not a great Chief Justice, he may have been just what the Supreme Court needed after the turbulence of *Dred Scott,* the war, Reconstruction, and the political maneuverings of Chase and some of his colleagues. The Court's tarnished reputation was largely refurbished during his tenure, and at his death in 1888, it was ready to take its place again as a fully coordinate department of government.

During the Waite tenure, the intellectual leadership in the Court remained with Justices Miller and Field, though an important role in this respect was also assumed by Justice Joseph P. Bradley. It was Bradley who, along with William Strong, had been appointed by President Grant in 1870 to ensure the overruling of *Hepburn v. Griswold.* Strong was a capable judge who had served with distinction on the Pennsylvania Supreme Court. He is largely forgotten today; his tenure was too short and most of his opinions dealt with matters of little current interest. The same is not true of Bradley, "whom I regard," Justice Frankfurter once wrote, "as one of the keenest, profoundest intellects that ever sat on that bench."[78]

Bradley was one of the very few men who became a Justice without having held prior public office (Justice Miller was the only other Justice in this category before the appointment of Justice Louis D. Brandeis in 1916).[79] Instead, Bradley's career had been entirely at the Bar; he had been an eminent railroad attorney in New Jersey for many years. Yet, Frankfurter tells us, though Bradley was thus "a corporation lawyer par excellence when he went on the Court . . . his decisions on matters affecting corporate control in the years following the Civil War were strikingly free of bias in favor of corporate power."[80]

Justice Bradley's importance on the Waite Court was greater than is apparent from only his reported opinions. Soon after Waite succeeded to his position, Bradley struck up a close relationship with him and the new Chief Justice relied on Bradley in his work. Waite once wrote that he respected Bradley's advice above all others and he was quite willing to acknowledge the Justice's help.[81] As he wrote to Bradley about an opinion he was working on, "I will take the credit, and you shall do the work, as usual."[82] Indeed, as we shall see, Bradley's help in the most important Waite opinion—that in the *Granger Cases*—was so extensive that he has been characterized as a virtual coauthor of the famous opinion there.[83]

Less needs to be said about the other Waite Court Justices. When the new Chief Justice was appointed, there were four other holdover Justices from the Chase Court: Justices Clifford, Swayne, Davis, and Hunt. The latter, lately Chief Judge of the New York Court of Appeals, had been appointed in 1872 on Justice Nelson's resignation. His tenure was brief and marred by illness. He suffered a stroke in 1878, but he held on until

1882 when Congress voted a special pension bill (he had been ineligible before) and Hunt finally resigned. In his place, President Arthur selected Circuit Judge Samuel Blatchford of New York, who turned out to be, in Waite's phrase, "a good worker,"[84] though hardly an outstanding jurist.

The year earlier, Justice Clifford had died and President Arthur appointed Horace Gray, Chief Justice of the Supreme Judicial Court of Massachusets, who had served on that bench since 1864. According to a letter by Justice Miller, "Gray . . . is the choice of our Court"[85] and his nomination was generally approved. Gray was to serve for twenty years and be succeeded by another Massachusets Chief Justice, Oliver Wendell Holmes.

Two other Waite Court appointees served even briefer terms than Justice Hunt. On Justice Strong's resignation in 1880, President Hayes appointed Circuit Judge William B. Woods of Georgia. Woods was the first judge from the South since the appointment of Justice Campbell twenty-eight years earlier. Woods was a hard worker, but a man of average capabilities, who served only seven years; his tenure was cut short by his death in 1887.

On Justice Swayne's retirement in 1881, his place was taken by Stanley Matthews, a railroad lawyer and former judge and Senator from Ohio. Matthews, too, served briefly (only eight years) and, as *The Nation* put it on his death, "[H]is service as judge has been without special distinction."[86]

Much more significant had been the appointment in 1877 of John Marshall Harlan to succeed Justice Davis, who had resigned after being elected to the Senate. Harlan was, in Justice Frankfurter's pithy description, a six-foot-three Kentuckian[87] who had been Attorney General of his state. In 1956, Frankfurter wrote to a close friend. "The present fashion to make old Harlan out a great judge is plumb silly."[88] Despite the Frankfurter disparagement, most students of the Court list Harlan as one of the most important Justices. He was certainly one of the great dissenters in Supreme Court history; his frequent challenges to the majority led his colleagues, as he once wrote to Chief Justice Waite, to suggest that he suffered from "dis-sent-ery."[89] Most important, Harlan's key dissents have generally been affirmed in the court of history. A century later, his rejection of the narrow view toward civil rights adopted by the Court majority has been generally approved.

On the bench, Harlan was the most serious of judges, who, Justice Frankfurter tells us, "wielded a battle-ax";[90] his opinions were vigorous, often impatient, sometimes bitter. With his colleagues, however, he was most sociable, being one of the leaders in the whist parties which, enlivened by the usual rounds of bourbon, were a traditional recreation of the Justices.[91] Harlan displayed a sense of humor that must have been a pleasant relief in a group of men who, from their photographs, were as somber, if not always as sober, "as a judge." When the Chief Justice sent Harlan a photograph, the Justice responded, "You look natural and life-like as you would look if I were to say that a gallon of old Bourbon was on the way from Kentucky for you."[92]

A few years later, a Harlan letter to the Chief Justice contained a witty sketch on the vacation activities of some of the Justices:

> The last I heard from Bro Woods he was at Newark. Bros Matthews and Blatchford will, I fear, get such lofty ideas in the Mountains that there will be no holding them down to mother Earth when they return to Washington. Bro Bradley, I take it, is somewhere studying the philosophy of the Northern Lights, while Gray is, at this time, examining into the Precedents in British Columbia. Field, I suppose has his face towards the setting sun, wondering, perhaps, whether the Munn case or the essential principles of right and justice will ultimately prevail.[93]

Granger Cases

The *Munn* case to which Justice Harlan referred was the principal case in a series of companion cases collectively known as the *Granger Cases*.[94] They were decided in 1877 and were the most important cases in the Waite Court. It was during Chief Justice Waite's term that the Supreme Court was first called upon to respond to the modern current of social legislation. His was the beginning of the epoch when due process served as the most fertile source of constitutional lawmaking. Before the trend toward due process as the basic restriction upon state power became established in the Supreme Court, the Court decided the *Granger Cases*. "Judged by any standards of ultimate importance," says Justice Frankfurter, Waite's ruling in the *Granger Cases* "places it among the dozen most important decisions in our constitutional law."[95] It upheld the power of the states to regulate the rates of railroads and other businesses—a holding, never since departed from, which has served as the basis upon which governmental regulation in this country has essentially rested.

The *Granger Cases* arose out of the abuses that accompanied the post/-Civil War growth of railroads. Highly speculative railroad building, irresponsible financial manipulation, and destructive competitive warfare resulted in monopolies, fluctuating and discriminatory rates, and inevitable public outcry. The grievances against the railroads were especially acute in the Midwest, where the farmer was dependent upon them for moving his crops, as well as on the grain elevators in which those crops were stored. The farmers' resentment led to the Granger movement, which swept through the Midwest in the early 1870s. The Grangers sought to correct these abuses through state regulation. They secured laws in Illinois, Wisconsin, Minnesota, and Iowa regulating railroads and grain elevators and limiting the prices they could charge. These were the laws at issue in the *Granger Cases*. In the principal case before the Court, an Illinois law fixed the maximum prices to be charged by grain elevators in Chicago; four companion cases involved state statutes regulating railroad rates.

The Court sustained all these laws against due process attacks on the ground that "property . . . become[s] clothed with a public interest when used in a manner to make it of public consequence, and affect the community at large. When, therefore, one devotes his property to a use in which the public has an interest, he, in effect, grants to the public an interest in that use, and must submit to be controlled by the public for the common good, to the extent of the interest he has thus created."[96]

Chief Justice Waite's opinion in the *Granger Cases* was greatly influenced by an outline prepared by the Court's leading legal scholar, Justice Bradley.[97] In particular, it was Bradley who called Waite's attention to the common law on the subject, especially Lord Hale's seventeenth-century statement that when private property is "affected with a publick interest, it ceases to be juris privati[98] only."[99] But the *Granger* opinion was more than a rehash of the Bradley outline. Waite articulated his opinion in language broad enough to transform the whole course of the law of business regulation. In the words of a contemporary, "Suffice it, that the decision itself in its general breadth and purpose has no precedent."[100]

Waite was only following the time-honored judicial technique of pouring new wine into old bottles. He read and expounded Lord Hale in the spirit of the industrial era. He tore a fragment from the annals of the law, stripped away its limited frame of reference, and re-created it in the image of the modern police power.[101] Under the *Granger* approach, for a business to be subject to regulation, it need only be one which affects the community. "Waite's reference to property 'clothed with a public interest' surely meant no more than that the Court must be able to attribute to the legislature the fulfillment of a public interest."[102] In this sense, a business affected with a public interest becomes nothing more than one in which the public has come to have an interest.[103] This rationale becomes a means of enabling governmental regulatory power to be asserted over business far beyond what was previously thought permissible. As a member of the highest bench once pointed out, "There is scarcely any property in whose use the public has no interest."[104] The public is concerned about all business because it contributes to the welfare of the community.[105]

Waite's rationale did not really reveal its potential until over half a century later. In the years immediately following *Granger,* it was virtually neutralized by judicial adoption of the Field-Bradley notion of due process. It was revived when the due process current was reversed, starting in 1934.[106] Since that time, it has been the doctrine that has furnished the constitutional foundation for the ever-broader schemes of business regulation that have become so prominent a feature of the present-day society.

Civil Rights Cases

The Reconstruction period saw not only the adoption of the postbellum amendments but also the enactment of significant civil rights statutes by Congress. The most important of these laws was the Civil Rights Act of 1875.[107] That statute constituted the culmination of the postbellum Republican program and a decade of efforts to place the ideal of racial equality upon the legal plane. One may go further and see in the 1875 law the last victory for the egalitarian ideal of the Reconstruction period. From 1875 to the middle of the next century, there were to be no further legal gains for racial equality. On the contrary, the civil rights legislation enacted during the Reconstruction decade was soon to be virtually emasculated by both Congress and the Supreme Court.

To the present-day observer, the Civil Rights Act of 1875 is of particular interest, for it provides the historical nexus between the Fourteenth Amendment and the Civil Rights Act of 1964. The goal of equality in public accommodations, which Congress sought to attain by enactment of the 1964 statute, was what had been intended by the 1875 law. The legislative history of the 1875 law demonstrates that the legislators of the post– Civil War period were intimately concerned with many of the key problems that are still with us in the field of civil rights: integration versus segregation (particularly in education), legal versus social equality, and the crucial question of whether an ideal such as racial equality can be achieved by legislative action, especially in a society opposed to practical implementation of that ideal.

The key figure in the enactment of the 1875 Civil Rights Act was Senator Charles Sumner. Whatever may be said against Sumner, he was, throughout his career, a sincere believer in the cause of equal rights. Well before emancipation, he gave substance to Whittier's economium, "He saw a brother in the slave, With man as equal man he dealt." His crucial position in the history of the act is not affected by the fact that he died in March 1874, a year before it became law. As the sponsor of the measure in the House pointed out, the bill was originated by Sumner and he regarded it as his main legacy to his country.[108] On his death bed, he is said to have told Judge Ebenezer R. Hoar, then a Congressman from Massachusetts, "You must take care of the civil-rights bill,—my bill, the civil rights bill,— don't let it fail!"[109]

Sumner's bill reached the statute book as the Civil Rights Act of 1875. It contained a prohibition against racial discrimination in inns, public conveyances, and places of amusement. The prohibition was, however, ruled invalid by the Waite Court in the 1883 *Civil Rights Cases*[110] on the ground that it sought to reach discriminatory action that was purely private in nature and consequently not within the scope of the Equal Protection Clause. "Can the act of a mere individual," asked Justice Bradley for the Court, "the owner of the inn, the public conveyance, or place of amusement, refusing the accommodation, be justly regarded as imposing any badge of slavery or servitude upon the applicant, or only as inflicting an ordinary civil injury, properly cognizable by the laws of the state, and presumably subject to redress by those laws until the contrary appears?" Answering this query in favor of the latter construction, the opinion asserted, "Individual invasion of individual rights is not the subject-matter of the amendment."[111]

Only Justice Harlan—a former slaveholder—dissented from this narrow construction of the Fourteenth Amendment. He declared that the majority's narrow concept of "state action" reduced the amendment to "splendid baubles, thrown out to delude those who deserved fair and generous treatment at the hands of the nation."[112] Similarly, Frederick Douglass attacked the decision as "a concession to prejudice" and contrary to Christianity, the Declaration of Independence, and the spirit of the age.[113]

Despite the attraction of the Harlan dissent for present-day jurists, the Court has continued to follow the rule laid down in the *Civil Rights Cases*. As stated in the leading modern case, "The principle has become firmly embedded in our constitutional law that the action inhibited by the first section of the Fourteenth Amendment is only such action as may fairly be said to be that of the States. That Amendment erects no shield against merely private conduct, however discriminatory or wrongful."[114]

Critics of the Court, especially in recent years, have contended that the decision in the *Civil Rights Cases* amounted to virtual judicial usurpation, that the Justices emasculated the post–Civil War amendments to nullify the broad remedial intent of their framers. The congressional debate on the 1875 civil rights law demonstrates that such a view is unfounded. There is ample indication in the debates that a substantial number of legislators considered the bill before them unconstitutional, many of them[115] for the very reason later stated by the Supreme Court—that is, that it sought to reach individual, rather than state, action.

The congressional discussions also bear directly upon more recent developments in the field of civil rights. In the Senate debate in February 1875, Senator Matthew H. Carpenter (one of the outstanding lawyers of the day) stated that Congress might try to accomplish the public accommodation purposes of the 1875 act under its commerce power. "Such provision in regard to theaters," he asserted, "would be somewhat fantastic as a regulation of commerce."[116] In 1964, the Congress did enact a civil rights act based upon the commerce power of the very type which Senator Carpenter had termed "fantastic" in 1875,[117] and, as Carpenter also prophesied,[118] such a statute was upheld by the Supreme Court.[119]

To the observer today, the congressional debate on the 1875 Act is also most pertinent for its discussion of discrimination in education. The very problems that have become so important since the landmark decision in *Brown v. Board of Education*[120]—integration versus segregation, the threat of the South to close down the public school system rather than have "mixed" schools, and the claim that all that is really needed is "separate but equal" facilities for the two races—appear here. The sharpest controversy arose over the original Sumner bill's prohibition of racial discrimination not only in public accommodations but also in all "common schools and public institutions of learning or benevolence supported in whole or part by general taxation." Though a strong effort was made to strike out this prohibition in the Senate, it was defeated, and the bill as passed by the upper chamber in 1874 contained the provision drafted by Sumner.

The situation was different when the House considered the bill a year later. Public opposition (as shown by the 1874 election results) led it to strike out the clause prohibiting racial discrimination in educational institutions, as well as one covering cemeteries. The actual vote on the amendment to strike out the school provision was overwhelmingly in favor. The consensus against the school prohibition was now so great that no effort was made in the Senate to reinsert it in the bill. Instead, that chamber speedily passed the House bill.

The debate on the proposed prohibition of racial discrimination in schools is directly relevant to the intent of those who wrote the Fourteenth Amendment with regard to segregation in education. One who has read it cannot help but conclude that the Congress that sat less than a decade after the Fourteenth Amendment was sent to the states for ratification did not think that the amendment had the effect of prohibiting school segregation. If it had, the whole debate would have been irrelevant, since integration would have been constitutionally required, regardless of any congressional provision in the matter. It is fair to say that no participant in the congressional debate took such a view (which was, of course, that ultimately taken by the Supreme Court in *Brown v. Board of Education*).

This does not, however, mean that the decision in *Brown* was wrong. The Court there was interpreting the Constitution to meet society's needs in 1954—needs which were not necessarily the same as those of 1875. Only those who would make the Constitution as inflexible as the laws of the Medes and Persians will object to such constitutional construction. Stability and change are the twin sisters of the law and together make the Constitution a document enduring through the ages.

Corporate Protection

The Waite Court's narrow construction of the Fourteenth Amendment in the *Civil Rights Cases* should be contrasted with its extension of constitutional protection to corporations.

One of the key questions for constitutional historians has been whether the framers of the Fourteenth Amendment intended to include corporations within the scope of its protection. A decade and a half after it was adopted, Roscoe Conkling, a former member of the committee which drafted it, implied, in argument before the Supreme Court, that he and his colleagues, in framing the Due Process and Equal Protection Clauses, had deliberately used the word "person" in order to include corporations.[121] "At the time the Fourteenth Amendment was ratified," he averred, "individuals and joint stock companies were appealing for congressional and administrative protection against invidious and discriminating State and local taxes."[122] The implication was that the committee had taken cognizance of such appeals and had drafted its text to extend the organic protection to corporations: "The men who framed . . . the Fourteenth Amendment must have known the meaning and force of the term 'persons.'"[123]

Most historians reject the Conkling insinuation.[124] From a purely historical point of view, it is clear that Conkling, influenced by the advocate's zeal, overstated his case. Yet even if his argument on the real intent of the draftsmen was correct, that alone would not justify the inclusion of corporations within the word "person." As Justice Hugo L. Black once put it, "a secret purpose on the part of the members of the committee, even if such be the fact, . . . would not be sufficient to justify any such construction."[125] After all, what was adopted was the Fourteenth Amendment and not what Roscoe Conkling or the other members of the draftng committee thought about it.[126]

What stands out to one concerned with the meaning of the amendment is the deliberate use in its Equal Protection and Due Process Clauses of the same language employed in the Fifth Amendment. It is surely reasonable to assume that, when Congressman John A. Bingham, "the Madison . . . of the Fourteenth Amendment,"[127] deliberately used that language,[128] he intended to follow the same approach as his predecessors with regard to the applicability of the new safeguard. By the middle of the nineteenth century, the corporate entity had become an established part of the economy. If corporate "persons" were to be excluded from the new constitutional protections, it is difficult to see why the unqualified generic term "persons"[129] was employed.

It must be emphasized that corporate personality antedated the Fourteenth Amendment. Its protection had, by the time of the postbellum amendments, become a vital concern of the law. The end of the Civil War saw a vast expansion in the role of the corporation in the economy,[130] but even before that conflict, the corporate device was recognized as an indispensable adjunct of the nation's growth. This realization had already led to decisions favorable to the corporate personality.[131] When the ultimate protection of person and property was transferred by the Fourteenth Amendment from the states to the nation, the judicial trend in favor of the corporation also became a national one. The role of the corporate person in the post/-Civil War economy made the use of the Fourteenth Amendment to safeguard such persons a natural development, whatever may have been the subjective goals of its framers.

At any rate, the Waite Court speedily gave corporate protection its imprimatur and it did so in a manner that indicated that it had no doubt on the matter. In 1877, in the *Granger Cases,*[132] the Court had ruled on the merits of state regulatory laws which were asserted to violate due process and equal protection without even considering whether plaintiff corporations were "persons" capable of invoking the organic guarantees. Nine years later, in *Santa Clara County v. Southern Pacific Railroad Company,*[133] the question of whether corporations were "persons" within the meaning of the Fourteenth Amendment was extensively briefed by counsel. At the beginning of oral argument in the case, however, Chief Justice Waite tersely announced: "The court does not wish to hear argument on the question whether the provision in the Fourteenth Amendment to the Constitution, which forbids a State to deny to any person within its jurisdiction the equal protection of the laws, applies to these corporations. We are all of opinion that it does."[134]

The Court in the *Santa Clara* case was apparently so sure of its ground that it wrote no opinion on the point.[135] Be that as it may, nevertheless, the Waite pronouncement definitively settled the law on the matter. In the words of Justice Douglas, "It has been implicit in all of our decisions since 1886 that a corporation is a 'person' within the meaning of the . . . Fourteenth Amendment."[136] Countless cases since *Santa Clara* have proceeded upon the assumption that the Fourteenth Amendment assures corporations, as well as individuals, both due process and equal protection.[137]

The Justices and the Election Crisis

Over a century later, the Waite Court may appear unduly restrictive in its interpretation of civil rights guaranties. To its contemporaries, however, the Court appeared as a prime instrument of the conciliation needed to signal the end of Reconstruction and a desired return to "normalcy." The Court composition itself reflected that changing era when, just a few months before Chief Justice Waite's death in 1888, President Cleveland appointed Lucius Quintus Cincinnatus Lamar of Mississippi to replace Justice Woods, who had died. Lamar, then Secretary of the Interior and a former Senator, was the first Democrat appointed since Justice Field in 1862 and, more important, the first Justice who had served in the Confederate Army. To have an ex-Confederate on the highest Court was, to most, a welcome symbol of the post-Reconstruction reconciliation. Now the Court, like the country itself, could return to "normalcy" and put the trauma of the war and Reconstruction behind it.

The movement for returning stability, both for the Court and the country, however, had been rudely interrupted by the election crisis of 1876–1877. The Supreme Court was involved in that crisis, not as an institution, but through the appointment of five Justices to the commission set up by Congress to resolve the dispute. "Divided seven to seven on recognized party lines, the decisive vote and opinion was that of the member appointed for judicial impartiality, Justice Joseph P. Bradley."[138]

Coming so soon after the traumatic experiences of civil conflict and Reconstruction, the disputed Hayes–Tilden election of 1876 might well have resulted in a permanent breakdown of the constitutional system— bringing about, in the contemporary terminology, the "Mexicanization" of American politics. A study published thirty years after the event asserts, "Few of the generation which has grown up since then will have any but the faintest conception of the gravity of the situation existing during the winter of 1876–1877."[139] Such a comment is, of course, even more true today. It is all but impossible now to appreciate the gravity of the crisis presented by the dispute. At the time, "[M]ore people dreaded an armed conflict than had anticipated a like outcome to the secession movement of 1860–1861."[140] This sentiment was expressed in a letter from Senator Sherman to Hayes himself: "The same influence now rules . . . as did in 1860–1861, and I feel that we are to encounter the same enemies that we did then."[141] Well might Chief Justice Waite, in a letter to a federal judge, characterize the situation as a "great trial."[142]

As so often in our history, an essentially political controversy was converted into a legal battle, with the disputed issues argued and resolved in constitutional terms. Perhaps, ultimately, that is why the crisis could be settled without bloodshed. Justice Jackson's striking claim—"struggles over power that in Europe call out regiments of troops, in America call out battalions of lawyers"[143]—was given dramatic corroboration in this peaceful resolution.

Legally speaking, the conflict arose because of a lacuna in the Constitution with regard to the process of electing the President. Article II provides for the selection of presidential electors under state laws, for the casting of their votes, and for the certification of the electors to the President of the Senate. It then goes on. "The President of the Senate shall, in the Presence of the Senate and House of Representatives, open all the Certificates, and the Votes shall then be counted." But counted by whom? The President of the Senate (who in 1876–1877 was Thomas W. Ferry, a leading Republican), the two Houses separately (leading to a deadlock, since the Senate had a Republican and the House a Democratic majority), or the Houses jointly (in which case the Democratic House would outvote the Republican Senate)?

The Constitution did not answer these questions, which assumed critical importance in the face of conflicting electoral certificates from four states. The returning boards in Florida, Louisiana, and South Carolina (solidly Republican) had certified the electors for Hayes, converting Tilden majorities by disallowing thousands of Democratic votes. The Democrats, claiming fraud, had conflicting certificates, certifying their electors, sent to Washington. In Oregon, where Hayes had received a clear majority, one of his electors was ineligible under the Constitution because he was a federal officer. The Governor certified the other two Republican electors and the Democrat who had lost to the ineligible official. The two Repulican electors chose a third Republican to fill the vacancy and sent a certificate, accompanied by a certification of the election results by the Oregon Secretary of State. In these circumstances, the crucial question was, of course, who counted the electoral votes, and it could not be decided by the normal political machinery. That the situation was finally resolved by extra-constitutional means may be attributed to the overwhelming popular desire, particularly in the South, for a peaceful solution.[144]

When the crisis was at its height, early in 1877, Chief Justice Waite wrote that "the good sense of the people is exerting its influence upon the leaders."[145] Two weeks later Congress set up an Electoral Commission to decide which of the disputed electoral votes to count.[146] It was to be composed of five Senators and five Congressmen, equally divided between the two parties, and four Supreme Court Justices, designated by circuits. These four (two from each party) were then to select a fifth member of the Court, who would be the commission's key man, if, as was expected, the other fourteen members divided evenly along party lines. It was hoped that the partisan element in the commission's work would be neutralized by the selection of David Davis as the fifth Justice (he was the only member of the Court not formally affiliated with either party). This expectation was frustrated by Davis's sudden resignation from the Court after his election to the Senate by the Illinois legislature. That left only Republican members of the Court to choose from, and Justice Bradley, supposedly the least partisan among them, was chosen.

When it came to decide the disputed returns, the Electoral Commission divided, in every instance, strictly along party lines. Justice Bradley's vote, added to those of the other Republicans, meant an eight-to-seven division in Hayes's favor on every disputed elector, and the Republican candidate was declared elected by the margin of one electoral vote. By then it was March 2, 1877—only two days before inauguration day.

The vital legal question before the Electoral Commission was that of whether it could properly go behind the returns certified by the relevant state officials. The Republican majority gave a negative answer on the disputed returns from Florida, Louisiana, and South Carolina. "It seems to me," declared Justice Bradley for the majority, "that the two Houses of Congress, in proceeding with the count, are bound to recognize the determination of the State board of canvassers as the act of the State, and as the most authentic evidence of the appointment made by the State."[147]

This position appears valid and was, in fact, the one ultimately adopted when Congress, a decade later, finally provided a permanent procedure for counting the electoral vote.[148] Any other rule leaves it open for a majority in Congress—even though repudiated at the polls—to perpetuate its candidate in the highest office. It has, however, been claimed that the commission did not follow its own ruling when, in Oregon, it refused to accept the Governor's certification of the one Tilden elector. Though a close question, it seems that the claim misconceives the nature of what the commission did in this case. Under the Oregon statutes it was the Secretary of State who had the authority to canvass the returns, and his certificate as to those chosen was accepted in view of his exclusive authority under state law.[149]

Legally justified or not, a storm of controversy resulted from the Electoral Commission's rulings. In particular, Justice Bradley was subjected to vituperative attacks for allegedly changing his original opinion in favor of the Democrats after pressure from leading Republicans and railroad interests, to which he was supposedly beholden. The charge against Bradley's integrity severely tarnished the remaining career of one who (from the point of view of legal ability) was one of the best men ever to sit on the high tribunal.

More important, as in every instance in which Justices have performed nonjudicial duties, the judicial descent into the political arena reflected unfavorably on the Supreme Court itself. According to James A. Garfield, "All the judges, save one, were very sorry to be called to this commission."[150] The spectacle of the Justices casting their votes on partisan lines cannot but have had a deleterious effect upon the Court's reputation. At the same time, it must be recognized that without them, it is doubtful that the Electoral Commission could ever have been approved—much less had its decision accepted by the country. Had they refused to serve, they would have upset the carefully worked-out compromise and plunged the country into a crisis which it might not have been possible to settle peacefully.

"Just at present," wrote Chief Justice Waite after the Electoral Commission had decided, "our judges are severely criticized, but I feel sure time will bring us out all right."[151] And so it turned out. With Henry Adams, the country "still clung to the Supreme Court, much as the churchman still clings to his last rag of Right. Between the Executive and the Legislature, citizens could have no Rights; they were at the mercy of Power. They had created the Court to protect them from unlimited Power."[152] The need for an impartial umpire in a working federation was too great. For the balance properly to be kept, judicial power could not long be kept in repose.

Resolution of the dispute took place soon after the centennial of the nation. Writing at the time, Henry Adams mourned that "the system of 1789 had broken down, and with it . . . the fabric of . . . moral principles. Politicians had tacitly given it up."[153] Adams was (as all too often) wide of the mark. The constitutional system set up a century earlier had endured the test of fire. Reconstruction and the bitterness it engendered was at last ended. For the first time since the war, it was governmental policy that "the flag should wave over states, not provinces, over freemen and not subjects."[154] The focus of the Court's concern could now shift from the great issues that had almost destroyed the nation to those more appropriate to a less troubled era.

7

Fuller Court, 1888–1910

The law is both a mirror and a motor. It is, as Holmes tells us, a mirror wherein we see reflected the society which it serves.[1] But the law also helps to move the society in the direction that is perceived to serve best the "felt necessities of the time."[2] The jurisprudence of the Supreme Court inevitably reflects both aspects of the law's role.

This was particularly true as the Supreme Court began its second century. Despite the celebrated Holmes animadversion, the law then did virtually "enact Mr. Herbert Spencer's *Social Statics.*"[3] The Court's decisions reflected the Spencerean laissez faire that had become dominant in the society as a whole at the time. However, the Court also helped to mold the society and economy in the Spencerean image. It furnished the legal tools to further the period's galloping industrialism and ensure that public power would give free play to the unrestrained capitalism of the era.

The end of the last century saw the beginning of a period of negative Supreme Court jurisprudence, when government was denied the essential powers which it was to assume during the next century. As the Court itself recently conceded, the Court then "imposed substantive limitations on legislation limiting economic autonomy . . . , adopting [instead] the theory of *laissez faire.*"[4] The Court became a primary pillar of the dominant jurisprudence of the day—its overriding theme that of the individualism of the law applied with mechanical rigor, with abstract freedom of the individual will as the crucial factor in social progress. This was the time when the Court apparently believed in everything we now find it impossible to believe in: the danger of any governmental interference with the economy, the danger of subjecting corporate power to public control, the danger of any restriction upon the rights of private property, the danger of disrupting the social and economic status quo—in short, the danger of making anything more, the danger of making anything less.

Chief Justice Fuller

In the Court's history, Melville Weston Fuller remains the prime example of that luckiest of all persons known to the law—the innocent third party without notice. Fuller was a little-known Chicago lawyer when the lightning struck. On Chief Justice Waite's death in March 1888, President Cleveland decided to appoint someone from Illinois in his place. His first choice, Judge John Shoffield of the state's Supreme Court, refused the appointment because he did not want to raise his nine children in Washington. Cleveland then chose Fuller, who had become a successful attorney—though, in his own phrase at the time, not someone "already in the public eye" or even "publicly known to a greater or less extent."[5]

Fuller was the first Chief Justice who had never held any federal office prior to his appointment. In fact, as his biographer concedes, his prior public service was "pitifully small,"[6] consisting only of brief membership in a constitutional convention and a state legislature. One newspaper called him the most obscure man ever appointed Chief Justice. Fuller's new colleagues apparently agreed. Writing to Justice Field, who had hoped for the appointment, Justice Bradley complained that though Fuller was "a very estimable man and a successful practitioner; [this] hardly fills the public expectation for the place of Chief Justice of the United States."[7]

Fuller was an extremely small man—so small, indeed, that it was necessary to elevate his chair on the bench and provide a hassock to prevent his feet from swinging in the air. As Justice Frankfurter tells it, "Fuller came to a Court that wondered what this little man was going to do. There were titans, giants on the bench. They were powerful men, both in experience and in force of conviction, and powrful in physique, as it happened." A number of the Justices were big men: Justices Miller, Harlan, and Gray were great, stocky men, over six feet; Justice Field was also well above average size. "Those were the big, powerful, self-assured men over whom Melville Fuller came to preside."[8]

"Oh, but there were Giants on the Court in those days," Fuller used to say of his early years as Chief Justice. Nor was he referring only to the great stature of his colleagues. The Court at the time contained some of the greatest Justices in its history; Justices Miller, Field, Bradley, and Harlan were among the Justices who have left an enduring mark on our law. Well could the new Chief Justice write soon after his appointment, "No rising sun for me with these old luminaries blazing away with all their ancient fires."[9]

As it turned out, Fuller was one of the most effective Chief Justices. His new colleagues, says Justice Frankfurter, "looked upon him . . . with doubt and suspicion, but he soon conquered them." Fuller had a remarkable capacity for leading the Court conferences. In Frankfurter's phrase, "He had gentle firmness, courtesy, and charm."[10] At the core of the Fuller success as Chief Justice was what his biographer calls his "remarkable capacity for mediation; he liked to 'tinker a compromise,' as Justice Holmes put it." Holmes himself used to say "that there never was a better presiding officer, or rather, and more important in some ways, a better moderator inside the conference chamber, than this quiet gentleman from Illinois."[11] Indeed, according to Holmes, "I think he was extraordinary. He had the business of court at his fingers' ends; he was perfectly courageous, prompt, decided. He turned off the matters that daily called for action easily, swiftly with the least possible friction, with imperturbable good humor, and with a humor that relieved any tension with a laugh."[12]

Humor, in fact—what Justice Frankfurter termed "lubricating humor"—was Fuller's secret weapon. Frankfurter gives an example that he heard from Justice Holmes:

> Justice Harlan, who was oratorical while Justice Holmes was pithy, said something during one of the Court's conferences that seemed to Holmes not ultimate wisdom. Justice Holmes said he then did something that ordinarily isn't done in the conference room of the Supreme Court. Each man speaks in order and there are no interruptions, because if you had that you would soon have a Donnybrook Fair instead of orderly discussion. But Holmes afterward said, "I did lose my temper at something that Harlan said and sharply remarked, 'That won't wash. That won't wash.'" Tempers flared and something might have happened. But when Holmes said, "That won't wash," the silver-haird, gentle little Chief Justice said, "Well, I'm scrubbing away. I'm scrubbing away."[13]

Fuller presided, in Frankfurter's phrase, "with great but gentle firmness."[14] After returning from a Saturday conference, Justice Gray, the first to hire a law clerk, used to tell Samuel Williston, his clerk when Fuller took his seat (and later the leading authority on the law of contracts), how the discussion had been heated but the Chief Justice had kept the Court together. It was Fuller who inaugurated the practice still followed of having each Justice greet and shake hands with every other Justice each morning. "This practice," wrote Fuller's biographer, "tends to prevent rifts from forming."[15]

The Court and the Justices

A former Fuller classmate at Bowdoin described the courtroom as the new Chief Justice prepared to take the oath: "A small white-walled room . . . a long massive desk, behind which, seated in their massive chairs ranged side by side, sit eight figures . . . clothed to their feet in flowing robes of 'solemn black,' apparently unmovable with stern fixed gaze, and seemingly almost as emotionless as effigies of departed greatness which are clustered in 'Statuary Hall' not far distant."[16]

The Fuller Court in operation fully followed the practice and traditions established by its predecessors, since, as Justice Holmes once put it, "Fuller hated to change anything—even to adjourn half an hour for luncheon instead of two."[17] One thing, however, was substantially changed during Fuller's tenure—the burden of circuit court duty that had been the bane of the Justices for a century.

A bill for the relief of the Supreme Court had been pending for some years in Congress. None had been passed and the Court's situation continued in its unsatisfactory state, because of both the circuit obligation and increasing docket congestion. Fuller's biographer tells how the aging Chief Justice was compelled to go on circuit to South Carolina in June in very hot weather. Though the burden may have been less great than it was before the railroad, it was still substantial. "I am so weary I can hardly sit up," wrote Fuller at the end of the 1889 Term.[18]

In 1881, former Justice Strong published an article which called attention to the unsatisfactory situation. He also urged "the only possible adequate remedy for the existing evil . . . the establishment of a court of appeals in each of the circuits into which the country is now divided—a court intermediate between the Supreme Court and the circuit courts."[19] Such a measure had been proposed in a bill introduced in the Senate by former Justice Davis after he had resigned to accept a seat in the upper House. It was also the solution preferred by the Justices themselves. As early as 1872, Justice Miller had written "of the plan which has always had my preference, an intermediate appellate court in each circuit, or such a number of intermediate courts of appeal as may be found useful."[20]

Now Chief Justice Fuller put the prestige of his office behind the proposal. But he did it informally at a dinner to which he invited the important members of the Senate Judiciary Committee and by cultivation of Senator George F. Edmunds, the key member of the committee. The Fuller effort was soon to bear fruit. The Senate Judiciary Committee wrote Fuller that it would "be agreeable to the Committee to receive . . . the views of the Justices."[21] The latter, in a report prepared by Justice Gray, recommended the establishment of circuit courts of appeal. Congress then set up the proposed courts in 1891, under a bill sponsored by Senator Edmunds. Though it did not specifically relieve the Justices of circuit duty, it did so in practice, since they were not expected to sit in the new courts. The old circuit courts were finally abolished by a 1911 statute.

During Chief Justice Fuller's tenure, there was an almost complete change in the Court's composition. Indeed, when the Chief Justice died in 1910, only Justice Harlan remained of those who had been on the bench when Fuller was appointed. The first of the new Justices was David J. Brewer, appointed in 1889 to fill the seat of Justice Matthews, who had died. Brewer, a former Kansas Supreme Court Justice and federal circuit judge, was Justice Field's nephew. On the Court, Brewer proved a strong supporter of his uncle's due process posture. Soon after his appointment, Brewer wrote an opinion in which he both attacked the *Granger Cases*[22] doctrine as "radically unsound" and issued the clarion of the coming era of

Court jurisprudence: "The paternal theory of government is to me odious."[23] Instead, the theme was to be that of Brewer's 1891 address at Yale Law School: "Protection to Private Property from Public Attack."[24]

Most of the other Justices appointed during the Fuller tenure can be dealt with more briefly. On Justice Miller's death in 1890, after twenty-eight years on the Court, Henry B. Brown of Michigan, a federal district judge for fifteen years, was chosen. He was to serve until 1906—an average Justice whose work on the bench is all but forgotten.

In the same mold was George Shiras, Jr., of Pennsylvania, appointed in 1892 in the place of Justice Bradley, upon the latter's death. Shiras had no previous judicial experience; he was a successful practitioner whose clients included large steel corporations and the Baltimore and Ohio Railroad. Shiras served only until 1903, when he retired. "His retirement," writes a biographer, "attracted little notice, and his death even less, for Shiras appeared to have been an undistinguished jurist."[25]

Other Justices on the Fuller Court who were buried in what a biographer of one of them terms "the shroud of anonymity"[26] were Howell E. Jackson of Tennessee, a former Senator and federal circuit judge—the first Democrat appointed by a Republican President since Justice Field in 1861—who died in 1895, only two years after he was chosen; Joseph McKenna of California, a former Congressman, circuit judge, and Attorney General, chosen to succeed Justice Field, who had resigned in 1897; William R. Day of Ohio, who had been Secretary of State and a circuit judge, when Justice Shiras resigned in 1903; William H. Moody of Massachusetts, a former Congressman, Secretary of the Navy, and Attorney General, chosen when Justice Brown resigned in 1910; and Horace H. Lurton, a circuit judge, appointed in place of Justice Rufus W. Peckham, who died in 1909 (and whose appointment will be discussed shortly).

Three Justices appointed during Fuller's term stand out from this group of less then mediocre Justices. They are Edward Douglass White, Rufus Wheeler Peckham, and Oliver Wendell Holmes. White was a Senator from Louisiana when he was appointed in 1894 to succeed Justice Blatchford, who had died. Owner of a large plantation, White had served in the Confederate Army and briefly in his state's highest court. White was a more than competent Justice, who served as an Associate for sixteen years, but he is remembered primarily as Fuller's successor as Chief Justice and will be discussed more fully in that position in Chapter 9. Justice Holmes also will be dealt with later (in Chapter 8), since his contribution to both the Court and the law was so important that it deserves separate treatment.

That leaves Justice Peckham, a member of New York's highest court, appointed in 1895 to the seat of Justice Jackson, who had died after serving only two years. Holmes wrote that his Supreme Court colleagues were "enthusiasts for liberty of contract."[27] Foremost among these liberty-of-contract enthusiasts was Justice Peckham, who was to be Holmes's principal antagonist in *Lochner v. New York*,[28] the leading case on the subject.

More than any other judge, Peckham was the exemplar of the conservative jurist early in this century. His decisions were prime applications of the dominant legal thought of the day—using the law as the barrier against interferences with the operation of the economic system. If laissez faire was read into the Due Process Clause, that was true in large part because of Justice Peckham's opinions.

Like so many of the Justices of his day, Peckham was a strong earthy character; Holmes termed him "a master of Anglo-Saxon monosyllabic interjections."[29] A young law clerk once asked Holmes, "What was Justice Peckham like, intellectually?" "Intellectually?" Holmes replied, puzzlement in his voice. "I never thought of him in that connection. His major premise was, 'God damn it!'"[30] A few years later, after making the same comment, Holmes explained that he meant "thereby that emotional predilections governed him on social themes."[31]

Peckham's opinions bear witness to the acuteness of the Holmes observation. The "emotional predilections" that governed Peckham's decisions were based upon the fear of changes in the existing order. "When socialism first began to be talked about," Holmes tells us, "the comfortable classes of the community were a good deal frightened. I suspect that this fear has influenced judicial action both here and in England."[32] It certainly influenced the Peckham jurisprudence.

At the same time, few can doubt Justice Peckham's importance in helping to translate the prevailing jurisprudence into the law of the land. It was Peckham who wrote the opinions in both the case in which liberty of contract was first relied upon by a majority of the Supreme Court[33] and the case where the liberty-of-contract tide reached its crest.[34] In both cases, Justice Frankfurter tells us, "Mr. Justice Peckham wrote Mr. Justice Field's dissents into the opinions of the Court."[35] It was because of the Peckham opinions that Holmes could say that the Fourteenth Amendment may have begun with "an unpretentious assertion of the liberty to follow the ordinary callings," but "[l]ater that innocuous generality was expanded into the dogma, Liberty of Contract."[36]

Due Process and Liberty of Contract

That "dogma" became the foundation of Fuller Court jurisprudence. Justice Frankfurter once said that, to judges like those on the Fuller Court, "Adam Smith was treated as though his generalizations had been imparted to him on Sinai."[37] Such judges were bound to look upon the law as though it were intended to suppy a legal sanction to those generalizations. To them, regulatory legislation presented itself as a clear infringement upon the economic laws posited by Adam Smith and Herbert Spencer, and the progressive evolution of the society which was supposed to be based upon them:[38] "[A]ny legislative encroachment upon the existing economic order [was] infected with unconstitutionality."[39]

The Fuller Court jurisprudence, Justice Frankfurter tells us, was ulti-

mately based upon "misapplication of the notions of the classic econo-
mists. . . . The result . . . was that economic views of confined validity
were treated by lawyers and judges as though the Framers had enshrined
them in the Constitution."[40]

To the Justices of the day as to most of their contemporaries, the law
existed above all to protect freedom of contract. Whatever the law might
do in other respects, it might not limit contractual capacity,[41] because that
capacity was an essential element of what a Peckham opinion termed "the
faculties with which [man] has been endowed by his Creator."[42] In
another opinion, Justice Peckham referred to "the general rule of absolute
liberty of the individual to contract" and urged that "no further violation"
of that rule "should be sustained by this court."[43] If a 1900 American Bar
Association paper could proclaim "there is . . . complete freedom of contract; competition is now universal, and as merciless as nature and natural
selection,"[44] that was true largely because of the Fuller Court opinions in
the matter.

It was Justice Peckham who spoke for the majority in the important
freedom-of-contract cases decided by the Fuller Court. In particular, he
wrote the landmark opinion in *Allgeyer v. Louisiana* (1897),[45] where freedom of contract was established as an essential element of the "liberty"
protected by due process. At issue in *Allgeyer* was a state law that prohibited an individual from contracting with an out-of-state insurance company for insurance of property within the state. Such a law was ruled
violative of the Due Process Clause, which protects liberty to contract:
"Has not a citizen of a State, under the provisions of the Federal Constitution above mentioned, a right to contract outside of the State for insurance
on his property—a right of which state legislation cannot deprive him?"[46]

The key passage of the Peckham opinion gave a broad construction to
the "liberty" protected by the Fourteenth Amendment. "The liberty mentioned in that amendment," Peckham wrote, "means not only the right of
the citizen to be free from the mere physical restraint of his person, as by
incarceration, but the term is deemed to embrace the right of the citizen to
be free in the enjoyment of all his faculties; to be free to use them in all
lawful ways; to live and work where he will; to earn his livelihood by any
lawful calling; to pursue any livelihood or avocation, and for that purpose
to enter into all contracts which may be proper, necessary and essential to
his carrying out to a successful conclusion the purposes above mentioned."[47]

Yet *Allgeyer* did more than enshrine liberty of contract in the constitutional pantheon. It made due process dominant as the doctrine virtually
immunizing economic activity from regulation deemed contrary to the
laissez-faire philosophy of the day. Justice Peckham once said that to uphold
government regulation of the *Granger Cases* or *Allgeyer* type "is to take a
long step backwards and to favor that class of paternal legislation,
which . . . interferes with the proper liberty of the citizen."[48] Economic
abuses should be dealt with not by a law which "will not, as seems to me

plain, even achieve the purposes of its authors," but by "the general laws of trade [and] the law of supply and demand."[49]

To the Fuller Court, a law which interfered with the free operation of the market was a violation of substantive due process. In this respect, *Allgeyer* marked the culmination of the trend away from the *Slaughter-House Cases*,[50] which marked Chief Justice Fuller's tenure. Indeed, much of the substance of Fuller Court jurisprudence involved the elevation of the dissents in *Slaughter-House* into the law of the land. The development starts with cases involving railroad regulation. In the *Granger Cases*,[51] we saw, the Court followed the strict *Slaughter-House* approach, ruling that the Due Process Clause did not subject the legislative judgment in fixing rates to judicial review: "For protection against abuses by Legislatures, the people must resort to the polls, not to the courts."[52]

The restrictive approach to due process was, however, abandoned during the Fuller Chief Justiceship. The Court soon held that the power to regulate was not the power to confiscate;[53] whether rates fixed were unreasonable "is eminently a question for judicial investigation, requiring due process of law."[54] The rule laid down was that the Due Process Clause permits the courts to review the substance of rate-fixing legislation—at least to determine whether particular rates are so low as to be confiscatory.[55]

During this period, the state courts were moving even faster in developing substantive due process. Even before Chief Justice Fuller took his seat, they had used substantive due process to strike down regulatory laws on the ground that such a law "arbitrarily deprives him of his property and some portion of his personal liberty."[56] In an 1885 decision, the New York court declared that the "liberty" protected by due process meant one's right to live and work where and how he will; laws that limit his choice or place of work "are infringements upon the fundamental rights of liberty, which are under constitutional protection."[57]

These state decisions directly anticipated the Supreme Court's adoption of the substantive due process concept: "[A]ll that happened," writes a federal judge who came to the Bar at the time, "was that the Supreme Court joined hands with most of the appellate tribunals of the older states."[58]

The joining of hands occurred in the *Allgeyer* case, where, Justice Frankfurter tells us, "Mr. Justice Field's [*Slaughter-House*] dissent in effect established itself as the prevailing opinion of the Supreme Court."[59] In *Allgeyer*, for the first time, a state law was set aside on the ground that it violated substantive due process. The "liberty" referred to in the Due Process Clause, said Justice Peckham's opinion, embraces property rights, including that to pursue any lawful calling. A state law that takes from its citizens the right to contract outside the state for insurance on their property deprives them of their "liberty" without due process.

Between the dictum of the *Granger Cases,* that for protection against legislative abuses "the people must resort to the polls, not to the courts," and *Allgeyer v. Louisiana* and its progeny lies the history of the emergence of modern large-scale industry, of the consequent public efforts at control of business, and of judicial review of such regulation.[60] Thenceforth, all governmental action—whether federal or state—would have to run the gantlet of substantive due process; the substantive as well as the procedural aspect of such action would be subject to the scrutiny of the highest Court: "[T]he legislatures had not only domestic censors, but another far away in Washington, to pass on their handiwork."[61]

The Court's utilization of substantive due process was not mere control of state legislation in the abstract. Court control was directed to a particular purpose, namely, the invalidation of state legislation that conflicted with the doctrine of laissez faire which dominated thinking at the turn of the century. What Justice Frankfurter termed "the shibboleths of a pre-machine age . . . were reflected in juridical assumptions that survived the facts on which they were based. . . . Basic human rights expressed by the constitutional conception of 'liberty' were equated with theories of *laissez-faire*."[62] The result was that due process became the rallying point for judicial resistance to the efforts of the states to control the excesses and relieve the oppressions of the rising industrial economy.[63]

In the Fuller Court jurisprudence, the "liberty" protected by due process became synonymous with governmental hands-off in the field of private economic relations. "For years," Justice William O. Douglas tells us, "the Court struck down social legislation when a particular law did not fit the notions of a majority of Justices as to legislation appropriate for a free enterprise system."[64]

Substantive due process now became the businessman's first line of defense. Behind it, corporate power could operate free from legal interference. In the Fuller Court, the negative conception of law reached its judicial climax. The Court now saw its task as one not of further innovation but of stabilization and formalization. The law itself had become the great bulwark against economic and social change.

Shackling Federal Power

What the Fuller Court's interpretation of due process did to state regulatory power, its interpretation of the Commerce Clause did to federal regulatory power. Toward the end of the century, John Marshall's broad conception of the commerce power[65] gave way to a much more restricted view. To Marshall, the reach of the Commerce Clause extended to all commerce that concerned more states than one. Now this concept of "effect upon commerce" gave way to emphasis upon transportation across state lines. While commerce remained within the confines of a state, it did not come within federal regulatory power even though it had impacts which radiated beyond the state's borders.

This narrow conception of commerce subject to the Commerce Clause was adopted in *Kidd v. Pearson*,[66] decided a few months after Chief Justice

Fuller took office in 1888. At issue was a state statute which prohibited the manufacture of intoxicating liquors, even though they were made for sale outside the state. The law was upheld on the ground that it involved the regulation, not of commerce, but of manufacturing. The power to regulate commerce, according to the decision, does not include any authority over manufacturing, even though the products are the subject of commercial transactions in the future. To hold otherwise, the Court asserted, would be to vest in Congess control over all productive industries.

In *Kidd v. Pearson*, the challenged power was that of a state. What was said there about congressional authority was thus technically only obiter. But it soon became the basis of decision in *United States v. E. C. Knight Co.* (1895),[67] more popularly styled the *Sugar Trust Case*. That case arose out of the first important prosecution brought by the Government under the Sherman Anti-Trust Act. Defendant company had obtained a monopoly over the manufacture of refined sugar. The complaint charged that defendant had violated the Sherman Act by its acquisition of the stock of competing sugar refining companies. The Court, in an opinion by the Chief Justice, held that such an acquisition could not be reached by the federal commerce power. The monopolistic acts alleged related only to manufacturing, which was not within the scope of the Commerce Clause.

The *Knight* decision, like that in *Kidd v. Pearson* upon which it was based, was a logical consequence of the Court's changed approach to commerce. Once the Marshall conception of commerce as an organic whole gave way to the crossing of state lines as the criterion of congressional power, *Kidd-Knight* followed naturally. The result was the artificial and mechanical separation of "manufacturing" from "commerce," without regard to their economic continuity or the effects of the former upon the latter. Manufacture was treated as a "purely local" activity and hence beyond the compass of the Commerce Clause.

The motivating consideration that led the Court to this restricted conception of commerce was stated by Chief Justice Fuller in his *Knight* opinion: "Slight reflection will show that if the national power extends to all contracts and combinations in manufacture, agriculture, mining, and other productive industries, whose ultimate result may effect external commerce, comparatively little of busines operations and affairs would be left for state control."[68]

But the converse of this assertion was also true. If the commerce power did not extend to manufacture, agriculture, mining, and other productive industries, comparatively little of business operations and affairs in this country would really be subject to federal control.

Almost needless to say, the Fuller Court's restricted conception of the commerce subject to federal regulatory power fitted in perfectly with the laissez-faire theory of governmental function that dominated political and economic thinking at the time. To bar federal intervention, as the Court did in these cases, was all but to exclude the possibility of any effective regulation in them. This was, of course, exactly what was demanded by the advocates of laissez faire; to them, the economic system could function properly only if it was permitted to operate free from governmental interference.

The Constitution, states Justice Holmes in a celebrated passage, "is not intended to embody a particular economic theory, whether of paternalism and the organic relation of the citizen to the state or of laissez faire."[69] At the same time, it was most difficult for judges not to assume that the organic document was intended to embody the dominant economic beliefs of their own day. The Constitution may not, under the famous Holmes phrase, enact Mr. Herbert Spencer's *Social Statics*.[70] But the Court's narrow notion of commerce was a necessary complement to the translation of Spencerean economics into the keystone of the American polity.

Income Tax Case

Chief Justice Fuller's most famous and most criticized opinion[71] was that in the 1895 case of *Pollock v. Farmer's Loan & Trust Co.*[72]—usually known as the *Income Tax Case*—in which the Court ruled the Income Tax Act of 1894 unconstitutional. This was the case which demonstrated dramatically how the Court would use the Constitution to protect private property from governmental infringements.

The issue in the *Income Tax Case* was relatively simple: Was the income tax a "direct tax" which, under article I, section 2, "[s]hall be apportioned among the several States . . . according to their respective numbers"? If so, the 1894 statute was invalid, since it was levied on all persons who had incomes of over $4,000, with no provision for any apportionment. Both counsel and the Justices, however, did not limit themselves to the relatively uncomplicated issue before them. Instead, the case was used as the vehicle for a broadside attack upon governmental interferences with private property.

The attack began when former Senator Edmunds warned the court that such a tax "imposed by those who pay nothing, upon a very small minority" would lead to "communism, anarchy, and then, the ever following despotism."[73] Then Joseph Choate, considered the leading advocate of the day, declared that the challenged tax "is communistic in its purposes and tendencies, and is defended here upon principles as communistic, socialistic—what shall I call them—populistic as ever have been addressed to any political assembly in the world."[74]

Addressing the Justices directly, Choate warned, "If you approve this law . . . and this communistic march goes on," the law would thenceforth be helpless: "There is protection now or never." The Court "will have nothing to say about it if it now lets go its hold upon this law."[75]

To uphold the tax, Choate argued, would be "to enunciate . . . a doctrine worthy of a Jacobin Club . . . of a Czar of Russia," the "new doctrine of this army of 60,000,000—this triumphant and tyrannical majority—who want to punish men who are rich and confiscate their property."[76]

The Court rendered two decisions in the *Income Tax Case*. In the first, it held that the tax on income from land was invalid, but it was evenly divided (four-to-four) on the tax on other income. After reargument, the Court, in a five-to-four decision, decided that the entire income tax law was invalid. Chief Justice Fuller's opinions were largely technical in seeking to show that the income tax was a "direct tax" within the meaning of the Constitution. The constitutional provision, according to Fuller, was "put in antithesis to . . . duties, imposts, and excises,"[77] which were to be the principal source of federal revenue. Reliance was placed upon a work by Albert Gallatin to show that direct taxes were "those which are raised on the capital or revenue of the people; . . . indirect such as are raised on their expense."[78] The income tax was a tax on the capital or revenue of the people rather than on their expense. Hence it was a direct tax and invalid because not apportioned among the states in proportion to their population, as required by the Constitution.

The decisions in the *Income Tax Case* were severely criticized because they appeared contrary to earlier decisions which the Chief Justice failed to distinguish adequately. In addition, it was widely reported that the second decision resulted from a switched vote by one of the Justices, who had been in favor of sustaining the tax on nonrent income at the time of the first decision, to the antitax side, which gave the latter its bare majority. At the time it was believed that it was Justice Shiras who had changed his mind, and he was widely blamed for wrecking the income tax. More recent commentators have questioned whether it was Justice Shiras who had switched, or even whether there was any switch at all.[79] At any rate, if there was what has been termed "The Mystery of the Vacillating Jurist,"[80] it remains to this day essentially unsolved.

Instead, the *Income Tax Case* remains as the Fuller Court's response to warnings such as those raised in the Choate argument. Justice Harlan delivered an impassioned dissent rejecting what he called the argument urging "this court . . . to stand in the breach for the protection of the just rights of property against the advancing hosts of socialism." Instead, Harlan declared that the decisions were "a disaster to the country" because their "interpretation of the Constitution . . . impairs and cripples the just powers of the National Government." The Court's action, Harlan asserted, invests property owners "with power and influence that may be perilous to that portion of the American people upon whom rests the large part of the burdens of the government, and who ought not to be subjected to the dominion of aggregated wealth."[81]

Justice Holmes called Justice Harlan "the last of the tobacco-spittin' judges."[82] As Harlan delivered his *Income Tax* dissent, his face grew visibly redder, and his fist banged the bench as he glared defiantly at the Chief Justice. *The Nation* called Harlan's dissent "the most violent political tirade ever heard in a court of last resort."[83]

But the fervent rhetoric was not all on one side. Choate's ardent oratory found a receptive ear in the *Income Tax* majority. This may be seen from the concurring opinion of Justice Field, whose due process philosophy was now to be dominant in the Fuller Court. "The present assault upon capital," declared Justice Field, "is but the beginning. It will be but the stepping-stone to others, larger and more sweeping, till our political contests will become a war of the poor against the rich; a war constantly growing in intensity and bitterness." If the Court were to sanction the income tax law, "it will mark the hour when the sure decadence of our present government will commence."[84]

The judges who felt this way about a tax of 2 percent on annual incomes above $4,000 now had at their disposal the newly fashioned tool of substantive due process. How they would use it became clear in the *Lochner* case, to be discussed in Chapter 8.

Insular Cases

The so-called *Insular Cases*,[85] decided in 1901 by the Fuller Court, are remembered today largely because of Mr. Dooley's famous aphorism that the Supreme Court follows the election returns.[86] All but forgotten is the fact that the *Insular Cases* were great cases at the time which decided one of the crucial constitutional issues of the day. That issue was also well stated in Mr. Dooley's summary of the Court's holding: "I see," said Mr. Dooley, "Th' supreme court has decided th' constitution don't follow th' flag."[87]

The question of whether the Constitution follows the flag was an issue of the relationship between the organic instrument and conquered territory. The problem was posed in acute form by the war with Spain and the territorial acquisitions that were its outcome. Soon after the war ended, the question of whether government in the new territories was subject to those constitutional limitations which apply in the continental United States came before the Supreme Court in the series of 1901 decisions that have come to be known as the *Insular Cases*.

They arose out of the 1900 statute providing a government for conquered Puerto Rico. Among that law's provisions were sections requiring customs duties to be paid upon goods imported into the United States from the island. It was contended that such a provision was invalid, since Puerto Rico had become a part of the United States within the constitutional requirement that "all Duties, Imposts and Excises shall be uniform throughout the United States." This, in turn, said the Court, posed the broader question of whether the provisions "of the Constitution extend of their own force to our new acquired territories."[88]

According to the *Insular Cases*, whether the Constitution follows the flag depends, in the particular case, upon the type of territory that is involved. The Court drew a distinction between "incorporated" and "unincorporated" territories. The former are those territories which Congress has incorporated into and made an integral part of the United States.

Without express provision by Congress, territory acquired by the nation remains unincorporated. Applying the distinction between incorporated and unincorporated territories to Puerto Rico, the Court found that that island belonged to the latter category: "[W]hilst in an international sense Porto Rico was not a foreign country, since it was subject to the sovereignty of and was owned by the United States, it was foreign to the United States in a domestic sense, because the island had not been incorporated into the United States, but was merely appurtenant thereto as a possession."[89]

Justice Frankfurter tells us that the "over two hundred pages of opinions . . . were illuminatingly summarized by that great philosopher, Mr. Dooley, when he said that so far as he could make out, 'the Supreme Court decided that the Constitoosh'n follows the flag on Mondays, Wednesdays, and Fridays.'"[90] In other words, the Constitution sometimes follows the flag and sometimes does not. Whether it does depends on whether the territory concerned is incorporated or unincorporated. In the former, all the constitutional rights and privileges must be accorded. But the same is not true in unincorporated territory like Puerto Rico. In them, a constitutional provision like that governing duties and imposts does not restrict governmental authority.

Chief Justice Fuller, joined by Justices Harlan, Brewer, and Peckham, delivered what his biographer calls a "calm and restrained" dissent.[91] Justice Harlan also delivered his more typical impassioned dissent, in which he urged, "The idea that this country may acquire territores anywhere upon the earth, by conquest or treaty, and hold them as mere colonies or provinces—the people inhabiting them to enjoy only such rights as Congress chooses to accord to them—is wholly inconsistent with the spirit and genius as well as with the words of the Constitution."[92]

Three years after the *Insular Cases*, its holding was applied in *Dorr v. United States*.[93] The question there was whether trial by jury became a necessary incident of justice in the Philippine Islands when they became American territory. The Supreme Court held that it did not. Since the Philippines had not been incorporated into the United States, the jury-trial requirement of the Sixth Amendment was not a limitation upon the power to provide a government for that territory.

Mr. Dooley may have satirized the Court that decided the *Insular Cases*. But that holding has remained the basic principle in dealing with the relationship of the Constitution to overseas territories. It was to be of particular pertinence in a later age when the United States was to become a leader of the international community. What the Supreme Court said in those cases on the government provided in territories acquired as a result of the Spanish War was to apply as well to the military governments set up in conquered territory almost half a century later.

At the same time, there is an anomaly in the holding that our Government may act in violation of constitutional guarantees. Justice Harlan indicated how troubling such a result was in a letter he sent to the Chief

Justice. "The more I think of these questions," Harlan wrote, "the more alarmed I am at the effect upon our institutions of the doctrine that this country may acquire territory inhabited by human beings anywhere upon the earth and govern it at the will of Congress and without regard to the restraints imposed by the Constitution upon governmental authority. There is a danger that commercialism will sweep away the safeguards of real freedom."[94]

Separate but Equal

Chief Justice Fuller's biography does not even mention the 1896 case of *Plessy v. Ferguson*.[95] Yet the decision there is now one of the most criticized decisions of the Fuller Court. It was the seminal decision which, for more than half a century, made equal protection no more than a hortatory slogan for African-Americans. While the Fuller Court developed the Fourteenth Amendment's Due Process Clause as the principal safeguard of property rights, its *Plessy* decision ensured that the amendment was of little value to the blacks for whose benefit it had primarily been adopted.

Homer Plessy, a Louisiana resident, was one-eighth black. In 1892, while riding on a train out of New Orleans, he was ejected by the conductor from a car for whites and directed to a coach assigned to nonwhites. The conductor acted under a Louisiana statute that provided for "equal but separate [railroad] accommodations for the white and colored races."

Plessy claimed that the statute was contrary to the Fourteenth Amendment's requirement of equal protection of the laws and took his case to court. As recently summarized by the Court, *Plessy* held that "racial segregation . . . works no denial of equal protection, rejecting the argument that racial separation enforced by the legal machinery of American society treats the black race as inferior."[96] Instead, the *Plessy* Court stated, "Laws permitting, and even requiring their separation in places where they are liable to be brought into contact do not necessarily imply the inferiority of either race to the other."[97]

The *Plessy* decision gave rise to what has been termed one of the most vigorous dissents in Supreme Court history[98]—certainly the greatest delivered by Justice Harlan. The dissent attacked the very basis of the Court's decision, that is, that segregation alone was not discriminatory. To anyone familiar with the techniques of racial discrimination, the Court's view is completely out of line with reality. The device of holding a group of people separate—whether by confinement of Jews to the Ghetto, by exclusion of untouchables from the temple, or by segregation of the black—is a basic tool of discrimination. "The thin disguise of 'equal' accommodations for passengers in railroad coaches," movingly declared the Harlan dissent, "will not mislead anyone, nor atone for the wrong this day done."[99]

Plessy v. Ferguson gave the lie to the American ideal, so eloquently stated in the Harlan dissent: "Our Constitution is color-blind, and neither knows nor tolerates classes among citizens."[100] Upon the "separate but equal" doctrine approved by the Court was built a whole structure of racial discrimination. Jim Crow replaced equal protection, and legally enforced segregation became the dominant fact in Southern life.

To the present-day observer, the merits of the case are all with Justice Harlan's dissent. All the same, perhaps, one should not be too harsh in judging the Fuller Court for a decision that mirrored its own time and place. Living in an era that has witnessed a virtual egalitarian revolution, we find it all too easy to censure the Fuller Court for its reliance upon doctrine that we now deem outmoded. But that Court—a reflection of the less tolerant society in which it sat—could hardly hope to lift itself (by its own bootstraps as it were) above the ingrained prejudices of its day.

In addition, we should recognize that, inadequate though the *Plessy* decision may have been in its failure to implement the guaranty of equality, it was the language in *Plessy* requiring equality of treatment for the separate races that was to prove decisive in the movement for legal equality in the racial field. Indeed, it was to be precisely the requirement of equality of treatment articulated in the *Plessy* opinion that was half a century later to provide the opening wedge for the ultimate overruling of the *Plessy* holding itself.

8

Watershed Cases:
Lochner v. New York, 1905

Aside from *Dred Scott* itself, *Lochner v. New York*[1] is now considered the most discredited decision in Supreme Court history. When commentators discuss the case at all, they use it as a vehicle to illustrate the drastic change in jurisprudence during the twentieth century, which has seen the Holmes dissent in *Lochner* elevated to established doctrine.

Yankee from Olympus

The paradigmatic early twentieth-century conflict over constitutional construction was that in the *Lochner* case between Justice Rufus W. Peckham and Justice Oliver Wendell Holmes. The appointment of Holmes was a capital event in the history of the Supreme Court. Our discussion of John Marshall quoted Holmes: "[I]f American law were to be represented by a single figure, skeptic and worshipper alike would agree without dispute that the figure could be one alone, and that one, John Marshall."[2] If American law were to be represented by a second figure, most jurists would say that it should be Holmes himself. For it was Holmes, more than any other legal thinker, who set the agenda for modern Supreme Court jurisprudence. In doing so, he became as much a part of American legend as law: the Yankee from Olympus[3]—the patrician from Boston who made his mark on his own age and on ages still unborn as few have done. To summarize Holmes's work is to trace the development from nineteenth-century law to that of the present day. The younger Holmes came from what his father, the famous New England writer, termed "the Brahmin caste of New England"—the "untitled aristocracy" of early America.[4] After graduation from Harvard and Civil War service in the Union Army, Holmes decided to study law. He told his father that he was going to Harvard Law School, and Dr. Holmes is said to have asked, "What is the use of that? A lawyer can't be a great man."[5] Holmes's career showed how mistaken his father was.

After his graduation from law school, Holmes was admitted to the Bar, joined a law firm, and became a part-time lecturer at Harvard. He wrote articles for legal periodicals and edited the twelfth edition of Kent's *Commentaries*. Then, in 1880, came the invitation to deliver a series of lectures. He chose as his topic *The Common Law* and the lectures were published in a book of that name in 1881. This book was to change both Holmes's life and the course of American law.

As a state judge tells us, "The book propounds an idea audacious and even revolutionary for the time."[6] The Holmes theme has become so settled in our thinking that we forget how radical it was when it was announced over a century ago. The very words used must have appeared strange to the contemporary reader: "experience," "expediency," "necessity," "life." Law books at the time used far different words: "rule," "precedent," "logic," "syllogism."[7] As Holmes's biographer tells us, "The time-honored way was to deduce the *corpus* from *a priori* postulates, fit part to part in beautiful, neat logical cohesion."[8] Holmes rejected "the notion that a given [legal] system, ours, for instance, can be worked out like mathematics."[9] Instead, he declared, "The law embodies the story of a nation's development through many centuries, and it cannot be dealt with as if it contained the axioms and corollaries of a book of mathematics."[10]

But the great Holmes theme was stated at the very outset of *The Common Law*: "The life of the law has not been logic: it has been experience. The felt necessities of the time, the prevalent moral and political theories, intuitions of public policy, avowed or unconscious, even the prejudices which judges share with their fellow-men, have had a good deal more to do than the syllogism in determining the rules by which men should be governed."[11]

When Holmes wrote these words, he was pointing the way to a new era of jurisprudence that would, in former Attorney General Francis Biddle's words, "break down the walls of formalism and empty traditionalism which had grown up around the inner life of the law in America."[12] The courts, Holmes urged, should recognize that they must perform a legislative function, in its deeper sense. The secret root from which the law draws its life is consideration of "what is expedient for the community." The "felt necessities of the time," intuitions of what best serve the public interest, "even the prejudices which judges share with their fellow-men"—all have much more to do than logic in determining the legal rules that govern the society.

The Holmes lectures purported to be only a descriptive statement of what the law was and how, historically, it came to be that way. In fact, however, Holmes was making a prescriptive statement of what the law ought to be—a statement that was ultimately to set the theme for the jurisprudence of the coming century. The success of *The Common Law* led to a Harvard law professorship in 1882. But Holmes taught there only a term, for he was appointed in December 1882 to the Supreme Judicial Court of Massachusetts. He served on that tribunal for twenty years (from 1899 as Chief Justice), when he was elevated to the U.S. Supreme Court. Though he was already sixty-one when he took his seat on that Court, he still had his greatest judicial years to serve. He did not leave the Supreme Court until his retirement in January 1932. During the thirty years he spent in Washington, he made the greatest contribution since Marshall to American constitutional law.

Peckham Precursor

Holmes's great antagonist in the *Lochner* case was Justice Peckham, who had already been discussed as the exemplar of the conservative Justice at the turn of the century.[13] To us today, Peckham's *Lochner* opinion of the Court is a prime example of the reactionary jurisprudence of the day. We should, however, realize that Peckham himself saw his opinion quite differently. To him it was the culmination of the progressive trend away from the paternalistic theory of the state that was as "odious" to him as it was in Justice Brewer's already quoted statement.[14]

Justice Peckham best explained his posture in this respect in *People v. Budd,*[15] decided in 1889, while Peckham was still on the highest New York court. At issue was a state law fixing maximum rates for grain elevators. The New York court followed the *Granger Cases*[16] and upheld the law. Judge Peckham delivered a vigorous dissent, which exemplifies the conception of law that he was soon to elevate to Supreme Court doctrine.

In his dissent, Peckham flatly rejected the businesses-affected-with-a-public-interest doctrine adopted in the *Granger Cases,* which had been established as the common-law rule by Lord Hale in the seventeenth century. Lord Hale, Peckham stressed, had written "when views of governmental interference with the private concerns of individuals were carried to the greatest extent." Indeed, "in those days the theory of a paternal government . . . was to watch over and protect the individual at every moment, to dictate the quality of his food and the character of his clothes, his hours of labor, the amount of his wages, his attendance upon church, and generally to care for him in his private life."[17]

Two centuries later, Peckham went on, a different view prevailed "as to how far it is proper to interfere in the general industrial department of the country." There was now "no reason [to] go back to the seventeenth or eighteenth century ideas of paternal government." "State interference in matters of private concern" was no longer considered proper, since it was contrary to "the later and, as I firmly believe, the more correct ideas which an increase of civilization and a fuller knowledge of the fundamental laws of political economy, and a truer conception of the proper functions of government have given us at the present day."[18]

The economic orthodoxy of the day told Judge Peckham that such legislative attempts "to interfere with what seems to me the most sacred rights of property and the individual liberty of contract" were bound to be ineffective. Such a law "will result either in its evasion or else the work will not be done, and the capital employed will seek other channels where such [unregulated] rate can be realized, or the property will become of little or no value." In the latter case, the law "may ruin or very greatly impair the value of the property of wholly innocent persons." To Peckham, it was clear that such a law was "wholly useless for any good effect, and only powerful for evil."[19] Instead, the New York law was an unjustified interference with "the law of supply and demand."

Illustrating the Holmes comment quoted in the last chapter on the fear of socialism and judicial action, Peckham warned that "[t]o uphold legislation of this character is to provide the most frequent opportunity for arraying class against class." The challenged law was not only, "in my belief, wholly inefficient to permanently obtain the result aimed at," but it was "vicious in its nature, communistic in its tendency."[20]

The Peckham jurisprudence was, of course, to attain its apogee in the *Lochner* case. However, "The Peckham of *Lochner v. New York* surely needs no introduction after *People v. Budd*."[21] The Peckham concept of law was fully developed in his *Budd* dissent. His important Supreme Court opinions, particularly that in *Lochner*, were foreshadowed by what he wrote in *Budd*.

Facts and Opinion

Joseph Lochner had been convicted for violating a New York law by requiring a worker in his bakery to work more than sixty hours in one week. The statute prohibited bakery employees from working more than ten hours a day, or sixty hours a week. The law was challenged as a violation of due process: "The Statute in Question is Not a Reasonable Exercise of the Police Power."[22]

The case gave the Supreme Court great difficulty. The Justices first voted by a bare majority to uphold the law. The case was assigned to Justice Harlan, who wrote a draft opinion of the Court. Justice Peckham wrote a strong draft dissent. Before the case came down, however, there was a vote switch. The Peckham dissent became the opinion of the Court and the Harlan opinion a dissent.[23] It is not known who changed his vote, though the probability is that it was Chief Justice Fuller. Though Fuller had voted to uphold other maximum-hour laws,[24] his biographer tells us that "the ten-hour law for bakers seemed to him to be 'featherbedding,' paternalistic, and depriving both the worker and employer of fundamental liberties."[25]

The other cases in which laws regulating hours of labor had been sustained were treated by the Court as health measures. In *Lochner,* it has been suggested, Justice McKenna, whose father had owned a bakery, may have persuaded Fuller and others in the majority that bakery work was not dangerous and that the health rationale was a sham.[26] Justice Peckham himself needed no persuading. His *Lochner* opinion of the Court was a natural product of the judge who had elevated liberty of contract to the constitutional plane in *Allgeyer v. Louisiana,*[27] ruled against interferences with the operation of the market in his *Budd* opinion, and even voted against a maximum-hours law for miners.[28] If Justice Peckham refused to consider the law protecting miners a legitimate health measure, it was obvious that he would not accept the health justification for a similar law regulating bakery work.

For Justice Peckham, *Lochner* was essentially a reprise of *Allgeyer* and *Budd.* Here, too, the crucial factor was the violation of freedom of contract which *Allgeyer* had ruled "part of the liberty of the individual." The *Lochner* "statute necessarily interferes with the right of contract between the employer and employees, concerning the number of hours in which the latter may labor in the bakery of the employer."[29]

In this case, the state has clearly limited "the right of the individual to labor for such time as he may choose." It then becomes "a question of which of two powers or rights shall prevail—the power of the State to legislate or the right of the individual to liberty of person and freedom of contract." To answer that question, the Court must answer the further Peckham query: "Is this a fair, reasonable and appropriate exercise of the police power of the State, or is it an unreasonable, unnecessary and arbitrary interference with the right of the individual to his personal liberty or to enter into those contracts in relation to labor which may seem to him appropriate or necessary for the support of himself and his family?"[30]

The Peckham opinion had no doubt about the answer. In the first place, Justice Peckham asserted, the argument that "this act is valid as a labor law, pure and simple, may be dismissed in a few words." There is no reason for treating bakers differently from other employees and "the interest of the public is not in the slightest degree affected by such an act." Hence, "There is no reasonable ground for interfering with the liberty of person or the right of free contract, by determining the course of labor, in the occupation of a baker."[31]

In consequence, said Justice Peckham, if the law is to be upheld, it "must be . . . as a law pertaining to the health of the individual engaged in the occupation of a baker."[32] But the mere assertion that the subject relates to health is not enough. The relationship to public health must be direct enough for the Court to deem it reasonable.

The Peckham opinion found the required relationship to be lacking.

The trade of baker, in the Court's view, was not an unhealthy one, which would justify legislative interference with the right of contract. Of course, "almost all occupations more or less affect the health." Yet that alone does not mean that such occupations must be subject to any police power regulation on public health grounds. "There must be more than the mere fact of the possible existence of some small amount of unhealthiness to warrant legislative interference with liberty." Labor, in and of itself, may carry with it the seeds of unhealthiness. "But are we all, on that account, at the mercy of legislative majorities?"[33]

The Peckham conclusion is that the *Lochner* law "is not, within any fair meaning of the term, a health law, but is an illegal interference with the rights of individuals, both employers and employees, to make contracts regarding labor upon such terms as they may think best." Its real purpose was, not to protect health, but to have the state again assume "the position of a supervisor, or *pater familias,* over every act of the individual."[34]

To Peckham, the *Lochner* law was a reversion to the paternal role of government which he had condemned in his *Budd* opinion. "Statutes of the nature of that under review," he declared, "limiting the hours in which grown and intelligent men may labor to earn their living, are mere meddlesome interferences with the rights of the individual."[35]

Unfortunately, "This interference on the part of the legislatures of the several States with the ordinary trades and occupations of the people seems to be on the increase." Under the Peckham jurisprudence, however, these interferences cannot pass muster. Such attempts "simply to regulate the hours of labor between the master and his employees (all being men, *sui juris*), in a private business" must give way before "the right of free contract and the right to purchase and sell labor upon such terms as the parties may agree to." Hence the *Lochner* law must fall: "[T]he freedom of master and employee to contract with each other in relation to their employment, and in defining the same, cannot be prohibited or interfered with, without violating the Federal Constitution."[36]

Holmes Dissent

Aside from its use as the horrible example of what we now consider the wrong kind of judicial activism, *Lochner* is remembered today for its now-classic dissent by Justice Holmes, celebrated for its oft-quoted aphorisms. Indeed, the Holmes *Lochner* opinion is probably the most famous dissent ever written. "There is a famous passage," Justice Cardozo tells us, "where Matthew Arnold tells us how to separate the gold from the alloy in the coinage of the poets by the test of a few lines which we are to carry in our thoughts."[37] The flashing epigrams[38] in Holmes's *Lochner* dissent do the like for those who would apply the same test to law.

The *Lochner* dissent contains one of Justice Holmes's most famous statements: "General propositions do not decide concrete cases." Yet the Holmes dissent is based more upon general propositions than upon con-

crete rules or precedents. Indeed, Holmes begins with a broad proposition: "This case is decided upon an economic theory which a large part of
the country does not entertain."[39] Holmes neither explains nor elaborates
the charge.[40] Instead, he goes on to point out that the decision on economic grounds is not consistent with his conception of the judicial function. "If it were a question whether I agreed with that theory, I should
desire to study it further and long before making up my mind. But I do not
conceive that to be my duty, because I strongly believe that my agreement
or disagreement has nothing to do with the right of a majority to embody
their opinions in law."[41]

The dissent then strikes directly at the dominant conception, which
equated the law with laissez faire. That conception is stated by Holmes as a
paraphrase of Herbert Spencer's first principle:[42] "The liberty of the citizen
to do as he likes so long as he does not interfere with the liberty of others to
do the same." That may have been "a shibboleth for some well-known
writers," but it "is interfered with by school laws, by the Post Office, by
every state or municipal institution which takes his money . . . whether
he likes it or not." Indeed, it is settled that "laws may regulate life in many
ways which we as legislators might think as injudicious . . . as this, and
which equally with this interfere with the liberty to contract."[43]

This leads Holmes to his best-known aphorism: "The Fourteenth
Amendment does not enact Mr. Herbert Spencer's Social Statics."[44] This
"general proposition" is supported by the "decisions cutting down the
liberty to contract." Cited without discussion are the cases upholding a
maximum-hours law for miners and prohibiting sales of stock on margin,
as well as compulsory vaccination laws.[45] The lack of discussion is explained by the Holmes conviction that it is irrelevant whether the judges
share the "convictions or prejudices" embodied in these laws:

> [A] constitution is not intended to embody a particular economic theory,
> whether of paternalism and the organic relation of the citizen to the State or of
> *laissez faire*. It is made for people of fundamentally differing views, and the
> accident of our finding certain opinions natural and familiar or novel and even
> shocking ought not to conclude our judgment upon the question whether stat-
> utes embodying them conflict with the Constitution of the United States.[46]

This proposition, Holmes indicates, may be an exception to his warning against "general propositions." As such, it supports the general
Holmes approach to judicial review. "I think that the word liberty in the
Fourteenth Amendment is perverted when it is held to prevent the natural
outcome of a dominant opinion, unless it can be said that a rational and fair
man necessarily would admit that the statute proposed would infringe
fundamental principles as they have been understood by the traditions of
our people and our law."[47]

A law such as that at issue is not be be invalidated unless it fails to meet
this standard. "No such sweeping condemnation can be passed upon" the
Lochner law. On the contrary, "A reasonable man might think it a proper

measure on the score of health." That is all that is necessary for the conclusion that the law should be sustained. "Men whom I certainly could not pronounce unreasonable would uphold it."[48]

Judge Richard Posner asserts that the Holmes *Lochner* dissent would not have received a high grade in a law school examination: "It is not logically organized, does not join issue sharply with the majority, is not scrupulous in its treatment of the majority opinion or of precedent, is not thoroughly researched, does not exploit the factual record."[49] Certainly, the Holmes opinion is utterly unlike the present-day judicial product—all too often the work of law clerks for whom the acme of literary style is the law review article. The standard opinion style has become that of the reviews: colorless, prolix, platitudinous, always erring on the side of inclusion, full of lengthy citations and footnotes—and above all dull.[50]

The Holmes dissent is, of course, anything but dull. In fact, as Posner sums it up, it may not be "a *good* judicial opinion. It is merely the greatest judicial opinion of the last hundred years. To judge it by [the usual] standards is to miss the point. It is a rhetorical masterpiece."[51]

Scope of Review

There are indications that Justices Holmes and Peckham did not have high opinions of each other. Reference has been made to Holmes's deprecating comment when asked what Peckham was like intellectually.[52] On his side, Peckham was once asked by Chief Justice Fuller whether he was "willing to part with" the opinion in a case[53] to Holmes. Peckham answered, "I will part with it in spite of _____ _____ as Brother Harlan would say!" Harlan's well-known vituperation usually had a subject, and that of Peckham's blanks was undoubtedly Holmes.[54]

The difference between the two Justices in *Lochner* was, however, based on more than their possible personal antagonism. The Peckham and Holmes *Lochner* opinions represent two opposed conceptions of jurisprudence—the one that of the late nineteenth century, the other that of the coming legal era. In this respect there were two essential differences between the two antithetic approaches: (1) on the proper scope of judicial review; and (2) on the reliance upon economic theory by the reviewing court.

According to Justice Peckham, the question to be determined in cases involving challenges to *Lochner*-like legislation is: "Is this . . . fair, reasonable and appropriate . . . or is it an unreasonable, unnecessary and arbitrary interference with the right of the individual?"[55] Judge Posner asserts that the Holmes dissent does not really "take issue with the fundamental premise of the majority opinion, which is that unreasonable statutes violate the due process clause of the Fourteenth Amendment."[56] Instead, by his conclusion that a "reasonable man might think it a proper measure,"[57] Posner asserts, "Holmes seems to concede the majority's conclusion that the due process clause outlaws unreasonable legislation."[58]

Though Justices Peckham and Holmes both state a test of reasonableness, there is all the difference in the world between the Peckham and Holmes manner of applying the reasonableness test. The Peckham opinion indicated that the reasonableness of the challenged statute must be determined as an objective fact by the judge upon his own independent judgment. In holding the *Lochner* law invalid, the Court in effect substituted its judgment for that of the legislator and decided for itself that the statute was not reasonably related to any of the social ends for which the police power might validly be exercised. This interpretation of reasonableness, as an objective criterion to be determined by the judge himself, permeates the *Lochner* opinion.

The *Lochner* Court, in striking down a law whose reasonableness was, at a minimum, open to debate, in effect determined upon its own judgment whether such legislation was desirable. Such an approach was utterly inconsistent with the basic Holmes doctrine, which was one of judicial restraint. Justice Holmes himself once wrote: "On the economic side, I am mighty skeptical of hours of labor . . . regulation."[59] His personal opinion about the desirability of the law was, however, irrelevant under his theory of review. The "criterion," Holmes stated in a 1923 case, "is not whether we believe the law to be for the public good," but whether "a reasonable man reasonably might have that belief."[60] If the "character or effect" of a law "be debatable, the legislature is entitled to its own judgment, and that judgment is not to be superseded . . . by the personal opinion of judges, 'upon the issue which the legislature has decided.'"[61]

Under the Peckham approach, the desirability of a statute was determined as an objective fact by the Court on its own independent judgment. For Justice Holmes, a more subjective test was appropriate: Could rational legislators have regarded the statute as a reasonable method of reaching the desired result?[62] In the words of Holmes's leading judicial disciple, "It can never be emphasized too much that one's own opinion about the wisdom or evil of a law should be excluded altogether when one is doing one's duty on the bench. The only opinion of our own even looking in that direction that is material is our opinion whether legislators could in reason have enacted such a law."[63]

Reliance on Economic Theory

The other fundamental Peckham–Holmes difference was on the reliance on economic theory by the Court in its review of the *Lochner* law. The Holmes assertion that the case was "decided upon an economic theory" was the opening salvo in the twentieth-century approach to review of regulatory action. According to Sir Frederick Pollock, what Holmes was saying here was "that it is no business of the Supreme Court of the United States to dogmatize on social or economic theories."[64] The *Lochner* Court struck down the statute as unreasonable because a majority of the Justices disagreed with the economic theory on which the state legislature had acted. This was precisely the approach to judicial review that Justice Holmes rejected. There may, in the given case, be economic arguments against a challenged regulatory law. To Holmes, however, such arguments were properly addressed to the legislature, not to the judges.

As the Court put it half a century later, after the Holmes posture had become the accepted one, it is improper for the courts "to strike down state laws, regulatory of business and industrial conditions, because they may be unwise, improvident, or out of harmony with a particular school of thought."[65] It is not for the judge to intervene because he disagrees with the economic theory upon which a law is based. "Whether the legislature takes for its textbook Adam Smith, Herbert Spencer, Lord Keynes, or some other is no concern of ours."[66]

According to Justice Peckham, however, that is exactly what should be the concern of a reviewing court. If a law is based upon what the judge considers an unsound economic theory, the judge should hold the law invalid. And there is no doubt that Peckham considered the *Lochner*-type law to be, at the least, unsound.

Lochner has become so discredited that we forget that, in his own day, Justice Peckham was considered a good judge (he had been recommended for Chief Justice in a letter to President Cleveland by Melville W. Fuller, before the latter was appointed to the position).[67] In addition Peckham was representative of the dominant legal thought at the beginning of the century. To such a jurist, *Lochner* was correct both in its rationale and result, since the statute violated the fundamental proposition upon which his jurisprudence was grounded: that of law designed to ensure the unfettered operation of the market and the freedom of contract that was its foundation.

The law to Peckham was based upon what the New York court had stressed as "the unceasing struggle for success and existence which pervades all societies of men." But the operation of the evolutionary struggle must not be interfered with by government. "Such governmental interferences disturb the normal adjustments of the social fabric, and usually derange the delicate and complicated machinery of industry and cause a score of ills while attempting the removal of one."[68]

Justice Peckham followed a similar approach. As he saw it, the "liberty" protected by the law "is deemed to embrace the right of man to be free in the enjoyment of the faculties with which he has been endowed by his Creator, subject only to such restraints as are necessary"—here, too, a paraphrase of Herbert Spencer's first principle.[69] This principle was plainly violated by the *Lochner* law.

To Justice Peckham, the regulation of bakery hours was a deviation from the evolutionary progress that underlay his conception of law. It was a reversion to the paternalism of an earlier day. "The paternal theory of government" was as odious to the *Lochner* majority as it was to Justice Brewer.[70] "The utmost possible liberty to the individual, and the fullest possible protection to him and his property, is both the limitation and duty of government."[71] To judges who adopted this philosophy, the "liberty" protected by law became synonymous with no interference with economic activities. "For years," Justice Douglas was later to explain, "the Court struck down social legislation when a particular law did not fit the notions of a majority of Justices as to legislation appropriate for a free enterprise system."[72] Any other posture, to judges like Justice Peckham, would have meant a reversion to the paternalistic theory of government that had been repudiated by "the more correct ideas [of] the present day."[73]

Rehabilitating *Lochner?*

A word remains to be said about a surprising sequel to the *Lochner* story that, not too long ago, few would have anticipated. Those who were to change the course of jurisprudence as the twentieth century went on "repeated to one another, as creed, . . . Justice Holmes's dissenting opinion in *Lochner.*"[74] By midcentury, the once-heretic creed had become accepted doctrine; *Lochner* appeared to be as repudiated as *Dred Scott.* Indeed, not long ago, Justice John Paul Stevens declared, "When the Court repudiated the line of cases that is often identified with *Lochner v. New York,* it did so in strong language that . . . seemed to foreclose forever any suggestion that the due process clause of the fourteenth amendment gave any power to federal judges to pass on the substance of the work product of state legislatures."[75]

To Stevens, as to other commentators, the post-*Lochner* jurisprudence "seemed to foreclose" any possible *Lochner* revival. And so it appeared until recently, when a number of jurists have sought to accomplish the result stated in the title of a 1985 article: "Rehabilitating Lochner."[76] They argue that even if Justice Holmes was correct in his claim that *Lochner* was decided upon the Court's own economic theory, that need not mean that the decision was wrong. On the contrary, these jurists urge, the *Lochner* Justices were doing only what judicial review requires when they invalidated the law because it was based upon what they considered an incorrect economic theory.

Perhaps the most influential writer to support the economic theory behind *Lochner* is Judge Posner. He notes that the prevailing view in recent years has been that *Lochner* and similar decisions earlier in the century "reflected a weak grasp of economics."[77] According to Posner, however, it is the economic analysis that led to the *Lochner* statute that was seriously flawed.

To Judge Posner, laws like that in *Lochner* "were attempts to suppress competition under the guise of promoting the general welfare."[78] Such attempts are all but heresy to advocates of present-day Chicago School economics, which Posner himself has done so much to translate into legal doctrine. That school has never reconciled itself to the fact that, in this century, the "invisible hand" of Adam Smith has increasingly been replaced by the "public interest" as defined in regulatory legislation and administration. To the Chicago School, the overriding goal of law, as of economics, should be that of efficiency. The law should intervene "to reprehend only that which is inefficient," and even then the law's role should be limited, since the "market punishes inefficiency faster and better than the machinery of the law."[79]

Such an approach would turn the legal clock back to *Lochner*. The primary criterion for those, like Judge Posner, who see economics as the foundation of law is efficiency and to them efficiency is best promoted by the free operation of the market. Thus they are drawn inevitably to the *Lochner* rationale—that governmental interference with the market promotes inefficiency and must normally be considered arbitrary. We are thus brought back to the law at the turn of the century, when cases like *Lochner* set the pattern for judicial reception of laws that attempted to curb the excesses and abuses of a completely unrestrained market.

Judge Posner writes that he "was the first to suggest that the discredited 'liberty of contract' doctrine could be given a solid economic foundation and as good a jurisprudential basis as the Supreme Court's aggressive modern decisions protecting civil liberties." Nevertheless, Posner denies advocating the *Lochner* approach, declaring, "I have never believed, however, that such a restoration of the '*Lochner* era' . . . would be, on balance, sound constitutional law."[80] The denial is disingenuous. The Posner approach lends direct support to the effort to take our public law back to *Lochner*. With efficiency and wealth maximization as its end and the market as the instrument through which it is achieved, Posnerian jurisprudence leads to what Posner himself terms the constitutionalization of laissez faire.[81]

Though it was uniformly thought that the verdict of history was resoundingly against *Lochner,* it thus now appears that the jury may still be out. Yet the accepted judgment on the Peckham–Holmes disagreement has surely been correct. The primary effect of the Peckham approach was to immunize the economy from interference by the machinery of law. To a judge like Peckham, *Lochner* did not involve control of legislation in the abstract. Court control was directed to a particular purpose—the invalidation of legislation that conflicted with the doctrine of laissez faire that dominated thinking at the turn of the century.[82] What Justice Frankfurter termed "the shibboleths of a pre-machine age . . . were reflected in juridical assumptions that survived the facts on which they were based. . . . Basic human rights expressed by the constitutional conception of 'liberty' were equated with theories of *laissez-faire*."[83] The result was that the law became the rallying point for judicial resistance to the efforts to control the excesses and relieve the oppressions of the rising industrial economy.[84]

Almost a century later, we tend to forget how inadequate the law was at the time of *Lochner*. To return to the Peckham conception of law is to return to a time when "it was unconstitutional to intrude upon the inalienable right of employees to make contracts containing terms unfavorable to themselves, in bargains with their employers." In those days, "[a]n ordinary worker was told, if he sought to avoid harsh contracts made with his employer . . . that he had acted with his eyes open, had only himself to blame, must stand on his own feet, must take the consequences of his own folly."[85] And if, as in *Lochner*, a law sought to equalize the situation, it was ruled an invalid interference with freedom of contract. To return to *Lochner* is to return to the abuses that inevitably accompany unrestricted laissez faire. Few today will agree that such a return is desirable.

From a broader point of view, Justice Holmes was surely correct in repudiating the Peckham concept that the judge should decide on the basis of the economic theory that he deems correct. The judge qua economist will inevitably write his own economic views into the Constitution, and, as Holmes once put it, all too often on the basis "of the economic doctrines which prevailed about fifty years ago."[86]

If we have learned anything in this century, however, it is that judges should not substitute their economic judgments for those of the legislature. What Justice Holmes told his fellow judges is still as valid as it ever was—that the Constitution was not "intended to give us *carte blanche* to embody our economic . . . beliefs in its prohibitions."[87] "Otherwise," as Holmes put it in his first Supreme Court opinion, "a constitution, instead of embodying only relatively fundamental rules of right, . . . would become the partisan of a particular set of . . . economical opinions, which by no means are held *semper ubique et ab omnibus.*"[88] The economic theory behind a law should continue to be primarily a question for the legislator, not the judge. Provided that they have a rational basis, it is the judge's duty to enforce even "laws that I believe to embody economic mistakes."[89]

"Judges," reads another famous Holmes passage, "are apt to be Naif, simpleminded men."[90] This was particularly true of a judge like Justice Peckham when he used his conception of economics as his legal compass. The result in *Lochner* was that the Constitution was virtually treated as a legal sanction of the Survival of the Fittest.

That result will be avoided only if the courts follow the Holmes approach and reject the view that their notions of economics should override the economic theory upon which the legislature acted. Today, as in Holmes's day, the proper posture is for the judge to say that even though "speaking as a political economist, I should agree in condemning the law, still I should not be willing to think myself authorized to overturn legislation on that ground, unless I thought that an honest difference of opinion was impossible, or pretty nearly so."[91]

9

White and Taft Courts, 1910–1930

William Howard Taft, the only man who was both President and Chief Justice, once said that the Supreme Court was his notion of what heaven must be like. This led Justice Frankfurter to say that "he had a very different notion of heaven than any I know anything about."[1]

If the Supreme Court early in this century was not "heaven," it certainly had attained a preeminent rank in the American polity. In this respect, there was a quantum difference between the Supreme Court and all other courts in the country. Before his elevation to the supreme bench, Benjamin N. Cardozo was for many years a distinguished member and chief of New York's highest tribunal. With the perspective gained from judicial service in both Albany and Washington, he could note acutely a basic difference between the highest courts of state and nation. The New York Court of Appeals, he said, "is a great common law court; its problems are lawyer's problems. But the Supreme Court is occupied chiefly with statutory construction—which no man can make interesting—and with politics."[2] Of course, as Justice Robert H. Jackson pointed out, in this statement, the word "politics" is used in no sense of partisanship but in the sense of policy-making.[3]

The transition from a state court, eminent though it may be, to the supreme tribunal in Washington is more than mere promotion to a higher judicial body. When Cardozo came to Washington, he left what was, without a doubt, the greatest common-law court in the land; it dealt essentially with the questions of private law that are the preoccupation and delectation of most lawyers. The Court to which he came was not, and has not been, since the time of John Marshall, the usual type of law court. Public, not private, law is the stuff of Supreme Court litigation. Elevation to that tribunal requires the judge to make the adjustment from preoccupation with the restricted, however novel, problems of private litigation to the most exacting demands of judicial statesmanship.

It is the fact that the Supreme Court is more than the usual law court that makes its work of vital significance to more than a relatively small number of jurists. The Court is primarily a political institution, in whose keeping lies the destiny of a mighty nation. Its decrees mark the boundaries between the great departments of government; upon its action depend the proper functioning of federalism and the scope to be given to the rights of the individual.

A judge on such a tribunal has an opportunity to leave his imprint upon the life of the nation as no mere master of the common law possibly could. Only a handful of men in all our history have made so manifest a mark on their own age and on ages still to come as did Justice Holmes.[4] The same cannot be said of even the greatest of modern English judges. To be a judge, endowed with all the omnipotence of justice, is certainly among life's noblest callings; but the mere common-law judge, even in a preeminently legal polity like that in Britain, cannot begin to compare in power and prestige with a Justice of our Supreme Court. A judge who is regent over what is done in the legislative and executive branches—the deus ex machina who has the final word in the constitutional system—has attained one of the ultimates of human authority.

"Mr. Taft Has Rehabilitated the Supreme Court"

President Taft devoted more attention to the choice of Justices than any other President. Taft had been Solicitor General and a circuit judge and was fully aware of the importance of judicial appointments. He himself appreciated the need for changes in the now-aged Fuller Court. "The condition of the Supreme Court is pitiable," Taft wrote soon after his election, "and yet those old fools hold on with a tenacity that is most discouraging."[5]

The new President was soon given the opportunity virtually to remake the Court. He was given his first chance a half year after he took office in 1909 when Justice Peckham died. Taft chose his old friend and former circuit court colleague, Horace H. Lurton of Tennessee. It is, as Taft's biographer points out, illogical that, having deplored the senility of the highest Court only four months earlier, the President should have appointed a judge who was almost sixty-six.[6] But the friendship, born of years of service together on the circuit court, overrode the fact that Lurton was the oldest ever to ascend to the Court. Taft himself wrote that "there was nothing that I had so much at heart in my whole administration as

Lurton's appointment."[7] Lurton was a mediocre judge who turned out to be a mediocre Justice. He served only five years and contributed little to Supreme Court jurisprudence.

The same was not true of Taft's second appointee, Charles Evans Hughes. His is, of course, one of the great names in Supreme Court history, for he was to preside over the Court during one of its most crucial periods. That was, however, to be two decades later, when he was to be appointed a second time to fill the Court's center chair. In 1910, when he took the place of Justice Brewer on the latter's death, Hughes was Governor of New York. He had gone into politics after a successful career at the Bar and as a legislative investigator. He had been a possible presidential candidate and had declined the vice presidential nomination in 1908. Hughes resigned from the Court in 1916 to accept the Republican nomination for the Presidency.

When he appointed Hughes, President Taft indicated in a letter to his nominee that he would choose him as Chief Justice should that position become vacant. The vacancy occurred only two months later, when Chief Justice Fuller died even before Hughes had taken his seat on the bench. But Hughes was not to receive the appointment. Taft gave "prayerful consideration" to the matter for months and then, swayed in part by the preference of the Justices as reported to him by the Attorney General, promoted Justice Edward D. White to the Chief Justiceship.[8] "Hughes is young enough to wait," said the President, "and if he makes good on the bench I may yet be able to appoint him.[9]

The new Chief Justice had been a more than competent Justice for sixteen years. As Justice Frankfurter later recalled, White "looked the way a Justice of the Supreme Court should look. . . . He was tall and powerful. I think a jowl also helps a Justice of the Supreme Court, and White had an impressive jowl."[10]

White had served on the Southern side in the Civil War. Frankfurter tells us that this was a factor in his appointment: "Taft was glad to appoint . . . White as Chief Justice because White had been a Confederate . . . to make a Confederate, an ex-Confederate—are Confederates ever 'ex'?—Chief Justice was something that could contribute much, even then, so Taft thought, and I believe rightly, to the cohesion of our national life."[11]

Despite the Frankfurter aside, White was an "ex-Confederate" who had completely changed his views during what he was to call "the anguish, more appalling than the calamity of the war"[12] of the Reconstruction period. "My God," he later said in horror-stricken tones of the Confederate cause, "my God, if we had succeeded."[13] At any rate, it was Chief Justice White, the ex-Confederate plantation owner from the Deep South, who was to write the first opinions vindicating the black right to vote and to live in predominantly white areas.[14]

After he had been in office a year and a half, President Taft wrote with satisfaction to his half-brother, "I shall have the appointment of probably a majority of the Supreme Court before the end of my term, which, in view of the present agitation in respect to the Constitution, is very important."[15] As it turned out, Taft was able to appoint five Justices and the new Chief Justice.

When Chief Justice White's nomination went to the Senate it was accompanied by two nominations for Associate Justice: Willis Van Devanter of Wyoming and Joseph R. Lamar of Georgia, chosen to succeed White as a Justice and Justice Moody, whom ill health forced to resign in 1910. Van Devanter had been the first Chief Justice of his state's highest court, though he soon resigned to return to a lucrative practice. He later served as Assistant Attorney General and a circuit judge. Van Devanter was to be on the Court for twenty-six years. He is remembered now primarily for his role as one of the "Nine Old Men" who struck down much of Franklin D. Roosevelt's New Deal legislation. To his associates in the White and Taft Courts, however, he was a most valuable colleague. Chief Justice Taft was later to say his "mainstay in the court is Van Devanter."[16]

Van Devanter illustrates how a Justice can play an important role in the Court, belying a poor public reputation. "Mr. Justice Van Devanter," says Justice Frankfurter, "is a man who plays an important role in the history of the Court, though you cannot find it adequately reflected in the opinions written by him because he wrote so few."[17] Van Devanter was afflicted with what one of his colleagues describes as "pen paralysis." On the other hand, Chief Justice Hughes was to recall years later, "his careful and elaborate statements in conference . . . were of the greatest value. If these statements had been taken down stenographically, they would have served with little editing as excellent opinions. His perspicacity and common sense made him a trusted advisor in all sorts of matters. Chief Justice White leaned heavily upon him and so did Chief Justice Taft."[18]

Justice Lamar, appointed at the same time as White and Van Devanter, had been a successful corporation lawyer and member of Georgia's highest court. A tall Southern patrician with silver hair, Lamar was said by the *New York Times* to look both the scholar and the judge.[19] He died in 1916; his tenure of less than five terms was too short to make a mark. Today, in Supreme Court history, Lamar remains a cipher who, according to one commentator, "is remembered, if at all, only by his grandchildren."[20]

The last of the Taft appointees was Mahlon Pitney, then Chancellor of New Jersey, chosen in 1912 on the death of Justice Harlan. His appointment was opposed in the Senate because of allegedly antilabor decisions in the state court. Justice Pitney served until 1922, when a stroke forced him to retire. On the Court, Pitney was, according to one summary, "a respectable judge . . . without sign or promise of distinction."[21] As Holmes later put it, "He had not wings and was not a thunderbolt, but he was a very honest hard working Judge."[22]

President Taft expected his appointments to confirm the Court's conservative balance. His new Court majority consoled him with the realization that the six staunch conservatives would protect the Constitution from the attacks leveled by Theodore Roosevelt and his new Progressive party. The duty of the Court, Taft had written to Justice Moody, was to "preserve the fundamental structure of our government as our fathers gave it to us."[23] The sound men he had appointed would ensure the needed preservation. Well might Joseph H. Choate, the leading conservative lawyer of the day, boast at a dinner in the President's honor, "Mr. Taft has rehabilitated the Supreme Court."[24]

That it did not quite work out as the President had planned was due in large part to circumstances beyond anyone's control. Speaking of his Court appointees, Taft told reporters, "I have said to them, 'Damn you, if any of you die, I'll disown you.'"[25] But die or retire most of them did. Except for Justice Van Devanter, Taft's appointees as Justices served only four, six, and ten years and Chief Justice White sat for only ten years.

Under the new Chief Justice, the atmosphere within the Court became noticeably less rigid. Before his Court appointment, White's career had been more in politics than law and he remained, in Holmes's phrase, "naturally a politician and a speaker."[26] Chief Justice Hughes later recalled that White "was most considerate and gracious in his dealings with every member of the Court, plainly anxious to create an atmosphere of friendliness and to promote agreement in the disposition of cases."[27] In particular, the new Chief Justice, perhaps influenced by his senatorial experience, conducted conferences with much less restraint—allowing freer discussion and being, as Holmes was to note, less inclined "to stop the other side when the matter seems clear to all of us."[28] The result was a more relaxed Court, which was greatly appreciated by its members, who, as Hughes put it, "became a reasonably happy family."[29]

However, the family's material condition continued to be that of genteel poverty. The Court continued to sit in the old Senate Chamber that had been its home since 1860. As time went on, it became ever more apparent that the courtroom was scarcely suitable for the highest bench in the land. First of all, as Justice Frankfurter once said, "It was a small room, you know,"[30] badly ventilated and poorly lighted. The conference room in the Capitol basement was even worse. As Chief Justice Hughes remembered it. "[T]he room became overheated and the air foul . . . not conducive to good humor." Hughes complained, "I suppose that no high court in the country had fewer conveniences."[31]

Yet many of the Justices were attached to the old courtroom and deemed its location in the Capitol an important plus. The Court may have been barren in physical facilities, but its atmosphere was rich in dignity and tradition. President Taft wanted to initiate construction of a Supreme Court building. Chief Justice White resisted the idea; he feared that the public would lose interest in the Court if it left the Capitol.[32] Indeed, after the Court finally moved out of its cramped Capitol quarters in 1935 to the marble temple in which it now sits, a Justice was to refer longingly to "that wonderful old courtroom in the Capitol, which I think it was almost a desecration of tradition to leave."[33]

At the time of the White Court, no office space was provided for the Justices. They all had to write their opinions and do their research at home, providing their own working space for themselves and their secretaries. Since 1886, the Justices had been provided with $2,000 a year to pay a secretary or clerk. Most of the Justices hired stenographer-typists. Thus Justice Hughes hired three young lawyers during his first brief Court tenure. He later recalled, "I kept them busy with dictation, hating to write in longhand," and, referring to research, "whatever was necessary in that line I did myself." In a prescient passage, Hughes also noted "that if we had experienced law clerks, it might be thought that they were writing our opinions."[34] Though some Justices, notably Gray and his successor Holmes, hired law clerks rather than stenographers, it was not until 1919 that Congress provided specifically for law clerks in addition to secretarial assistance.

The improved atmosphere in the White Court could not lighten what Chief Justice Taft was to call the "exhausting character"[35] of the ever-growing docket. The cases docketed showed a steady augmentation, and each term ended with the Court increasingly in arrears. Justice Louis D. Brandeis asked former Justice Hughes, who had returned to practice, whether he did not envy Brandeis's coming summer vacation. "Not at the price you have to pay for it," was Hughes's answer.[36]

Rule of Reason

When Chief Justice White took his seat, there was the usual mass of matters on the docket. Among them were cases on what a commentator called "the main question of the day"—"the regulation by law of corporate activity in its relation to the country at large."[37] To the public and the profession, the greatest of these were the landmark Sherman Act cases.

Harper's noted that a number of Supreme Court decisions had "been awaited with country-wide suspense and attention." Nevertheless, none of them had "caused the market and the whole industrial and commercial world to pause more perceptibly than have the cases of the Government against the Standard Oil Company and the American Tobacco Company. They came to be known simply as the Trust Cases. For months the financial markets have virtually stood still awaiting their settlement."[38]

The key decision in the *Trust Cases*[39] arose out of the Government's prosecution of the Standard Oil Company under the Sherman Act. The lower court had found the great oil monopoly in violation of the Act and had ordered its dissolution. A unanimous Supreme Court upheld the dissolution ruling. Yet, ironically, though the Government won the case, it had to accept an interpretation of the Sherman Act which greatly reduced that law's effectiveness. The Court ruled that Standard Oil's practices constituted "unreasonable" restraint of trade prohibited by the Sherman Act, rather than the type of "reasonable" restraint which the Act permitted. Thus was born "the most curious *obiter dictum* ever indulged in by the

Supreme Court"[40]—the so-called *rule of reason* in antitrust cases: "[I]n every case where it is claimed that an act or acts are in violation of the statute the rule of reason . . . must be applied."[41]

The rule of reason was almost entirely the handiwork of Chief Justice White; Justice Holmes called White's "invention of the rule of reason . . . [t]he Chief Justice's greatest dialectical coup."[42] Certainly it was the principal legal legacy of White's Court tenure. In practice, it greatly restricted the scope of the Sherman Act and made prosecutions under it more difficult. At the same time, it greatly increased the judicial role in Sherman Act cases and ensured that our antitrust law, based though it might be on enactments by Congress, would still be primarily judge-made law.

Justice Harlan delivered a typically strong dissent, which Chief Justice Hughes later recalled as "a passionate outburst seldom if ever equaled in the annals of the Court."[43] To Harlan, the Court's action was an example of "the tendency to judicial legislation . . . [t]he most alarming tendency of this day, in my judgment, so far as the safety and integrity of our institutions are concerned." Here, the Court was led "to so construe the Constitution or the statutes as to mean what they want it to mean." The result was "that the courts may by mere judicial construction amend the Constitution . . . or an Act of Congress."[44]

All the same it was difficult to disagree with a decision that asserted that it was only interpreting a law "by the light of reason."[45] Justice Holmes pointed this out in his comment to a law clerk about the White *Standard Oil* opinion: "The moment I saw that in the circulated draft, I knew he had us. How could you be against that without being for a rule of unreason?"[46]

A Progressive Court?

Writing in 1912, Felix Frankfurter asserted that "the fate of social legislation in this country rests ultimately with our judges."[47] In the White Court there were conflicting currents in the judicial attitude to what a magazine termed "the main question" of the day—"the regulation by law of corporate activity."[48] To supporters of the laissez-faire jurisprudence of the Fuller Court, there were disturbing tendencies in some of its successor's decisions. In a 1913 article, "The Progressiveness of the United States Supreme Court,"[49] Charles Warren noted that few recent social and economic laws had been invalidated by the Court. Instead, the Court upheld laws drastically changing the rules governing employer liability as well as wage and hour laws.[50]

By the beginning of this century, the failure of the common law to provide adequately for workers' injuries had made reform efforts inevitable. In 1908, a federal statute abolished the fellow-servant rule and restricted the defenses of contributory negligence and assumption of risk. An opinion by Justice Van Devanter for a unanimous Court upheld the statute[51]—though, as a Van Devanter letter informs us, "there had theretofore been pronounced differences of opinion" among the Justices.[52]

The 1908 statute, however, was at most a partial remedy. As Justice Brandeis later wrote, "[N]o system of indemnity dependent upon fault on the employers' part could meet the situation."[53] What was needed was a complete transformation of tort law, with its essential element the elimination of the rule "that fault is requisite to liability."[54] The needed reform was provided by workmen's compensation laws, enacted in almost half the states by the time White became Chief Justice. Workmen's compensation involved a rejection of fault as the basis for liability; instead, liability was based on the concept of protection of individuals from the consequences of industrial accidents, regardless of culpability on their part. The fundamental principle is that of liability without fault, with recovery provided for all injuries arising out of and in the course of employment. The result, as Justice Holmes pointed out, is that the pain and mutilation incident to production is thrown upon the employer in the first instance and the public in the long run.[55]

The Court, most of whose members were still set in conservative laissez-faire beliefs, found the *Arizona Employers' Liability Cases (1919)*,[56] in which a state workmen's compensation law was challenged, difficult to decide. Liability without fault by legislative decree was something the Chief Justice along with Justices McKenna, Van Devanter, and McReynolds could not accept. In the words of the latter Justice's dissent, "As a measure to stifle enterprise, produce discontent, strife, idleness, and pauperism, the outlook for the enactment seems much too good."[57]

Nevertheless, a bare majority voted to uphold the law. The opinion was assigned to Justice Holmes, but he wrote one of such "sweeping generality"[58] that he lost the majority. According to Holmes, Justices Pitney and Day "thought there was danger in this op. and P wrote what none of his majority could disagree with."[59] Pitney's draft became the opinion of the Court and Holmes's, which "was thought too strong by some of the majority,"[60] became a concurrence.

Despite his famous aphorism against general propositions,[61] the Holmes opinion was based upon the "general proposition that immunity from liability when not in fault is [not] a right inherent in free government."[62] The Pitney opinion of the Court was more pedestrian, stressing the limitations of review power: the courts should not consider the wisdom of legislation, and the novelty of the statutory scheme "is not a constitutional objection."[63] The important thing, however, is not the lack of sparkle in the Pitney opinion but the fact that the Court upheld a law which sought to protect labor by making such drastic inroads into the principle that was still widely seen as "of the very foundation of right—of the essence of liberty as it is of morals—to be free from liability if one is free from fault."[64]

Even more significant in the White Court's "progressive" decisions were decisions upholding maximum-hour laws. First came a 1917 decision on the Adamson Act, which Congress had enacted to avert a railroad strike. Among its provisions was the requirement of an eight-hour day for railroad workers. The provision was sustained in an opinion by Chief Justice White that stressed the plenary congressional power over interstate commerce—particularly its power to deal with a strike that would interrupt commerce.[65] As Justice Holmes summarized it, the Court "went the whole unicorn as to the power of Congress."[66] The case was thus one on the commerce power, not the police power to regulate hours. "I think," Holmes wrote explaining the decision, "if Congress can weave the cloth it can spin the thread."[67]

Bunting v. Oregon,[68] decided later the same year, did decide the police power issue, for it involved the constitutionality of an Oregon law limiting the hours of work in manufacturing establishments to ten hours a day. Felix Frankfurter, who presented the case for the law, later recalled a dramatic moment in the argument: "During the course of the argument McReynolds said to me, 'Ten hours! Ten hours! Ten! Why not four?' . . . in his snarling, sneering way. . . . I moved down towards him and said, 'Your honor, if by chance I may make such a hypothesis, if your physician should find that you're eating too much meat, it isn't necessary for him to urge you to become a vegetarian.'"

According to Frankfurter, "Holmes said, 'Good for you!' very embarrassingly right from the bench. He loathed these arguments that if you go this far you must go further. 'Good for you!' Loud. Embarrassingly."[69]

To the Court that had decided *Lochner, Bunting* would have been an even easier case. But two things had happened since *Lochner* was decided. In 1908, *Muller v. Oregon*[70] had upheld a maximum-hours law for women. The Justices had been persuaded by Louis D. Brandeis's famous argument and even-more-famous brief that, unlike the situation in *Lochner,* there was a direct correlation between hours worked by women and their health. Hence a law limiting the hours worked by women was valid, even though, under *Lochner,* a maximum-hours law for all bakery employees was not. Yet, though *Muller* was thus based upon the need to protect women, it was but a short step for its reasoning to be used to justify similar laws to protect all workers.

In addition, the personnel of the Court had changed substantially since *Lochner.* Only the Chief Justice and Justices McKenna, Holmes, and Day remained, and all but McKenna had been dissenters in *Lochner.* As it turned out, McKenna was the key Justice in *Bunting.* It has been said that McKenna was all but "mesmerized" by the brilliant Brandeis presentation in *Muller v. Oregon.*[71] Though, as seen,[72] he had played a key role in the *Lochner* decision process, Justice McKenna now voted to uphold maximum-hours regulation under the Adamson Act and in *Bunting.* More than that, he delivered the opinion of the court upholding the Oregon law.

The *Bunting* opinion ignored *Lochner.* It did not cite the case or even refer to the freedom of contract issue that had been so crucial to the *Lochner* decision. Instead, the opinion assumed the correctness of the *Muller* finding that a law regulating the hours of service was a health law, since "it is injurious to their health for them to work . . . more than ten hours in any one day." The argument the other way was summarily rejected: "The record contains no facts to support the contention."[73]

Bunting in effect overruled *Lochner,* though sub silentio. Why it did so was intimated in the only pregnant portion of the otherwise pedestrian McKenna opinion: "New policies are usually tentative in their beginnings, advance in firmness as they advance in acceptance. . . . Time may be necessary to fashion them to precedent customs and conditions, and as they justify themselves or otherwise they pass from militancy to triumph."[74] Maximum-hour laws had gone through this process; the "new policies" upon which they were based had gone from the *Lochner* rejection to the *Bunting* triumph.

Child Labor Case

The decisions just discussed may have demonstrated, as Charles Warren called it in his already-mentioned article, the "progressiveness" of the White Court. But they were more than balanced by decisions restricting the rights of labor[75] and, most of all, by that in *Hammer v. Dagenhart* (1918),[76] usually known as the *Child Labor Case.* Most commentators today place that case with *Lochner* on the list of discredited Supreme Court decisions.

In the *Child Labor Case,* a federal statute prohibited transportation in interstate commerce of goods made in factories that employed children. Though the statute did not in terms interfere with local production or manufacturing, its real purpose was to suppress child labor. With goods produced by children denied their interstate market, child labor could not continue upon a widespread scale. To the majority of the Court, the congressional purpose rendered the law invalid. Congress was seeking primarily to regulate the manner in which manufacturing was carried on; such manufacturing under the restrictive meaning of the Court in cases like the already-discussed *Sugar Trust Case*[77] was not commerce which could be reached by federal authority. Congress could not, even by an act whose terms were specifically limited to regulation of commerce, use its commerce power to exert authority over matters like manufacturing, which were not, within the Court's restricted notion, commerce.

Justice Holmes wrote just after the decision that he thought it "ill timed and regrettable."[78] This led him to deliver a dissent that stands second only to that delivered by him in *Lochner.* Holmes explained his dissent in a letter: "I said that as the law unquestionably regulated interstate commerce it was within the power of Congress no matter what its indirect effect on matters within the regulation of the states, or how obviously that effect was intended."[79]

The Holmes dissent emphasized the distressing effect of the Court's decision. "If there is any matter," asserted Holmes, "upon which civilized countries have agreed . . . it is the evil of premature and excessive child labor."[80] Yet the practical result of the decision was to render effective regulation of child labor all but impossible. In a country like the United States, if a practice like child labor is to be dealt with effectually, it must be by national regulation. By rigidly excluding Congress from exercising regulatory authority, the *Child Labor Case* virtually decreed that child labor should be left only to whatever controls were afforded by the workings of an unrestrained system of laissez faire. The United States alone, among nations, was precluded from taking effective action against an evil so widely censured by civilized opinion.

Chief Justice Taft and His Court

"There is nothing I would have loved more than being chief justice of the United States," President Taft told his Attorney General as he signed Chief Justice White's commission. "I cannot help seeing the irony in the fact that I, who desired that office so much, should now be signing the commission of another man."[81]

However, Chief Justice White died in May 1921. A month later, President Harding appointed Taft to what he had called only two months earlier "the position, which I would rather have than any other in the world."[82] Taft began his new judicial career at an age when most men are about to retire; he was sixty-four when he was appointed. "I am older and slower and less acute," Taft wrote.[83] Despite this, he was an effective Chief Justice. Even those who criticize the conservatism that permeated his jurisprudence concede that he brought a needed leadership to the highest Court.

Felix Frankfurter tells how Justice Brandeis once said to him, "It's very difficult for me to understand why a man who is so good a Chief Justice . . . could have been so bad as President. How do you explain that?" Frankfurter replied, "The explanation is very simple. He loathed being President and being Chief Justice was all happiness for him."[84]

There is no doubt that Taft was, in his biographer's phrase, "a happy man" as Chief Justice.[85] The Chief Justiceship was to him the ultimate compensation for the unhappiness of his years in the White House. Of course, Taft was the very paradigm of the genial politician—in Frankfurter's phrase, "instinctively genial, with great warmth, and a capacity to inspire feelings of camaraderie about him."[86] More than that, he proved in the Court what he had failed to be in the White House—a strong leader. As Chief Justice, Taft wielded greater authority than any of his predecessors since Taney. He met the criterion for a Chief Justice stated in one of his letters of "leadership in the Conferences, in the statement of the cases, and especially with respect to applications for certiorari."[87]

Justice Holmes summed up the situation under the new Chief Justice: [W]ith the present C.J. things go as smoothly as possible." That was particularly true of the Court conferences. According to Holmes, "The meetings are perhaps pleasanter than I have ever known them—thanks largely to the C.J." Before, Holmes wrote, the Justices at conference "regarded a difference of opinion as a cockfight," but now "the Chief's way of conducting business" results in "keeping things moving."[88] Taft was stricter than his predecessors in controlling the conference discussion. As Holmes said, "[T]he C.J. inclines more than Fuller or White to stop the other side when the matter seems clear . . . , which I think is a good thing."[89] At any rate, Holmes wrote to Taft in 1925, "never before . . . have we gotten along with so little jangling and dissension."[90]

A large part of the Chief Justice's success in this respect was due to the fact that he had colleagues on the Court who were congenial to his conservative views. The majority who had decided the *Child Labor Case* were still on the bench, except for Chief Justice White, whom Taft himself had succeeded. It is true that Justices Day, Pitney, and Clarke were soon replaced, since they resigned or retired from the Court in 1922. But they were replaced by men who were, if anything, even more conservative. Day's seat was taken by Pierce Butler of Minnesota, Pitney's by Edward T. Sanford of Tennessee, and Clarke's by Senator George Sutherland of Utah.

Justice Sanford was, in the characterization of a biographer, an undistinguished conservative Justice who generally followed the lead of the Chief Justice.[91] To such an extent was this carried that Sanford died on the same day as Taft. Justice Sutherland replaced the more liberal Justice Clarke, who had dissented in the *Child Labor Case*. After Sutherland's appointment, Woodrow Wilson wrote to Clarke about his successor, "In my few dealings with Mr. Justice Sutherland I have seen no reason to suspect him of either principles or brains."[92] This was not entirely fair. Sutherland was a better than average spokesman for his conservative views. Both he and Justice Butler are today remembered primarily as two of the four Justices who opposed the New Deal measures during the 1930s.

The other members of what Justice Frankfurter called the "Four Horsemen,"[93] Justices Van Devanter and McReynolds, also sat on the Taft Court. James C. McReynolds of Kentucky had been Wilson's Attorney General; he had been appointed to the Court in 1914 to succeed Justice Lurton, who had died. McReyonlds was perhaps the least lovable person ever to sit on the Court. *Time* called him "intolerably rude," "savagely sarcastic," "incredibly reactionary," and "anti-Semitic."[94] The latter characteristic was as pronounced as it has been in any American public figure. McReynolds once refused to accompany the Court on a ceremonial occasion because Justice Brandeis would be there. "As you know," McReynolds wrote to the Chief Justice, "I am not always to be found when there is a Hebrew abroad."[95] Taft wrote that McReynolds was "fuller of prejudice than any man I have ever known."[96] Harold J. Laski summed him up: "McReynolds and the theory of a beneficent deity are quite incompatible."[97]

McReynolds's bête noire, Louis D. Brandeis, was the first Jewish Justice on the Supreme Court. Brandeis is, of course, another of the great names in the Court's history. Brought up in Louisville, he attended Harvard Law School and became a successful attorney in Boston. Brandeis soon became active as a reformer; he became known as the "People's Attorney," since he worked without fee for a great number of public causes. Brandeis was not, however, the typical turn-of-the-century progressive—content only to expose and deplore. While the muckrakers of the day dealt in invective and generalities, he sought remedies achieved through social legislation, particularly in his reforms of gas rates, insurance, and railroads.

When Justice Joseph Lamar died in 1916, President Wilson chose Brandeis in his place. The nomination led to a bitter confirmation battle (Brandeis himself wrote that "[t]he dominant reasons for the opposition . . . are that he is considered a radical and is a jew").[98] Brandeis was confirmed despite the unprecedented opposition of the American Bar Association and ex-President Taft, who asserted in a letter, "Mr. Louis D. Brandeis . . . is not a fit person to be a member of the Supreme Court."[99]

Brandeis at the Bar is best remembered for the new dimension that he added to legal thought—one that emphasized the facts to which the law applied. The Brandeis method was inaugurated by the brief submitted by him in *Muller v. Oregon*[100]—the generic type of a new form of legal argument, ever since referred to as the "Brandeis Brief." To persuade the Court to uphold an Oregon law prohibiting women from working in factories more than ten hours a day, Brandeis marshaled an impressive mass of statistics to demonstrate "that there is reasonable ground for holding that to permit women in Oregon to work . . . more than ten hours in one day is dangerous to the public health, safety, morals, or welfare."[101] The Brandeis Brief and argument in *Muller* were devoted almost entirely to the facts—and to facts not in the record, but which the Court was asked to accept as "facts of common knowledge." The *Muller* brief contains 113 pages. Only two of them are devoted to argument on the law.

The *Muller* Court upheld the Oregon law and it did so by relying upon the Brandeis approach, expressly noting in its opinion how the Brandeis Brief had supplied the factual basis for its decision. *Muller* thus was, as Justice Frankfurter termed it, "'epoch making,' . . . because of the authoritative recognition by the Supreme Court that the way in which Mr. Brandeis presented the case—the support of legislation by an array of facts which established the *reasonableness* of the legislative action, however it may be with its wisdom—laid down a new technique for counsel."[102] The Brandeis Brief still remains the model for constitutional cases.

Brandeis on the bench was only a more exalted version of Brandeis in the forum. If the Brandeis Brief replaced the black-letter judge with the man of statistics and master of economics,[103] Justice Brandeis himself was the prime exemplar of the new jurist in action. Above all, as in his brief, the Justice was a master of the facts in his opinions. For him the search of the legal authorities was the beginning, not the end, of research. He saw that the issues which came to the Court were framed by social and economic conditions unimagined even a generation before. Hence "the judicial weighing of the interests involved should, he believed, be made in the light of facts, sociologically determined and more contemporary than those which underlay the judicial approach to labor questions at the time."[104] When Brandeis marshaled his usual mass of facts in one of his opinions, Holmes characterized a Justice who had written the other way: "I should think [he] would feel as if a steam roller had gone over him."[105]

The Brandeis method on the bench was used for a particular purpose: to reject the prevailing notion that the law was to be "equated with theories of *laissez faire*."[106] If twentieth-century law has enabled the society to move from laissez faire to the welfare state, that has been true in large part because it has accepted Justice Brandeis's approach.

The Brandeis technique helped persuade jurists that the legal conception of "liberty" should no longer be "synonymous with the laissez faire of Herbert Spencer."[107] Instead, the law has come to believe with Brandeis that "[r]egulation . . . is necessary to the preservation and best development of liberty."[108] This in turn has led to acceptance of the Brandeis rejection of laissez faire as the foundation of our jurisprudence.

It was the Brandeis fact-emphasis technique as much as anything that heralded the end of the turn-of-the century concept of law. Compare the Brandeis opinion in a case involving a regulatory law, with its emphasis throughout on the economic and social conditions that called forth the challenged statute, with that in *Lochner,* where those factors were all but ignored. The difference is as marked as that between the poetry of T. S. Eliot and Alfred Austin.

There was another appointment to the Taft Court that should be mentioned. In 1925, Justice McKenna retired and Attorney General Harlan F. Stone of New York recieved his seat. Stone had been a professor and dean at the Columbia Law School and had indicated adherence to the laissez-faire economics that still dominated jurisprudence. On the Court, however, Chief Justice Taft complained in a letter, Stone became "subservient to Holmes and Brandeis. I am very much disappointed in him; he hungers for the applause of the law-school professors and the admirers of Holmes."[109]

Despite Taft's disappointment with Stone (he had urged his appointment to President Coolidge), the majority continued to follow the Chief Justice's conservative lead in most cases. He led the Court to decisions defending private property and laissez faire that reaffirmed the trend established by the Fuller Court. On the whole, he was one of the most successful Chief Justices—if success is measured by the extent to which his Court mirrored his own juristic philosophy.

The frustration that had encumbered Taft in the White House was now largely forgotten. "The truth is," the Chief Justice wrote in 1925, "that in my present life I don't remember that I ever was President."[110]

Taft has been called "the most 'political' of Chief Justices in American history."[111] He never hesitated to "lobby" Presidents and Congress to secure judicial appointments and legislative action. It was Taft's "politicking" that led to the legislation that finally allowed the Supreme Court, in Justice Frankfurter's phrase, "to be master in its own household."[112] The Judges Bill of 1925, which gave the Court virtual control over its own docket, was largely the result of the Chief Justice's own lobbying on Capitol Hill. "Van Devanter, McReynolds and I," he wrote, "spent two full days at the Capitol, and Van and I one full day more to get the bill through."[113]

The 1925 law's remedy for Court congestion was increased resort to certiorari, which was to be granted or denied in the Court's discretion. Henceforth, as Chief Justice Taft wrote, it was to "be reserved to the discretion of the Supreme Court to say whether the issue between [litigants] is of sufficient importance to justify a hearing of it in the Supreme Court."[114] The result is, Justice Frankfurter tells us, "that the business which comes to the Supreme Court is the business which the Supreme Court allows to come to it. Very few cases can come up without getting its prior permission."[115] The Court was thus now able "to confine its adjudication to issues of constitutionality and other matters of essentially national importance."[116]

Labor and the Court

"The only class which is arrayed against the court . . . is organized labor," wrote Chief Justice Taft to his brother in 1922.[117] Labor had good cause for complaint against the Court, for the Justices continued to serve as the censors of legislative attempts to enact regulatory laws, particularly those protecting the rights of labor.

The Taft Court's antilabor jurisprudence began soon after the Chief Justice took his seat with the 1921 decision in *Truax v. Corrigan*.[118] The Court there struck down an Arizona law barring courts from issuing injunctions in most cases growing out of a labor dispute. The union in *Truax* had picketed a restaurant, which caused its business to fall dramatically. This, said Taft for the majority, was an intentional injury that constituted a tort. A statute which made such a legal wrong remediless deprived the person suffering the loss of due process.

To Chief Justice Taft, the case was an example of how "we have to hit [organized labor] every little while, because they are continually violating the law and depending on threats and violence to accomplish their purpose."[119] But the practical result of a decision such as *Truax* was clear—in the phrase of Justice Holmes's dissent, "to prevent the making of social experiments that an important part of the community desires."[120]

Chief Justice Taft also delivered the opinion in the 1922 *Child Labor Tax Case*[121]—a sequel to the White Court's most controversial decision. That decision had invalidated a direct congressional prohibition of child labor on the ground that the national commerce power did not extend so far.[122] Congress then enacted a law imposing a 10 percent tax upon the profits of all persons employing children. The purpose of the law was, of course, to achieve precisely the result aimed at by the statute previously held unconstitutional, that is, to eliminate child labor. This motive, said the Taft opinion of the Court, rendered the exercise of the taxing power invalid. Congress was trying to suppress an activity that, under the *Child Labor Case,* could be controlled only by the states. When the taxing power is utilized for regulatory purposes in an area where Congress has no direct regulatory power, the tax itself must be invalidated because of its improper underlying purpose.

In his opinion, the Chief Justice conceded that the law at issue was "legislation designed to promote the highest good."[123] To his brother he wrote that the statute was an "effort of good people, who wish children protected." Nevertheless, he went on, "Unfortunately we cannot strain the Constitution . . . to meet the wishes of good people."[124] At the same time, the decision was ruinous in its impact on child labor regulation. In effect, in the words of a biographer, "Taft's opinion had effectively driven a second spike into congressional efforts to regulate child labor."[125]

The *Truax* and *Child Labor Tax* decisions were, however, only the beginning in the Taft Court's attack on regulatory legislation. "Since 1920," Professor Frankfurter noted in 1930, "the Court had invalidated more legislation than in fifty years preceding. Views that were antiquated twenty-five years ago have been resurrected in decisions nullifying minimum-wage laws for women in industry, a standard-weight-bread law to protect buyers from short weights and honest bakers from unfair competition, a law fixing the resale price of theater tickets by ticket scalpers in New York, laws controlling exploitation of the unemployed by employment agencies, and many tax laws."[126]

Of these decisions, the most extreme was that in the 1923 minimum-wage case—*Adkins v. Children's Hospital.*[127] So extreme was it, indeed, that Chief Justice Taft himself, despite his frequently expressed aversion toward dissents, could not go along with the majority and issued a dissenting opinion. The majority opinion was by Justice Sutherland, who thus began to assume the intellectual leadership of the most conservative Justices— soon to become the "Four Horsemen" in the Court that invalidated New Deal legislation.

In *Adkins* itself, as the Court recently summarized, "this Court held it to be an infringement of constitutionally protected liberty of contract to require the employers of adult women to satisfy minimum wage standards."[128] The *Adkins* opinion, like that in *Lochner,* was based upon extreme reliance upon the doctrine of freedom of contract. The law in question, said the Court, "forbids two parties having lawful capacity . . . to freely contract with one another in respect of the price for which one shall render service to the other in a purely private employment where both are willing, perhaps anxious, to agree." Nor did the economic inequality of women make them a proper subject for such protective legislation. In the

Court's view, such inequality did not justify the added burden imposed upon employers: "[T]o the extent that the sum fixed exceeds the fair value of the services rendered, it amounts to a compulsory exaction from the employer for the support of a partially indigent person, for whose condition there rests upon him no peculiar responsibility, and therefore, in effect, arbitrarily shifts to his shoulders a burden which, if it belongs to anybody, belongs to society as a whole."[129]

In *Adkins* freedom of contract reached its apogee. As stated by Justice Sutherland: "[F]reedom of contract is, nevertheless, the general rule and restraint the exception; and the exercise of legislative authority to abridge it can be justified only by the existence of exceptional circumstances."[130]

This, said Justice Frankfurter years later, was "the most doctrinaire view about 'liberty of contract.'" What it meant was "that presumptively every encroachment on 'liberty of contract' is unconstitutional, and you had to show some very good reason why there should be a curtailment of the freedom of contract—freedom of contract between a great, big laundry and Bridget McGinty. They were at arm's length, etcetera, and therefore you shouldn't interfere with their contracting equality."[131]

The effect of *Adkins* on social legislation was even more devastating than that of *Lochner*. The decision, Frankfurter tells us, "struck the death knell not only of this legislation, but of kindred social legislation because it laid down as a constitutional principle that any kind of change by statute has to justify itself, not the other way around." As such, *Adkins* had a serious inhibiting effect on legislative action. "It prevented legislation from being introduced, and it made still-born legislation which was by way of being introduced."[132]

Holmes and Judicial Restraint

In *Adkins,* Justice Holmes delivered another now-classic dissent. A Holmes letter states that his dissent "was intended . . . to dethrone Liberty of Contract from its ascendancy into the Liberty business."[133] But his *Adkins* dissent also sounded the theme that was most prominent in the Holmes jurisprudence. His English correspondent, Sir Frederick Pollock, summarized Holmes's dissent in a letter to the Justice: "The wisdom of the Act of Congress may well be open to grave doubt: but that, as you have said, was not the question before the Court."[134]

This was a pithy summary of the doctrine of judicial restraint that was Justice Holmes's great contribution to Supreme Court jurisprudence. Holmes was led to the doctrine by his innate skepticism, which made him dubious of dogma and decisions based upon dogmatic clichés. "Lincoln for government and Holmes for law," Justice Frankfurter once wrote, "have taught me that the absolutists are the enemies of reason— that . . . the dogmatists in law, however sincere, are the mischief-makers."[135] For Holmes, the only absolute was that there were no absolutes in law. His philosopher's stone was "the conviction that our . . . system rests upon tolerance and that its greatest enemy is the Absolute."[136] It was not at all the judicial function to strike down laws with which the judge disagreed. "There is nothing I more deprecate than the use of the Fourteenth Amendment . . . to prevent the making of social experiments that an important part of the community desires . . . even though the experiments may seem futile or even noxious to me."[137] Not the judge but the legislator was to have the primary say on the policy considerations behind a regulatory measure. The judge's business was to enforce even "laws that I believe to embody economic mistakes."[138]

The same theme was to be repeated many times by Justice Holmes. He continually reiterated that, as a judge, he was not concerned with the wisdom of the social policy behind a challenged legislative act. The responsibility for determining what measures were necessary to deal with economic and other problems lay with the people and their elected representatives, not the judges. The Constitution, Holmes declared, was not "intended to give us *carte blanche* to embody our economic or moral beliefs in its prohibitions."[139] The Constitution was never intended to embody absolutes. Instead, "Some play must be allowed for the joints of the machine, and it must be remembered that legislatures are ultimate guardians of the liberties and welfare of the people in quite as great a degree as the courts."[140]

Justice Holmes recognized, with the majority in cases such as *Lochner v. New York,* that the question at issue was whether the challenged law was a *reasonable* exercise of the police power of the state. But if Holmes, too, started with the test of reasonableness, he applied it in a manner very different from the *Lochner* majority. The Holmes approach was based upon the conviction that it was an awesome thing to strike down an act of the elected representatives of the people, and that the power to do so should not be exercised save where the occasion was clear beyond fair debate.[141] The Constitution was not to be treated "as prohibiting what 5 out of 9 old gentlemen don't think about right."[142]

In the Holmes view, the test to be applied was whether a reasonable legislator—the legislative version of the "reasonable man"—could have adopted a law like that at issue. Was the statute as applied so clearly arbitrary that legislators acting reasonably could not have believed it necessary or appropriate for public health, safety, morals, or welfare?[143]

In the individual case, to be sure, the legislative judgment might well be debatable. But that was the whole point about the Holmes approach. Under it, the opposed views of public policy, as respects business, economic, and social affairs, were considerations for the legislative choice,[144] to which the courts must defer unless it was demonstrably arbitrary or irrational.[145] "In short, the judiciary may not sit as a super-legislature to judge the wisdom or desirability of legislative policy determinations . . . in the local economic sphere, it is only the . . . wholly arbitrary act which cannot stand."[146]

In his *Lochner* dissent, as we saw, Holmes asserted, "This case is decided upon an economic theory which a large part of the country does not entertain." It may now be fairly said that both the economic and the legal theories upon which *Lochner* rested have been repudiated. While the Supreme Court at the beginning of this century was increasingly equating the law with laissez faire, men turned to Holmes's dissents as the precursors of a new era. The at-first-lonely voice soon became that of a new dispensation which wrote itself into American public law.[147]

This was true because the Holmes doctrine of judicial restraint was the necessary legal foundation for the soon-to-emerge welfare state. The Holmes approach meant that the courts would uphold laws that coincided with changing views on the proper scope of governmental regulation. American judges were soon to follow Holmes when he rejected legal shibboleths that equated "the constitutional conception of 'liberty' . . . with theories of *laissez faire*."[148] They came to recognize that the rule of restraint was essential if the law was to enable the society to make the necessary transition from laissez faire to the welfare state.

Holmes and Free Speech

The theme of judicial restraint was, however, overridden by another Holmes theme in cases involving the freedom of expression guaranteed by the First Amendment. Restraint may have been the proper posture for the Justice in cases like *Lochner v. New York,* where economic regulation was an issue. But a different situation was presented in First Amendment cases. Here, says Justice Frankfurter, history had taught Holmes that "the free play of the human mind was an indispensable prerequisite"[149] of social development.

More than that, the Bill of Rights itself, as Holmes recognized, specifically enshrines freedom of speech as its core principle. "If there is any principle of the Constitution that more imperatively calls for attachment than any other it is the principle of free thought," he asserted in a 1928 dissent.[150] Because freedom of speech was basic to any notion of liberty, "Mr. Justice Holmes was far more ready to find legislative invasion in this field than in the area of debatable economic reform."[151]

The Holmes concept of freedom of speech is a direct descendant of John Milton and John Stuart Mill. It found its fullest expression in the Justice's dissent in the 1919 case of *Abrams v. United States,*[152] which has been termed "the greatest utterance on intellectual freedom by an American."[153] Milton's *Areopagitica* argues for "a free and open encounter" in which "[Truth] and Falsehood grapple."[154] The *Abrams* dissent sets forth the foundation of the First Amendment as "free trade in ideas,"[155] which through competition for their acceptance by the people would provide the best test of truth. Or as Holmes put it in a letter, "I am for aeration of all effervescing convictions—there is no way so quick for letting them get flat."[156]

Like Milton and Mills, Justice Holmes stressed the ability of truth to win out in the intellectual marketplace. For this to happen, the indispensable sine qua non was the free interchange of ideas.[157] In the crucial passage of his *Abrams* dissent, Holmes tells us that those who govern too often seek to "express [their] wishes in law and sweep away all opposition," including "opposition by speech." They forget that time may also upset their "fighting faiths" and that government itself is an experimental process. The Constitution also "is an experiment, as all life is an experiment." To make the experiment successful, room must be found for new ideas which will challenge the old, for "the ultimate good desired is better reached by free trade in ideas." And "the best test of truth is the power of the thought to get itself accepted in the competition of the market, and that truth is the only ground upon which their wishes safely can be carried out."[158]

In *Abrams* defendants had been convicted under the Espionage Act of 1917 for the publishing of leaflets which incited resistance to the American war effort by encouraging "curtailment to cripple or hinder the United States in the prosecution of the war." Written in lurid language, the leaflets contained a bitter attack against the sending of American soldiers to Siberia and urged a workers' general strike in support of the Russian Revolution. The Supreme Court affirmed the convictions, holding that, even though the defendants' primary intent had been to aid the Russian Revolution, their plan of action had necessarily involved obstruction of the American war effort against Germany.

As already noted, Justice Holmes issued a strong dissent in *Abrams*, setting forth his conception of the "free trade in ideas" as the foundation of the right of expression. The Holmes dissent argued that the "silly" leaflets thrown by obscure individuals from a loft window presented no danger to the American war effort. Not enough, he said, "can be squeezed from these poor and puny anonymities to turn the color of legal litmus paper."[159]

According to Justice Holmes, "Only the emergency that makes it immediately dangerous to leave the correction of evil counsels to time warrants making any exception to the sweeping command, 'Congress shall make no law . . . abridging the freedom of speech.'"[160] But when does such an "emergency" arise? Holmes himself had provided the answer a few months earlier in another case: when "the words used are used in such circumstances and are of such a nature as to create a clear and present danger that they will bring about the substantive evils that Congress has a right to prevent."[161]

Under this Clear and Present Danger Test, speech may be restricted only if there is a real threat—a danger, both clear and present, that the speech will lead to an evil that the legislature has the power to prevent. In the *Abrams* case, the legislature had the right to pass a law to prevent curtailment of war production; but, said Holmes, there was no danger, clear and present, or even remote, that the leaflets would have had any effect on production.

Justice Brandeis tells us that the Clear and Present Danger Test, as stated by Justice Holmes, "is a rule of reason. Correctly applied, it will preserve the right of free speech both from suppression by tyrannous, well-meaning majorities and from abuse by irresponsible, fanatical minorities."[162]

The Holmes test is above all a test of degree. "Clear and present" danger is a standard, not a mathematical absolute. "It is a question of proximity and degree," said Holmes, after the passage stating the test quoted above.[163] As such, its application will vary from case to case and will depend upon the particular circumstances presented.

That the Holmes test is sound can be seen from the analogy of the law of criminal attempts. Just as a criminal attempt must come sufficiently near completion to be of public concern, so there must be an actual danger that inciting speech will bring about an unlawful act before it can be restrained.

Thus, if I gather sticks and buy some gasoline to start a fire in a house miles away and do nothing more, I cannot be punished for attempting to commit arson. However, if I put the sticks against the house and pour on some gasoline and am caught before striking a match, I am guilty of a criminal attempt. The fire is the main thing, but when no fire has occurred, it is a question of the nearness of my behavior to the outbreak of a fire. So under the Constitution, lawless acts are the main thing. Speech is not punishable as such, but only because of its connection to lawless acts.

But more than a remote connection is necessary, just as with the attempted fire. The fire must be close to the house; the speech must be close to the lawless acts. So long as the speech is remote from action, it is protected by the Constitution.[164] But if the speech will result in action that government can prohibit, then the speech itself can constitutionally be reached by governmental power, provided there is a clear and present danger that the action will result from the speech.

It is true that the Holmes doctrines discussed—both that of judicial restraint and that of Clear and Present Danger—were stated almost entirely in dissents. But those dissents were prime examples of Chief Justice Hughes's characterization of a dissent as "an appeal to . . . the intelligence of a future day."[165] When Harold J. Laski was told by an English lawyer that he was amazed that Holmes was not speaking for the Court in *Abrams,* he wrote to the Justice, "I explained that you were speaking for the Court of the next decade."[166] And so it turned out (though Laski was overoptimistic on the time it would take for the Holmes approach to become accepted doctrine).

Today, more than ever, we can see that it was Justice Holmes who set Supreme Court jurisprudence on its coming course. When Holmes asserted in his *Common Law,* "The life of the law has not been logic: it has been experience," and that the law finds its philosophy in "consideration of what is expedient for the community concerned," he was sounding the clarion of twentieth-century law. If the law reflected the "felt necessities of the time,"[167] then those needs rather than any theory should determine what the law should be. These were not, to be sure, the views followed by American judges and lawyers at the beginning of this century—or even by the majority of the Supreme Court during Holmes's tenure. But the good that men do also lives after them. If the late nineteenth century was dominated by the passive jurisprudence of the Fuller Court, the twentieth was, ultimately, to be that of Mr. Justice Holmes.

10

Hughes Court, 1930–1941

In 1935 the Supreme Court moved into its new building across the plaza from the Capitol. For the first time in its history, the Court had a home of its own—a magnificent Marble Palace symbolizing its role as the ultimate embodiment of Equal Justice Under Law—the motto inscribed on the frieze above the front entrance.

The new building at last presented the Court with a physical setting that matches the splendor of its constitutional role. Whatever purely architectural criticisms may be directed at the Court building, it cannot be denied that it is one of the most imposing structures in Washington. Its design is modeled on a classic Greek temple and is intended as a shrine to the majesty of the law. Its physical scale is truly impressive. Both longer and wider than a football field, it is four stories high and is constructed almost entirely of marble. At the main entrance are sixteen huge Corinthian columns, topped by a pediment; in its center, Liberty Enthroned, holding the scales of justice, guarded by Order on one side and Authority on the other.

The interior of the Court building is as impressive as its exterior. As a Smithsonian pamphlet describes it, "Six kinds of marble, three domestic and three foreign, along with thousands of feet of clear-grained white oak, give the building a feeling of polished, dust-free smoothness. Openwork gates, elevator doors, even the firehose cupboards are of gleaming bronze. Two spiral marble staircases, self-supporting marvels of engineering, soar from garage to the top floor, five levels up. For security's sake, these are unused—but only the Vatican and the Paris Opera can boast similar ones."

The building also has abundant office space, both for Justices' chambers and Court staff, and a magnificent courtroom—as well as ample amenities, including a minigymnasium and basketball court on the top floor, which Court personnel like to call "the highest court in the land."

When Justice Harlan F. Stone visited the new Court building, he wrote to his sons, "The place is almost bombastically pretentious, and thus it seems to me wholly inappropriate for a quiet group of old boys such as the Supreme Court of the United States."[1] When the Court was preparing to move into the new edifice, one of the Justices sardonically asked, "What are we supposed to do, ride in on nine elephants?"

But traditions give way slowly at the Supreme Court. There are still spittoons behind the bench for each Justice, pewter julep cups (now used for their drinking water), and, not long ago, there were quill pens at counsel tables. After the Court formally moved from the cramped quarters in the old Senate Chamber into the new building, most of the Justices resented the change and declined to use the handsome new suites provided for them. With Chief Justice Hughes, they continued to work in their homes, as they had been accustomed to do. One result of this was that Hugo L. Black, the first "new" Justice to occupy chambers in the building, and at that time the junior member of the Court, got his pick and was able to choose a splendid corner suite, which he kept for the remaining thirty-four years of his life (instead of the least desirable space normally left to the junior Justice).

Chief Justice Hughes and His Court

The new Court building was actually the result of Chief Justice Taft's efforts. "We ought to have a building by ourselves . . . as the chief body at the head of the judiciary," Taft told a key Senator and his efforts bore fruit when Senator Reed Smoot wrote to him that the Court would be included in an appropriation for new buildings.[2] Taft's lobbying ultimately ensured funding both for a building site and the new Court building itself. At the 1932 cornerstone ceremony, Chief Justice Hughes declared that the building was "the result of his [predecessor's] intelligent persistence."[3]

"My prayer," Taft wrote to his nephew in 1927, "is that I may stay long enough on the Court to see that building constructed."[4] Taft was not to be granted his wish. A complete breakdown in health forced his resignation on February 3, 1930, and he died a month later.

On the day Chief Justice Taft resigned, President Hoover nominated Charles Evans Hughes to succeed him. Hoover later wrote, "It was the obvious appointment."[5] Despite Hughes's eminent qualifications, his nomination led to the bitterest confirmation fight over a Court head since Chief Justice Taney was named a century earlier. The chief opponent was Senator George Norris, who asserted, "No man in public life so exemplifies the influence of powerful combinations in the political and financial world as does Mr. Hughes."[6] Despite the opposition, Hughes was confirmed by fifty-two to twenty-six.

Chief Justice Hughes was sixty-eight when he was appointed—the oldest man chosen to head the Court. However, he undertook his new duties with the vigor of a much younger person. In addition, his more than distinguished career endowed him with prestige which few in the highest judicial office have had. "He took his seat at the center of the Court," Justice Frankfurter was to write, "with a mastery, I suspect, unparalleled in the history of the court."[7]

As a leader of the Court, Hughes must be ranked with Marshall and Taney. Whatever the test of leadership, in Frankfurter's words, "Chief Justice Hughes possessed it to a conspicuous degree. In open court he exerted this authority by the mastery and distinction with which he presided. He radiated this authority in the conference room."[8]

The Hughes manner of conducting the Court's business has been described by Frankfurter, who had served as a Justice under him. "In Court and in conference he struck the pitch, as it were, for the orchestra. He guided discussion by opening up the lines for it to travel, focusing on essentials, evoking candid exchange on subtle and complex issues, and avoiding redundant talk. He never checked free debate, but the atmosphere which he created, the moral authority which he exerted, inhibited irrelevance, repetition, and fruitless discussion."[9]

In another passage Frankfurter writes that the "word which for me best expresses the atmosphere that Hughes generated . . . was taut. Everything was taut. He infected and affected counsel that way. Everybody was better because of Hughes, the leader of the orchestra."[10]

The Hughes conferences were militarylike models of efficiency. According to Justice Brandeis these "lasted for six hours and the Chief Justice did virtually all the speaking."[11] This was an exaggeration, but Hughes is still noted for his tight control over the conference discussion. Rarely did any Justice speak out of turn and Hughes made sure that the discussion did not stray from the issues he had stated in his incisive presentation. For him, Justice Frankfurter later told an interviewer, "the conference was not a debating society."[12] Frankfurter concludes, "To see [Hughes] preside was like witnessing Toscanini lead an orchestra."

The new Chief Justice's leadership abilities were precisely what the Court needed to enable it to confront its most serious crisis in almost a century. Some years earlier, Justice Holmes had written that, while things at the Court were quiet then, it was really only "the quiet of a storm centre."[13] The same could be said of the Court when Hughes took its center chair.

From outward appearances, the Hughes Court also seemed quiet—ready to continue along the conservative path taken by its predecessors. Yet from almost the beginning of the new Chief Justice's tenure, it was "enclosed in a tumultuous privacy of storm."[14] In decisions like that in the

Adkins case,[15] the Court's conservative core had carried their laissez-faire interpretation of the Constitution to the point where there was, in Holmes's phrase, "hardly any limit but the sky to the invalidating of [laws] if they happen to strike a majority of the Court as for any reason undesirable."[16] But changing conditions in the society itself were about to render the jurisprudence of the day obsolete.

When Chief Justice Hughes ascended the bench early in 1930, the country was already deep in the most serious economic crisis in our history. The crisis only became worse as the Hughes term went on—putting the entire leadership of the country, and not least the Court itself, to perhaps its most severe test.

The new Chief Justice had to meet that test with a Court composed almost entirely of Justices who had served under his predecessor. Four of the five Justices who had made up the *Adkins* majority (Van Devanter, McReynolds, Sutherland, and Butler) were still on the bench, serving as the nucleus for decisions extending the Taft Court's laissez-faire jurisprudence. It is true that the Court also contained Justices Holmes, Brandeis, and Stone as its liberal bloc. They could act as a brake upon the tendency to further treat the Constitution as a legal sanction for laissez faire—particularly if they were joined by the new Chief Justice and Justice Roberts, who came to the Court at the same time as Chief Justice Hughes.

Owen J. Roberts had been a successful practicing lawyer in Philadelphia who attracted national attention as prosecutor in the Teapot Dome scandal. Roberts had been appointed to fill Justice Sanford's seat upon the latter's death in 1930, after the Senate had turned down Circuit Judge John J. Parker's nomination. Roberts was a better than average Justice—but one who, as he himself recognized, would never be placed in the judicial pantheon. "Who am I," he wrote after leaving the bench, "to revile the good God that he did not make me a Marshall, a Taney, a Bradley, a Holmes, a Brandeis or a Cardozo."[17]

Justice Roberts was nevertheless to play a crucial role as the "swing vote" in the Hughes Court. It was his key vote that enabled the conservative bloc to prevail in the decisions striking down the important New Deal measures during the first part of Chief Justice Hughes's tenure. It was Justice Roberts's vote as well that enabled the Chief Justice to bring about the great jurisprudential reversal that took place in 1937.

Before we discuss the Hughes Court's jurisprudence, a word should be said about another personnel change, for it involved two of the outstanding Justices in the Court's history. On January 3, 1932, after the Justices had heard oral arguments, Justice Holmes casually announced, "I won't be here tomorrow," and he submitted his resignation later that day. On that day coincidentally, Earl Warren, then a California district attorney, had argued his first case before the Court. Warren used to say that his friends accused him of driving Holmes from the bench. They used to tease him— "one look at you and he said, 'I quit.'"

The departure of Holmes meant that the Court had lost one of its giants—the Justice who more than any other had set the theme for twentieth-century jurisprudence. Then a most unusual thing happened. The all but unanimous national consensus was that only one man deserved the succession to Holmes—Benjamin Nathan Cardozo. From a political point of view, his appointment seemed impossible. Cardozo was a New Yorker and there were already two on the Court from New York (Hughes and Stone); he was a Jew and there was already a Jewish Justice (Brandeis); and a conservative Republican President could scarcely name a liberal not a member of his party.

The sentiment in Cardozo's favor overrode these objections. Justice Stone told President Hoover that he was willing to resign to overcome the geographical objection and the powerful Senator from Idaho, William E. Borah, told the President that Cardozo belonged as much to Idaho as to New York. Borah went on, "Just as John Adams is best remembered for his appointment of John Marshall to the Supreme Court, so you, Mr. President, have the opportunity of being best remembered for putting Cardozo there."[18] And so it turned out. One of the few positive things for which President Hoover is remembered is his appointment of Cardozo to fill the Holmes seat.

Except for Holmes himself, Justice Cardozo was the preeminent judge of the first half of the twentieth century. Indeed, Cardozo was the outstanding common-law jurist of the twentieth century. It was he who led the way in adapting the common law to the requirements of the postindustrial society. Cardozo showed how traditional principles and techniques could be used to effect a complete change in the relationship between the law and individual rights of substance.

In American law, Cardozo remains the consummate legal craftsman, the master of the principles, ideals, and techniques of Anglo-American law. More than any other judge, Roscoe Pound once pointed out, "he has known the tools of his craft and his known how to use them."[19]

There was little drama in Cardozo's life. He was born in New York City, the son of a judge who was besmirched by his association with the notorious Tweed Ring. During his youth, he had a number of tutors, including Horatio Alger. After study at Columbia College and Law School and admission to the New York Bar in 1891, Cardozo became a lawyer's lawyer, to whom other attorneys referred difficult cases. He became a judge in 1914, serving eighteen years (five of them as chief judge) on the New York Court of Appeals and then six years on the United States Supreme Court. Though he was to be granted less than six full terms on the highest bench, Cardozo was able to play an important role as a member of the liberal wing which first resisted the Hughes Court's early reactionary decisions and then led the great changeover in the Court's constitutional jurisprudence.

Cardozo's major contribution to our law was his use of traditional judicial techniques to adapt the law to society's changing requirements. To Cardozo, the job of the judge was to adapt the experience of the past so that it would best serve the needs of the present. More than any judge, he showed how the common-law technique could be adapted for contemporary use. Reasoning by analogy, he showed how existing doctrines could be adapted to new needs. His mastery of judicial technique made the emerging law appear to be the logical product of established doctrines; in his hands the changing common law was made a blend of both continuity and creativeness.[20]

It was Cardozo who, next to Holmes, most marked the transition from the concept of law that prevailed a century ago. Cardozo was the first judge who explained systematically how judges reason and who made the first serious effort to articulate a judicial philosophy.[21] But he was not just an academic legal theorist; his judicial tenure gave him ample opportunity to apply his philosophy to the needs of the changing law.

Cardozo did more than demonstrate that law was Heraclitean rather than Newtonian in nature. To him, the law was neither an *is* nor an *ought;* it was also an endless *becoming.* Far from being the static system posited by turn-of-the-century jurists, "even now there is change from decade to decade." Since Cardozo, we have not doubted that "the end of the law" should "determin[e] the direction of its growth."[22] Cardozo showed how the common law could be freed to serve present needs—how the judge could be truly innovative while remaining true to the experience of the past. By doing so, he helped to move the law closer to the goal of making the law an effective instrument of social welfare.[23]

First Hughes Court

Justice Frankfurter once said that Chief Justice Hughes "was, in fact, the head of two courts, so different . . . was the supreme bench in the two periods of the decade during which Hughes presided over it."[24] The first Hughes Court sat from the Chief Justice's appointment in 1930 until 1937. The period was dominated by the decisions which both nullified the most important New Deal legislation and restricted state regulatory power. In both respects, the Court confirmed the laissez-faire jurisprudence of its predecessors. But the Court was sharply divided in its controversial decisions.

Just before Hughes's appointment, Chief Justice Taft, failing in health, gave voice to his concerns about the continued conservatism of the Court. The most that could be hoped for, he wrote Justice Butler, "is continued life of the present membership . . . to prevent disastrous reversals of our present attitude. With Van [Van Devanter] and Mac [McReynolds] and Sutherland and you and Sanford, there will be five to steady the boat . . . we must not give up at once."[25] But Chief Justice Taft and Justice Sanford both died a few months later. Now the Court's conservative majority itself might be in danger.

Taft's stricture about President Hoover—"that it is just as well for him to remember the warning in the Scriptures about removing landmarks"[26]—appeared borne out by some of the early Hughes Court decisions. The first of them, *Home Building & Loan Association v. Blaisdell* (1934),[27] upheld the governmental power to enact moratory laws for the relief of debtors during the Great Depression. The Hughes opinion found that the economic emergency furnished legitimate occasion for the exercise of the police power to protect both debtors and the community against the collapse of values that had occurred. To the four dissenters, the law was a patent violation of the categorical prohibition against "impairing the obligation of contracts."[28] Yet, as Justice Cardozo put it in an undelivered concurrence, while the moratory law "may be inconsistent with things" which the Framers believed, "their beliefs to be significant must be adjusted to the world they knew. It is not . . . inconsistent with what they would say today."[29]

Another case in which the early Hughes Court adjusted the intent of the Framers to contemporary needs was *Nebbia v. New York* (1934),[30] where a New York law fixing milk prices was ruled valid. In his opinion of the Court, Justice Roberts transformed the Court's attitude toward the legality of price regulation by doing away with the limited category of "businesses affected with a public interest" upon which the price-fixing power had until then been based. In doing so, Justice Frankfurter tells us, "Roberts wrote the epitaph on the misconception, which had gained respect from repetition, that legislative price-fixing as such was at least presumptively unconstitutional."[31] Instead, price fixing, like other types of business regulation, was to be sustained when there was a reasonable relationship between it and the social interests which may be vindicated by the police power.

In the *Blaisdell* and *Nebbia* cases, Chief Justice Hughes and Justice Roberts had joined with Justices Brandeis, Stone, and Cardozo. Liberal commentators expressed hope that the constitutional tide had now turned in their direction. That hope, however, proved short-lived when Justice Roberts joined the four *Blaisdell-Nebbia* dissenters in striking down a federal law requiring railroads to contribute to a pension fund for aged employees.[32] The decision was based upon a restricted conception of the commerce power that would also form the basis for the invalidation in the next two years of the most important New Deal measures—particularly when the Chief Justice joined Justice Roberts in his shift to the conservative wing of the Court.

New Deal Cases

"During the past half century," asserted President Franklin D. Roosevelt in 1937, "the balance of power between the three great branches of the Federal Government, has been tipped out of balance by the Courts in direct contradiction of the high purposes of the framers of the Constitution."[33] What has been called government by judiciary was dramatically illustrated by the Supreme Court reception of the early New Deal laws.

Based on the need to resuscitate the depressed economy by extended governmental intervention, the New Deal program involved the negation of
laissez faire; it meant a degree of economic regulation far greater than any
previously attempted. If the country was to go forward, said President
Roosevelt in his First Inaugural Address, "we must move as a trained and
loyal army willing to sacrifice for the good of a common discipline, because
without such discipline no progress is made, no leadership becomes effective."[34]

The effort to move the nation forward came up against the restricted
view of governmental power still held by the Supreme Court. The result
was a series of decisions that invalidated most of the important New Deal
legislation. In 1935 and 1936 cases, the Supreme Court struck down the
two key New Deal antidepression measures, the National Industrial Recovery Act and the Agricultural Adjustment Act.[35] Both measures were
held beyond the reach of the federal commerce power.

The NIRA was ruled invalid as applied to small wholesale poultry
dealers in Brooklyn. The business done by them was purely local in character, even though the poultry handled by them came from outside the state.
And, under the Court's approach, it did not make any difference that there
was some effect upon interstate commerce by the business being regulated.
Similarly, in holding the AAA unconstitutional, the Court replied upon
the proposition that agriculture, like manufacturing, is not commerce and
hence is immune from federal control.

Nor were these decisions all. Still other cases struck down measures
providing for regulation of the bituminous coal industry, municipal bankruptcy relief, and farm debtors' relief.[36]

The decisions alone were bad enough. Even worse, however, was the
manner in which the decrees of invalidity were delivered. Speaking in 1928
of Supreme Court decisions setting aside statutes enacted by Congress,
Chief Justice Hughes had asserted that few of these cases had been of great
importance in shaping the course of the nation.[37] The same could not be
said of the decisions nullifying the New Deal measures. As far as the
measures themselves were concerned, even their proponents had to concede that many of them were imperfectly conceived and crudely executed.[38] There is little doubt that they were subject to legitimate constitutional attack. Had the Court in its decisions confined itself to these limited
constitutional issues, at the same time leaving the way open for Congress
to remedy the defects by tighter draftsmanship, it would hardly have been
subjected to such bitter controversy.

But the Court deliberately did not choose the more prudent course.
To paraphrase Justice Robert H. Jackson,[39] in striking at the New Deal
legislation, the Court allowed its language to run riot. It sought to engraft
its own nineteenth-century laissez-faire philosophy into the Constitution.
In invalidating the National Industrial Recovery Act, the Court did not
limit itself to criticism of the obviously objectionable features of that law;
the rationale of its decision was, instead, so broad that it struck at all

national efforts to maintain fair industrial and labor standards. Similarly, in overthrowing the Agricultural Adjustment Act, the Court cast doubt upon all federal aid to agriculture, as well as upon any extensive use of the congressional power to tax and spend in order to promote the general welfare.

The Court's action in this respect came to its culmination just before the 1936 election, when it followed *Adkins* and ruled that there was no power in either states or nation to enact a minimum-wage law.[40] In the words of a contemporary critic, "The Court not merely challenged the policies of the New Deal but erected judicial barriers to the reasonable exercise of legislative powers, both state and national, to meet the urgent needs of the twentieth-century community."[41]

The narrow interpretation of governmental power in these decisions was catastrophic. "We have . . . reached the point as a Nation," President Roosevelt declared, "where we must take action to save the Constitution from the Court."[42] Elimination of manufacturing, mining, and agriculture from the reach of federal power had rendered Congress powerless to deal with problems in those fields, however pressing they might become.

Obviously the rejection of governmental power in the New Deal decisions fitted perfectly with the restricted theory of governmental function that had dominated American thinking for the previous half century. The Constitution, stated Justice Holmes in a noted passage, "is not intended to embody a particular economic theory, whether of paternalism and the organic relation of the citizen to the state or of laissez faire."[43] But it was most difficult for judges not to assume that the basic document was intended to embody the dominant economic beliefs on which they had been nurtured.

The grim economic background behind the New Deal measures, however, indicated how totally unrealistic was reliance on laissez faire. Giant industries prostrate, nationwide crises in production and consumption, the economy in a state of virtual collapse—if ever there was a need for exertion of federal power, it was after 1929. The market and the states had found the crisis beyond their competence. The choice was between federal action and chaos. A system of constitutional law that required the latter could hardly endure. The New Deal decisions, the Supreme Court was later to concede, "produced a series of consequences for the exercise of national power over industry conducted on a national scale which the evolving nature of our industrialism foredoomed to reversal."[44]

Constitutional Revolution, Ltd.

President Roosevelt's answer to the judicial decisions was his "Court-packing" plan of February 5, 1937. Under it, the President could appoint another judge for every federal judge who was over seventy and had not retired. This would have given the President the power to appoint six new Supreme Court Justices.

After lengthy hearings and public discussion, the Senate Judiciary Committee rejected the plan. Yet, if the President lost the Court-packing battle, he was ultimately to win the constitutional war, for the Supreme Court itself was soon to abandon its restrictive approach to the proper scope of governmental power. Hence, in Justice Jackson's summary of the Court-packing fight, "Each side of the controversy has comforted itself with a claim of victory. The President's enemies defeated the court reform bill—the President achieved court reform."[45]

A remarkable reversal in the Supreme Court's attitude toward the New Deal program took place early in 1937. From 1934 through 1936, the Court rendered twelve decisions declaring New Deal measures invalid; starting in April 1937, that tribunal upheld every New Deal law presented to it, including some that were basically similar to earlier nullified statutes. It is, in truth, not too far-fetched to assert that in 1937 there was a veritable revolution in the Court's jurisprudence, which one commentator characterized as "Constitutional Revolution, Ltd."[46]

It is too facile to state that the 1937 change was merely a protective response to the Court-packing plan, to assert, as did so many contemporary wags, that "a switch in time saved Nine." The furor over the President's proposal obviously had repercussions within the Court's marble halls. As FDR himself expressed it, "It would be a little naive to refuse to recognize some connection between these decisions and the Supreme Court fight."[47] At the same time, it misconceives the nature of the Supreme Court and its manner of operation as a judicial tribunal to assume that the 1937 change in jurisprudence was solely the result of the Court-packing plan. The 1937 reversal reflected changes in legal ideology common to the entire legal profession. The extreme individualistic philosophy upon which the Justices had been nurtured had been shaken to its foundations. If laissez-faire jurisprudence gave way to judicial pragmatism, it simply reflected a similar movement that had taken place in the country as a whole.

In an unpublished 1934 draft opinion, Justice Cardozo had asserted, "A gospel of laissez faire . . . may be inadequate in the great society that we live in to point the way to salvation, at least for economic life."[48] The conception of the proper role of government upon which the 1934–1936 decisions were based was utterly inconsistent with an era which demanded ever-expanding governmental authority. "Leviathan hath two swords: war and justice," states Thomas Hobbes in a famous passage. The need effectively to deal with the great economic crisis of the 1930s had, nevertheless, made it plain that the armory of the state had to include much more than these two elementary weapons. Before the New Deal, government was chiefly negative; its main task, apart from defense, was to support the status quo and maintain some semblance of fair play while private interests asserted themselves freely. Under the Roosevelt Administration, government became positive in a new sense.

For the Supreme Court, Canute-like, to attempt to hold back indefinitely the waves of ever-increasing governmental authority was to set itself an impossible task. "Looking back," declared Justice Roberts (the man whose switch is, more than anything else, said to have "saved the Nine" in 1937) in 1951, "it is difficult to see how the court could have resisted the popular urge for uniform standards throughout the country—for what in effect was a unified economy."[49] The laissez-faire doctrine, upon which the operation of American government had been essentially based since the founding of the Republic, had by then proved inadequate to meet pressing economic problems. As the Court recently put it, "the Depression had come and with it the lesson that seemed unmistakable to most people by 1937, that the interpretation of contractual freedom protected in *Adkins* rested on fundamentally false factual assumptions about the capacity of a relatively unregulated market to satisfy minimum levels of human welfare."[50]

The national economy could be resuscitated only by extended federal intervention. For the Government in Washington to be able to exercise regulatory authority upon the necessary national scale, it was essential that the Supreme Court liberalize its construction of the Constitution. To quote Justice Roberts again, "An insistence by the court on holding federal power to what seemed its appropriate orbit when the Constitution was adopted might have resulted in even more radical changes in our dual structure than those which have been gradually accomplished through the extension of the limited jurisdiction conferred on the federal government."[51]

In any event, there *was* a real conversion in a majority of the Supreme Court and its effects do justify the "constitutional revolution" characterization. And it is usually overlooked that the decision first signaling the reversal in jurisprudence was reached before the President introduced his Court-packing plan. On March 29, 1937, Chief Justice Hughes announced a decision upholding a state minimum-wage law, basically similar to one the Court had held to be beyond the power of both states and nation to enact only nine months previously.[52] The Court's confession of error was announced a month after the President's proposal, but the case itself was decided in conference among the Justices about a month before the publication of the Court-packing plan. This circumstantial evidence strongly bears out the statement made some years later by Chief Justice Hughes to his authorized biographer: "The President's proposal had not the slightest effect on our decision."[53]

It should, however, be noted that, even with the majority change in jurisprudence, a hard core of the 1934–1936 majority utterly refused to alter its views—which shows how narrow the actual margin of change was in the Court. In reality, it was the recognition by two Justices (primarily Justice Roberts and, to a lesser extent, Chief Justice Hughes) of the need for increased national governmental power that made for the switchover in the high tribunal. "Years ago," wrote an eminent professor of constitutional law of the Court in the mid-thirties, "that learned lawyer John

Selden in talking of 'Council' observed: They talk (but blasphemously enough) that the Holy Ghost is President of their General Councils when the truth is, the odd Man is still the Holy Ghost.'"[54] It was the conversion of "odd men" Roberts and Hughes that made the constitutional revolution of 1937 possible.

301 U.S.

Narrow though the margin for change may have been, there is little doubt that there was a real conversion among the new majority of the Supreme Court. On March 29, 1937, the Chief Justice announced the opinion upholding a state minimum-wage law,[55] in the process overruling both *Adkins* and a decision following that case made only nine months earlier.[56] According to one who sat at the Government counsel table that day, "[T]he spectacle of the Court that day frankly and completely reversing itself and striking down its opinion but a few months old was a moment never to be forgotten."[57]

March 29, 1937, as already indicated, saw the upholding of a state minimum-wage law. Though not directly concerned with national power, the Court did expressly overrule *Adkins,* which had denied congressional authority to fix wages; hence the decision was a substantial step forward in the movement toward increased national power. On the same day, the Court dealt squarely with federal statutes similar to several annulled in the 1934–1936 period. This time the Court upheld laws providing for farm debtors' relief, collective bargaining in the nation's railroads, and a penalizing tax on firearms analogous to that which it had struck down under the Agricultural Adjustment Act.[58] Well could a leading New Dealer chortle, "What a day! To labor, minimum-wage laws and collective bargaining; to farmers, relief in bankruptcy; to law enforcement, the firearms control. The Court was on the march!"[59]

These cases, in Volume 300 of the *Supreme Court Reports,* were to prove but the prelude to an even more drastic revolution in consitutional jurisprudence. To demonstrate the extent of the judicial revolution, one has, to use the method stated by Edward S. Corwin, only to "turn to Volume 301 of the *United States Supreme Court Reports,* a volume which has a single counterpart in the Court's annals. I mean Volume 11 of Peters's *Reports,* wherein is recorded the somewhat lesser revolution in our constitutional law precisely 100 years earlier, which followed upon Taney's succession to Marshall."[60]

On page 1 of 301 U.S., there is printed the April 12, 1937, decision of the Court in *National Labor Relations Board v. Jones & Laughlin Steel Corp.*[61] In it, the constitutionality of the National Labor Relations Act of 1935 was upheld. Robert H. Jackson termed the decision there the most far-reaching victory ever won on behalf of labor in the Supreme Court.[62] This was no overstatement, for the 1935 act was the Magna Carta of the American labor movement. It guaranteed the right of employees to organize collectively in unions and made it an unfair labor practice prohibited by law for employers to interfere with that right or to refuse to bargain collectively with the representatives chosen by their employees.

The Labor Act was intended to apply to industries throughout the nation, to those engaged in production and manufacture as well as to those engaged in commerce, literally speaking. But this appeared to bring it directly in conflict with the decisions drastically limiting the scope of the Federal Government's authority over interstate commerce, including some of the decisions of the 1934–1936 period on which the ink was scarcely dry. This was particularly true of the Court's decision nullifying the National Industrial Recovery Act, which had denied power in Congress to regulate local business activities, even though they affected interstate commerce. In the *Jones & Laughlin* case, these precedents were not followed: "These cases," laconically stated the Court, "are not controlling here."[63]

Instead, the Court gave the federal power over interstate commerce its maximum sweep. Mines, mills, and factories, whose activities had formerly been decided to be "local" and hence immune from federal regulation, were now held to affect interstate commerce directly enough to justify congressional control. There is little doubt that, as the dissenting *Jones & Laughlin* Justices protested, Congress in the Labor Act exercised a power of control over purely local industry beyond anything theretofore deemed permissible.

The *Jones & Laughlin* case was followed six weeks later by three equally significant decisions, also printed in Volume 301 of the *Supreme Court Reports,* upholding the constitutionally of one of the most important of the New Deal innovations, the Social Security Act of 1935. That law, which for the first time brought the Federal Government extensively into the field of social insurance, had been held unconstitutional by the lower court. The Supreme Court, however, in a precedent-making opinion by Justice Cardozo, reversed, holding that the scheme of old-age benefits provided for by the federal law did not contravene any constitutional prohibition.[64] In so doing, the Court gave the broadest possible scope to the congressional power to tax and spend for the general welfare, even though its reasoning on this point was inconsistent with its 1936 decision invalidating the Agricultural Adjustment Act.

In addition, the Court upheld the unemployment compensation schemes established under the Social Security Act.[65] The decisions sustaining that law put an end to fears that unemployment insurance and old-age benefit laws might prove beyond the power of either states or nation, as minimum-wage regulation had been held to be under the pre-1937 Court. Henceforth the United States was not to be the one great nation powerless to adopt such measures.

These decisions in 301 U.S. formed the heart of the constitutional revolution of 1937. Breaking with its previous jurisprudence, the Supreme Court upheld the authority of the Federal Government to regulate the entire economy under its commerce power and to use its power to tax and spend to set up comprehensive schemes of social insurance. And, it should be noted in light of later criticisms of the Court whose members were subsequently appointed by President Roosevelt, because of its claimed cavalier discard of established precedents, the cases in 301 U.S. were decided before a single Roosevelt-appointed Justice took his seat upon the bench. The most important of the old precedents which so restricted the scope of governmental authority were repudiated by the identical Court that had previously invoked them. There was no change in the Court's personnel until after it provided new precedents that served as a basis for much that the new judges were later to decide.

Second Hughes Court

The 1937 reversal marked the accession of what may be considered the second Hughes Court—so different was its jurisprudence from that of the Hughes Court that had preceded it. As just stressed, the new decisions were the product of the same Justices who had sat in the first Hughes Court. But the composition of the Court was now to change drastically as President Roosevelt was given the opportunity to "pack" the Court through the retirement or death of most of its members. Among the new Roosevelt-appointed Justices were some of the greatest names in modern Supreme Court history.

The first of them was Hugo L. Black of Alabama, a leading New Deal Senator, who was chosen to fill the first vacancy in five years, after Justice Van Devanter retired after the 1936 Term. The Black appointment was a controversial one because he had been a member of the Ku Klux Klan. In a short time, however, the furor over the disclosure that he had once been a member of the Ku Klux Klan seemed an echo from another world. "At every session of the Court," a *New York Times* editorial thundered, after Black's Klan membership had been revealed, "the presence on the bench of a justice who has worn the white robe of the Ku Klux Klan will stand as a living symbol of the fact that here the cause of liberalism is unwittingly betrayed." Within a few years, Justice Black became a leader of the Court's liberal wing.

Justice Black never forgot his origins in a backward Alabama rural county. But his Alabama drawl and his gentle manners masked an inner firmness found in few men. "Many who know him," wrote Anthony Lewis when Black turned seventy-five, "would agree with the one-time law clerk who called him 'the most powerful man I have ever met.'"[66] Though of only slight build, Black always amazed people by his physical vitality. He is quoted in *The Dictionary of Biographical Quotation* as saying, "When I was forty my doctor advised me that a man in his forties shouldn't play tennis. I heeded his advice carefully and could hardly wait until I reached fifty to start again."

Black's competitive devotion to tennis became legend. Until he was eighty-three, he continued to play several sets every day on the private court of his landmark federal house in the Old Town section of the Washington suburb of Alexandria. He brought the same competitive intensity to his judicial work. According to his closest colleague, Justice William O. Douglas, "Hugo Black was fiercely intent on every point of law he presented."[67] Black was as much a compulsive winner in the courtroom as on the tennis court. "You can't just disagree with him," acidly commented his great Court rival, Justice Jackson, to a columnist. "You must go to war with him if you disagree."[68] Black would fight bitterly on the issues that concerned him, such as the First Amendment.

If impact on the law is a hallmark of the outstanding judge, few occupants of the bench have been more outstanding than Black. It was Justice Black who fought for years to have the Court tilt the Constitution in favor of individual rights and liberties and who was a leader in what Justice Fortas once termed "the most profound and pervasive revolution ever achieved by substantially peaceful means."[69] Even where Justice Black's views have not been adopted literally, they have tended to prevail in a more general, modified form. Nor has his impact been limited to the Black positions that the Court has accepted. It is found in the totality of today's judicial awareness of the Bill of Rights and the law's newly intensified sensitivity to the need to apply its protection to all.

More than anything, Justice Black brought to the Supreme Court a moral fervor rarely seen on the bench. A famous passage by Justice Holmes has it that the black-letter judge will be replaced by the man of statistics and the master of economics.[70] Justice Black was emphatically a judge who still followed the black-letter approach in dealing with the constitutional text. "That Constitution," he said, "is my legal bible. . . . I cherish every word of it from the first to the last."[71] The eminent jurist with a dog-eared copy of the Constitution in his right coat pocket became a part of contemporary folklore. In protecting the sanctity of the organic word, Justice Black displayed all the passion of the Old Testament prophet in the face of graven idols. His ardor may have detracted from the image of the "judicial," but if the Justice did not bring to constitutional issues that "cold neutrality" of which Edmund Burke speaks,[72] his zeal may have been precisely what was needed in the Supreme Court. Anything less might have been inadequate to make the Bill of Rights the vital center of our constitutional law.

Justice Black's principal supporter on the Court was William O. Douglas, who succeeded Justice Brandeis, forced to retire in 1939 after a heart attack. To outside observers, Justice Douglas seemed the personification of the last frontier—the down-to-earth Westerner whose granite-hewed physique always seemed out of place in Parnassus. More than that, Douglas was the Court's Horatio Alger, whose early life was a struggle against polio and poverty. Told that he would never walk, he became a noted sportsman. He came east on a freight car to enroll at Columbia Law School, with six cents in his pocket, and went on to become an eminent law professor, chairman of the Securities and Exchange Commission, and at the age of forty a Supreme Court Justice—the youngest Supreme Court appointee in over a century.

On the bench, Justice Douglas was known as Justice Black's chief ally. "If any student of the modern Supreme Court took an association test," wrote Hugo Black, Jr., in his book about his father, "the word 'Black' would probably evoke the response 'Douglas' and vice versa."[73] Justice Black himself recognized this. Declining a 1958 invitation to write an article about Justice Douglas, Justice Black wrote, "our views are so nearly the same that it would be almost like self praise for me to write what I feel about his judicial career."[74]

Yet, even though he was normally to be found in the Black column, on the bench, as in his personal life, Justice Douglas always was a maverick, who went his own way regardless of the feelings of the other Justices. Justice Douglas would stick to his own views, and he was quick to use his own method of deciding in concurrence or dissent. It made little difference whether he carried a majority or stood alone. He would rarely stoop to lobbying for his position and seemed more interested in making his own stand public than in working to get it accepted. As a law clerk once put it to me, "Douglas was just as happy signing a one-man dissent as picking up four more votes."

If Justice Douglas was the strongest Black supporter, Justice Black's greatest intellectual adversary on the Court was Felix Frankfurter, who became a member of the Court just before Douglas did in 1939, replacing Justice Cardozo. Few members of the Supreme Court have been of greater interest than Frankfurter both to the public and to Court specialists. In large measure, this has been true because his career poses something of a puzzle. Before his appointment to the bench, he was known for his interest in libertarian causes. He was also closely connected in the public mind with the New Deal, and it was generally expected that, once on the Court, he would continue along a liberal path. Yet if one thing is certain, it is that it is risky to make predictions of how appointees will behave after they don the robe. "One of the things," Frankfurter once said, "that laymen, even lawyers, do not always understand is indicated by the question you hear so often: 'Does a man become any different when he puts on a gown?' I say, 'If he is any good, he does.'"[75] Frankfurter himself seemed an altogether different man as a Justice than he had been off the bench. From academic eminence behind the New Deal to leader of the conservative court cabal—thus did press and public tend to tag Justice Frankfurter.

Frankfurter's career was another legal version of the Horatio Alger success story. He arrived at Ellis Island as a twelve-year-old immigrant from Vienna in 1894, had outstanding records at City College in New York and Harvard Law School, and then divided his time between private practice and government work. In 1914, he was appointed to the Harvard law faculty. He became one of the best-known law professors in the country, specializing in the emerging field of administrative law. His primary contribution was not, however, academic, but his work in defense of liberal causes (he argued successfully in defense of maximum-hour and minimum-wage laws and was a leader in the opposition to the Sacco-Vanzetti convictions) and as governmental adviser. He was the intellectual force behind much of the New Deal program and many of its most effective administrators were recruited by him. The pervasiveness of Frankfurter's "Happy Hot Dogs," as his protégés in Washington were called, led the National Recovery Administration's director to label Frankfurter "the most influential individual in the United States."[76]

His friendship with Franklin D. Roosevelt and his role as an intimate adviser to the President led to Frankfurter's 1939 Supreme Court appointment. News of the appointment led to a champagne celebration in Harold Ickes's Department of the Interior office, attended by leading New Deal liberals. "We were all very happy," Ickes wrote, "there will be on the bench of the Supreme Court a group of liberals under aggressive, forthright, and intelligent leadership."[77]

It did not turn out that way. There would be a cohesive liberal majority on the Court, but it would not be led by Frankfurter. Instead, the liberal leadership was assumed by Frankfurter's two judicial rivals, Justice Black and later by Chief Justice Earl Warren. Frankfurter became the leader of the Court's conservative core, particularly during the Vinson and early Warren years.

Mention should also be made of two lesser appointees to the Hughes Court. The first was Stanley Reed of Kentucky, then Solicitor General, who was appointed in 1938 upon Justice Sutherland's retirement. Justice Reed was the most conservative of the Roosevelt-appointed Justices as well as the least intellectually gifted of them. He was a solid, plodding worker, who considered law in the traditional terms that had prevailed before the Court was recast by eight Roosevelt appointments. More than any other of the New Deal Brethren, Reed voted to uphold federal power, whether directed against property or personal rights. "How sure I was," reads a handwritten note from Reed to Frankfurter, "in the innocent days when law to a country lawyer seemed automatic—no two sides to any legal issue."[78] Reed tended throughout his judicial tenure to view often complex issues in simplistic terms.

Altogether different was Frank Murphy of Michigan, who was President's Roosevelt's Attorney General when he was appointed in 1940 in place of Justice Butler, who had died. Justice Murphy turned out to be one of the most liberal Justices on the modern Court—noted for voting with his heart rather than his head in cases involving racial minorities and the poor. To many his overriding emphasis on reaching the *right* result regardless of technicalities smacked of cant or sanctimony. "Justice tempered by Murphy" was the way wags described his record on the Court. Justice Frankfurter sarcastically called Murphy "St. Frank" in his correspondence. "The short of the matter," he once wrote Justice Reed, "is that today you would no more heed Murphy's tripe than you would be seen naked at Dupont Circle at high noon tomorrow."[79]

National Power Extended

The second Hughes Court contained some of the most eminent Justices in Supreme Court history: the Chief Justice himself and Justices Stone, Black, Frankfurter, and Douglas. The latter three did their most important work under Hughes's successors. In the later Hughes Court, they and their Roosevelt-appointed colleagues confirmed and extended the "constitutional revolution" that had taken place just before their appointment.

The *Jones & Laughlin* decision marked a definite break with the restricted view that productive industries are beyond the federal commerce power. *Jones & Laughlin* removed the immunity from the commerce power that manufacturing had come to enjoy. But the Court soon extended the *Jones & Laughlin* approach to other productive industries. *Sunshine Anthracite Coal Co. v. Adkins*[80] upheld a 1937 congressional act regulating the coal industry, similar in many ways to that which had been annulled in 1936. *Mulford v. Smith*[81] sustained the Agricultural Adjustment Act of 1938, whose basic features were not unlike those of the law of the same name that had previously been condemned.

It is true that *Mulford* did not expressly hold that agriculture came within the definition of commerce. This omission led a lower court as late as 1954 to assert that agriculture is not commerce and, hence, "Federal regulation of agriculture invades the reserved rights of the states."[82] The Supreme Court quickly laid to rest this ghost, declaring categorically, in reversing the lower-court decision, that regulation of agriculture was within the commerce power.[83]

Since *Jones & Laughlin,* productive industries have no longer been removed from the commerce power. No longer must production be treated as purely "local" activity, immune from congressional control regardless of the impact that it may have. The law has come back to the Marshall conception of commerce as an organic whole, with the Commerce Clause embracing all commerce that concerns more than one state. Under the conception, federal power is not limited to commerce that actually moves across state lines. It includes all activities that affect interstate commerce, though such activities, taken alone, might be considered "local." The whole point about the post-1937 law is that these activities can no longer be considered alone. If they have an effect upon interstate commerce, they concern more than one state and come within the Commerce Clause.

The key question thus becomes: Does the subject of regulation affect interstate commerce? An affirmative answer compels the conclusion that it is within federal power. Mines and mills, factories and farms—all engaged in production, rather than commerce in the literal sense—are brought within the sweep of the Commerce Clause, provided only that they exert some effect upon interstate commerce.

"Almost anything," caustically declared the *Jones & Laughlin* dissent "—marriage, birth, death—may in some fashion affect commerce."[84] Under contemporary conditions, the economic system has become so interconnected that there are few local business activities that may not have at least some repercussions upon commerce that extends beyond state lines. If centripetal forces are elevated, to the exclusion of the forces that counteract them, there is practically no economic activity that is immune from congressional control. If effect upon commerce is the test, irrespective of degree, the radius of federal power becomes as broad as the economic life of the nation.

In addition, reference should be made to *United States v. Darby* (1941),[85] one of the last cases in the Hughes Court. The decision there removed another pre-1937 limitation upon federal power when it overruled the *Child Labor Case*.[86] At issue in *Darby* was the Fair Labor Standards Act, which provides for the fixing of minimum wages and maximum hours by a federal agency. It goes on to prohibit the shipment in interstate commerce of goods manufactured by employees whose wages are less than the prescribed minimum or whose hours of work are more than the prescribed maximum. The Act was attacked on the ground that, while the prohibition was nominally a regulation of commerce, its motive was really regulation of wages and hours.

The Court candidly recognized that that was in fact the case. But the whole point about *Darby* is that, under it, the end toward which a congressional exercise of power over commerce is directed is irrelevant. This is, of course, contrary to the *Child Labor Case* thesis that the motive of the prohibition or its effect to control production within the states could operate to deprive the congressional regulation of its constitutional validity. The Court in *Darby* expressly disowned the *Child Labor Case* approach. The *Child Labor Case,* the *Darby* opinion declared, "should be and now is overruled."[87]

The overruling of the *Child Labor Case* returned the Court to Marshall's view of the federal power to regulate under the Commerce Clause as a complete one. In *Darby,* the Court relied directly upon Marshall's definition of the power to regulate commerce as the power "to prescribe the rule by which commerce is governed."[88] Under *Darby,* once again, the sole question is whether a challenged law does prescribe a governing rule for commerce. If it does, it is valid, regardless of the ends that may have induced its enactment.

The *Darby* decision marks the culmination in the development of the Commerce Clause as the source of a natural police power. The *Child Labor* decision constituted a significant rebuff to that development. *Darby* removed whatever limitation it imposed. Under *Darby,* Congress can utilize its commerce power to suppress any commerce contrary to its broad conception of public interest. The national police power (as this aspect of the commerce power may be termed) is the plenary power to secure any social, economic, or moral ends, so far as they may be obtained by the regulation of commerce.

Due Process and Judicial Restraint

The decisions of the second Hughes Court signaled a significant change in the Supreme Court's role in the constitutional structure. Where the Court had previously set itself up as virtual supreme censor of the wisdom of challenged legislation, it now adopted the view formerly expressed in dissent by Justice Holmes. Under the earlier approach, the desirability of a statute was determined as an objective fact on the Court's independent judgment. Now the more subjective Holmes test was applied: Could rational legislators have regarded the statute as a reasonable method of reaching the desired result?[89]

The change in the Court's approach had a seismic impact on the doctrine of substantive due process. Few today doubt that the high tribunal went too far before 1937 in its application of the doctrine or that the Court after that time was correct in deliberately discarding the extreme due process philosophy. The Due Process Clause was not intended to prevent legislatures from choosing whether to regulate or leave their economies to the blind operation of uncontrolled economic forces, futile or even noxious though the choice might seem to the judges. Economic views of confined validity are not to be treated as though the Framers had enshrined them in the constitution.[90]

In his dissent in *Lochner,* Justice Holmes had asserted, "This case is decided upon an economic theory which a large part of the country does not entertain."[91] After 1937, both the economic and legal theories on which *Lochner* rested were repudiated by the Supreme Court. In the 1937 decision overruling its earlier holding that a minimum-wage law violated due process by impairing freedom of contract between employers and employees, the Court asked: "What is this freedom? The Constitution does not speak of freedom of contract."[92] The liberty safeguarded by the Constitution is liberty in a society that requires the protection of law against evils which menace the health, safety, morals, or welfare of the people. Regulation adopted in the interests of the community, the Court concluded, is due process.

After 1937, the Hughes Court had only one occasion directly to overrule other due process decisions of its predecessors. That occurred in 1941, when a 1928 decision voiding a state statute regulating the fees charged by employment agencies as inconsistent with due process had been relied on by a lower court to invalidate a similar Nebraska law; the Supreme Court speedily reversed, holding that the earlier case could no longer be deemed controlling authority.[93]

Although the Hughes Court had no occasion directly to repudiate other specific due process decisions of the pre-1937 period, its later decisions were clearly inconsistent with the earlier due process philosophy. From 1890 to 1937 the high bench used the Due Process Clause as a device to enable it to review the desirability of regulatory legislation. Since that time, "The day is gone when this Court uses the Due Process Clause of the

Fourteenth Amendment to strike down state laws, regulatory of business and industrial conditions, because they may be unwise, improvident, or out of harmony with a particular school of thought."[94]

The view that due process authorized judges to hold laws unconstitutional because they believe the legislature has acted unwisely was definitely discarded. Instead, under the post-1937 jurisprudence, it is not for courts to judge the correctness of the economic theory behind a regulatory law. Not only was the Holmes view that the Constitution does not enact Spencer's *Social Statics* emphatically adopted; in the Court's more recent words, "Whether the legislature takes for its textbook Adam Smith, Herbert Spencer, Lord Keynes, or some other is no concern of ours."[95]

The decline of substantive due process was now firmly ingrained in our public law. And, the Court would say, such rejection is entirely consistent with the role of the judiciary in a representative democracy. To draw the pre-1937 due process line as a limit to regulatory action is, in Holmes's phrase, to make the criterion of constitutionality only what the judges believe to be for the public good.[96]

From this point of view, the decisions of the second Hughes Court can be characterized as constituting a constitutional revolution not just because they recognized significant governmental powers that had theretofore been denied. They also inaugurated a drastic shift in the balance that had previously existed between the Court and the other branches. The pre-1937 interpretation of the doctrine of judicial supremacy had been dominated by the primacy of the Supreme Court, culminating in the Court's review of the desirability of the early New Deal legislation. After 1937, the Court receded to a much more subdued position.

The new restrained posture was, nevertheless, one that was ultimately to give way to a more activist role. Writing as a contemporary who participated at the Government counsel table in the 1937 cases that have been discussed, Robert H. Jackson noted acutely that the new decisions of the Court did not necessarily establish immutably the constitutional law of the future: "No doubt another day will find one of its tasks to be correction of mistakes that time will reveal in this structure in which we now take pride. As one who knows well the workmen and the work of this generation, I bespeak the right of the future to undo our work when it no longer serves acceptably."[97]

The future was to do exactly what Justice Jackson had predicted. Judical restraint was to prove inadequate to meet the needs of the society during the second half of the century. Starting with the accession of Chief Justice Warren, the Court was once again to assume a primary role in the constitutional structure. That was to occur, however, only after the more subdued interlude of the Stone and Vinson Courts.

11

Stone and Vinson Courts, 1941–1953

While it may be the custom to designate the Supreme Court by the name of its head, one who looks only to the bare legal powers of the Chief Justice will find it hard to understand this underscoring of his preeminence. Aside from his designation as Chief of the Court and the attribution of a slightly higher salary, his position is not superior to that of his colleagues— and certainly is not legally superior. In Justice Tom C. Clark's words, "The Chief Justice has no more authority than other members of the court."[1]

The Chief Justiceship should not, however, be approached only in a formalistic sense. Starting with Marshall, the greatest of the Chief Justices have known how to make the most of the extralegal potential inherent in their position. Although perhaps only primus inter pares, the Chief Justice is *primus*. Somebody has to preside over a body of nine, and it is the Chief Justice who directs the business of the Court.

Charles Evans Hughes is the prime modern example of an effective Chief Justice. As Justice Frankfurter once put it, "Chief Justice Hughes radiated authority, not through any other quality than the intrinsic moral power that was his. He was master of the business."[2] The same was not true of Hughes's two successors. In terms of their leadership abilities, in fact, Chief Justices Harlan F. Stone and Fred M. Vinson were the least effective Court heads during the present century.

Chief Justice Stone and His Colleagues

Harlan Fiske Stone on Olympus shows that the qualities that make for an outstanding Justice are not necessarily the qualities that make for a good Chief Justice. There is no doubt that, since his appointment to the Court in 1925, Stone had made his mark as a superior Justice. He had been a leader in the fight against equating the Constitution with laissez faire, as well as in the movement to conform the Court's jurisprudence to the "felt necessities" of the changing times. By the time of Chief Justice Hughes's retirement in June 1941, Stone was recognized as the intellectual leader of the Court. He was, therefore, the natural choice to succeed to the Chief Justiceship—acknowledged as such by both his colleagues and the country. Hughes himself told President Roosevelt that Stone's record gave him first claim upon the honor.[3]

Time wrote that "it liked the idea of a solid man as Chief Justice to follow Charles Evans Hughes. And solid is the word for Chief Justice Stone—200 lb., with heavy, good-natured features and a benign judicial air."[4] Yet, impressive though the new Chief Justice may have been as a figure of justice, he proved anything but a leader in the Hughes mold. Indeed, Stone at the head of the Court was the very antithesis of the Hughes model of dynamism and efficiency. According to Stone's biographer, indeed, "Stone's techniques stood in bold contrast with those of his predecessor."[5]

Justice Frankfurter recalls that while the Hughes passion for efficiency made everything in his Court "taut . . . Stone was an 'easy boss.'"[6] After he had attended the new Chief Justice's first conference, Frankfurter wrote to Stone about "the relaxed atmosphere and your evident desire to have our conferences an exchange of . . . views of nine men."[7]

Chief Justice Hughes had conducted the conferences in the manner of a strict teacher in the classroom. The Hughes conference was normally a four-hour affair;[8] discussion was brief and to the point and woe to the Justice who spoke out of turn. In the Stone conferences, on the other hand, Frankfurter could write about "the deviations from the tradition against speaking out of turn."[9] Like the law professor that he once was, Stone was slow to cut off debate in his eagerness to have all issues thoroughly explored. The result was a freewheeling discussion in which the Chief Justice was more a participant than a leader. "The Chief Justice," commented Justice Reed, "delighted to take on all comers around the conference table and . . . to battle . . . for his views."[10] Stone, says Frankfurter, had "the habit . . . of carrying on a running debate with any justice who expresses views different from his."[11]

Justice Potter Stewart confirmed to me that he heard that "Stone's problem was that, at a conference, he himself always insisted upon having the last word, and that's not the way you preside—always arguing with the person that had spoken." Discussion became wrangling and the Justices emerged from these interminable meetings irritated and exhausted, their differences inflamed from excessive argument.[12] The Stone method only exacerbated the personal and professional differences in what soon became one of the most divided Courts in our history.

The fragmentation in the Stone Court was, in part at least, caused by personal antipathy among the Justices. That, in turn, was aggravated by the appointment to the Stone Court of Robert H. Jackson. But first, a brief word about two other appointments to that Court.

Just before the Stone nomination to head the Court, President Roosevelt selected Senator James F. Byrnes of South Carolina for the vacancy created by Justice McReynolds's retirement. The new Justice contributed little to the Court, since he resigned in 1942, only a year after he had taken his seat, to become, in Roosevelt's own term, "Assistant President"—in charge of domestic policies during the war.

To fill the Byrnes seat, the President selected Wiley B. Rutledge, a former dean of the Iowa and Washington University (St. Louis) Law Schools—at the time a judge on the Court of Appeals for the District of Columbia. Justice Rutledge became a member of the Court's liberal bloc and is noted for his opinions in the field of civil liberties, though he was to serve only six years until his death in 1949.

Between the Byrnes and Rutledge appointments, there was the selection of Robert H. Jackson, then Attorney General, to the seat held by Stone as a Justice. Jackson had been a successful practitioner in upstate New York and had gone to Washington to hold various legal positions in the Roosevelt Administration.

A book on the differences between Justices Black and Frankfurter is titled *The Antagonists*.[13] Yet the differences between the two were primarily intellectual. Those between Justices Black and Jackson were intensely personal. In fact, the two became as bitter personal antagonists as ever sat together on the Supreme Court.

Justice Harold H. Burton noted in his diary his surprise that Justices Black and Jackson exchanged "Good mornings" and even "joined in a brief discussion" at that morning's conference. "I mention this," he wrote, "because of the popular idea . . . that they could not speak to each other." Burton did, however, observe, "These were the first instances of their speaking to each other that I have seen this fall."[14]

The Black–Jackson feud racked both the Stone and Vinson Courts and contributed to the ineffectiveness of both Chief Justices. It also played a major part in Jackson's personal failure as a Justice. One of the most gifted men ever to serve on the Court, Jackson harbored ambitions to be President or, at the least, to occupy the Court's central chair. His lack of success in this regard poisoned his outlook and made him, in his last years on the bench, increasingly embittered.

Justice Jackson was termed by his closest colleague, Justice Frankfurter, "by long odds the most literarily gifted member on the Court."[15] As a stylist and phrase-maker, Jackson can be compared only with Justice Holmes. It was Justice Jackson who aphorized the reality of the Supreme

Court's position: "There is no doubt that if there were a super-Supreme Court, a substantial proportion of our reversals of state courts would also be reversed. We are not final because we are infallible, but we are infallible only because we are final."[16]

The Court and the Law of War

During Chief Justice Stone's tenure, the change from judicial supremacy to the judicial restraint of the post-1937 period gained added emphasis from American participation in World War II. To the Holmes restraint canon was added the truth contained in a striking passage from Burke's *Reflections on the Revolution in France:* "Laws are commanded to hold their tongues amongst arms; and tribunals fall to the ground with the peace they are no longer able to uphold." One familiar with the practical working of a constitutional system realizes that Burke's dictum all too often accords with the realities of wartime.

During the second global conflict, as during the Civil War period, the Supreme Court did little more than confirm the action taken by the Government to deal with the war emergency. Executive primacy is an inevitable concomitant of full-scale war, and it is perhaps unfair to expect the Justices to do more than stamp with their imprimatur measures deemed necessary by those wielding the force of the nation. In practice the Court could do no more than ratify the plenary power vested in government to meet the needs of global war.

In many ways the most dramatic of the Court's war decisions was rendered in the case of the eight German saboteurs who had been landed on our shores from submarines in June 1942. The eight had been arrested by FBI agents soon after their landings and tried by a military commission specially appointed by President Roosevelt for offenses against the law of war and the Articles of War. The commission had found them guilty of violating the law of war by attempting sabotage of our war facilities and had ordered death sentences for six of them and prison terms for the other two. The officers who had been appointed to defend the saboteurs before the military tribunal then sought habeas corpus.

To deal with the cases "without any avoidable delay," after the lower courts had refused relief, the Supreme Court convened in June 1942, in special term. After hearing argument for two days, it handed down a brief per curiam opinion denying habeas corpus. A formal opinion by Chief Justice Stone, setting forth the reasoning of the Court, was not filed until three months later[17]—weeks after the death sentences ordered by the military commission had been carried out.

Before the Supreme Court, the German saboteurs had contended that they could not validly be tried by a military tribunal, asserting that they were entitled to be tried in the civil courts with the safeguards, including trial by jury, which the Fifth and Sixth Amendments guarantee to all persons tried in such courts for criminal offenses. The Court rejected their contention, holding that the consitutional safeguards did not apply to offenses against the law of war.

Japanese Evacuation

Few, it is believed, will take issue with the Court's decision in the case of the German saboteurs. But the same is scarcely true of the high bench's handling of what a *Harper's* article was to term "America's Greatest War-time Mistake"—the evacuation of those of Japanese ancestry from the West Coast.

Acting upon their belief that those of Japanese heritage posed a security threat after Pearl Harbor, the military moved to eliminate the danger by a number of restrictive measures. The most important of them was a series of Civilian Exclusion Orders, issued early in 1942, excluding "all persons of Japanese ancestry, both alien and non-alien," from the westernmost part of the country. Those so excluded were gathered together in so-called assembly centers and then evacuated to what were euphemistically termed Relocation Centers in interior states, where they were detained until almost the end of the war. Under this evacuation program, over 112,000 persons of Japanese ancestry were herded from their homes on the West Coast into the relocation centers, which, had they been set up in any other country, we would not hesitate to call by their true name—concentration camps.

The record of his government in dealing with the West Coast Japanese during the war is hardly one that an American can contemplate with satisfaction. As the Court eloquently declared in 1943, "Distinctions between citizens solely because of their ancestry are by their very nature odious to a free people whose institutions are founded upon the doctrine of equality."[18] Despite this, the Court did uphold the evacuation of the Japanese in *Korematsu v. United States* (1944).[19]

Korematsu had been convicted for remaining in a military area contrary to the Civilian Exclusion Order of the military commander. Such an order, said the Court, could validly be issued by the military authorities in light of the particular situation confronting them on the West Coast after Pearl Harbor. In the face of a threatened Japanese attack, citizens of Japanese ancestry could rationally be set apart from those who had no particular associations with Japan; in time of war residents having ethnic affiliations with an invading enemy may be a greater source of danger than those of a different ancestry. That being the case, it could not be said that the exclusion order bore no reasonable relation to the demands of military necessity.

In *Ex parte Endo* (1944),[20] it is true, the Court did order the release of the Japanese-Americans from the Relocation Centers, on the ground that, though the original evacuations might have been justified by necessity, such necessity did not exist three years after Pearl Harbor, during which time the Government had had ample opportunity to separate the loyal from the disloyal among those detained. There was no evidence of disloyalty against Miss Endo. In her case, the authority to detain her as part of a program against espionage or sabotage was exhausted as soon as her loyalty was conceded. Consequently, said the Court, "Mitsuye Endo is entitled to an unconditional release by the War Relocation Authority."[21]

The *Endo* decision demonstrates both the strength and weakness of judicial review of exercises of the war power. Certainly, the Court's grant of habeas corpus to Miss Endo vindicates the rule of law even in wartime. It shows that, even though military authorities can take whatever measures may be demanded by the exigencies of war, the military's ipse dixit is not of itself conclusive of the necessity for the measures taken. The test of necessity is, with us, a judicial question, and the *Endo* case illustrates how it can be applied by the Court to condemn measures that bear no reasonable relation to military needs.

At the same time, the *Endo* case shows the limitations of judicial power as a practical check on military arbitrariness. By its very nature, judicial justice is dispensed slowly, though it may be dispensed exceedingly well. Mitsuye Endo was evacuated from her home and placed in a Relocation Center early in 1942. In July 1942, she filed a petition for a writ of habeas corpus in a federal district court; yet it was not until December 1944 that she was ordered released by the Supreme Court. But the Court's decision did not and could not affect her three-year deprivation of liberty, illegal though such deprivation might have been.

Despite the Supreme Court's decision, to quote from the concurring opinion of Justice Roberts, "An admittedly loyal citizen has been deprived of her liberty for a period of years. Under the Constitution she should be free to come and go as she pleases. Instead, her liberty of motion and other innocent activities have been prohibited and conditioned."[22] Mitsuye Endo may have had the satisfaction of being immortalized in the *Supreme Court Reports;* but she was most unlikely to consider that an adequate substitute for her loss of liberty during her illegal confinement in the Relocation Center.

Property Rights

The Stone Court also did little to restrain the drastic restrictions upon property rights that took place during World War II. The war power has, of course, always had a drastic impact upon property as well as persons. Nor is it surprising that this should be so. At a time when it is the undoubted law of the land that citizens may by conscription be compelled to give up their life for their country's cause, it should occasion no astonishment that they may similarly be required to yield their earthly possessions, if that be demanded. We may resist a planned economy and the coming of the omnipotent state in peacetime; at the same time, complete control of both person and property (what Justice Jackson once termed "military socialization")[23] is accepted by all as a patriotic necessity in time of war.

Total mobilization of both the work force and the economy reached its peak during World War II. Indeed, under the war statutes delegating authority to him, President Franklin D. Roosevelt can be said to have been vested with more arbitrary power over persons and property than any English-speaking statesman since Oliver Cromwell. But such total power in the Executive was deemed necessary to meet the demands of global war. And the Supreme Court was in the forefront of those recognizing this necessity.

The power fully to mobilize workers had been recognized in the Government during World War I;[24] thus, as the Court put it, "The constitutionality of the conscription of manpower for military service is beyond question."[25] The Court's decisions arising out of World War II recognized governmental authority over property rights as extensive as that over the work force. Said the Court, in the case just quoted, on the impact of "total global warfare" upon our system: "With the advent of such warfare, mobilized property in the form of equipment and supplies became as essential as mobilized manpower. Mobilization of effort extended not only to the uniformed armed services but to the entire population. . . . The language of the Constitution authorizing such measures is broad rather than restrictive."[26] Congress can, of course, clearly draft men for battle service. According to a 1942 opinion, "Its power to draft business organizations to support the fighting men who risk their lives can be no less."[27]

The most important World War II decision upholding governmental authority over property rights was *Yakus v. United States* (1944),[28] which involved the validity of price control. It was contended that the Emergency Price Control Act of 1942 unconstitutionally delegated the authority to control prices to the Price Administrator. Under the Act, the Administrator was empowered to fix maximum prices for all commodities. Under this law the most extensive scheme of price-fixing ever attempted in our system was carried out. The prices of almost all goods and services were directly controlled from Washington in a manner which was unprecedented. Yet the Court had no difficulty in sustaining the congressional assertion of such broad authority under the war power. Under that power governmental authority was upheld over the most important property right, namely, the right to dispose of property at the highest obtainable price.

In the realities of a war economy, the sanctions available to the government through its power over priorities and allocations are even more important as a means of securing obedience to its orders than the traditional methods of judicial enforcement provided for in the Price Control Act. Thus the power given to the President to allocate materials during World War II was used extensively to withhold materials from those who disobeyed the economic control orders of governmental agencies. This was true even though there was no statutory authority for such coercive use of the allocation power.

A good example of the manner in which the power was used is found in a 1944 case.[29] A retail fuel oil dealer had been found by the Office of

Price Administration to have violated its rationing order by obtaining large quantities of fuel oil without surrendering ration coupons and by delivering thousands of gallons of fuel oil to consumers without receiving ration coupons. As a punishment, the OPA suspended the right of the dealer to receive any fuel oil for resale for a year. This suspension order was upheld by the Supreme Court as against the claim that it imposed a penalty not provided for by Congress. The President's statutory power to allocate materials, said the Court, included the power to issue suspension orders against retailers and to withhold rationed materials from them where it was established they had violated the ration regulations.

The extreme sanction utilized by the OPA (which amounted, in effect, to an economic death sentence against the dealer concerned, at least for the period of the suspension order) was sustained, although the power to impose such a sanction was nowhere conferred by Congress. The normal reluctance of our courts to imply penal powers in the administration gave way in the face of what the Court felt to be the necessities inherent in a scheme of effective wartime rationing.

The Least Effective Chief

Fred M. Vinson may have been the least effective Court head in the Supreme Court's history. Appointed in 1946 after Chief Justice Stone's sudden death, the Kentuckian had been a Congressman and circuit judge, and had served in high executive positions, ending in his pre-Court tenure as Secretary of the Treasury. His appointment was due primarily to his close friendship with President Truman, who hoped that his skill at getting along with people would enable him to restore peace to a Court that had become splintered under Chief Justice Stone.

The new Chief Justice was (as his predecessor had been) a large man— in the Frankfurter phrase, "tall and broad and [with] a little bit of a bay window."[30] Throughout his career, Vinson had been known for his skill at smoothing ruffled feathers. But his hearty bonhomie was not enough to enable him to lead the Court effectively. The Justices looked down on the new Chief as the possessor of a second-class mind. Even Justice Reed, the least intellectually gifted of the Roosevelt appointees, could dismiss the dour-faced Chief, in a comment to Justice Frankfurter, as "just like me, except that he is less well-educated."[31] Frankfurter himself could characterize Vinson in his diary as "confident and easy-going and sure and shallow . . . he seems to me to have the confident air of a man who does not see the complexities of problems and blithely hits the obvious points."[32]

Phillip Elman, who had clerked for Frankfurter and was one of the most knowledgeable Court watchers, wrote to the Justice about "the C.J." from the Solicitor General's Office: "What a mean little despot he is. Has there ever been a member of the Court who was deficient in so many respects as a man and as a judge. Even that s.o.b. McReynolds, despite his defects of character, stands by comparison as a towering figure and power-

ful intellect . . . this man is a pygmy, morally and mentally. And so un-
couth."[33]

The new Chief Justice was even more inept than his predecessor in
leading the conference. According to Frankfurter's diary, Chief Justice
Vinson presented cases in a shallow way. He "blithely hits the obvious
points . . . disposing of each case rather briefly, by choosing, as it were,
to float merely on the surface of the problems raised by the cases."[34]
Certainly, the Vinson conference management was anything but masterful.

When Vinson supported Justice Tom Clark's appointment, Washington wags said it was because he wanted someone on his Court who knew
less law than he did.[35] Throughout his tenure, the Justices were openly to
display their contempt for their Chief. As a law clerk recalls it, several of
Vinson's colleagues "would discuss in his presence the view that the Chief's
job should rotate annually and . . . made no bones about regarding
him—correctly—as their intellectual inferior."[36]

Such a Chief Justice could scarcely be expected to lead the gifted prima
donnas who then sat on the Court. If anything the division among the
Justices intensified and, all too often, degenerated into personal animosity
during Vinson's service.

As it turned out, the Vinson Court was the most fragmented in the
Court's history. During the last Vinson term, only 19 percent of the cases
were decided unanimously—a record low. The situation was worsened by
the fact that the antagonism between Justices Black and Jackson became
even more bitter in the Vinson Court. The rancor between the two was
exacerbated by the Jackson belief that Justice Black had frustrated his long
ambition to become Chief Justice by intervening with President Truman
after Chief Justice Stone's death in 1946. A widely circulated column
quoted the President as saying, "Black says he will resign if I make Jackson
Chief Justice and tell the reasons why. Jackson says the same about
Black."[37] Some time later Justice Frankfurter wrote to a close friend, quot-
ing "a most reliable witness," that Truman had planned to appoint Jackson
until he was "threatened with resignations" from the Court.[38] Jackson,
then chief American prosecutor at the Nuremberg war crimes trials, held a
well-publicized press conference, in which he ventilated his dispute with
Black. Then he wrote bitterly to Frankfurter, "Black is now rid of the
Chief. . . . Now if he can have it understood that he has a veto over the
promotion of any Associate, he would have things about where he wants
them."[39]

The bitterness between these two Justices continued until Jackson's
death in 1954. Not long before, Jackson sent Frankfurter a note in which,
referring to Black, he declared, "I simply give up understanding our colleague and begin to think he is a case for a psychiatrist."[40] Black, on the
other side, wrote a former law clerk that a copy of the *Macon Telegraph* had
been sent him that contained an editorial entitled "Jackson Is an Unmiti-
gated Ass." On the same page, Black went on, "appeared an article by John

Temple Graves on the same subject. I have nothing but sympathy for John Temple."[41]

Vinson Court Appointments

During the early Vinson years, the Court was sharply divided between two blocs. The first, led by Justice Frankfurter, joined by the new Chief Justice and Justices Jackson and Reed, urged judicial restraint as the only criterion of constitutional adjudication. The other, led by Justice Black, joined by Justices Douglas, Murphy, and Rutledge, asserted that restraint was not the judicial be-all-and-end-all and that it should give way in cases involving civil liberties.

The rather even division between the two blocs was changed dramatically by the three Court appointments made by President Truman. The first was made just before Chief Justice Vinson was chosen, when Justice Roberts left the Court in 1945. The Truman appointee was Harold H. Burton, a Senator from Ohio, who had become a friend of the President as a member of the famous Truman Committee which investigated defense contracts during World War II. Burton's path to the high bench had followed the traditional pattern of practice as a local leader of the Bar, a political career (state legislator, Mayor of Cleveland, U.S. Senator), then elevation to the Court. In appearance, he has been compared to a small, neat version of a village storekeeper.[42] His biographer concedes that "Burton, an average justice, was not a bright, witty intellectual like a Frankfurter or a Black."[43] Yet, if he was quiet and unassuming, he was also hardworking. Frankfurter wrote Burton in 1947 that he had known every Justice since 1906, and "it is on that basis that I can say to you what I have said behind your back, that this court never had a Justice who was harder working or more conscientious."[44]

Burton, like the other Truman-appointed Justices, was usually to be found in the Frankfurter wing of the Court. Fully appreciative of his modest abilities, Burton was most deferential toward the former professor. Praise from the celebrated Harvard scholar led Burton to write, "I feel as though you have awarded me a mythical grade 'A' in your course on Consitutional Law, of which I may be proud."[45]

Truman's second Court appointee was Tom C. Clark, chosen to fill the seat of Justice Murphy, who died in 1949. Clark had started his legal career as a prosecutor, moved on to the Department of Justice, and then been raised to Attorney General when Truman became President. Clark remained the glad-handing politician, whose flamboyant bow tie (worn even under the judicial robe) gave him a perpetual sophomoric appearance. Clark continued to flaunt his Texas background—his normal way of signifying agreement to an opinion was to write "Okey." "The duck was delicious," Clark wrote to Chief Justice Earl Warren in 1955. "It was big enough to be from Texas."[46]

The Justices, on their side, constantly teased Clark about his state. In

1959, Justice John M. Harlan sent Clark a postcard from Australia: "Some of their sheep stations are nearly the size of the Sovereign State of Texas."[47] President Truman himself was quoted as saying that "Tom Clark was my biggest mistake. No question about it." According to the former President, "[I]t isn't so much that he's a bad man. It's just that he's such a dumb son of a bitch. He's about the dumbest man I think I've ever run across."[48]

To one familiar with Justice Clark's work, the Truman comment is ludicrous. Clark may not have been the intellectual equal of his more brilliant Brethren, but he developed into a more competent judge than any of the other Truman appointees. In fact, Clark has been the most under-rated Justice in recent Supreme Court history.

Under Chief Justice Vinson, Justice Clark almost always followed the conservative pro-Government Vinson position. The same was true of Sherman Minton, the last Truman-appointed Justice, who succeeded Justice Rutledge upon the latter's death in 1949. Like President Truman and Justice Burton, Minton had been a midwestern Senator, elected from Indiana. During his one term, he was a strong Roosevelt supporter and served as Democratic whip. This probably cost him reelection, but FDR consoled him with a seat on the U.S. Court of Appeals for the Seventh Circuit, where he served without distinction for eight years before being named to the high bench.

Minton was square-faced, with the build of a heavyweight boxer, and characterized by the saltiness of his tongue. He may have been the last Justice to use the spittoon provided for him behind the bench, which always upset the fastidious Justice Burton next to him. Minton, the worst of the Truman appointees, ranks by any standard near the botton on a list of Justices. Yet his fellows considered the earthy Hoosier most congenial to work with. "Minton," wrote Justice Frankfurter to a close friend, "will not go down as a great jurist, but he was a delightful colleague." Then he plaintively asked, "Why are most lawyers dull company?"[49]

The Court and the Cold War

The appointment of Chief Justice Vinson and Justices Burton, Clark, and Minton drastically tilted the Court balance in favor of the Frankfurter bloc. To Frankfurter, the judge who tried most to be a conscious Holmes disciple, the Holmes canon had become the only orthodox doctrine. For Justice Black, the restraint doctrine may no longer have been an appropriate response to the new constitutional issues presented to the Court. Frankfurter, on the contrary, whose attitude toward Holmes smacked as much of veneration as agreement, remained wholly true to the approach of the mentor whom he called "My Master." To Frankfurter, the Holmes canon remained the judicial polestar throughout his career on the Court. Judicial restraint became the established order so long as the Frankfurter bloc constituted the Court majority.

To Justice Frankfurter and his followers, restraint was the proper posture for a nonrepresentative judiciary, regardless of the nature of the asserted interest in particular cases. To refuse to defer to the legislative judgment did violence to the basic presuppositions of representative democracy.

Adherence to the restraint doctrine was carried to its logical extreme in the Stone and Vinson Court decisions deferring to the Government demand for security. It may be unfair to expect the Justices to have done more than confirm measures deemed necessary by those wielding the force of the nation. But the same was surely not true of the "cold war" period that followed the cessation of hostilities. Doubtless some of the excesses committed during the postwar years are still too close in time to be able to deal impartially with them. These excesses did, however, reveal that security, like the patriotism of which Dr. Johnson speaks, might also come to be the last refuge of a scoundrel: many of the things done in the postwar era in security's name would not have been tolerated in less tense times.

A basic problem for the American system of law is that of reconciling the antinomy between liberty and security. Both have, to be sure, always been essential elements in the polity, whose coexistence has had to be reconciled by the law. In the postwar period, nevertheless, security tended to dominate. The response to the tensions of the cold war made American law security-conscious as it had never been before.

The governmental demand for security was articulated in laws and other measures restricting rights normally deemed fundamental. For the first time since the notorious Alien and Sedition Acts of 1798, a peacetime sedition law (making subversive speech criminal) put people in prison. The law in question—the Smith Act—was enacted in 1940; but the first significant prosecutions, those brought in 1948 against the leaders of the American Communist party, were a direct fruit of the postwar confrontation with the Soviet Union. The Communist prosecutions were upheld in *Dennis v. United States* (1951)[50] with the decision turning on the "clear and present danger" presented by Communist advocacy during the tense postwar period.

The *Dennis* decision, however, involved a clear watering-down of the Clear and Present Danger Test as it was stated by Justice Holmes. According to Chief Justice Vinson's opinion of the Court, the correct interpretation of the test was: "In each case [courts] must ask whether the gravity of the 'evil,' discounted by its improbability, justifies such invasion of free speech as is necessary to avoid the danger."[51]

No substantive "evil" is as grave as the forcible overthrow of government; hence the danger that that will occur need not be nearly as "clear and present" as would be required if a lesser "evil" was involved. To the Court, the "evil" in *Dennis* was so overwhelming that it had to defer to Congress on whether any nexus between defendants' teaching and advocacy and that "evil" did exist.

Restraint and deference were, indeed, carried to their extreme in *Dennis*. Of course, as Justice Douglas pointed out in dissent, Communists in this country were miserable merchants of unwanted ideas, whose wares remained unsold.[52] To the majority, however, that did not justify a judicial tribunal in substituting its judgment for that of the elected representatives of the people that the Communist party constituted a "clear and present" danger to national security. In a democratic system, under the restraint canon, the primary responsibility for determining when there is danger that speech will induce the substantive evil that the Government has a right to prevent—in this case, attempts to overthrow government by force and violence—must lie with the elected representatives of the people. In this, as in other fields of judicial review, the Vinson Court saw its function as one to serve as a brake upon arbitrary extremes. But its task was exhausted when there was found to be a rational basis for the particular exercise of congressional judgment.

Dennis was not the only decision sustaining cold war restrictions under the restraint doctrine. In addition, the Court upheld other significant restrictions, ranging from drastic restraints upon aliens to the loyalty–security programs instituted by governments in this country. The restrictions on Communist aliens were ruled within the plenary power of Congress over citizens of other lands,[53] and the federal loyalty program was upheld under the settled principle that "[t]he Constitution does not guarantee public employment."[54] In addition, the Court refused to strike down restrictions on the procedural rights of notice and full hearing in cases where national security was involved.[55]

These decisions, occurring during the immediate postwar period, may be understandable as a continued reaction from the excesses of pre-1937 judicial supremacy. Justices who had repudiated those excesses continued to display the same Holmes-like approach of deference to the legislator. Nevertheless, one may wonder whether the Supreme Court did not go too far in standing aside in the face of the extreme restrictions imposed in security's name.

A Court overimbued with the dominant demand for security may tend to accede to that demand, even if the cost be distortion of accepted principles of constitutional law. Yet this can hardly be done without important effects on general jurisprudence. A supreme tribunal that molds its law only to fit immediate demands of public sentiment is hardly fulfilling its proper role. As Justice Frankfurter once put it, "The Court has no reason for existence if it merely reflects the pressures of the day."[56] The doctrine of deference to the legislature may require abnegation on the part of the Court but hardly abdication by it of the judicial function. Whatever may be said about the strains and stresses of the cold war period, the enemy was not so near the gates that the nation had to abandon respect for the organic traditions that had theretofore prevailed in the American system.

Decisions against Government

The ledger was not, however, all in favor of governmental power during the Stone and Vinson years. There were also significant cases in which the Court refused to accept claims of public authority vis-à-vis individual rights. The most important of them was the *Steel Seizure Case* (1952),[57] in which the Court (over Chief Justice Vinson's dissent) invalidated President Truman's seizure of the nation's steel mills during the Korean War. According to the Government, a strike disrupting steel production for even a brief period would have so endangered the well-being and safety of the nation that the President had "inherent power" to do what he had done.

The Court denied that the President had any such prerogative to exercise what amounted to lawmaking authority: "In the framework of our Constitution, the President's power to see that the laws are faithfully executed refutes the idea that he is to be a lawmaker. . . . The Founders of this Nation entrusted the lawmaking power to the Congress alone in both good and bad times." The President, in Justice Black's opinion, had no authority, either as Chief Executive or as Commander in Chief, to take possession of private property in order to keep labor disputes from stopping production: "This is a job for the Nation's lawmakers."[58]

If, at first glance, the *Steel Seizure* decision seems inconsistent with the judicial restraint doctrine, the inconsistency is more apparent than real. The case was not one involving only the question of whether the Court should defer to the President's action. On the contrary, the Court here, in Justice Frankfurter's phrase, had "to intervene in determining where authority lies as between the democratic forces in our scheme of government."[59] In the *Steel* case, the Court was acting not as overseer of the Chief Executive but as holder of the constitutional balance between the two political branches of government, to both of which the Court, under its post-1937 doctrine, owed deference.

In such a situation, to which branch should the Court yield? Where power to seize private property for public use (clearly within legislative authority) is involved, the supremacy of the congressional policy is the only principle consistent with a polity in which "[a]ll legislative powers herein Granted" have been vested in Congress.

There were also Stone and Vinson Court decisions that were favorable to the protection of civil liberties. The most dramatic of them well demonstrated the emotional trials presented by a rigorous employment of the Frankfurter judicial restraint technique. At issue was the constitutionality of a state law making it compulsory for school children to salute the flag. In *Minersville School District v. Gobitis,*[60] decided in 1940, Justice Frankfurter had delivered the opinion of the Court sustaining the flag-salute requirement, with Justice Stone alone dissenting. In a letter to Stone explaining his opinion, Frankfurter claimed that "nothing has weighed as much on my conscience, since I have come to this Court, as has this case."[61] Then, only three years later, in *West Virginia Board of Education v. Barnette*,[62] the Court reversed itself and ruled the compulsory salute unconstitutional.

In *Barnette*, Justice Jackson delivered perhaps his best opinion for the Court. "The Bill of Rights," he declared, "denies those in power any legal opportunity to coerce" allegiance. "Authority here is to be controlled by public opinion, not public opinion by authority."[63] Justice Frankfurter stood his original ground and delivered a sharp *Barnette* dissent, which emphasized the need for judicial restraint even in such a case. As the Justice wrote to Chief Justice Stone explaining his opinion upholding the flag salute, "What weighs with me strongly in this case is my anxiety that, while we lean in the direction of the libertarian aspect, we do not exercise our judicial power unduly, and as though we ourselves were legislators by holding with too tight a rein the organs of popular government."[64]

The *Barnette* case well illustrates that no matter how he tried to clothe his opinions with the Holmes mantle, there was an element of shabbiness in the results reached by Justice Frankfurter in too many cases. After Frankfurter delivered his opinion upholding the compulsory flag salute, he was talking about the opinion over cocktails at Hyde Park. Eleanor Roosevelt impulsively declared that regardless of the Justice's learning and legal skills, there was something wrong with an opinion that forced little children to salute a flag when such a ceremony was repugnant to their conscience.[65]

In *Barnette*, the Court also stated the preferred-position doctrine that was to prove seminal in First Amendment cases. The Jackson opinion emphasized the difference in approach which must be followed as between First Amendment cases and other cases: "The right of a State to regulate, for example, a public utility may well include, so far as the due process test is concerned, power to impose all of the restrictions which a legislature may have a 'rational basis' for adopting. But freedoms of speech and of press, of assembly, and of worship may not be infringed on such slender grounds."[66]

In later Courts, the view that the First Amendment freedoms are in a preferred position led to renewed emphasis upon the need to protect personal rights in preference to those rights primarily economic in nature. Certainly, as Justice Frankfurter stated it, "those liberties of the individual which history has attested as the indispensable conditions of an open as against a closed society come to this Court with a momentum for respect lacking when appeal is made to liberties which derive merely from shifting economic arrangements."[67]

The judicial tendency to find legislative invasion more readily where personal rights are involved than in the sphere of economics[68] began in the Stone and Vinson Courts. According to its proponents, this approach was not really inconsistent with the Holmes restraint canon. This was pointed out by Justice Stone in a letter just before he became Chief Justice: "Justice Holmes' opinions indicate that he thought that the judge should not be too rigidly bound to the tenet of judicial self-restraint in cases involving civil liberties, although so far as I know he never formulated the distinction. You will find my formulation of it in a footnote in *United States v. Carolene Products Company.*"[69]

In his *Carolene Products* footnote[70]—the second most famous footnote in Supreme Court history[71]—Justice Stone also asked whether legislation aimed at "discrete and insular minorities" should not be subject to "more searching judicial inquiry."[72] Then in the *Korematsu* case, the Stone Court stated that "all legal restrictions which curtail the civil rights of a single racial group are immediately suspect . . . courts must subject them to the most rigid scrutiny."[73] These statements by Stone and his Court were to be the foundation of the "suspect class" concept that was to prove of such significance in the Warren and Burger Courts.

Perhaps the most important Stone and Vinson Court decisions protecting individual rights were those involving racial discrimination. It was those Courts that began to make equal protection more than a mere slogan for minority groups. For virtually the first time since their adoption, the gulf between the letter of the Fourteenth and Fifteenth Amendments and their practical effect began to be significantly narrowed. In its decisions the Stone-Vinson Court removed the legal prop from some of the most important manifestations of racial discrimination in this country. In a 1944 case, the so-called white primary, upon which Southern efforts to disfranchise blacks were based, was ruled unconstitutional.[74] In 1948 the enforcement of racial restrictive covenants was stricken down.[75]

It is true that neither the Stone nor the Vinson Court questioned the "separate but equal" doctrine of *Plessy v. Ferguson*,[76] which served as the legal foundation for racial segregation—particularly in the field of education. But those Courts were, nevertheless, able to make important progress by placing increasing emphasis upon judicial implementation of the requirement of equality in facilities. Their decisions were aimed at ensuring that the separate facilities provided for blacks were in fact substantially equal to those afforded to whites.

In 1948, the Court held that black applicants must be admitted to law schools reserved to whites unless equivalent facilities were provided for them.[77] In Texas, a separate black law school was, indeed, opened. A first-rate law school cannot, however, come into being full-grown, like Minerva from the head of Jove. The facilities afforded by the newly opened black law school could not, in fact, equal those available to white students at the University of Texas. Under these circumstances, the Court ruled that blacks in Texas had been denied a legal education equivalent to that offered to students of other races: "We hold that the Equal Protection Clause of the Fourteenth Amendment requires that petitioner be admitted to the University of Texas Law School."[78]

The Court, to be sure, did not directly overrule *Plessy* during the Stone and Vinson years. But its decisions came very close to holding that segregation as such, at least in higher education, was contrary to the Constitution. What was said of the University of Texas Law School applied with equal force to all long-established white educational institutions. Though *Plessy* permitted segregation, provided the requirement of equality in facilities was met, the Court's approach meant, in practice, that that requirement could never be met in the field of higher education.

But what is true of segregation in higher education is, in reality, also true of segregation as such. There can never be equality in separated facilities, for the mere fact of segregation makes for discrimination. The arbitrary separation of African-Americans solely on the basis of race is a "badge of servitude"[79] that must generate in them a feeling of inferior social status, regardless of the formal equality of the facilities provided for them.

In effect, then, the Stone and Vinson Courts were not far from the holding that segregation as such was discriminatory and hence a denial of equal protection—a holding but a step away from the decision in the Texas law school case. As we shall see, however, Chief Justice Vinson strongly opposed taking that step and it was left to the Court under his successor to do so in the *Brown* case, to be discussed in Chapter 13.

But it was not only the *Brown* decision that was to signal the arrival of a Court head so different from Chief Justice Vinson himself. The cases discussed in this section, in which the claims of the individual vis-à-vis government were upheld, were but a small part of the work of the Stone and Vinson Courts. Apart from them, those Courts, and particularly that under Chief Justice Vinson, were most receptive to claims of governmental authority as against claims of individual right. Judicial restraint and deference to the legislature had carried the day in the Supreme Court.

All this was, however, soon to change. A new Chief Justice was to turn the Court in a new direction—one which adapted Supreme Court jurisprudence to the changing needs of the second half of the twentieth century.

12

Warren Court, 1953–1969

There have been two great creative periods in American public law. The first was the formative era, when the Marshall Court laid down the foundations of American constitutional law, giving specific content to the broad general terms in which the Constitution is written. The judicial task at that time was to work out from the constitutional text a body of legal doctrines adapted to the needs of the new nation and the new era into which it was entering.

The second great creative period was the Warren Court era. The judicial task then was to keep step with the twentieth century's frenetic pace of societal change. To do this, the Warren Court had to perform a transforming role, usually thought of as more appropriate to the legislator than the judge. In the process it rewrote much of the corpus of American constitutional law. Indeed, in terms of creative impact on the law, the Warren Court's tenure can be compared only with that of the Marshall Court.

Chief Justice Warren and His Background

Of course, the Supreme Court is a collegiate institution whose collegiate nature is underscored by the custom the Justices have had of calling each other "Brethren." But each of the Brethren can only be guided, not directed. As Justice Frankfurter once stated in a letter to Chief Justice Vinson, "[G]ood feeling in the court, as in a family, is produced by accommodation, not by authority—whether the authority of a parent or a vote."[1]

The Court "family" is composed of nine individuals, who have borne out James Bryce's truism that "judges are only men."[2] "To be sure," Frankfurter once wrote Justice Reed, "the Court is an institution, but individuals, with all their diversities of endowment, experience and outlook determine its actions. The history of the Supreme Court is not the history of an abstraction, but the analysis of individuals acting as a Court who make decisions and lay down doctrines."[3]

Foremost among the individuals who made up the Warren Court was, of course, the Chief Justice. In many respects Earl Warren could have been a character out of Sinclair Lewis. Except for his unique leadership abilities, he was a rather typical representative of the Middle America of his day, with his bluff masculine bonhomie, his love of sports and the outdoors, and his lack of intellectual interests or pretensions.

In an interview Justice Potter Stewart, a member of the Warren Court, told me that the Chief Justice had a "simple belief in the things we now laugh at: motherhood, marriage, family, flag, and the like." These furnished the foundation for Warren's scale of values throughout his professional life. If there was something of the Babbitt in this, it was also, as Justice Stewart put it to me, "a great source of strength, that he did have these foundations on which his thinking rested."

The early Warren was a direct product of his upbringing and surroundings. Born and raised in California, he grew up in a small town that was a microcosm of the burgeoning West. From a last vestige of the American frontier—with cowboys on horses, saloons, and gunfights—the town and state quickly came to be the paradigm of twentieth-century America. "All changed, changed utterly,"[4] as growth became the prime element of California life.

Like his state, Warren displayed a capacity for growth throughout his career. The popular conception of Warren's judicial career has, indeed, been one of a virtual metamorphosis—with the political grub suddenly transformed into the judicial lepidopteran. Certainly, Warren as Chief Justice appeared an entirely different person than he had been before his elevation to the Court. As his state's leading law-enforcement officer, Warren had been perhaps the foremost advocate of the forced evacuation of persons of Japanese ancestry from the West Coast after the Japanese attack on Pearl Harbor in December 1941. As Chief Justice, Warren was the foremost proponent of racial equality. From his crucial role in the *Brown* segregation case[5] to the end of his Court tenure, he did more than any other judge in American history to ensure that the law, in W. H. Auden's phrase, "found the notion of equality."

As Governor, Warren strongly opposed reapportionment of the California legislature, even though, as he later conceded, "My own state was one of the most malapportioned in the nation." As Chief Justice, Warren led the movement to bring the apportionment process within the equal protection guaranty, a movement that culminated in the Chief Justice's own opinion laying down the "one person, one vote" principle.[6]

Like John Marshall, Earl Warren had a politicl background. Soon after he had obtained his law degree from the University of California, Warren worked in the office of the District Attorney of Alameda County, across the bay from San Francisco. Five years later, in 1925, he was elected District Attorney, serving in that position until 1938. A 1931 survey of American district attorneys by Raymond Moley (later famous as a member of President Franklin D. Roosevelt's so-called Brain Trust) "declared without hesitation that Warren was the best district attorney in the United States."[7]

In 1938 Warren was elected Attorney General of California and became Governor in 1942. He was a most effective chief executive; he reorganized the state government and secured major reforming legislation—notably measures for a modern hospital system, improving the state's prisons and its correction system, providing an extensive highway program, and improving old-age and unemployment benefits. Warren proved an able administrator and was the only Governor of his state to be elected to three terms.

On September 8, 1953, Chief Justice Vinson suddenly died. President Eisenhower appointed Warren to the seat. The California Governor resigned his position and took up his new duties as Chief Justice at the beginning of the 1953 Term.

Leadership Not Scholarship

According to a famous Macaulay statement: "There were gentlemen and there were seamen in the navy of Charles II. But the seamen were not gentlemen, and the gentlemen were not seamen." There have been scholars and there have been great Justices on the Supreme Court. But the scholars have not always been great Justices, and the great Justices have not always been scholars.

To be sure, outstanding scholars did sit on the Warren Court; among them, Felix Frankfurter stands out. Frankfurter was as learned a Justice as ever sat on the Court. His scholarship far exceeded the bounds of legal arcana, unlike so many juristic scholars. The range of the Justice's scholarly interests is illustrated not only in his opinions and published writings, but also in his amazingly varied correspondence with the leading intellectual figures of the day—ranging from Alfred North Whitehead to John Dewey to Albert Einstein. Publication of Frankfurter's best letters would serve not only law, but scholarship and literature as well.

Yet Frankfurter may have been a better letter writer than he was a judge. With all his intellect and scholarly talents, Frankfurter's judicial career remained essentially a lost opportunity. As far as public law was concerned, he may well have had more influence as a law professor than as a Supreme Court Justice. Although Frankfurter may have expected to be the intellectual leader of the Court, as he had been of the Harvard law faculty, the Chief Justice himself performed the true leadership role in the Warren Court.

But Warren was never a legal scholar in the Frankfurter sense. "I wish that I could speak to you in the words of a scholar," the Chief Justice once told an audience, "but it has not fallen to my lot to be a scholar in life."[8] The Justices who sat with him all stressed that Warren may not have been an intellectual like Frankfurter, but then, as Justice Stewart observed to me, "he never pretended to be one."

In assessing the importance of scholarship as a judicial attribute, one should distinguish sharply between a member of the Supreme Court and its Chief Justice. Without a doubt, Justice Story was the greatest legal scholar ever to sit on the Court. His scholarship enabled him to make his outstanding judicial record, and his legal expertise supplied the one thing that Chief Justice Marshall lacked. Indeed, Marshall is reputed to have once said: "Brother Story here . . . can give us the cases from the Twelve Tables down to the latest reports."[9] It is safe to assume that Story's learning often fleshed out the Chief Justice's reasoning with the scholarly foundation needed to support some Marshall opinions.

Still, no one conversant with American law will conclude that Story was a greater judge than Marshall. Story's scholarship could scarcely have produced the constitutional landmarks of the Marshall Court. When Marshall died, Story's admirers hoped that he would become the new Chief Justice. "The Supreme Court," Harvard President Josiah Quincy toasted, "may it be raised one Story higher."[10] But Story's appointment could have been a disaster; the scholar on the bench would have been a misfit in the Court's center chair.

This hypothesis is not mere conjecture. It is supported by the Court's experience under Chief Justice Stone, who, like Frankfurter, had been a noted law professor. From an intellectual viewpoint, Stone was an outstanding judge. Yet, as we saw, he failed as Chief Justice; his lack of administrative ability nearly destroyed the Court's effectiveness. The Stone Court presented a spectacle of unedifying atomization wholly at variance with its functioning as a collegiate tribunal.

Warren clearly did not equal Stone as a legal scholar. But his leadership abilities and skill as a statesman enabled him to be the most effective Chief Justice since Hughes. Those Justices who served with him stressed Warren's leadership abilities, particularly his skill in conducting the conference. "It was incredible," said Justice Brennan just after Warren's death, "how efficiently the Chief would conduct the Friday conferences, leading the discussion of every case on the agenda, with a knowledge of each case at his fingertips."[11]

A legal scholar such as Stone treated the conference as a law school seminar, "carrying on a running debate with any justice who expresse[d] views different from his."[12] At conference, Chief Justice Warren rarely contradicted the others and made sure that each of them had his full say. Above all, he stated the issues in a deceptively simple way, reaching the heart of the matter while stripping it of legal technicalities. As the *Wash-*

ington Post noted, "Warren helped steer cases from the moment they were first discussed simply by the way he framed the issues."[13]

In his first conference on *Brown v. Board of Education*,[14] Warren presented the question before the Court in terms of racial inferiority. He told the Justices that segregation could be justified only by belief in the inherent inferiority of blacks and, if *Plessy v. Ferguson*[15] was followed, it had to be upon that basis. A scholar such as Frankfurter certainly would not have presented the case that way. But Warren's "simplistic" words went straight to the ultimate human values involved. In the face of such an approach, arguments based on legal scholarship would have seemed inappropriate, almost pettifoggery.

The work of a Chief Justice differs greatly from that of other members of the Court as far as legal scholarship is concerned. A person without scholarly interest would find the work of an Associate Justice most unrewarding, since an Associate Justice spends time in Court only hearing and voting on cases and writing opinions. Thus, while considering the appointment of a successor to Chief Justice Vinson, President Eisenhower asked a member of Governor Warren's staff whether Warren would really want to be on the Court after his years in high political office: "Wouldn't it be pretty rarified for him?" "Yes," came back the answer, "I frankly think he'd be very likely to be bored to death [as an Associate Justice]." But, the response went on: "My answer would be emphatically different if we were talking about the Chief Justiceship. He could run the place."[16]

The staff member's answer gets to the heart of the matter. The essential attribute of a Chief Justice is not scholarship but leadership. One who can "run the place" and induce the Justices to follow will effectively head the Court.

The Chief Justice must still write opinions backed by the traditional indicia of legal scholarship: discussion of complicated technical issues, citation and consideration of precedents, and learned-looking footnotes. But a lack of scholarly attainments does not necessarily preclude the production of learned opinions. The necessary scholarship can be supplied by the bright, young, ex-law review editors who serve as the Justices' law clerks. It did not take *The Brethren*[17] to make students of the Court aware of how much of the opinion-writing process has been delegated to the clerks. "As the years passed," wrote Justice Douglas of his own Court years, "it became more and more evident that the law clerks were drafting opinions."[18] The first drafts of the opinions that Chief Justice Warren assigned to himself were almost all prepared by his law clerks.

The Chief Justice would outline the way he wanted the opinion drafted, leaving the clerk with a great deal of discretion to flesh out the details of the opinion. Warren never pretended to be a scholar interested in research and legal minutiae. He left the reasoning and research supporting the decision to his clerks, as well as the task of compiling extensive footnotes, an indispensable component of the well-crafted judicial opinion.

Perhaps the most famous footnote in any Supreme Court opinion appeared in *Brown v. Board of Education*.[19] Noted footnote 11 listed seven works by social scientists to support the statement that segregation meant black inferiority. Yet one of Warren's law clerks inserted the footnote into the opinion, and, as will be seen, neither the Chief Justice nor the Associate Justices paid much attention to it at the time.

It *Was* the *Warren* Court

There are those who claim that although Chief Justice Warren may have been the nominal head of the Court that bears his name, the actual leadership was furnished by other Justices. Thus a biography of Justice Black is based on the proposition that the Alabaman was responsible for the "judicial revolution" that occurred during the Warren years.[20] More recently, one review of my biography of the Chief Justice asserts that, more than anything, it shows that the proper title of the Court while Chief Justice Warren sat in its center chair would be the *Brennan* Court.[21]

Justice Black himself always believed that he had led the judicial revolution that rewrote so much of the corpus of our constitutional law. Black resented the acclaim the Chief Justice received for leading what everyone looked on as the Warren Court. As Justice Black saw it, the Court under Chief Justice Warren had only written into law the constitutional principles that Justice Black had been advocating for so many years. When Warren retired as Chief Justice, the Justices prepared the traditional letter of farewell. The draft letter read, "For us it is a source of pride that we have had the opportunity to be members of the Warren Court." Justice Black changed this to "the Court over which you have presided."[22]

Nevertheless, the other Justices who served with Chief Justice Warren all recognized his leadership role. Justice Douglas, closest to Justice Black in his views, ranks Warren with Marshall and Hughes "as our three greatest Chief Justices."[23] Another member of the Warren Court told me that it was the Chief Justice who was personally responsible for the key decisions during his tenure. The Justices who sat with him have all stressed to me that Chief Justice Warren may not have been an intellectual like Justice Frankfurter, but then, to quote Justice Stewart again, "he never pretended to be one." More important, says Stewart, he possessed "instinctive qualities of leadership." When I asked Stewart about claims that Justice Black was the intellectual leader of the Court, he replied, "If Black was the intellectual leader, Warren was the *leader* leader."

Chief Justice Warren brought more authority to the Chief Justiceship than had been the case for years. The most important work of the Supreme Court, of course, occurs behind the scenes, particularly at the conferences where the Justices discuss and vote on cases. The Chief Justice controls the conference discussion, his is the prerogative to call and discuss cases before the other Justices speak. All those who served with him stressed Chief Justice Warren's ability to lead the conference.

Justice Stewart told me that at the conferences, "after stating the case, [Warren] would very clearly and unambiguously state his position." The Chief Justice rarely had difficulty in reaching a decision, and once his mind was made up, he would stick tenaciously to his decision. As Justice White expressed it to me in an interview, Warren "was quite willing to listen to people at length . . . but, when he made up his mind, it was like the sun went down, and he was very firm, very firm about it." The others never had any doubt about who was the head of the Warren Court.

A reading of the conference notes of Justices on the Warren Court reveals that the Chief Justice was as strong a leader as the Court has ever had. As the *Washington Post* summarized my Warren biography, it "shows Warren as even more a guiding force in the landmark opinions of his court than some have previously believed. Chief Justice Warren helped steer cases from the moment they were first discussed simply by the way he framed the issues."[24] In almost all the important cases, the Chief Justice led the discussion toward the decision he favored. If any Court can properly be identified by the name of one of its members, this Court was emphatically the *Warren* Court and, without arrogance, he, as well as the country, knew it. After an inevitable initial period of feeling his way, Chief Justice Warren led the Supreme Court as effectively as any Chief Justice in our history. When we consider the work of the Warren Court, we are considering a constitutional corpus that was directly a product of the Chief Justice's leadership.

The Other Justices

Of course, the Supreme Court is inevitably more than the judge who sits in its center chair. In many ways, the individual Justices operate, as some of them have said, like "nine separate law firms." Plainly, Chief Justice Warren was not going to be able to deal with the Justices in the way he had directed matters as Governor of California. "I think," Justice Stewart affirmed to me, "he came to realize very early, certainly long before I came here [1958], that this group of nine rather prima donnaish people could not be led, could not be told, in the way the Governor of California can tell a subordinate, do this or do that."

What Justice Stewart says is true of any Supreme Court, but it was particularly true of the Justices on the Warren Court. The Court in the mid-1940s was characterized by Yale law professor Fred Rodell as "the most brilliant and able collection of Justices who ever graced the high bench together."[25] The stars of that Court were Justices Black, Frankfurter, Douglas, and Jackson—four of the greatest judges ever to serve on the highest bench, brilliant jurists, each possessed of a peculiarly forceful personality. In addition, Justices Black and Douglas, on the one side, and Justices Frankfurter and Jackson, on the other, represented polar views on the proper role of the Court in interpreting the Constitution. Their doctrinal differences, fueled by increasing personal animosity, erupted into the most bitter feud in Supreme Court history. All four were still serving when Warren was appointed, though Jackson died before he could play an important role in the development of the Warren Court's jurisprudence.

Justices Black and Douglas were the leading advocates among the Justices of the activist approach that became the Warren Court's trademark. But the Chief Justice's principal ally was Justice William J. Brennan, Jr., appointed in 1956 on Justice Minton's retirement. Before then, Brennan had been a judge in New Jersey for seven years, rising from the state trial court to its highest bench. He was the only Warren Court Justice to have served as a state judge.

"One of the things," Justice Frankfurter once said, "that laymen, even lawyers, do not always understand is indicated by the question you hear so often: 'Does a man become any different when he puts on a gown?' I say, 'If he is any good, he does.'"[26] Certainly Justice Brennan on the supreme bench proved a complete surprise to those who saw him as a middle-of-the-road moderate. He quickly became a firm adherent of the activist philosophy and a principal architect of the Warren Court's jurisprudence. Brennan had been Frankfurter's student at Harvard Law School; yet if Frankfurter expected the new Justice to continue his pupilage, he was soon disillusioned. After Brennan had joined the Warren Court's activist wing, Frankfurter supposedly quipped, "I always encourage my students to think for themselves, but Brennan goes too far!"

Justice Brennan soon became Chief Justice Warren's closest colleague. The Chief Justice would turn to Brennan when he wanted to discuss a case or some other matter on which he wanted an exchange of views. The two would usually meet on Thursdays, when the Chief Justice would come to the Brennan chambers to go over the cases that were to be discussed at the Court's Friday conference.

Justice Brennan's unassuming appearance and manner mask a keen intelligence. He was perhaps the hardest worker on the Warren Court. Unlike some others (notably Justice Douglas), Brennan was always willing to mold his language to meet the objections of some of his colleagues, a talent that would become his hallmark on the Court and one on which the Chief Justice would rely frequently. It was Justice Brennan to whom the Chief Justice was to assign the opinion in some of the most important cases to be decided by the Warren Court.

A year before the Brennan appointment, in 1955, Justice John Marshall Harlan was chosen to fill the vacancy created by Justice Jackson's death. Harlan was the grandson of the Justice with the same name who had written the dissent in *Plessy v. Ferguson*.[27] As soon as he took his place on the bench, Harlan had taken a place in the Court's history, as the only descendant of a Justice to become one. The first Harlan had been a judicial maverick, who had been an outspoken dissenter. The second Harlan took a more cautious approach to the judicial function that reflected his background. Educated at Princeton, he had been a Rhodes Scholar and had had a successful career with a leading Wall Street law firm prior to his appointment in 1954 to the U.S. Court of Appeals for the Second Circuit.

On the Supreme Court, Harlan, like Justice Jackson before him, became a firm adherent of Justice Frankfurter's judicial restraint philosophy. After Frankfurter's retirement, it was Harlan who came to be looked on as the conservative conscience of an ever-more-activist Court. But Harlan had none of the acerbity that made Frankfurter distasteful to some of his colleagues.

Harlan looked like the contemporary vision of a Supreme Court Justice. Tall and erect, with sparse white hair, conservatively dressed in his London-tailored suits, with his grandfather's gold watch chain across the vest under his robe, he exuded the dignity associated with high judicial office. Yet underneath was a warm nature that enabled him to be close friends with those with whom he disagreed intellectually, notably Justice Black. Visitors could often see the two Justices waiting patiently in line in the Court cafeteria. The two were a study in contrasts: the ramrod-straight patrician with his commanding presence and his slight, almost wispy colleague who always looked like the lively old Southern farmer.

If Harlan at the Bar had been known as a "lawyer's lawyer," on the Court he soon acquired the reputation of a "lawyer's judge." Soon after Justice Harlan took his seat (he was confirmed by the Senate on March 17, 1955), Justice Frankfurter wrote a friend that "Harlan both on the bench and in the Conference is as to the manner born."[28] The new Justice was plainly one of the best, if not the best, lawyer on the Court and, next to Frankfurter, the Justice most interested in the technical aspects of the Court's work. He became a sound, rather than brilliant, Justice who could be relied on for learned opinions that thoroughly covered the subjects dealt with, though they degenerated at times into law review articles of the type Frankfurter too often wrote.

The term most frequently used to describe Justice Harlan by those who knew him is "gentleman"—though some say that beneath the veneer was a bland grayness that permeated both his life and his work. To the other Justices, Harlan always appeared the quintessential patrician, with his privileged upbringing and Wall Street background. "I hear 'mi lord,'" reads a note from Justice Clark to Harlan, "that you have been under the weather. . . . Your Lordship should be more careful of your whiskey and your habits."[29]

Soon after Justice Brennan's appointment, President Eisenhower selected Charles E. Whittaker in place of Justice Reed, who retired in 1957. The new Justice had grown up on a small farm in Kansas. He had quit high school at sixteen and had never gone to college. He began work in a Kansas City law firm as a messenger, office boy, and "bottle washer," going to an unaccredited law school at night. He eventually became a prosperous lawyer and was appointed a district judge and then promoted to the U.S. Court of Appeals for the Eighth Circuit.

Whittaker may have been the least talented Justice appointed during this century. Justice Whittaker, a member of the Warren Court told me, "used to come out of our conference literally crying. You know Charlie had gone to night law school, and he began as an office boy and he'd been a farm boy and he had inside him an inferiority complex, which . . . showed and he'd say, 'Felix used words in there that I'd never heard of.'"

A more important Warren Court member was Potter Stewart of Ohio, a federal circuit judge, appointed in 1958 after Justice Burton retired. Stewart was one of the youngest Justices (only forty-three when selected by President Eisenhower); at sixty-six he also was one of the youngest to retire. When he first took his seat, Stewart's youth and handsome appearance added an unusual touch to the highest bench, showing that it need not always be composed of nine old men. To those who knew him in those days, it was painful to see his physical decline before he died in 1985.

Before then, people who met Justice Stewart were surprised by his vigor and clearly expressed views, which contrasted sharply with his public image as an indecisive centrist without clearly defined conceptions. Unlike Justices Black and Frankfurter, Stewart never acted on the basis of a deep-seated philosophy regarding the proper relationship between the state and its citizens. When asked if he was a liberal or a conservative, he answered, "I am a lawyer," and went on to say, "I have some difficulty understanding what those terms mean even in the field of political life. . . . And I find it impossible to know what they mean when they are carried over to judicial work."[30]

Stewart was a moderate with a pragmatic approach to issues that polarized others. In his early years on the Court, he tended to be the "swing man" between the activist and judicial restraint blocs. During that time, the two blocs were evenly divided and Stewart cast the key vote in many cases, voting now with the one and now with the other group. With Justice Frankfurter's 1962 retirement, Chief Justice Warren and his activist supporters gained the upper hand. Justice Stewart remained in the center as the Court moved increasingly to the left. At the end of Warren's tenure, he continued as the Court's leading moderate—though, according to Justice Douglas, by that time "Stewart and Harlan were the nucleus of the new conservatism on the Court."[31]

Justice Stewart was best known for his comment in a 1964 obscenity case, "I know it when I see it"[32]—a phrase that he later lamented might well become his epitaph. Stewart's aptness for the pungent phrase helped make him the press's favorite Justice, and he was more accessible to reporters than any of his colleagues.

Another moderate on the Warren Court was Byron R. White of Colorado, who was Deputy Attorney General when chosen in 1962 after Justice Whittaker was forced to retire because of ill health. On graduation from Yale Law School after the war, White had clerked for Chief Justice Vinson; he was the first former law clerk to become a Justice.

Justice White was certainly not the typical Supreme Court appointee. He was known to most Americans as "Whizzer" White—the all-American back who became the National Football League rookie of the year in 1938. Physically, White is most impressive. At six feet two and a muscular 190 pounds, he has maintained the constitution that made him a star football player. Even as a Justice, White retained his athletic competitiveness, never hesitating to take part in the clerks' basketball games in the gymnasium at the top of the Court building.

On the Supreme Court, White, like Stewart, defied classification. He, too, tended to take a lawyerlike approach to individual cases, without trying to fit them into any overall judicial philosophy. He did tend to be one of the more conservative Justices in the Court's center, particularly in criminal cases. "In the criminal field," as Justice Harry Blackmun has said, "I think Byron White is distinctly a conservative."[33] After his appointment to the Court, White was never close to Chief Justice Warren. "I wasn't exactly in his circle," White told me. Certainly White never became a member of the Warren Court's inner circle. He went his own way, voting against the Chief Justice as often as not.

During the first part of Chief Justice Warren's tenure, the Court was split between the activist wing led by Warren and the advocates of judicial restraint led by Frankfurter. With the 1962 retirement of Justice Frankfurter and his acolyte, Justice Whittaker, the Court balance was completely altered. For the remainder of his tenure the Chief Justice had a solid majority. The Warren majority was secured when Arthur J. Goldberg succeeded Justice Frankfurter, who wrote to the President in 1962 that he was "left with no choice but to" retire because of severe heart attacks.

A newspaper once characterized Justice Goldberg's success story as "almost too corny."[34] The last of eleven children of an immigrant Russian fruit peddler, Goldberg rose from one of the poorest sections of Chicago. He worked his way through Northwestern Law School, where he graduated first in his class. He developed a labor practice and became counsel to the AFL-CIO. He became a Kennedy supporter in early 1960 and was appointed Secretary of Labor when the new Administration took office. In that position, he became the Kennedy Administration's troubleshooter. In fact, as Robert Kennedy wrote about his discussions with the President on the Court vacancy, "[T]he only reservation about Arthur Goldberg was the fact that he was so valuable to the Administration and could handle labor and management in a way that could hardly be equalled by anyone else."[35]

Physically and emotionally Goldberg was the antithesis of the first Kennedy appointee. Where White was dispassionate and reserved, Goldberg was warm and ebullient. He was of average build, with wavy grey hair, and none of the athletic propensities of his colleague. Goldberg noted the contrast in a speech the winter after his appointment. "As a new brother," he said, "I think I should tell you some of the innermost secrets of the Supreme Court. The Court is sorely divided. There is one group that believes in an active judicial philosophy; Justices Black, Douglas, and White are addicted to physical exercise. The middle wing, which believes in moderate exercise, consists of the Chief Justice and Justices Clark and

Brennan. And then there's the third group—Justices Harlan and Stewart and myself."[36]

Goldberg soon became Warren's firm ally on the Court. "There is nobody," Goldberg told me, "I felt closer to than Warren." In fact, states the former Justice, "if you look at the *Harvard Law Review* annual rack-up of how we voted during my tenure, he and I voted together more than any other Justice, as I recall it." In some ways, indeed, Goldberg was even more activist than the Chief Justice. Wrote former Justice Minton to retired Justice Frankfurter in 1964, "Mr. J. Goldberg is a walking Constitutional Convention. Wow what an activist he is!"[37]

Justice Goldberg, however, served only three years; he resigned in 1965 to become Ambassador to the United Nations. His place was taken by Abe Fortas, who proved to be an equally firm Warren supporter. Fortas was no ordinary junior Justice. He was one of the few Court appointees who had to be virtually drafted for his seat. He also had one of the best minds of anyone appointed to the Supreme Court. Justice Harlan, himself an outstanding legal craftsman, told his law clerks that he considered Fortas the most brilliant advocate to appear in his time.

Before his appointment, Fortas listed his business address in *Who's Who in the South and Southwest* as "c/o White House, 1600 Pennsylvania Avenue, Washington, D.C." He came to the Court with the celebrity and status of the President's close adviser. After John F. Kennedy's assassination, the first telephone call the new President made in Dallas was to Fortas.

The Warren majority on the Court was strengthened toward the end of the Chief Justice's tenure when the seat of Justice Clark, who retired in 1967, was taken by Thurgood Marshall. Though Clark had backed the Chief Justice in important cases, he could not be counted on as a consistent Warren supporter. On the other hand, from the time he took his seat, Justice Marshall became a firm vote for Warren. Marshall gave an equal protection dimension to the Horatio Alger legend. Great-grandson of a slave and son of a Pullman car steward, Marshall was the first black appointed to the highest Court. Solicitor General at the time of his appointment, he had headed the N.A.A.C.P. Legal Defense Fund's staff for over twenty years and had been chief counsel in the *Brown* segregation case.

The appointments of Justices Goldberg, Fortas, and Marshall gave the Chief Justice a firm majority that enabled him to lead the Justices in what became the key decisions of the Warren Court jurisprudence. Of course, there were still opposing views among the Justices, but the principal Warren opponent had become the courtly Justice Harlan, who had neither Justice Frankfurter's abrasive personality nor his leadership abilities. Harlan's voice calling for judicial restraint was but a faint echo of that of his mentor and, in one reporter's phrase, "was now heard in a symbolic wilderness without the assured support of Frankfurter."[38] The balance of power had definitely shifted to the Chief Justice. Now the Warren Court jurisprudence could accelerate the onward rush of the "constitutional revolution" for which the Chief Justice, more than any person, was responsible.

Activism versus Judicial Restraint

There is an antinomy inherent in every system of law: the law must be stable and yet it cannot stand still.[39] It is the task of the judge to reconcile these two conflicting elements. In doing so, jurists tend to stress one principle or the other. Indeed, few judges can keep an equipoise between stability and change.

Chief Justice Warren never pretended to try to maintain the balance. As soon as he had become established on the Court, he came down firmly on the side of change, leading the Supreme Court's effort to enable our public law to cope with rapid societal change. Warren strongly believed that the law must draw its vitality from life rather than precedent. What Justice Holmes termed "intuitions" of what best served the public interest[40] played the major part in Warren's jurisprudence. He did not sacrifice good sense for the syllogism. Nor was he one of "those who think more of symmetry and logic in the development of legal rules than of practical adaptation to the attainment of a just result."[41] When symmetry and logic were balanced against considerations of equity and fairness, he normally found the latter to be weightier.[42] In the Warren hierarchy of social values, the moral outweighed the material.[43]

Throughout his tenure on the Court, the Chief Justice tended to use "fairness" as the polestar of his judicial approach. Every so often in criminal cases, when counsel defending a conviction would cite legal precedents, Warren would bend his bulk over the bench to ask, "Yes, yes—but were you fair?"[44] The fairness to which the Chief Justice referred was no jurisprudential abstraction. It related to such things as methods of arrest, questioning of suspects, and police conduct—matters that Warren understood well from his earlier years as District Attorney in Alameda County, California. Decisions like *Miranda v. Arizona*[45] were based directly upon the Warren fairness approach.

The Chief Justice's emphasis upon fairness and just results led him to join hands with Justices Black and Douglas and their activist approach to constitutional law. Their activism led to Warren's break with Justice Frankfurter—the foremost advocate on the Court of the Holmes doctrine of judicial restraint. To Justice Holmes, as we saw, the legislator was to have the primary say on the considerations behind laws; the judge's duty was to enforce "even laws that I believe to embody economic mistakes."[46] Justice Frankfurter remained true to the Holmes approach, insisting that self-restraint was the proper posture of a nonrepresentative judiciary, regardless of the nature of the asserted interests in particular cases. Warren followed the canon of judicial restraint in the economic area, but he felt that the Bill of Rights provisions protecting personal liberties imposed more active enforcement obligations on judges. When a law allegedly infringed upon personal rights guaranteed by the Bill of Rights, the Chief Justice refused to defer to the legislative judgment that had considered the law necessary.

Warren rejected the Frankfurter philosophy of judicial restraint because he believed that it thwarted effective performance of the Court's constitutional role. Judicial restraint, in the Chief Justice's view, all too often meant judicial abdication of the duty to enforce constitutional guarantees. "I believe," Warren declared in an interview on his retirement, "that this Court or any court should exercise the functions of the office to the limit of its responsibilities." Judicial restraint meant that "for a long, long time we have been sweeping under the rug a great many problems basic to American life. We have failed to face up to them, and they have piled up on us, and now they are causing a great deal of dissension and controversy of all kinds." To Warren, it was the Court's job "to remedy those things eventually," regardless of the controversy involved.[47]

The Warren approach in this respect left little room for deference to the legislature, the core of the restraint canon. Warren never considered constitutional issues in the light of any desired deference to the legislature. Instead, he decided those issues based on his own independent judgment, normally giving little weight to the fact that a reasonable legislator might have voted for the challenged law.

For Chief Justice Warren, the issue on judicial review was not *reasonableness* but *rightness*. If the law was contrary to his own conception of what the Constitution demanded, it did not matter that a reasonable legislator might reach the opposite conclusion. When Warren decided that the Constitution required an equal population apportionment standard for all legislative chambers except the United States Senate, the fact that no American legislature had followed the new requirement did not deter him from uniformly applying the standard.[48] Justice Harlan's dissent may have demonstrated that the consistent state practice was, at the least, reasonable. For the Chief Justice, however, legislative reasonableness was irrelevant when the practice conflicted with his own interpretation of the Constitution.

The Court and Civil Rights

Not too long ago, legal observers expected the post-1937 "constitutional revolution" to signal a permanent decline in the Supreme Court's position. The subdued role played by the Court in the later New Deal period, during World War II, and in the postwar years led many to expect the Court to wither away, much as the state was supposed to do in Marxist theory. Yet one thing is clear: the Marxist state may have disappeared, but America's high tribunal has anything but withered away in the second half of the century.

Still, the work of the Supreme Court under Chief Justice Warren differed from earlier periods. In the Warren Court, the emphasis in the

Court completed the shift from the safeguarding of property rights to the protection of personal rights. In enforcing the liberties guaranteed by the Bill of Rights, the Supreme Court forged a new and vital place for itself in the constitutional structure. More and more the Court came to display its solicitude for individual rights. Freedom of speech, press, and religion, the rights of minorities and those accused of crime, those of individuals subjected to legislative and administrative inquisitions—all came under the Warren Court's fostering guardianship.

Foremost in the Warren Court's catalogue of individual rights was that to racial equality. The landmark decision here was, of course, that in the *Brown* school segregation case, to which the next chapter is devoted. As we shall see there, *Brown* not only outlawed school segregation but also served as the foundation for a series of decisions outlawing racial segregation in all public institutions. By 1963, the Court could declare categorically, "It is no longer open to question that a State may not constitutionally require segregation of public facilities."[49]

In its *Brown* decision, the Court had held that school segregation was unconstitutional, and a year later, in its second *Brown* decision, the Court ordered the district courts to take such action as was necessary and proper to ensure the nondiscriminatory admission of plaintiffs to schools "with all deliberate speed."[50] The enforcement of *Brown* was left to the lower courts, which met massive resistance in many Southern school districts. The Warren Court intervened in several cases to force desegregation, particularly in *Griffin v. County Board of Prince Edward County* (1964),[51] in which a Virginia county had closed down its school system rather than have blacks attend schools with whites. Then, in *Green v. County School Board* (1968),[52] the Court invalidated "freedom of choice" plans, under which most Southern school districts then operated. The Court held that a school board had a duty to come forward with a desegregation plan that "promises realistically to work now." If the board did not do so, the federal court had authority to issue any order deemed necessary to achieve immediate desegregation.

The Warren Court decisions furthering racial equality were an important catalyst for the civil rights protests of the 1960s and congressional action to protect civil rights. Both developments received support from the Warren Court decisions. Convictions of civil rights demonstrators in public places and at sit-in demonstrations at restaurants, lunch counters, and libraries were reversed in a number of cases.[53] Other cases established the right to use the streets and other public places as "public forums" for the dissemination of even unpopular views.[54]

Soon after the Court decided the last of the important sit-in cases, the Civil Rights Act of 1964 became law—the first time since Reconstruction that Congress had passed an important statute to protect civil rights. The 1964 law prohibited racial discrimination in hotels, restaurants, and other public accommodations. The decision in the *Civil Rights Cases*[55] of 1883 had invalidated a similar law (the Civil Rights Act of 1875) on the ground that the Fourteenth Amendment's guaranty of equal protection was limited to "state action" and did not reach the discrimination of private hotel and restaurant owners, but the Warren Court upheld the 1964 Act.

The Court ruled in *Heart of Atlanta Motel v. United States* (1964)[56] that Congress could pass the law under its commerce power and a law under that power was not subject to the "state action" limitation. The Court also upheld the Voting Rights Act of 1965, which contained far-reaching provisions protecting the right of blacks to vote. Though the Act provides for the supplanting of state election machinery by federal law and federal officials, the opinion of Chief Justice Warren found it a valid congressional measure to enforce the Fifteenth Amendment.[57]

Reapportionment

The *Brown* decision signaled the expansive attitude of the Warren Court toward the constitutional guaranty of equality. From the field of racial equality involved in *Brown*, the Court spread the equal protection mantle over an increasingly broad area, notably in the field of political rights. The key decision here was *Baker v. Carr* (1962),[58] in which the federal courts were ruled competent to entertain an action challenging legislative apportionments as contrary to equal protection. The problem arose from the fact that though American legislative districts had originally been apportioned on an equal population basis, over the years population shifts had made for extreme malapportionment. Thus the seats in the Tennessee legislature, where *Baker* arose, had last been apportioned in 1901. By the time the case was brought, a vote from the most populous country had only a fraction of the weight of one from the least populous: the population ratio of the most and least populous districts was then greater than nineteen to one.

Before *Baker*, the Supreme Court had held that the federal courts had no jurisdiction in such cases. The question of legislative apportionment was ruled a "political question" and therefore beyond judicial competence. The leading pre-*Baker* apportionment decision had declared, "Courts ought not to enter this political thicket."[59] In *Baker*, Chief Justice Warren and his colleagues overruled the earlier cases and held that attacks on legislative apportionments on equal protection grounds could be heard and decided by the federal courts.

Then, in *Reynolds v. Sims* (1964),[60] the Chief Justice's own opinion ruled that the Constitution lays down an "equal population" principle for legislative apportionment. Under this principle, substantially equal representation is demanded for all citizens. According to the most noted passage in the Warren opinion, "Legislators are elected by voters, not farms or cities or economic interests." It follows, Warren said, that "the Equal Protection Clause requires that the seats in both houses of a bicameral state legislature must be apportioned on a population basis."[61] Legislative districts must represent substantially equal populations, which means that the "one person, one vote" principle is now enshrined in the United States

Constitution: equal numbers of people are entitled to equal representation in their government.

Chief Justice Warren himself characterized the reapportionment cases as the most important cases decided by the Court during his tenure. In those cases, the Warren Court worked an electoral reform comparable to that achieved by the nineteenth-century Parliament in translating the program of the English Reform Movement into the statute book. The result has been a virtual transformation of the political landscape, with voting power shifted from rural areas to the urban and suburban areas in which most Americans have come to live.

The Chief Justice never had doubts about the correctness of the reapportionment decisions. He maintained that if the "one person, one vote" principle had been laid down years earlier, many of the nation's legal sores would never have festered. According to Warren, "Many of our problems would have been solved a long time ago if everyone had the right to vote, and his vote counted the same as everybody else's. Most of these problems could have been solved through the political process rather than through the courts. But as it was, the Court had to decide."[62]

Equality and Criminal Justice

If one great theme recurred in the jurisprudence of the Warren Court, it was that of equality before the law—equality of races, of citizens, of rich and poor, of prosecutor and defendant. The result was what Justice Fortas termed "the most profound and pervasive revolution ever achieved by substantially peaceful means."[63] More than that, it was the rarest of all political animals: a judicially inspired and led revolution. Without the Warren Court decisions giving ever-wider effect to the right to equality before the law, most of the movements for equality that have permeated American society would have encountered even greater difficulties.

In addition to racial and political equality, the Warren Court moved to ensure equality in criminal justice. The landmark case was *Griffin v. Illinois* (1956).[64] Griffin had been convicted of armed robbery in a state court. He filed a motion for a free transcript of the trial record, alleging that he was indigent and could not get adequate appellate review without the transcript. The motion was denied. In the Supreme Court's conference on the case, Chief Justice Warren pointed out that the state had provided for full appellate review in such a case. A defendant who could pay for a transcript should not be given an advantage over one who could not. "We cannot," declared the Chief Justice, "have one rule for the rich and one for the poor." Hence he would require the state to furnish the transcript.

The Court followed the Warren lead and held that it violates the Constitution for a state to deny to defendants alleging poverty free transcripts of trial proceedings, which would enable them adequately to prosecute appeals from criminal convictions. According to Justice Black's opinion, "There can be no equal justice where the kind of trial a man gets depends on the amount of money he has. Destitute defendants must be afforded as adequate appellate review as defendants who have money enough to buy transcripts."[65]

As it turned out, *Griffin* was a watershed in the Warren Court's jurisprudence. In it the Court made its first broad pronouncement of economic equality in the criminal process. After *Griffin* the Warren Court appeared to agree with Bernard Shaw that "the worst of crimes is poverty," as it tried to equalize criminal law between those possessed of means and the less affluent. It was the *Griffin* approach that was the foundation of the landmark decision in *Gideon v. Wainwright* (1963),[66] which required counsel to be appointed for indigent defendants.

The *Gideon* case was one of the most famous decided by the Warren Court. Indeed, Clarence Gideon and his case have become part of American folklore. In *Gideon* the Court overruled an earlier case that had refused to hold that the right to counsel was so fundamental as to be included in the due process guaranty. Again following the Chief Justice's lead, the Court reversed Gideon's conviction, because his request to the trial judge to have a court-appointed lawyer to assist him was denied. "[R]eason and reflection," declares the *Gideon* opinion, "require us to recognize that in our adversary system of criminal justice, any person haled into court, who is too poor to hire a lawyer, cannot be assured a fair trial unless counsel is provided for him. This seems to us to be an obvious truth."[67]

Criminal Procedure

Gideon made plain that the Constitution requires public provision of counsel for criminal defendants who cannot afford to hire their own attorneys. The need for the assistance of counsel is not, however, limited to the courtroom. *Gideon* was based on the express constitutional guaranty of the right to counsel in "all criminal prosecutions." But the case only raised another critical question: When does the right to counsel begin?

The Warren Court answered this question in *Miranda v. Arizona* (1966).[68] Miranda had been convicted in a state court of kidnapping and rape. He had been arrested and taken to an interrogation room, where he was questioned without being advised that he had a right to have an attorney present. After two hours, the police secured a confession, which was admitted into evidence over Miranda's objection.

The state's highest court affirmed the conviction, but the Warren Court reversed. The majority agreed with Chief Justice Warren that for the police to be able to use any confession, they must show that they gave full effect to the defendant's right to remain silent and to the presence of an attorney, either retained or appointed. Warren had no doubt that the right to counsel began as soon as there was what his opinion termed "custodial interrogation"—that is, interrogation while an individual is in police custody. To the Chief Justice, the constitutional right came into play when Miranda was arrested. "I didn't know," he commented during the argument before the Court, "that we could arrest people in this country for investigation. Wouldn't you say it was accusatory when a man was locked in jail?"

The *Miranda* decision worked a drastic change in American criminal law and its application by police officers, prosecutors, and judges. The Warren Court in effect laid down the rule that an accused who wants a counsel should have one at any time after being taken into custody. Under Warren's decision, the police must give so-called *Miranda* warnings: that the person arrested has a right to remain silent, that anything he says may be used against him, that he can have a lawyer present, and that he can have counsel appointed if he cannot afford one.

Protection of the rights of criminal defendants had become a primary concern of the Warren Court. The *Miranda* decision, as much as anything, exemplified Warren's own concern in such cases. Indeed, *Miranda* was the ultimate embodiment of the Warren approach in criminal cases.

That approach was also illustrated by *Mapp v. Ohio* (1961),[69] where the Warren Court adopted the exclusionary rule, which bars the admission of illegally seized evidence, in state criminal cases. The Supreme Court had refused to follow that rule before *Mapp,* holding that the exclusionary rule was not required by the Constitution in state criminal cases. Now, under Chief Justice Warren, the *Mapp* state conviction was reversed because illegally seized evidence had been admitted at the trial. Such a holding, the Warren Court affirmed, closes "the only courtroom door remaining open to evidence secured by official lawlessness" in violation of the Constitution's guaranty against unreasonable searches and seizures.

From Property to Personal Rights

The dominant trend in the Court during the Warren tenure was a shift in emphasis from property rights to personal rights. "When the generation of 1980 receives from us the Bill of Rights," Chief Justice Warren declared in a 1955 article, "the document will not have exactly the same meaning it had when we received it from our fathers."[70] The Bill of Rights as interpreted by the Warren Court had a meaning much different from that handed down by its predecessors.

There were three principal developments in the Warren Court regarding the protection of personal rights: (1) acceptance of the preferred-position doctrine, (2) extension of the trend toward holding Bill of Rights guaranties binding on the states, and (3) broading of the substantive content of the rights themselves.

The preferred-position theory, we saw in Chapter 11, was first stated in *United States v. Carolene Products Co.,*[71] though only tentatively in a footnote. Under Chief Justice Warren it became accepted doctrine. The theory is based on the view that the Constitution gives a preferred status to personal, as opposed to property rights. The result is a double standard in the exercise by the Supreme Court of its review function. The tenet of judicial restraint does not rigidly bind the judge in cases involving civil liberties and other personal rights.[72] The presumption of validity for laws gives way far more readily in cases where life and liberty are restrained. In those cases, the legislative judgment must be scrutinized with much greater care.

Critics say that the preferred-position approach, with its elevation of personal rights, creates a hierarchy of rights not provided for in the Constitution. It should, however, be recognized that each generation must necessarily have its own scale of values. In nineteenth-century America, concerned as it was with the economic conquest of a continent, property rights occupied the dominant place. A century later, in a world in which individuality was dwarfed by concentrations of power, concern with the maintenance of personal rights had become more important. With the focus of concern on the need to preserve an area for the development of individuality, judges were naturally more ready to find legislative invasion when personal rights were involved than in the sphere of economics.[73]

One of the last decisions of the Marshall Court had ruled the Federal Bill of Rights binding only on the Federal Government, not on the states.[74] It has been urged that the Fourteenth Amendment changed that result, incorporating the entire Bill of Rights in the Fourteenth Amendment's due process clause.[75] The Court has never accepted this view, adopting instead a selective approach, under which only those rights deemed "fundamental" are included in due process. Yet if advocates of full incorporation seemingly lost the incorporation battle, after midcentury they came close to winning the due process war; for the Warren Court, without formally abandoning its selective incorporation approach, held almost all the Bill of Rights guaranties to be fundamental and hence absorbed by due process.

The key decisions were *Mapp* and *Gideon,* which reversed earlier refusals to hold the right against the use of illegally secured evidence and the right to counsel to be so fundamental as to be included in due process. Both *Mapp* and *Gideon* spoke in broad terms of the need to protect individual rights; they signaled a trend to include ever more of the Bill of Rights guaranties in the Fourteenth Amendment. In the following decade the Warren Court held these rights fundamental and hence binding upon the states: rights against double jeopardy[76] and self-incrimination[77] and rights to jury trial in criminal cases,[78] to a speedy trial,[79] and to confrontation.[80] Add to these rights those that had been held binding on the states before midcentury,[81] and they include all the rights guaranteed by the Federal Bill of Rights except the rights to a grand jury indictment and to a jury in civil cases involving more than twenty dollars. The two exclusions hardly alter the overriding tendency to make the Due Process Clause ever more inclusive.

The Warren Court did more than merely apply the Bill of Rights to the states: it also broadened the substantive content of the rights guaranteed, giving virtually all personal rights a wider meaning than they had theretofore had in American law. This was particularly true in two crucial areas: criminal justice and freedom of expression. The former has already been discussed, but it remains to say a word about the latter.

Two members of the Warren Court, Justices Black and Douglas, consistently urged an absolutist view of the freedom of speech guaranty.[82] Their view was based upon the unqualified language of the First Amendment, which says that "Congress shall make no law . . . abridging the freedom of speech." In the Black-Douglas view, when the amendment says that no laws abridging speech shall be made, it means flatly that *no* laws of that type shall, under any circumstances, be made. Under the Black-Douglas approach, no speech may ever be restricted by government action, even speech which is libelous, obscene, or subversive.

The Warren Court did not adopt the Black-Douglas absolutist view, but it did place increasing emphasis on freedom of expression as a preferred right. The primacy of the First Amendment was firmly established. The right to use the streets and other public forums was extended to those using them for civil rights protests and demonstrations.[83] Most criticisms of conduct by public officials and public figures were exempted from the law of libel.[84] Freedom of the press was broadened; censorship laws were struck down,[85] and the power to restrain publication on grounds of obscenity was drastically limited.[86]

In addition, the Warren Court began to recognize new personal rights not specifically guranteed in the Federal Bill of Rights. Foremost among these was a constitutional right of privacy. Justice Douglas urged the existence of such a right in a 1961 dissent: "This notion of privacy is not drawn from the blue. It emanates from the totality of the constitutional scheme under which we live."[87] The Douglas notion of a right to privacy as part of the area of personal rights protected by the Constitution was accepted by the Warren Court only four years later.

In *Griswold v. Connecticut* (1965),[88] defendants had been convicted of violating a state law prohibiting the use of contraceptive devices and the giving of medical advice in their use. In reversing the conviction the Supreme Court held that the law violated the right of privacy. The opinion expressly recognized the existence of a constitutionally protected right of privacy—a right said to be within the protected scope of specific Bill of Rights guaranties.

During Chief Justice Warren's tenure, protection of personal rights and liberties became the very focus of the Court's enforcement of the contemporary Constitution. With property rights constitutionally curtailed, compensatory scope had to be given to personal rights if the ultimate social interest—the individual life—was not to be lost sight of.

The need to broaden the constitutional protection of personal rights received added emphasis from the growth and misuse of governmental power in the twentieth century. Totalitarian systems showed dramatically what it meant for the individual to live in a society in which Leviathan had become a reality. The "Blessings of Liberty," which the Constitution's

Framers had taken such pains to safeguard, were placed in even sharper relief in a world that had seen so clearly the consequences of their denial.

Warren in the Pantheon

In 1966, Vice President Hubert Humphrey declared that if President Eisenhower "had done nothing else other than appoint Warren Chief Justice, he would have earned a very important place in the history of the United States."[89] By then, it was widely recognized that Earl Warren had earned a place in the front rank of the American judicial pantheon. That is true even though Chief Justice Warren will never rank with the consummate legal craftsmen who have fashioned the structure of Anglo-American law over the generations—each professing to be a pupil, yet each a builder who added his few bricks.[90] But Warren was never content to deem himself a mere vicar of the common-law tradition. Instead he was the paradigm of the "result-oriented" judge, who used his power to secure the result he deemed right in the cases that came before his Court.

In reaching what he considered the just result, the Chief Justice was not deterred by the demands of stare decisis. For Warren, principle was more compelling than precedent. The key decisions of the Warren Court overruled decisions of earlier Courts. Those precedents had left the enforcement of constitutional rights to the political branches. Yet the latter had failed to act. In Warren's view, this situation left the Court with the choice either to follow the precedent or to vindicate the right. For the Chief Justice, there was never any question as to which was the correct alternative.

Warren cannot be deemed a great juristic technician, noted for his mastery of the common law. But he never pretended to be a legal scholar or to profess interest in legal philosophy or reasoning. To him, the outcome of a case mattered more than the reasoning behind the decision.

The result may have been a deficiency in judicial craftsmanship that subjected Warren to constant academic criticism, both during and after his tenure on the bench. Without a doubt, Warren does not rank with Holmes or Cardozo as a master of the opinion, but his opinions have a simple power of their own; if they do not resound with the cathedral tones of a Marshall, they speak with the moral decency of a modern Micah. Perhaps the *Brown* opinion did not articulate the juristic bases of its decision in as erudite a manner as it could have. But the decision in *Brown* emerged from a typical Warren judgment, with which few today would disagree. The Warren opinion was so *right* in that judgment that one wonders whether additional learned labor in spelling out the obvious was really necessary.

When all is said and done, Warren's place in the judicial pantheon rests, not upon his opinions, but upon his decisions. If impact on the law is the hallmark of the outstanding judge, few occupants of the bench have been more outstanding than Chief Justice Warren.

Employing the authority of the ermine to the utmost, he never hesitated to do whatever he thought necessary to translate his own conceptions of fairness and justice into the law of the land. His Court's principal decisions have taken their place in the forefront of historic judicial decisions. Their impact on a whole society's way of life can be compared only with that caused by political revolution or military conflict.

"It is a delicious irony," reads a famous passage by Anthony Lewis, "that a President who raised inactivity to a principle of government . . . should have appointed a Chief Justice for whom action was all."[91] President Eisenhower said that one of the reasons he chose Warren was because he felt that Warren did not "hold extreme legal or philosophical views." Yet this criterion of moderation in legal and philosophical views is not necessarily valid as a measure by which to judge greatness on the bench. When people object to extremism in legal and philosophical views, they are really objecting to judicial activism. The judge who holds strong views is bound to take a more expansive view of the judicial function.

Perhaps the period of judicial activism that took place under the Warren Court was unprecedented in legal history. But almost all the outstanding judges in American law have been characterized by a more affirmative approach to the judicial role than that taken by their lesser colleagues. The great American judges have been jurists who used the power of the bench to the full. This was particularly true of Chief Justice Warren. The Marshall Court and the Warren Court have now become major parts of American legal history. The two men who sat at the center of those Courts were strong leaders who acquired their influence by the force of their character and their integrity.

We need not, in Justice Frankfurter's phrase, subscribe to the "hero theory of history" to recognize that great judges make a profound difference in the law. It did make a vast difference that Earl Warren and his colleagues rather than some other judges sat when they did. If Warren could not have made his judicial reputation without the opportunity that his position afforded him, it is also true that he forced the opportunity to make the creative contributions that escaped lesser Chief Justices. This, after all, has been the common characteristic of the greatest judges. Not all of them were masters of the common law or consummate craftsmen of judicial techniques. But all seized the occasion to make the creative contributions that eluded their lesser brethren.

13

Watershed Cases:
Brown v. Board of Education,
1954

"This is the first indication that I have ever had that there is a God." This caustic comment to two former law clerks was Justice Frankfurter's reaction to Chief Justice Vinson's death in September 1953, just before reargument was scheduled in the *Brown* school segregation case.[1] Had Vinson presided over the Court that decided *Brown,* the result would have been a divided decision. In a May 20, 1954, letter to his colleague Stanley Reed, three days after the unanimous *Brown* decision was announced, Frankfurter wrote, "I have no doubt that if the *Segregation* cases had reached decision last Term there would have been four dissenters—Vinson, Reed, Jackson and Clark—and certainly several opinions for the majority view. That would have been catastrophic."[2]

Instead, *Brown v. Board of Education* stands at the head of the cases decided by the Warren Court. In many ways, *Brown* was the watershed constitutional case of this century. Justice Reed told one of his clerks that "if it was not the most important decision in the history of the Court, it was very close."[3] When *Brown* struck down school segregation, it signaled the beginning of effective enforcement of civil rights in American law.

First *Brown* Conference

Chief Justice Vinson, notably gregarious and good natured, had grown increasingly irritable during his last Court term. According to a close associate, "[H]e was distressed over the Court's inability to find a strong, unified position on such an important case."[4] But Vinson was as much a cause as a critic of the *Brown* fragmentation. So ineffective a Chief Justice could scarcely be expected to lead the gifted prima donnas who then sat on the court. If anything, we saw, the division among the Justices intensified and, all too often, degenerated into personal animosity during Vinson's tenure. The first *Brown* conference, presided over by Vinson, saw the Court as divided as at any time during the Vinson tenure.

At that conference, on December 13, 1952, Chief Justice Vinson sounded what can most charitably be described as an uncertain trumpet, instead of the clarion needed to deal with the issue that had become an incubus on American society. His conference presentation totally ignored the fact that the "separate but equal" doctrine approved in *Plessy v. Ferguson*[5] had been the foundation of a whole structure of racial discrimination. Jim Crow replaced equal protection, and legally enforced segregation had become the dominant feature in Southern life.

Instead, at the *Brown* conference, Vinson's opening words indicated that he was not ready to overrule *Plessy v. Ferguson*. There was a whole "body of law back of us on separate but equal," he declared. Nor had the accepted law been questioned by Congress. "However [we] construe it," he said, "Congress did not pass a statute deterring and ordering no segregation." On the contrary, Congress itself had provided for segregation in the nation's capital. Since segregation in the District of Columbia had been imposed by a Congress including "men . . . who passed [the] amendments," he found it "hard to get away [from] that interpretation," particularly in light of the "long continued acceptance" of segregation in the District.[6]

On the other side, Justice Black asserted categorically that segregation rested on the idea of Negro inferiority; one did "not need books" or other sociological evidence "to say that." To Black, the "basic purpose" of the Fourteenth Amendment was "to protect [against] discrimination" and to abolish "such castes." For Justice Douglas also, the merits were "not in [the] realm of argument." The Constitution barred "classification on the basis of race." This principle was simple, "though the application of it may present difficulties." A similar view on the merits was stated by Justices Burton and Minton. As the latter stated it, "[C]lassification by race is not reasonable [and] segregation [is] per se unconstitutional."

The others, however, were closer to the Vinson position. The remaining Southerners on the Court, Justices Clark and Reed, indicated that they were following the Chief Justice's lead. Clark said that if the Court tried to impose desegregation immediately, "he would say we had led the states on to think segregation is OK and we should let them work it out."

Justice Reed went further, stating that he "approach[ed the case] from [a] different view," and would uphold the "separate but equal doctrine." In his view, the "states should be left to work out the problems for themselves."

Justices Frankfurter and Jackson were ambivalent. Frankfurter said that it was "intolerable" to have segregation in the District of Columbia. He found it difficult, however, to rule that segregation in the states was unconstitutional. He said that "he want[ed] to know why what has gone before is wrong." He could not accept a broad rule that "it's unconstitutional to treat a negro differently than a white."

Justice Jackson was also somewhat negative, though not as much as Justice Reed. Jackson declared that a decision to override *Plessy* could not be based on law as traditionally conceived. He said that he found nothing in the text, the opinions of the courts, or the history of the Fourteenth Amendment "that says [segregation] is unconstitutional," and asserted that "[Thurgood] Marshall's brief starts and ends with sociology." Yet that did not, as we shall see, mean that Justice Jackson would have voted in favor of segregation.

With the conference thus divided, any decision would have been sharply split. According to Justice Frankfurter, in his already-quoted letter to Justice Reed, there would have been only a bare majority in favor of striking down segregation. Justice Douglas read the conference consensus differently. In a *Memorandum for the File* on the day the *Brown* decision was delivered, Douglas wrote that if the cases had been decided during the 1952 Term, "there would probably have been many opinions and a wide divergence of views and a result that would have sustained, so far as the States were concerned, segregation of students."[7] Either way, such a fragmented decision would have been, in Frankfurter's phrase in his letter to Reed, "catastrophic."

Reargument Ordered

Fabian tactics are used as frequently in the Supreme Court as on the battleground. Justice Frankfurter, we have seen, made an ambiguous statement at the 1952 *Brown* conference. The Justice was not, however, entirely candid in his conference comments. Far from being in doubt about segregation in the states, he had concluded early in his consideration of the *Brown* case that segregation should now be stricken down. According to William R. Coleman, who had been the first black clerk at the Supreme Court when he clerked for Frankfurter during the 1948 Term, there was never any doubt about the Justice's *Brown* vote. "From the day the cases were taken," Coleman told an interviewer, "it was clear how he was going to vote. . . . I know . . . that he was for ending segregation from the start."[8]

More than any other Justice, Frankfurter was of the opinion that the close vote that would have resulted after the 1952 conference would, as he indicated in his letter to Reed, have been disastrous. With the Court as divided as it was, the only tactic that appeared at all promising was delay. The longer the Court's decision could be postponed, the more chance there was that the majority to end segregation could be increased. At the

December 1952 conference Frankfurter urged that the case be set for reargument. During the weeks that followed, the Justice sought to have reargument ordered for the next term, which would begin in October 1953.

To secure so long a postponement Justice Frankfurter would have to devise a plausible pretext, which would convince both the Court and the country that more than stalling for time was involved. Two factors made it easier to persuade the Justices. The first was that the conference had not taken any vote on the segregation issue. Justice Burton's diary entry on the 1952 conference records, "We discussed the segregation cases thus disclosing the trend but no even tentative vote was taken."[9] The same was true in the later conferences discussing the matter.

Just as important was the fact that, in Frankfurter's words at the time, "no one on the Court was pushing it."[10] Certainly Chief Justice Vinson was not eager to force a vote that he was going to lose; nor were the other Justices satisfied with the divided decision that would have come in the spring of 1953. According to Alexander Bickel, one of his 1952–1953 clerks, Justice Frankfurter returned from a conference toward the end of May 1953 in a euphoric state: "He said it looked as if we could hold off a decision that term . . . and that if we could get together some questions for discussion at a reargument, the case would be held over until the new term."[11]

With Bickel's help, Frankfurter formulated five questions "to which the attention of counsel on the reargument of the cases should be directed." At a conference on May 29, a majority agreed to put all the questions, with only minor changes from the Frankfurter draft. On June 8, the Court ordered the segregation cases restored to the docket for reargument in the next term, and the parties were asked to discuss the questions Frankfurter had drawn up, which related to the intent of the framers of the Fourteenth Amendment, as well as the remedial aspect of the case.[12]

Justice Jackson and Chief Justice Rehnquist

During the 1952 Term, one of Justice Jackson's law clerks was William H. Rehnquist, who was to become Chief Justice in 1986. Just before the 1952 conference Rehnquist wrote a memorandum on *Brown* which was headed *A Random Thought on the Segregation Case* and signed "whr."[13] It compared judicial action to invalidate segregation to the Court's "reading its own economic views into the Constitution" in cases such as *Lochner v. New York*[14]—"the high water mark in protecting corporations against legislative influence." According to the memo: "In these cases now before the Court, the Court is . . . being asked to read its own sociological views into the Constitution." For the Court to hold segregation invalid here, the memo asserted, would be for it to repeat the error of the *Lochner* Court. "If this Court, because its members individually are 'liberal' and dislike segregation, now chooses to strike it down, it differs from the McReynolds court only in the kinds of litigants it favors and the kinds of special claims it protects."

Rehnquist concluded his memo: "I think *Plessy v. Ferguson* was right and should be re-affirmed. If the Fourteenth Amendment did not enact Spencer's *Social Statics*,[15] it just as surely did not enact Myrdahl's [*sic*] *American Dilemma*."[16]

The Rehnquist memo became an important factor in the Senate debate on the nominations of Rehnquist both to the Supreme Court and as Chief Justice. Justice Rehnquist himself maintained that his *Brown* memo "was prepared by me at Justice Jackson's request; it was intended as a rough draft of a statement of *his* views at the conference of the justices, rather than as a statement of my views."[17] The Rehnquist explanation has been challenged, particularly during the hearings on his judicial nominations.[18] It is hard not to conclude, as Richard Kluger did in his monumental book on the *Brown* case, that "one finds a preponderance of evidence to suggest that the memorandum in question—the one that threatened to deprive William Rehnquist of his place on the Supreme Court—was an accurate statement of his own views on segregation, not those of Robert Jackson."[19]

To the evidence so convincingly summarized in the Kluger book,[20] one must add the draft concurrence which Justice Jackson prepared, but never issued, in *Brown*. It has long been known that Jackson worked on a draft which he intended as the basis for a concurring *Brown* opinion. That draft, which has become available in the Jackson papers at the Library of Congress,[21] appears inconsistent with Rehnquist's assertion that his memo was intended to state Jackson's rather than Rehnquist's view on the constitutionality of segregation.

The key sentence in the Jackson draft states categorically, "I am convinced that present-day conditions require us to strike from our books the doctrine of separate-but-equal facilities and to hold invalid provisions of state constitutions or statutes which classify persons for separate treatment in matters of education based solely on possession of colored blood." At the end of his draft, Justice Jackson repeated this conclusion, stating, "I favor enter[ing] a decree that the state constitutions and statutes relied upon as requiring or authorizing segregation merely on account of race or color, are unconstitutional."

The Jackson draft shows clearly that the Justice held the view that school segregation was unconstitutional. He may, as the other Justices did, have recognized that before *Brown* the law had been the other way. He also had no illusions about the difficulties involved in enforcing a desegregation decision. Still, his draft expressed no doubt on the correctness of the *Brown* decision. By the time of the case, he was plainly ready to announce the principle that segregation was unconstitutional. And he did so in the draft prepared but never issued by him.

It is hard to believe that the man who wrote the sentences holding segregation invalid in his draft held the view only a few months earlier attributed to him in the Rehnquist memo—that *"Plessy v. Ferguson* was right and should be re-affirmed." So inconsistent, indeed, is this view with the Jackson draft that one is tempted to ask what might have happened had Justice Jackson's unequivocal draft statements on the invalidity of segregation been available when the Senate voted on the Rehnquist nomination to the Supreme Court or his later nomination as Chief Justice.

Warren's *Brown* Conference

By the time the *Brown* case was reargued in December 1953, Chief Justice Vinson's seat had been taken by Warren. Neophyte though he was, the new Chief Justice was to play the crucial role in fashioning the unanimous *Brown* decision. It is not too much to say, indeed, that Frankfurter's Fabian strategy was ultimately justified by Warren's leadership in disposing of the segregation issue. As Justice Burton wrote in his diary just before the Court handed down its unanimous *Brown* decision, "This would have been impossible a year ago. . . . However the postponement then was with the hope of a better result later."

The Warren leadership became apparent when the Justices met December 12, 1953, for their first conference on the case under the new Chief Justice. Until then, his Brethren had found a bluff, convivial colleague, who disarmed them with his friendliness and lack of pretension. Now they were to learn that he was a born leader, whose political talents were to prove as useful in the Judicial Palace as they had been in the State House.

The Warren *Brown* conference set the tone both for disposition of the segregation issue and for the entire tenure of the Warren Court. All the Justices were present except Black, who had to be away because his sister-in-law was near death in Alabama. The Chief Justice opened the conference by proposing that the Justices discuss the cases informally without taking any votes. The others agreed.

Warren's plan to talk about the issues as long as necessary without any vote led to what Justice Frankfurter called "a reconnoitering discussion, without thought of a vote."[22] That discussion, at the December 12 conference, was made easier by a step that Frankfurter had taken after the *Brown* case had come to the Court. Early in the previous term, the Justice had given Alexander Bickel, then one of his law clerks, the job of doing intensive research on the original intent of the framers of the Fourteenth Amendment. Bickel's work, based on months of plowing through the musty, near century-old folio volumes of the *Congressional Globe,* was finished late in the summer of 1953. His lengthy memorandum, carefully revised by both the Justice and Bickel himself, was printed and sent to the other Justices just before the *Brown* reargument in early December. Frankfurter's covering *Memorandum for the Conference* summarized the result of Bickel's labors: "The [Bickel] memorandum indicates that the legislative history of the Amendment is, in a word, inconclusive."[23] The Thirty-ninth Congress, which voted to submit the amendment, with its guaranty of equal protection, to the states for ratification, did not indicate an intent to have it either outlaw or not outlaw segregation in public schools.

The conclusion that the legislative history of the Fourteenth Amendment was "inconclusive" on school segregation was significant enough to be repeated as the first important point in the *Brown* opinion.[24] It meant that those who argued that the framers of the amendment had intended not to abolish segregation had misread the legislative history. At the beginning of the December 1952 conference, Chief Justice Vinson had asserted that it was "hard to get away from that contemporary interpretation of the amendments." If there was no such clear interpretation, the Justices were not foreclosed by any supposed intent in favor of segregation by those who wrote the amendment.

At the December 12 conference, the new Chief Justice set a completely different tone from that of his predecessor. His first words revealed his position and furnished a clear lead for the Court.

"I can't escape the feeling," Warren declared, "that no matter how much the Court wants to avoid it, it must now face the issue. The Court has finally arrived at the place where it *must* determine whether segregation is allowable in public schools." To be sure, there was legitimate concern over whether the Court was called upon to reverse the older cases and lines of reasoning. "But the more I've read and heard and thought, the more I've come to conclude that the basis of segregation and 'separate but equal' rests upon a concept of the inherent inferiority of the colored race. I don't see how *Plessy* and the cases following it can be sustained on any other theory. If we are to sustain segregation, we also must do it upon that basis."

Warren then asserted that "if the argument proved anything, it proved that that basis was not justified." More than that, he went on, "I don't see how in this day and age we can set any group apart from the rest and say that they are not entitled to exactly the same treatment as all others. To do so would be contrary to the Thirteenth, Fourteenth, and Fifteenth Amendments. They were intended to make the slaves equal with all others. Personally, I can't see how today we can justify segregation based solely on race."

Having indicated his opinion on the constitutional issue, Warren interposed some cautionary words. "It would be unfortunate," he said, "if we had to take precipitous action that would inflame more than necessary. The condition in the different states should be carefully considered by the Court." Referring to "the Deep South," he said, "[I]t will take all the wisdom of this Court to dispose of the matter with a minimum of emotion and strife. How we do it is important."

To sum up, the Chief Justice concluded his presentation, "[M]y instincts and feelings lead me to say that, in these cases we should abolish the practice of segregation in the public schools—but in a tolerant way."

His opening statement was a masterly illustration of the Warren method of leading the conference. Justice Abe Fortas told me that "it was

Warren's great gift that in presenting the case and discussing the case, he proceeded immediately and very calmly and graciously, to the ultimate values involved—the ultimate constitutional values, the ultimate human values." Warren's *Brown* presentation clearly stated the question before the Court in terms of the moral issue of racial inferiority. Segregation, he told the Justices, could be justified only by belief in the inherent inferiority of blacks and, if we follow *Plessy,* we have to do it upon that basis. Warren's words went straight to the ultimate human values involved. In the face of such an approach, traditional legal arguments seemed inappropriate, almost pettifoggery. To quote Fortas again, "[O]pposition based on the hemstitching and embroidery of the law appeared petty in terms of Warren's basic value approach."

Warren's presentation put the proponents of *Plessy* in the awkward position of appearing to subscribe to racist doctrine. Justice Reed, who spoke most strongly in support of *Plessy,* felt compelled to assert that he was not making "the argument that the Negro is an inferior race. Of course there is no inferior race, though they may be handicapped by lack of opportunity." Reed did not, however, suggest any other ground upon which the Court might rely to justify segregation now. The gap here was particularly apparent, since Reed went out of his way to "recognize that this is a dynamic Constitution and what was correct in *Plessy* might not be correct now." He also conceded that, under *Plessy,* "equal protection has not been satisfactory. The result has been less facilities for Negroes."

But Warren was not engaged in gaining debating points against Justice Reed and any of the others who might still oppose the overruling of *Plessy.* Warren knew that with the four Justices—Black, Douglas, Burton, and Minton—who had indicated at the conference a year earlier that they were in favor of striking down school segregation, he had at least five votes. But for him to gain only a sharply split decision would scarcely be a ringing victory. He tailored his presentation to mollify those who might be persuaded to join the evolving majority. By stating the matter in terms of the moral issue of racial inferiority, he placed those who might still lean toward *Plessy* on the defensive. Then he reached out to them by his stress on the need to proceed "in a tolerant way."

He recognized and regretted that precedents would have to be overturned but noted that this was necessary if the Court was not to rely on outdated racist theory. His moral tone did not contain any accusations against the South (that would certainly have raised the hackles of Justices Reed and Clark, the two Southern members present), but said only that segregation was no longer justifiable "in this day and age."

Warren also sought to assuage Southern sensibilities by stressing the need for caution in effectuating the Court's decision. Wisdom and understanding were called for if the decision was to produce "a minimum of emotion and strife." Warren was firm in his position that the Court should abolish school segregation. His clear lead on the overriding issue set the theme for the Court's decision—a matter of crucial importance given the uncertainties with which some of the Justices may have still been plagued. But the Chief Justice indicated that the ultimate decision should take account of conditions in different states. The Court should be flexible in how it framed its decree.

When Warren completed his opening statement, he probably supposed that he had five, or at most six, votes to abolish segregation. That, at any rate, was the reasonable assumption, based upon the Court's discussion the year before. Yet the views expressed after Warren spoke indicated that there had been some movement among the Justices. Justices Douglas, Burton, and Minton repeated their strong views against segregation. Justice Frankfurter again delivered an ambiguous statement, yet from all we know the Justice had concluded the year before to vote against segregation. But Justices Clark and Jackson, who had expressed reluctance the previous year, now suggested that they might be willing to support the new Chief Justice's position.

Justice Clark, who had supported Chief Justice Vinson at the December 1952 conference, pointed out how difficult the issue was for one who was closer to the race problem than anyone on the Court except Justice Black. Nevertheless, Clark said he was willing to agree to a decision against segregation, but "it must be done carefully or it will do more harm than good." Above all, there "must be no fiat or anything that looks like a fiat." Justice Clark indicated that he was not enthusiastic about an antisegregation decision, "but will go along as said before, provided relief is carefully worked out . . . in such a way that will permit different handling in different places."

Justice Jackson stated that the segregation issue was largely "a question of politics." This "for me personally is not a problem, but it is difficult to make it other than a political decision. . . . Our problem is to make a judicial decision out of a political conclusion"—to find "a judicial basis for a congenial political conclusion." The implication was that he would support a properly written decision striking down segregation. "As a political decision, I can go along with it."

The conference discussion showed Warren that he could count on five, and probably six, solid votes to end segregation, and could secure two more votes if an opinion could be written that would satisfy Justices Clark and Jackson. That left Justice Reed, who alone still supported the *Plessy* doctrine. But even Reed had conceded that *Plessy* should not be followed merely because it was a precedent. He had also recognized that, in practice, *Plessy* had not produced anything like equal facilities for blacks and had had difficulty in meeting Warren's claim that to affirm *Plessy* meant to affirm black inferiority.

Chief Justice Warren now devoted all his efforts to eliminating the danger of dissenting and concurring opinions. The main thing, he knew, was to avoid polarizing those who had not indicated wholehearted agreement with his own forthright view. The last thing he wanted was a candid concurrence by Justice Jackson, a learned separate opinion by Justice

Frankfurter—much less the Southern breach in Court unity that would be worked by a dissent by Justice Reed.

The Chief Justice's instinct for effective leadership was demonstrated by the way in which he acted after the December 12 conference. Warren moved on two levels to secure agreement to a single, unanimous opinion: the personal level and the conference level. During the months following the December conference, Warren had frequent lunches with several of his colleagues. Most often, the luncheon group consisted of the Chief Justice and Justices Reed, Burton, and Minton. The three worked on Reed to induce him to go along with the majority. Among those who strongly favored striking down segregation, Burton and Minton were the two most congenial to Reed and the most likely to influence his vote, particularly by stressing the baneful effects of a split decision.

Warren also had many individual meetings with his colleagues. Justice Burton's diary notes several such sessions at which he discussed the segregation case with the Chief Justice. From all we know about Warren, he was most effective when he was able to operate in a one-on-one setting. That was the way he had been able to accomplish things back in California. The result in *Brown* showed that he had not lost any of his persuasive powers in the Marble Palace.

Second *Brown* Conference

On January 16, 1954, Warren presided over his second *Brown* conference. For the January conference the Chief Justice had developed the second prong of his *Brown* strategy, which he had revealed to Justice Burton a month earlier. It was, in Burton's words, a "plan to try [to] direct discussion of segregation cases toward the decree—as providing now the best chance of unanimity in that phase."[25]

S. Sidney Ulmer, a political scientist who also used Justice Burton's diary in writing about the *Brown* decision, concluded that the conference plan disclosed to Burton "reveals poor judgment on Warren's part since the subsequent processes by which a decree was produced proved to be much more complicated and difficult than the processes leading to the initial decision."[26] This comment misconceives the Warren strategy. Far from revealing poor judgment, the plan remains a striking illustration of the instinctive leadership that Warren displayed throughout his career. Continued discussion of the decision itself would only polarize potential concurrers and dissenters, solidifying their intention to write separate opinions. By concentrating instead on the remedy, all the Justices would work on the assumption that the decision itself would strike down segregation. They would endeavor jointly to work out a decree that would best effectuate such a decision. Those who, like Justice Reed, found it most difficult to accept a decision abolishing segregation would grow accustomed to what might at first have seemed too radical a step.

At any rate, the second *Brown* conference under Warren was devoted to the question of remedy. The Chief Justice led off the session along the lines he had indicated to Justice Burton. And he did so by setting the remedial theme in the manner most calculated to appeal to Justices Reed, Clark, and even Black, who feared the effects of ordering the South to admit the black plaintiffs immediately into white schools. Above all, the Southerners wanted a flexible decree that, as Justice Clark had stated at the December 12 conference, "will permit different handling in different places."

In addition the Chief Justice sought to reassure Justice Frankfurter and, to a lesser extent, Justice Jackson, who feared that for the Supreme Court to order relief in the individual school districts would bog the Justices in a political quagmire. On the day before the January 16 conference, Frankfurter had circulated a five-page memorandum devoted to "considerations [which] have arisen within me in regard to the fashioning of a decree." In it, he categorically asserted, "The one thing one can feel confidently is that this Court cannot do it directly."[27]

The Chief Justice's opening statement at the January 16 conference was aimed, first of all, at Justice Frankfurter. The Court, Warren said, should be "as little involved in administration as we can. We should turn to the district courts for enforcement. But we ought not to turn them loose without guidance of what paths are open to them." Warren then addressed himself to the Southerners' concern. The Chief Justice stressed that the enforcement problem was not the same in different states. Kansas would present no difficulty, he said. In Delaware, there would likewise be no trouble and, in the District of Columbia, it could also be done quickly. The South was another matter. As Justice Frankfurter's notes tell us, Warren expressed "lot of concern for S.C. and Va."

Justice Black, who had been absent at the conference a month earlier, spoke next. The Alabaman wanted the Court to do as little as possible in enforcement. "Leave it to the district courts," he urged, "let them work it out." "Would you give them any framework?" Warren interjected. "I don't see how you could do it," came back the reply, "we should leave it up to the district courts."

Justice Reed also felt that the Court should lay down only general enforcement principles. But, unlike Justice Black, he thought the Court must say a few things. Above all, it must ensure that enforcement was "not a rush job. The time they give, the opportunities to adjust, these are the greatest palliative to an awful thing that you can do." By his remarks, Reed indicated that he was beginning to feel that the "awful thing" of desegregation was becoming palatable, if enough time for adjustment was given.

The notes we have of the January 16 conference were taken by Justice Frankfurter.[28] They do not contain any summary of his own statement. Presumably, he repeated the position taken in his January 15 memorandum that the Court should not enforce desegregation directly. He may also have repeated a suggestion that he made in the memorandum—that a master should be appointed, either by the Supreme Court or the district court, to determine the facts and propose decrees in each of the cases before the Court. Justice Douglas agreed that the use of a master was a possible solution, since it would be difficult for the Court to decide anything concrete on enforcement. Douglas stressed that any decree should reflect generosity, as well as flexibility.

Justice Jackson said that if it was only a matter of personal right, there was no answer to the N.A.A.C.P. claim that the black plaintiffs were entitled to immediate admission to desegregated schools. "If what we're doing is to uproot social policy, what business is it of ours?" Jackson did not attempt to resolve the dilemma. Instead, he suggested that the Court needed more time to consider the remedial issue. As Frankfurter's notes quoted him, "let's have a reargument on terms of a decree!!"

Jackson's suggestion for reargument on the terms of the decree was supported by Justices Clark and Black, and ultimately adopted by the Court. Justices Burton and Minton restated the need for the Court itself not to perform the enforcement function. Burton said that "our contribution is to decentralize" enforcement and Minton that he was "not given to throwing the Court's weight around."

The conference closed with a strong statement by Justice Black on the need for flexibility in enforcement. The worst thing would be for the Court to try to resolve everything at once by its decree. "If necessary, let us have 700 suits" to work out the process. "Vagueness is not going to hurt." What was most needed was time. That was why Black had no objection to reargument. "Let it simmer. . . . Let it take time. It can't take too long." In the Deep South "any man who would come in [in support] would be dead politically forever. . . . In Alabama most liberals are praying for delay." Justice Black concluded the conference by warning that enforcement would give rise to a "storm over this Court."

Unanimity Secured

At his second *Brown* conference, Warren had stated his basic approach in the manner calculated to appeal to those in doubt: delegation of the decree power (to meet Justice Frankfurter's fear of embroiling the Court directly in the enforcement process) and flexibility (to meet the fears of the Southerners over immediate desegregation). The discussion indicated a consensus that was in accord with the Chief Justice's own position. The Southerners' alarm at precipitate action had led even the strongest proponents of overruling *Plessy* to concede the need for flexibility and time to work out the desegregation process.

The general agreement on Justice Jackson's suggestion for a reargument on the terms of the decree now made the Chief Justice's task of securing a unanimous opinion outlawing segregation less difficult. Justices Frankfurter and Clark would find it easier to agree, once it was clear that the Court itself would not immediately decree an end to segregation. Justice Reed might, it is true, prove more difficult to convince. But he, too, could be persuaded more readily, now that all had agreed to flexibility in enforcement and postponed the specifics on the remedial question to a later day.

After the January 16 session, there was one other conference discussion, several weeks later, before the Court took a vote on the matter. We do not know any details about the later conference or when the vote was taken. In his posthumously published memoirs, the Chief Justice stated that the Court voted in February and that the vote was unanimous against the separate-but-equal doctrine.[29] In later interviews Warren indicated that the vote came in March. The question of the exact date is less important than that of how the vote went. Despite Warren's statement in his memoirs, it is probable that the vote was eight to one, with Justice Reed voting to follow the *Plessy* case.

That Reed did vote to uphold segregation is indicated by his working on the draft of a dissent during February, asserting that segregation by itself did not violate equal protection.[30] In addition, there are indications that the Justice persisted in standing alone until the end of April. Warren continued to work on him to change his vote, both at luncheon meetings and in private sessions. Then, toward the end, the Chief Justice put it to Justice Reed directly: "Stan, you're all by yourself on this now. You've got to decide whether it's really the best thing for the country."[31] As described by Reed's law clerk, who was present at the meeting, Warren was typically restrained in his approach. "Throughout the Chief Justice was quite low-key and very sensitive to the problems that the decision would present to the South. He empathized with Justice Reed's concern. But he was quite firm on the Court's need for unanimity on a matter of this sensitivity."[32]

Ultimately Justice Reed agreed to the unanimous decision. He still thought, as he wrote to Justice Frankfurter, that "there were many considerations that pointed to a dissent." But, he went on, "they did not add up to a balance against the Court's opinion . . . the factors looking toward a fair treatment for Negroes are more important than the weight of history."[33]

Drafting the Opinion

At the conference that took the vote to strike down segregation, it was agreed that the opinion should be written by the Chief Justice. The writing of the *Brown* opinion was done under conditions of even greater secrecy than usual. The extreme secrecy was extended to the entire deliberative process in the segregation case. Thus the covering note to a Frankfurter memorandum on the fashioning of a decree in the case stated at the end: "I need hardly add that the typewriting was done under conditions of strictest security."[34]

The Justices also took steps to ensure that the way they voted would not leak out. No record of action taken in *Brown* was written in the docket book that was kept by each Justice and was available to his clerks. Warren tells us that, at the conference at which the opinion was assigned, "the importance of secrecy was discussed. We agreed that only my law clerks should be involved, and that any writing between my office and those of the other Justices would be delivered to the Justices personally. This practice was followed throughout and this was the only time it was required in my years on the Court."[35]

Toward the end of April, after he had secured Justice Reed's vote, the Chief Justice was ready to begin the drafting process. Warren's normal practice was to leave the actual drafting of opinions to his law clerks. He would usually outline verbally the way he wanted the opinion drafted. The outline would summarize the facts and how the main issues should be decided. The Chief Justice would rarely go into particulars on the details involved in the case. That was for the clerk drafting the opinion, who was left with a great deal of discretion, particularly on the reasoning and research supporting the decision.

It has been assumed that this procedure was also followed in the *Brown* drafting process. However, it was Warren himself who wrote the first *Brown* draft. Headed simply "Memorandum" and undated, the original draft, in Warren's handwriting, is in pencil on nine yellow legal-size pages.

Certain things about Warren's first *Brown* draft stand out. First, it definitely set the tone of the final opinion. It was written in the typical Warren style: short, nontechnical, well within the grasp of the average reader. The language is direct and straightforward, illustrating the point once made to me by one of his law clerks: "He had a penchant for Anglo-Saxon words over Latin words and he didn't like foreign phrases thrown in if there was a good American word that would do."

Warren's own draft was based on the two things he later stressed to Earl Pollock, the clerk primarily responsible for helping on the *Brown* opinion: the opinion should be as brief as possible, and it was to be written in understandable English, avoiding legalisms. This was repeated in Warren's May 7 memorandum transmitting the draft opinion to the Justices. The draft, wrote the Chief Justice, was "prepared on the theory that the opinions should be short, readable by the lay public, non-rhetorical, unemotional and, above all, non-accusatory."[36]

Even more important is the fact that the Warren draft is basically similar to the final *Brown* opinion. Changes were, to be sure, made in the draft—both by the Chief Justice and his clerks as well as by other Justices. In particular, the clerks supplied supporting authority in the text and in fleshing out the opinion with footnotes. But the essentials of the *Brown* opinion (and most of its language) were contained in the draft Warren wrote out in his own forcefully legible hand at the end of April 1954.

The Warren draft contains two of the three most famous passages in the *Brown* opinion. First, after referring to the decision facing the Court, the draft states, "In approaching it, we cannot turn the clock of education back to 1868, when the Amendment was adopted, or even to 1895 [*sic*] when *Plessy v. Ferguson* was decided."

The Warren draft also contains *Brown*'s striking passage on the baneful effect of segregation on black children: "To separate them from others of their age in school solely because of their color puts the mark of inferiority not only upon their status in the community but also upon their little hearts and minds in a form that is unlikely ever to be erased."

In addition, it was the Warren draft that stressed the changed role of education in the contemporary society, as contrasted with the situation when the Fourteenth Amendment was adopted ("No child can reasonably be expected to succeed in life today if he is deprived of the opportunity of an education"). And it posed the critical question presented to the Court— "Does segregation of school children solely on the basis of color, even though the physical facilities may be equal, deprive the minority group of equal opportunities in the educational system?"—as well as its answer— "We believe that it does."

An early draft of the memorandum transmitting the *Brown* draft to the Justices declared, "On the question of segregation in education, this should be the end of the line."[37] If that was true, it was mainly the Chief Justice's doing—even more than commentators on *Brown* have realized. The Warren *Brown* draft shows us that the Chief Justice was primarily responsible not only for the unanimous decision, but also for the opinion in the case. This was one case where the drafting was not delegated.[38] The opinion delivered was essentially the opinion produced when Warren himself sat down and put pencil to paper.

Final Opinion

"An opinion in a touchy and explosive litigation . . . is like a souffle—it should be served at once after it has reached completion." The *Brown* drafting process, after Warren had penciled his draft opinion, showed that the Chief Justice agreed with this sentiment contained in a May 15, 1954, note from Justice Frankfurter.

As soon as Warren had finished his *Brown* draft and had it typed, he called in his three law clerks and told them that the decision of the Court was to overturn *Plessy,* and that the decision was unanimous. He enjoined them to the strictest secrecy, saying that he had not told anyone outside the Justices what the decision was—not even his wife. The Chief Justice asked all three clerks to write drafts based on his own draft opinion. It was now the end of April, and the clerks worked the entire weekend in order to have their drafts ready on Monday, May 3. Earl Pollock, in particular, recalls working on his draft straight through the weekend, virtually without sleep. Pollock was the clerk with primary responsibility for the case, since he had written the *Brown* bench memo. Such a memo is prepared for each Justice after the Court votes to hear a case.

Of the three clerks' drafts, Pollock's was the most important, for he made the only significant changes in the Chief Justice's draft. The most important change was the separation of the companion case of *Bolling v.*

Sharpe[39] from the *Brown* case itself. The segregation cases presented to the Court involved schools in four states and the District of Columbia. Warren's *Brown* draft had dealt with all these cases together, as shown by its opening sentence: "These cases come to us from the States of Delaware, Virginia, Kansas and South Carolina and from the District of Columbia." The Warren draft went on to say that, though they were separate cases, "the basic law involved in their decision is identical to the point that they can on principle properly be considered together in this opinion."

In treating the state and D.C. cases together, the Warren draft was making a legal mistake. The rationale for striking down segregation in the states cannot be used to reach that result in Washington, D.C. The state action in *Brown* was invalidated under the Equal Protection Clause of the Fourteenth Amendment. Yet that amendment is binding only on the states, not the Federal Government. The latter is bound by the Fifth Amendment, which contains a Due Process Clause but no requirement of equal protection. Obviously, the Court would not decide that the states could not have segregated schools while the District of Columbia could. But the result had to be reached in terms of due process, rather than equal protection, analysis.

Following the example set in Warren's draft, the three law clerks dealt with the state and D.C. cases together in their drafts. However, in the May 3 covering memorandum attached to his draft, Pollock indicated that this was not the proper approach. Pollock stated that his draft was "along the lines of your memo of last week. Like the memo, this draft covers all five cases in one consolidated opinion." Pollock then wrote, "I am inclined to think, however, that the District of Columbia case should be treated independently in a short, separate opinion accompanying the other one. . . . The material relating to the equal protection clause of the 14th Amendment has no direct relevance to the District of Columbia case, which, of course, is based primarily on the due process clause of the 5th Amendment."

Warren accepted Pollock's suggestion and the details of the legal theory underlying the D.C. case were worked out by the other two Warren clerks. They also drafted the separate short opinion delivered in the case.

In addition to the May 3 drafts of the clerks, there were further *Brown* drafts on May 5 and 7, before the draft opinion was ready to be circulated to the Justices, together with the draft in the D.C. case. This typed *Brown* draft, composed of nine legal size pages, was titled "MEMORANDUM ON THE STATE CASES." The draft in *Bolling v. Sharpe,* four typed pages, was titled "MEMORANDUM ON THE DISTRICT OF COLUMBIA CASE." The draft opinions were then delivered personally to the Justices on Saturday, May 8. Justice Burton's diary entry for that date notes: "In AM the Chief Justice brought his draft of the segregation cases memoranda. These were in accord with our conversations."[40] Warren himself took the printed drafts to the Justices who were in their chambers. His clerks delivered copies to the others. Justice Black's copy was brought to him on the tennis court at his home in Alexandria, Justice Minton's at his Washington apartment.

As already indicated, a few significant alterations were made in the Warren draft. The Justices themselves made only minor stylistic changes.[41] In the Warren chambers, the further drafts were prepared by Pollock, who also made numerous cosmetic alterations in the Chief Justice's draft. There were only two important changes. The first was a discussion of the post–*Plessy v. Ferguson* education cases. This added some legal meat to the opinion, but did not really add anything of substance to it.

The second change was more significant, particularly when we add the contribution made to it by a second clerk. After Warren's "hearts and minds" sentence on the baneful effect of segregation, Pollock's first draft of May 3 added the passage from the opinion of the lower court in the Kansas case that now appears in the *Brown* opinion. The Pollock draft then asserted, "Whatever may have been the state of psychological knowledge when *Plessy v. Ferguson* was decided, this finding is amply supported by modern authority. To the extent that there is language in *Plessy v. Ferguson* to the contrary, and to the extent that other decisions of this Court have been read as applying the 'separate but equal' doctrine to education, those cases are overruled."

This statement was accepted by the Chief Justice and repeated in the drafts of May 5, 7, and 8. The passage was put into its final form by written corrections in Pollock's writing in the draft of May 12, the final draft in the case.

In an interview Pollock told me that he had been greatly troubled by the *Plessy v. Ferguson* opinion. "It seemed to me," he went on, "that the most noxious part of *Plessy* was the notion that, if Negroes found segregation a 'badge of inferiority,' that was sort of in the eye of the beholder. I was looking for a way to part company and to say that, whatever the situation was in the 1890's, we know a lot more about law in society than we did then."

It was this approach that was responsible for the most controversial part of the *Brown* opinion—that relying not on law but "psychological knowledge." In the *New York Times* of May 18, 1954, James Reston analyzed *Brown* in a column headed "A Sociological Decision." According to Reston, the Court had relied "more on the social scientists than on legal precedents. . . . The Court's opinion read more like an expert paper on sociology than a Supreme court opinion."[42]

The Reston characterization was based on the assertion that the *Brown* opinion rested less on legal reasoning than on the Pollock statement on "psychological knowledge . . . amply supported by modern authority." Even more suggestive to commentators was the fact that the statement was backed by a footnote—the celebrated footnote 11, which soon became the most controversial note in Supreme Court history.

Footnote 11 listed seven works by social scientists, starting with an article by Kenneth B. Clark and concluding with the massive two-volume study by Gunnar Myrdal, *An American Dilemma.*[43] Both supporters and critics of the Court have assumed, like Reston, that footnote 11 means that the *Brown* decision was based on the work of the social scientists cited. As Kenneth Clark, the first mentioned in the footnote, summarized it, "[T]he Court . . . appeared to rely on the findings of social scientists in the 1954 decision."[44] A plethora of learned commentary followed, analyzing in amazing detail the works cited, the significance of the order of citation, and the fact that other relevant writings were not cited—as well as the whole subject of using the methods and products of social science in deciding controversial legal issues. Even supporters of the decision have shown how vulnerable the studies and tests relied on in most of the cited works really were.

Those who have focused intensively on the controversial footnote have acted out of ignorance of the manner in which Supreme Court opinions are prepared. Warren himself stressed that it was the decision, not the citations, that were important. "It was only a note, after all," he declared to an interviewer.[45] The Chief Justice normally left the citations in opinions to his law clerks. He considered them among the minutiae of legal scholarship, which could in the main be left to others.

In the *Brown* case, the fleshing out of the opinion with footnotes was left primarily to one of Warren's clerks, Richard Flynn. As he recalls it, there was no specific method in the organization of footnote 11. When it came time to cite supporting authority, the works listed (particularly those by Myrdal and Clark) were, to Flynn, the "obvious" things to list. When asked by me if there was any method in the way he organized the note, he answered, "I don't recall any, that's just the way it fell."

It has been said that footnote 11 provoked concern among several Justices, particularly in the gratuitous citation of the Myrdal work, which had in the decade since its publication become a red flag to the white South. In 1971, Justice Clark, then retired from the Court, told an interviewer, "I questioned the Chief's going with Myrdal in that opinion. I told him—and Hugo Black did, too—that it wouldn't go down well in the South. And he didn't need it."[46]

Clark's recollection appears inaccurate. Over the years, he may have come to feel that, in view of the storm in the South caused by the Myrdal citation, he should have said something at the time, which may, years later, have become a blurred recollection that he and Black had expressed concern. But in all likelihood, if Justices Black and Clark had objected, the Myrdal reference would have been taken out. The citation was just not important enough to Warren to withstand even a mild expression of concern by any of the Justices.

Justice Clark did suggest one correction in footnote 11 that was speedily accepted by the Chief Justice—which indicates that the same thing might well have happened if Clark had asked for the Myrdal work to be removed. In the draft of the *Brown* opinion originally delivered to the Justices, the footnote began with the citation "Clark, Effect of Prejudice and Discrimination on Personality Development (Midcentury White House Conference on Children and Youth, 1950)." Justice Clark objected that this did not sufficiently identify the Clark who had written the article cited. He did not want people, particularly in his own South, to think that the Court was citing an antisegregation article that he—Tom Clark—had authored. He therefore asked that the citation be changed to "K. B. Clark," so that no one would confuse the Justice with the author. In the final opinion, as Justice Clark had suggested, the name of the first author cited appears with initials, the only author in the footnote not identified solely by last name.

There was thus no controversy over footnote 11 in the Court itself. So far as we know, the draft *Brown* opinion was speedily approved by the Justices—and without significant changes. The general attitude among Warren's colleagues was expressed in a handwritten note by Justice Douglas: "I do not think I would change a single word. . . . You have done a beautiful job."[47]

The final *Brown* draft was circulated on May 13 in printed form. The next day, Saturday, May 15, was a conference day. At lunch, the Justices were entertained by Justice Burton, with a large salmon provided by Secretary of the Interior Douglas McKay. Just before, Burton wrote in his diary, the "conference finally approved Segregation opinions and instructions for delivery Monday—no previous notice being given to office staffs etc so as to avoid leaks. Most of us—including me—handed back the circulated print to C.J. to avoid possible leaks."[48]

Decision Day

Monday, May 17, 1954, Supreme Court Chamber. At noon precisely, the curtain behind the bench parted and, led by the Chief Justice, the Justices filed to their places. "As we Justices marched into the courtroom on that day," Warren later recalled, "there was a tenseness that I have not seen equaled before or since."[49] For weeks before the *Brown* decision was announced, the courtroom had been jammed; anticipation mounted as the weeks passed. Now, to the acute Court watcher, there were definite signs that the day would not be a quiet one. For one thing, several of the Justices' wives were present, something that rarely happened except on historic Court occasions. Even more significant was Justice Jackson's presence. Early that morning Warren had gone to the hospital, where Jackson was critically ill, to show him a copy of the final opinion. Then, says Warren, the ailing Justice "to my alarm insisted on attending the Court that day in order to demonstrate our solidarity."[50]

Warren had advised his law clerks to be in the courtroom that morning. They had also realized from the intensive work on the final proofs that weekend that announcement of the decision was imminent. Justice Frankfurter had also told his clerks that *Brown* was coming down. The other Justices' clerks had not been informed in advance, but they all sensed that this was to be more than the usual decision day. On his way to the robing room, Justice Clark had stopped to say to the clerks, "I think you boys ought to be in the courtroom today."[51]

Surprisingly, the air of expectancy was absent from the Supreme Court press room. The reporters were told before the Court convened that it looked like a quiet day. Their expectation appeared borne out by the routine business that occupied the first part of the Court session. First there was the admission of 118 attorneys to the Supreme Court Bar. Then three unimportant opinions were read by Justices Clark and Douglas.

By now, fifty minutes of the Court session had passed. The reporters did not bother going upstairs to hear the routine opinions being delivered. But then, as they sat in the press room over their coffee, Banning E. Whittington, the Court's press officer, started putting on his coat and announced, "Reading of the segregation decisions is about to begin in the courtroom." Whittington led a fast-moving exodus as the reporters dashed up the long flight of marble steps just as the Chief Justice began reading in his booming bass. It was then 12:52 P.M.

"I have for announcement" said Warren, "the judgment and opinion of the Court in No. I—Oliver Brown et al. v. Board of Education of Topeka." Then, in a firm, colorless tone, he read the opinion in the state segregation case. First, the background of the case was given. Plaintiff Negroes sought admission to public schools on a nonsegregated basis. They had been denied relief on the basis of the *Plessy* "separate but equal" doctrine. They contended that segregated schools were not "equal" and could not be made "equal" and, by their very nature, deprived plaintiffs of the equal protection of the laws. The Chief Justice then summarized the reargument and concluded (following the Bickel-Frankfurter memorandum) that the legislative history of the Fourteenth Amendment was, at best, inconclusive. He also stressed the changed position in the society of both public education and the Negro.

The opinion next went into the origin of the separate-but-equal doctrine and the cases which had applied it. Here, unlike those cases, the lower courts had found that the Negro and white schools had been or were being equalized as to buildings, curricula, teacher qualifications, and other "tangible" factors. The decision therefore could not turn on comparing merely these tangible factors: "We must look instead to the effect of segregation itself on public education."[52]

In approaching this problem, the Court could not "turn the clock back to 1868 when the Amendment was adopted, or even to 1896 when *Plessy v. Ferguson* was written. We must consider public education in light of its full development and its present place in American life throughout the Nation." Then, in what for him was eloquent language, the Chief Justice affirmed that "[t]oday, education is perhaps the most important function of state and local governments. . . . In these days, it is doubtful that any child may reasonably be expected to succeed in life if he is denied the opportunity of an education. Such an opportunity, where the state has undertaken to provide it, is a right which must be made available to all on equal terms."

This brought the opinion to the crucial question: "Does segregation of children in public schools solely on the basis of race, even though the physical facilities and other 'tangible' factors may be equal, deprive the children of the minority group of equal educational opportunities?"

Until then, with the opinion two-thirds finished, the Chief Justice had not indicated the outcome of the decision. Now, in the next sentence, he did. Answering the critical question, he asserted: "We believe that it does."

The opinion reviewed earlier cases which had held that, in colleges and universities, segregated schools could not provide equal educational opportunities. The same was true for children in public schools. "To separate them from others of similar age and qualifications solely because of their race generates a feeling of inferiority as to their status in the community that may affect their hearts and minds in a way unlikely ever to be undone." This was the final version of the soaring sentence Warren had put in his original outline; it stated the baneful impact of segregation on black children better than a whole shelf of learned sociological tomes.

Plessy v. Ferguson had said that the assumption that segregation supposed colored inferiority was fallacious. Warren now declared that segregation did mean the inferiority of the Negro group. "Whatever may have been the extent of psychological knowledge at the time of *Plessy v. Ferguson*, this finding is amply supported by modern authority. Any language in *Plessy v. Ferguson* contrary to this finding is rejected." This statement was supported by footnote 11 of the opinion, which listed the seven works by social scientists and was to become the most famous note in a Supreme Court opinion.

The Chief Justice then stated the Court's far-reaching conclusion: "We conclude"—here Warren departed from the printed text to insert the word "unanimously"—"that in the field of public education the doctrine of 'separate but equal' has no place. Separate educational facilities are inherently unequal." Warren later recalled: "When the word 'unanimously' was spoken a wave of emotion swept the room; no words or intentional movement, yet a distinct emotional manifestation that defies description."[53] Other observers agree that, as Warren pronounced the word, barely suppressed astonishment swept around the courtroom. Until then, neither the press nor the public had any idea that the man from California had achieved a unanimous decision.

The *Brown* opinion concluded with the holding that plaintiffs "and others similarly situated" had been deprived of equal protection by the segregation complained of. But it did not decree any relief. Instead, the Jackson suggestion at the January 16 conference was followed. "Because these are class actions, because of the wide applicability of this decision, the formulation of decrees in these cases presents problems of considerable complexity." This decision dealt only with "the constitutionality of segregation in public education. We have now announced that such segregation is a denial of the equal protection of the laws." The question of appropriate relief was still before the Court. "In order that we may have the full assistance of the parties in formulating decrees," the Court was scheduling further argument on the matter for the next Court term, beginning in October 1954. The U.S. Attorney General was invited to participate. So were the Attorneys General of states with segregated schools.

Almost never before, wrote Arthur Krock the next day in the *New York Times,* had the "high tribunal disposed so simply and briefly of an issue of such magnitude."[54] The *Brown* opinion was strikingly short for an opinion of such consequence: only ten pages in the *United States Reports,* in which Supreme Court opinions are printed. The second segregation opinion, that in *Bolling v. Sharpe,* was even shorter, disposing of District of Columbia segregation in six paragraphs.

When Warren concluded the *Bolling* opinion, it was 1:20 P.M.; it had taken only twenty-eight minutes for the Chief Justice to read both opinions. After Warren finished, everyone in the courtroom looked to Justice Reed to see what the courtly Southerner was going to do. Reed, who had had tears in his eyes during the opinion reading, now filed out silently with the Justices. Despite Warren's use of the word "unanimously," it was hard for many to grasp that neither Reed nor any of the other Justices had filed a separate opinion. For weeks thereafter people would call the clerk's office and ask to see the dissenting opinion.

Discussing the *Brown* unanimity later, Warren said with a characteristic generosity of spirit that the real credit should go to the three Southern Justices. "Don't thank me," he said to a California friend who visited him in his chambers, "I'm not the one. You should see what those . . . fellows from the southern states had to take from their constituencies. It was absolutely slaughter. They stood right up and did it anyway because they thought it was right."[55] Justice Reed, in particular, came to believe that the *Brown* decision was wholly right. He told one of his clerks that "if it was not the most important decision in the history of the Court, it was very close."[56]

The other Justices, too, were well aware that they had participated in what Justice Frankfurter termed "a day that will live in glory."[57] A few days earlier, in a note to Warren joining the opinion, Frankfurter wrote: "When—I no longer say 'if'—you bring this cargo of unanimity safely to port it will be a memorable day no less in the history of the Nation than in that of the Court. You have, if I may say so, been wisely at the helm throughout this year's journey of this litigation. Finis coronat omnia."[58]

Other Segregation Cases

Chief Justice Warren had been careful to limit the *Brown* opinion to desegregation in schools. But the decision necessarily had a broader impact. In *Bolling v. Sharpe,* the District of Columbia companion case, the Court had declared, "In view of our decision that the Constitution prohibits the states from maintaining racially segregated schools, it would be unthinkable that the same Constitution would impose a lesser duty on the Federal Government."[59] It would similarly be unthinkable that the same Constitution would prohibit segregation in schools and impose a lesser duty on other public institutions.

Soon after the first *Brown* decision, the U.S. Court of Appeals for the Fourth Circuit was presented with the question of whether Baltimore could continue to segregate its public beaches. The court held that *Brown* required a negative answer. If segregation in schools generates a feeling of inferiority in Negroes, the same must be true, said the court, of segregation in recreational facilities.[60] In November 1955, the Supreme Court unanimously affirmed this decision without opinion.[61] At the same time, it unanimously reversed a decision upholding a segregated municipal golf course in Atlanta, citing in support only its decision in the Baltimore case.[62]

In 1956, the Court affirmed a decision striking down the segregated bus system in Montgomery, Alabama.[63] After the conference, Chief Justice Warren told his law clerks that they were going to affirm summarily, without hearing oral argument. He said they were going to cite three cases: *Brown* and the Baltimore and Atlanta cases of the previous year. The clerk who drafted the opinion told me that, since he had no further guidance, he took the prudent course and simply followed the Chief Justice literally. That is the way the per curiam opinion in the Montgomery bus case appears in the *United States Reports*—stating merely that the decision below is affirmed and citing only the three cases, without explaining what the Court was doing and the reasoning behind its decision.

The clerk involved says, "I thought at the time that it was a pretty casual way for the Court to advance a major proposition of constitutional law and still do." Legal scholars have voiced a similar criticism of the Court's failure to explain its extension of *Brown* to public facilities outside the field of education. The criticisms, however, are beside the point. Was there a need for explanation once the *Brown* opinion—with the broad sweep of its language striking down separation of the races—had been written?

In the years after *Brown,* the Court ruled segregation invalid in all public buildings,[64] housing,[65] transportation,[66] and recreational[67] and eating facilities.[68] The Court, following Warren's lead, said as little as possible but continued to strike down every segregation law challenged before it. By 1963, an opinion could declare categorically: "[I]t is no longer open to question that a State may not constitutionally require segregation of public facilities."[69]

Chief Justice Warren used to recall that, when he first arrived, the Supreme Court had a separate washroom for Negroes. One of the first things he did was to end the discrimination taking place in the Court building itself. By the end of Warren's tenure, as the cases just discussed show, the Court had virtually rooted out racial discrimination from American law.

Brown in Perspective

An undated note, written on a Supreme Court memo pad in Justice Frankfurter's handwriting, reads: "It is not fair to say that the South has always denied Negroes 'this constitutional right.' It was NOT a constitutional right till May 17/54."[70] The more restrained critics of the *Brown* decision have been disturbed by what they claim is its inadequacy in explaining its vindication of the new right. They allege a lack of legal craftsmanship in the Warren opinion that has tended to deprive it of the respect to which it would otherwise be entitled. They point to the opinion's laconic nature, its failure to rely upon legal precedent, and its literary weakness, as compared with those produced by past masters of the judicial art, such as Marshall, Holmes, and Cardozo.

It may be unfortunate that the *Brown* opinion was not written by a John Marshall or a Benjamin Cardozo. Great cases deserve nothing less than opinions by the consummate giants of legal craftsmanship. But the criticism brings to mind a comment of Warren E. Burger, shortly after he succeeded Warren as Chief Justice. Seated before the fire at the elegant Elizabethan manor where international conferences are held in Ditchley, Oxfordshire, swirling the fine liqueur in a glass in his hand, the new Chief Justice was distressed by critics who contended that he did not have the qualifications of a Cardozo. "Who is there today who does?" he wryly asked.

The same can be asked of those who have urged that *Brown* deserved a Holmes or a Cardozo for its creator. One can go further and wonder whether those virtuosos of the opinion could possibly have secured the unanimous decision that was the Warren forte. If we take the *Brown* opinion as it is written, it certainly ranks as one of the great opinions of judicial history—plainly in the tradition of Chief Justice Marshall's seminal dictum that the Court must never forget that it is a *Constitution* it is expounding.[71] Perhaps the *Brown* opinion did not demonstrate as well as it might have that the mere fact of segregation denies educational equality. But *Brown* is so clearly right in its conclusion in this respect that one wonders whether additional labor in spelling out the obvious would really have been worthwhile. Considerations of *elegantia juris* and judicial craftsmanship are largely irrelevant to the truism that segregation was intended to keep blacks in a status of inferiority.

Four decades after the *Brown* case, it has become apparent that the criticisms of the Court's performance there have lost their relevancy. What is plain is that *Brown* has taken its place in the very forefront of the pantheon of historic decisions. In the light of what Justice Goldberg once termed the American commitment to equality[72] and its part in helping to fulfill the commitment, *Brown* will occupy a paramount position long after the contemporary criticisms will have ceased to have any more continuing significance than those voiced against the great decisions of the Marshall Court at the very outset of the nation's constitutional development.

14

Burger Court, 1969–1986

Before Warren E. Burger became Chief Justice, all Supreme Court documents were typed with as many carbon copies as needed. Thus the memoranda on in forma pauperis (or Miscellaneous Docket) petitions sent to each Justice required many copies. Because it was necessary to use very thin paper, these memoranda were called flimsies. The junior Justices, who received the last copies, often had difficulty reading them.

This was changed when Chief Justice Burger sent around an August 7, 1969, memorandum: "The necessary steps are now being taken toward acquiring a Xerox machine in the building." The new copiers "will be utilized primarily in preparation and distribution of the Miscellaneous Docket memoranda ('flimsies' as they are commonly called.)" Since that time, ample copiers have been provided for the Justices and Court staff, and word processors and computers have been introduced.

But the availability of copiers and the other communications devices had a baneful, though unintended, effect upon the operation of the Burger Court. One privy to the working of the Warren Court quickly notes the crucial importance of personal exchanges among the Justices—both in conference discussions and, even more so, in the postargument decision process. Such exchanges became less significant in the Burger Court. Conference notes during the Burger years show less an interchange of views than flat statements of each Justice's position in the case. "Not much conferencing goes on" at the conference, a Justice recently confirmed. "In fact, to call our discussion of a case a conference is really something of a misnomer, it's much more a statement of the views of each of the nine Justices, after which the totals are added and the case is assigned."[1]

"When I first went on the Court," writes Chief Justice Rehnquist, "I was both surprised and dismayed at how little interplay there was between the various justices."[2] Rehnquist, too, was referring to the conference, but his remark is equally applicable to the entire decision process. The constant personal exchanges in the Warren Court (much of it one-on-one lobbying by the Chief Justice and his allies or by opponents intended to influence votes) gave way to mostly written contacts through notes and memoranda. It is after all so much easier to make copies and send them around than to engage in protracted personal efforts to persuade others to change their positions. In the Court, as in other institutions, technology intended to facilitate communication has made for less personal interchange.

It was, however, more than the photocopier that made for the difference between the Burger and Warren Courts. A few years ago, the present Chief Justice compared the two tribunals. Rehnquist stated that the impact of the Court had been diminished under Chief Justice Burger. "I don't think that the Burger Court has as wide a sense of mission. Perhaps it doesn't have any sense of mission at all."[3]

In large part, this was true because of the differences in the leadership role assumed by the two Chief Justices. Warren brought more authority to the position of Chief Justice than had been the case for years, and the Warren Court bore his image as unmistakably as the earlier Courts of John Marshall and Roger B. Taney. This was plainly not as true while Chief Justice Burger occupied the Court's central chair. To be sure, Chief Justice Warren was inevitably a tough act to follow. But even on its own terms, the Burger tenure was not marked by strong leadership in molding Supreme Court jurisprudence.

Burger on Olympus

Warren E. Burger was cast from a different mold than Earl Warren. Burger's background was mostly in a law firm in St. Paul. He had nothing like Warren's spectacular career and broad political experience, although he had been active in the Republican party. He worked in Harold E. Stassen's successful campaign for Governor, and in 1952 he was Stassen's floor manager at the Republican national convention when Minnesota's switch supplied the necessary votes for Dwight Eisenhower's nomination.

After the election, Burger was appointed Assistant Attoney General in charge of the Claims Division of the Department of Justice. This experience led directly to one of his most noted opinions as an appellate judge.[4] It held that a government contractor might not be debarred from further government contracts without notice and hearing. Burger told me that the ruling was directly influenced by the many debarment orders that had come across his desk in the Justice Department—"issued in the name of the

Secretary of the Navy; but the actual decision was made by some Lieutenant, J.G., way down the line."

In 1956, Burger was named to the U.S. Court of Appeals in Washington, D.C., where he developed a reputation as a conservative, particularly in criminal cases. Then, in 1969, came what Justice John M. Harlan once called his "ascendancy to the Jupitership of Mount Olympus."[5] Burger was sworn in as Chief Justice when Chief Justice Warren retired in 1969 and headed the Court until his own 1986 retirement.

Burger, a reporter recently wrote, "looked as though he had been cast by Hollywood for the part of chief justice."[6] With his snow-white hair and broad shoulders, Burger was an almost too-perfect symbol of the law's dignity. His critics contended that he stood too much on the dignity of his office and was, more often than not, aloof and unfeeling. Intimates, however, stress his courtesy and kindness and assert that it was the office, not the man, that may have suggested a different impression. "The Chief," says Justice Harry A. Blackmun, who knew Burger far longer than anyone else on the Court, "has a great heart in him, and he's a very fine human being when you get to know him, when the tensions are off. One has to remember, too, that he's under strain almost constantly."[7]

Burger was not a person to develop intimate relationships with those with whom he worked; yet he was as close to his clerks as to anyone in the Marble Palace. A novel tells of a "tiny kitchen installed in his office by Warren E. Burger who loved to cook and often prepared lunch for himself or an occasional guest."[8] Every Saturday at noon, Burger made soup in his kitchen for his clerks. They never knew what was in it, one of them told me, but they ate it without question or complaint. Then Burger would sit and talk informally with them for hours, usually with colorful reminiscences about his career. The one rule at these sessions was that no one would talk about the cases before the Court on which they were currently working.

Others picture the Chief Justice as a petty pedant, not up to the demands of his position and most concerned with minor details and the formal dignity of his office. Burger himself undertook the redecoration of the Supreme Court cafeteria and personally helped choose the glassware and china. He also redesigned the Court bench, changing it from a traditional straight bench to a "winged," or half-hexagon shape.

Burger was always sensitive to what he perceived to be slights to his office and to himself; throughout his tenure he had an almost adversarial relationship with the press. According to one reporter, "He fostered an atmosphere of secrecy around the court that left some employees terrified of being caught chatting with us."[9] When a network asked permission to carry live radio coverage of the arguments at the Court, Burger replied with a one-sentence letter: "It is not possible to arrange for any broadcast of any Supreme Court proceeding." Handwritten at the bottom was a postscript: "When you get the Cabinet meetings on the air, call me!"[10]

The Chief Justice was deeply hurt by derogatory accounts about his performance, particularly in the best seller *The Brethren,* and was gleeful when he told me that copies of the book were remaindered at ninety-eight cents in a Washington bookstore. But all the Justices were sensitive to the effect of *The Brethren* on the public perception of the Court. In a memorandum to the others, Rehnquist urged that it would be unwise for the Court to take certain action, "especially . . . in light of the microscopic scrutiny which our actions are apt to receive for a while."[11]

From his "Middle Temple" cheddar, made according to his own recipe, to the finest clarets, Burger was somewhat of an epicure. One of the social high points of a 1969 British-American conference at Ditchley, Oxfordshire, was the learned discussion about vintage Bordeaux between Burger and Sir George Coldstream, head of the Lord Chancellor's office and overseer of the wine cellar at his Inn of Court. The Chief Justice was particularly proud of his coup in snaring some cases of a rare Lafite in an obscure Washington wine shop.

The effectiveness of a Chief Justice is, of course, not shown by his epicurean tastes. Certainly, Burger in action cannot be compared with past virtuosos of the Court's center chair, such as Chief Justices Hughes and Warren. Burger came to the Court with an agenda that included a dismantling of the jurisprudential edifice erected by the Warren Court, particularly in the field of criminal justice. In large part, Burger owed his elevation to the highest judicial office to his reputation as a tough "law and order" judge. He had commented disparagingly on the Warren Court decisions on the rights of criminal defendants. As Chief Justice, he believed that he now had the opportunity to transform his more restrictive views into positive public law.

Burger expressed opposition during most of his tenure to *Mapp* and *Miranda*[12]—the two landmark criminal procedure decisions of the Warren Court—but he was never able to persuade a majority to cast those cases into constitutional limbo. The same was true of other aspects of Burger's anti-Warren agenda. No important Warren Court decision was overturned by the Burger Court. If Burger hoped that he would be able to undo much of the Warren "constitutional revolution," he was clearly to be disappointed.

Warren Court Holdovers

The Court that Chief Justice Burger was called upon to lead contained some of the most noted Justices in Supreme Court history. Two of them— Justices Black and Harlan—retired after the second Burger term. But they both played important parts in major cases before then.

The Black who sat on the Burger Court was no longer the leading liberal. The Alabaman's fundamentalist approach to the Constitution did not permit him to adopt an expansive approach toward individual rights. Justice Black stood his constitutional ground where the rights asserted rested on specific provisions, such as the First Amendment or the Fifth

Amendment privilege against self-incrimination; but when he could not find an express constitutional base, Black was unwilling to create one. Justice Black's opposition to busing at the conference on the *Swann* school busing case[13] was, Justice Brennan remarked, because the word bus is not found in the Constitution.

The Court's leading conservative was now Justice Harlan, a position he had held since Justice Frankfurter's retirement in 1962. When Burger took his seat, Harlan's was the Court's principal voice calling for judicial restraint, and the new Chief Justice had reason to expect support from him. He shared Burger's concern over "the 'horse and buggy' conditions under which the federal judiciary, and particularly this Court, are now operating"[14] and he had been the leading conservative in the last years of the Warren Court.

Justice Harlan's conservative philosophy did not, however, permit him to go along with the Chief Justice's agenda vis-à-vis the Warren Court decisions. "Respect for the Courts," Harlan once wrote, "is not something that can be achieved by fiat."[15] He applied this principle to reflexive refusal to follow prior decisions with which he disagreed. The true conservative, Harlan believed, adhered to stare decisis, normally following even precedents against which he had originally voted.

Harlan's antithesis in many respects was Justice Douglas, who never allowed jurisprudence the other way to deter him from reaching the liberal results that he favored. After Justices Black and Harlan retired in 1971, Douglas was the senior Associate Justice on the Burger Court, yet though the public considered him the leader of its liberal wing he was not the one to play the leadership role. He was a maverick, an idiosyncratic loner who was least effective in the give-and-take required in a collegial institution.

Instead, the leader of the Burger Court's liberal bloc was Justice Brennan, who had served as the catalyst for some of the most important decisions of the Warren Court. After Warren's retirement, Brennan was no longer the trusted insider. Instead, he became the Justice who tried above all to keep the Warren flame burning and, as a consequence, Burger's leading opponent on the Court. Both Brennan and the Chief Justice accepted this over the years the two sat together. One year the Brennan law clerks were guests at a convivial Burger luncheon. The Chief Justice telephoned Brennan to tell him what fine gentlemen his clerks were. When they returned, the Justice reproved them. "You turncoats," he said, "what did you do over there?"

The third member of the liberal bloc in the Burger Court was Justice Marshall. In the Burger Court he served as a virtual judicial adjunct to Justice Brennan. The law clerks, it is said, took to calling Marshall "Mr. Justice Brennan-Marshall."

The leading role in the Burger Court was not, however, assumed by its conservative or liberal wing. Instead, the most influential part was played by those Justices who were characterized by Justice Harry A. Blackmun as "in the middle." According to Blackmun, "Five of us" were in that position—Justices Stewart, White, Powell, Stevens, and himself.[16] Of these, Justices Stewart and White were holdovers from the Warren Court.

Justice Stewart continued as a moderate with a pragmatic approach to issues that polarized others. Under Chief Justice Warren, Stewart had remained in the center as the Court moved to the left. In the Burger Court he continued as the leading moderate. His center position enabled him to play a pivotal role. The others tended to turn to Justice Stewart because they trusted his judgment as a lawyer. Thus in the *Swann* school busing case,[17] where he was the senior Justice in the conference majority, Justice Douglas wrote at the time, "The case was obviously for me to assign, and I would have assigned it to Stewart."[18]

The other centrist Warren holdover, Justice White, did not play as influential a role in the Burger Court. As a Justice, White, like Stewart, has defied classification. He has tended to take a lawyerlike approach to individual cases without trying to fit them into any overall judicial philosophy, and he was considered one of the more conservative Justices in the Burger Court's center, particularly in criminal cases. "In the criminal field," as Justice Blackmun sees it, "I think Byron White is distinctly a conservative."[19]

It is fair to say that Justice White has been more respected among his colleagues than outside the Court—in part because of his gruff bluntness and no-nonsense manner. When loyalty-oath cases were still part of the Court's agenda, he curtly told a conference, these oath cases are a "pain in the neck."[20] When he did not think much of a case, he termed it "this pipsqueak of a case."[21] And when an attorney was doing a particularly bad job in oral argument, White was heard in a stage whisper, "This is unbelievable."[22]

Burger Court Appointments

Harry A. Blackmun, the first new Justice to be apppointed to the Burger Court, took his seat in 1970, replacing Justice Fortas, who had been forced to resign a year earlier. Blackmun had served eleven years on the U.S. Court of Appeals for the Eighth Circuit. He went to grade school with Warren Burger, and the two remained close friends thereafter. After graduation from Harvard Law School, Blackmun served as law clerk in the court of appeals, practiced with a Minneapolis law firm, and was counsel to the Mayo Clinic.

In his early years on the Court few expected Blackmun to be more than an appendage of the Chief Justice. He was then virtually Burger's disciple; they were on the same side in almost all cases. The press had typecast Blackmun as the subordinate half of the "Minnesota Twins," after the baseball team.

All this was to change and by the later years of the Burger tenure, Blackmun was completely his own man. His opinions became increasingly as liberal as any that Justices Brennan and Marshall might have written.

One can also see an improvement in Blackmun's work. If we compare his hesitant first draft in *Roe v. Wade*[23] with his more self-assured opinions in later cases, we see how the Justice grew with increasing experience and self-confidence. As a 1983 *New York Times* article put it, "Justice Blackmun's evolution as a jurist and prominence on the Court represent one of the most important developments in the judiciary's recent history."[24]

At the end of 1971, Lewis F. Powell and William H. Rehnquist were appointed to succeed Justices Black and Harlan, who had retired. Powell was, in many ways, the paradigm of the lawyer turned Justice. He came to the Court after private practice with one of Virginia's most prestigious law firms, a career that was capped by his term as President of the American Bar Association. Surprisingly few Supreme Court Justices have been drawn directly from the practicing bar. Indeed, during the past half century, only Powell and Justice Abe Fortas were in private practice when they were appointed.

On the bench, Powell was "a lawyer's Justice" and a pragmatic centrist who followed the measured approach developed by his thirty-five years of practice. He avoided doctrinaire positions and hard-edged ideological decisions and gained a reputation as a moderate, though he voted more often with Chief Justice Burger than some of the others in the center bloc. His quest for the middle ground is best illustrated by his most famous opinion—that in the *Bakke* case[25]—where he cast the deciding vote. Though the Powell *Bakke* opinion was joined by no other Justice, it is considered by most commentators as the authoritative statement of law.

Powell was essentially a conservative whose judicial approach was reminiscent of that followed by Justice Harlan during the early Burger years. Like Harlan, Powell believed in following precedents of which he may have disapproved until they were overruled. Thus he voted to decide cases in accordance with the *Mapp* and *Miranda* decisions,[26] even though he might well have voted against those decisions had he been on the Court. In the conference on a 1983 case,[27] Powell said, about a prior decision, "It's bad law. I would want to limit it to its own facts, without overruling it in so many words."[28]

Justice Rehnquist was, of course, the Burger Court's most conservative member—dubbed by *Newsweek* "The Court's Mr. Right."[29] During his early years on the Court, the majority remained largely unsympathetic to the Rehnquist entreaties from the right. It was then that he received a Lone Ranger doll as a gift from his law clerks, who called him the "lone dissenter" during that period. During his fourteen years as an Associate Justice, Rehnquist dissented alone fifty-four times—a Court record. By the time of his nomination as Chief Justice, however, Rehnquist had become the most influential Associate Justice in the Burger Court—anticipating the paramount role he was to assume when he was elevated to the Court's center chair.

The next appointment to the Burger Court was that of John Paul Stevens, a judge on the U.S. court of appeals when he was chosen to fill the vacancy created by Justice Douglas's retirement in 1975. Stevens was appointed as a moderate. It is, however, even harder to classify Stevens than the others on the Burger Court. As the *New York Times* wrote about him, "On a Court that everyone likes to divide into liberal and conservative, Justice Stevens has a list of labels all his own: enigmatic, unpredictable, maverick, a wild card, a loner."[30]

Statistically, Stevens was the Justice nearest the Burger Court's center, disagreeing equally with the Justices at the poles. Thus in the 1981 term, Stevens disagreed with Justice Rehnquist in 35 percent of the cases and with Justice Brennan 33 percent of the time.

Stevens as a loner reminds one of Justice Douglas, to whose seat Stevens was appointed. Like his predecessor, Stevens makes little effort to win over other members of the Court. What a law clerk once told me about Douglas applies equally to Stevens: "Douglas was just as happy signing a one-man dissent as picking up four more votes." Justice Stevens wrote more dissents than any other member of the Court; he was often a lone dissenter. A book on the Burger Court concludes that, while Justice Stevens was once viewed as a potential leader of the Court, "the effect of his independence of mind often has been to fragment potential majorities and leave the state of the law indeterminate."[31]

Over the years, Justice Stevens has acquired something of the reputation of an iconoclast—albeit an idiosyncratic one. Stevens is idiosyncratic in more than his decisions. He hires only two law clerks instead of the usual four and drafts more of his own opinions than any of the others. He also deviates from the Court's unwritten conservative dress code; his constant bow tie (worn under the judicial robe) gives him a perpetual sophomoric appearance. In 1986, the Justices were hearing argument on whether Orthodox Jews, with their religious duty to wear yarmulkes, should be exempt from the military dress code's ban on hats indoors. Counsel for the government told the Justices, "It's only human nature to resent being told what to wear, when to wear it, what to eat."

"Or whether you can wear a bow tie?" chimed in Justice Stevens.[32]

The First Sister

When the Supreme Court was first established, the author of an opinion was designated, "Cushing, Justice." In 1820, the form was replaced by "Mr. Justice Johnson" as opinion author. This style lasted over a century and a half. Then, in 1980, Justice White suggested to the conference that, since a woman Justice was bound to be appointed soon, they should avoid the embarrassment of changing the style again at that time. All the others agreed, and the manner of designating the author of an opinion became, simply, "Justice Brennan."[33]

Justice White's prescience was borne out the next year when Sandra Day O'Connor was appointed as what *Time* called "the Brethren's first sister."[34] Her career dramatically illustrates the changed place of women in the law. Though O'Connor graduated third in her class at Stanford Law School (Rhenquist had been first), only one California law firm would hire her; a Los Angeles firm offered her a job as a legal secretary. Ironically, Attorney General William French Smith, one of the partners in the firm that had refused to hire her as an associate, recommended O'Connor for the Supreme Court.

After law school, O'Connor returned to Arizona, where she combined legal work with political activity. She became assistant attorney general and then a member, and ultimately majority leader, of the state senate. She was elected to the Arizona Superior Court, where she served for five years. In 1979, she was appointed to the Arizona Court of Appeals. O'Connor was the second Burger Court member (after Justice Brennan) to have served as a state judge.

The Justices used to jest among themselves about the effect a woman Justice would have. Tradition says that the junior Justice answers the conference door (one of them used to quip that he was the highest-paid doorman in the world). As rumors of a female appointment gained ground, the Justices joked, first to Justice Rehnquist and then to the new junior Justice, Stevens, that when a woman came to the Court, he should be a gentleman and continue to answer the door. In the event, when O'Connor became the newest Justice, she assumed the doorkeeper's task without question. And the Justices continued to be called the Brethren even though a woman had joined their ranks.

Soon after O'Connor took her seat, a *Time* headline read "And Now the Arizona Twins; Justice O'Connor Teams Up with Court Conservative Rehnquist."[35] There is no doubt that Justice O'Connor has been more conservative than her predecessor, Justice Stewart. "I think it is fairly clear," said Justice Blackmun during the last Burger term, that O'Connor "is on the right."[36] She typically voted opposite Justice Rehnquist only about 10 percent of the time, while disagreeing with Justices Brennan and Marshall in over 45 percent of cases.

It is, however, not accurate to picture Justice O'Connor as only a Rehnquist clone on the Burger Court. She tended to be as conservative as the latter in criminal cases and was, in fact, the author of opinions limiting *Miranda* during the last two Burger terms.[37] She also sided with her fellow Arizonian on the importance of recognizing state powers[38] and the need for judicial restraint vis-à-vis the legislature.[39] But she was more moderate in a few areas—most notably (in view of her own experience with sex discrimination) in cases involving sexual bias,[40] but also in cases involving affirmative action[41] and the First Amendment.[42]

Even those who disagree with her recognize that O'Connor has been an above-average Justice, who has become an effective conservative voice. She has not been hesitant in expressing her views both in conference and from the bench. Her opinions have been characterized by clear analysis and focus upon the points at issue. But they have at times been lightened by a little-credited gift for language. In a case involving the right of a defendant to represent himself, the O'Connor opinion noted, "We recognize that a . . . defendant may wish to dance a solo, not a pas de deux."[43]

The Changed Judicial Agenda

Chief Justice Taft, we saw,[44] once said that the Supreme Court was his idea of what heaven must be like. More recently, however, Justice Blackmun said of being a Justice, "It's a rotten way to earn a living."[45]

In large part, the change in the judicial attitude is a reflection of the Court's ever-increasing caseload. In its early years, the Court had relatively few cases on its docket: in 1803, there were only 51 cases; seven years later, the number was 98. The numbers increased as the years went on—but not by much. There were 723 cases on the Court's docket in 1900 and 1,039 when Chief Justice Taft died. In recent years, however, the caseload increased dramatically. In the 1953 Term, Chief Justice Warren's first, 1,293 cases were disposed of; in the 1968 Term, Warren's last, 3,117 cases were disposed of.

The increase continued during Chief Justice Burger's tenure: the cases disposed of increased from 3,357 in the 1969 Term to 4,289 in the 1985 Term. The Court's caseload has more than tripled during the past four decades. The burgeoning workload has been responsible for advancing the Court's first conference from early October to the last week in September, for the inclusion of Wednesdays in the conference schedule, and for increasing the number of staff members. In the Warren Court, each Justice had two law clerks; in the Burger Court, the number was increased to four.

Chief Justice Burger warned repeatedly that unless the workload was curbed it would soon be "impossible" for the Justices "to perform [their] duties well and survive very long." Not all the Justices agreed. But the pressures of the docket certainly changed the Court atmosphere from what it was in Chief Justice Taft's day. In a speech, Justice Blackmun described the Court as weary and overworked. He asserted that he was "never so tired" as he was at the end of the 1984 Court Term.[46]

It is, however, not only the volume of cases that has changed in recent years. There has been an even-more-dramatic change in the quality of the cases decided in the Marble Palace. The Court has, of course, always played a pivotal role in the American polity—a role encapsulated in Tocqueville's famous statement that "scarcely any political question arises in the United States that is not resolved sooner or later into a judicial question."[47] From this point of view, the Supreme Court is primarily a political institution. Because its decrees mark the boundaries between the great departments of government and because on its actions depend the proper functioning of federalism and the scope to be given to the rights of the individual, a judge on such a tribunal has an opportunity to leave an imprint on the life of the nation as no mere master of the common law possibly could. Certainly, as an English writer noted not long ago, "In America, the Supreme Court is supreme. In no other democratic country do nine judges, none of them elected, tell the president and the legislature what each may or may not do."[48]

Yet that is precisely what the Burger Court did. In the *Nixon* case,[49] the Court not only told the President what he must do; its decision led directly to his forced departure from the White House. In *Roe v. Wade*,[50] the Justices not only told the legislatures they might not restrict the right to abortions; their decision caused a social schism that has remained a major divisive factor in our society. The Court's other decisions may not be as dramatic, yet each illustrates some facet of the Court's crucial place in the polity.

It may be said that the Court has decided landmark cases throughout its history. Here, too, there has been a difference. In the past, a *Dred Scott* case[51] or an *Income Tax* case[52] would come up every quarter century or so, but the time of the Court would be occupied mainly with lesser cases. During this century, however, the pace of landmark decision making has increased tremendously. This has been particularly true of decision making since Chief Justice Warren took his seat. From the *Brown* case[53] at the beginning of the Chief Justice's tenure to the *Powell* case[54] at its end, the Warren years saw a judicial revolution that transformed both the law and the society that it served. Leading case after leading case was decided at a pace that would have astounded Court observers in an earlier day.

The pace did not decrease under Warren's successor. Scarcely a term passed in the Burger Court without decisions that would be landmarks in all but a jaded age. Certainly, the Burger Court was as much in the spotlight of press attention as its precedent-shattering predecessor.

Since Chief Justice Warren's day, there has been a substantial change not only in the volume of leading cases, but also in their quality. In earlier Courts, the important cases involved property rights. In the first part of the century, the Justices used the Due Process Clause to strike down laws regulating economic activity. More recently, the Court came to accept governmental restrictions on the free exercise of property rights to an extent never before permitted in American law. But this has not led to a lessening of the Court's role in the constitutional system. Instead, there has been a shift in judicial emphasis from the protection of property rights to protection of personal rights. Such protection has, indeed, been elevated to the top of the judicial agenda during the past quarter century.

The development in this respect has been accompanied by a continuing rights explosion; as never before, new interests have pressed upon the law to seek recognition in the form of legal rights. The Burger Court recognized rights that had previously scarcely even pressed for recognition, raising ever more rights to the legally protected plane. The rights in question cut across the entire legal spectrum, ranging from extensions of traditional rights, such as those to freedom of expression and equal protection, to newly emerging rights not previously vindicated by law.

The Burger Court cases well illustrate the changed nature of the rights asserted before the Court. Among these rights are those to sexual equality,[55] to bodily autonomy and privacy,[56] to treatment by committed persons,[57] to access to news,[58] to notice and hearing before even a "privilege" may be taken away,[59] and to access to court proceedings.[60] The judicial agenda was more and more concerned with the new personal rights that increasingly pressed for legal recognition.

The shift in judicial emphasis to protection of personal rights reflected the Justices' concern with the changing nature of our society. With property rights constitutionally curtailed, compensatory scope had to be given to personal rights if the ultimate social interest—that in the individual life—was not to be lost sight of. The Justices, like the rest of us, were disturbed by the growth of governmental authority and sought to preserve a sphere for individuality even in a society in which individuals stand dwarfed by the power concentrations that confront them.

Desegregation and Racial Discrimination

Brown v. Board of Education[61] was, of course, the seminal case in ensuring that the Equal Protection Clause became more than a paper protection against racial discrimination. A second *Brown* decision ruled that the transition to desegregated schools should take place "with all deliberate speed."[62] The result was a gap in the *Brown* enforcement process; deliberate speed too often meant indefinite delay in vindicating the right to attend desegregated schools.

It was the Burger Court which laid to rest the "all deliberate speed" formula in *Alexander v. Holmes County Board of Education* (1969),[63] the first school desegregation case decided under the new Chief Justice. As explained in a letter by Justice White, what *Alexander* held was "that the deliberate speed formula has been abandoned . . . and that as soon as possible is [its] substitute."[64] Under *Alexander,* "[T]he obligation of every school district is to [desegregate] at once . . . and immediately to operate as unitary school systems."[65]

After abandoning the deliberate speed formula, the Burger Court dealt with the extent of remedial power possessed by the federal courts in desegregation cases. The key case here was *Swann v. Charlotte-Mecklenburg Board of Education* (1971).[66] The Court upheld the far-reaching desegregation order issued by the district court, which included what amounted to racial quotas (at least as starting points) in the schools and provision for extensive busing to help achieve integration. The case gave rise to a serious conflict between the new Chief Justice and the majority. The latter was ultimately able to frustrate the Burger effort to weaken the remedial power of the federal courts in desegregation cases.

The key issue in *Swann* was that of the remedial power of the courts in desegregation cases. As Justice Douglas stated at the conference: "The problem is what is the power of the court without the help of Congress to correct a violation of the Constitution." He pointed to the broad remedial powers of the courts in other areas. "If there is an antitrust violation, we give a broad discretion." The same should be true here.

Even though the Douglas view on remedial power was supported by the conference majority, the Chief Justice circulated a draft *Swann* opinion[67] that stated a most restricted view of remedial power, which the draft contrasted with what it termed "a classical equity case," for example, removal of an illegal dam or divestiture of an illegal corporate acquisition. Here, the situation was said to be different. Under the Burger draft, the implication was that federal courts could act only to bring about the situation that would have existed had there never been state-enforced segregation.

Chief Justice Burger's efforts to limit the remedial power of the district courts were frustrated by the other Justices. Their pressure led the Chief Justice to modify his restricted approach, and the final *Swann* opinion adopted a very broad conception of remedial power.

The final *Swann* opinion specifically rejected the view stated in the Burger draft that the remedial power in this type of case was somehow less than that in the classical equity case. According to it, "[O]nce a right and a violation have been shown, the scope of a district court's equitable powers to remedy past wrongs is broad, for breadth and flexibility are inherent in equitable remedies." Remedial discretion includes the power to reach a goal of racial distribution in schools comparable to that in the community and to order extensive busing if that is deemed appropriate to "produce an effective dismantling of the dual system."[68]

The most important thing about the *Swann* decision was its recognition of broadside remedial power. *Swann* goes far beyond the no-segregation principle laid down in *Brown*. Under *Swann* the federal courts have the power to issue whatever orders may be necessary to bring about an integrated school system—the broad and flexible power traditionally exercised by courts of equity. This includes extensive busing appropriate to attain the goal "of insuring the achievement of complete integration at the earliest practicable date."[69]

The Burger Court, however, made a sharp distinction between segregation required by law and that brought about by other factors, such as racially separate housing patterns. The de jure–de facto distinction continued as the dividing line in desegregation jurisprudence.[70] The mere fact that segregated schools exist in a district is not enough. Only where de jure segregation is shown may the *Swann* remedial power be used to do away with a dual school system.

In addition, the Court drew the line at interdistrict remedies. In *Milliken v. Bradley* (1974),[71] it refused to recognize a power in the federal courts to order the amalgamation of urban and suburban school districts to obtain a desirable racial mixture in the schools, where de jure segregation existed only in the urban district. As it was put by the Chief Justice in a *Milliken* draft, "Federal authority to impose cross-district remedies presupposes a fair and reasoned determination that there has been a constitutional violation by all of the districts affected by the remedy."[72]

In *Washington v. Davis* (1976),[73] the Court ruled that proof that a government act has a disproportionate impact upon racial minorities is not enough to show a violation of equal protection. A discriminatory racial purpose must be shown. Though criticized, this holding follows logically from the de jure–de facto distinction. A de facto segregation situation is not enough to make for a constitutional violation. In the absence of racially discriminatory intent or purpose, the mere fact that black and white children are treated differently does not prove a denial of equal protection. The same approach is the basis of the *Washington v. Davis* rule.

Affirmative Action

The Warren Court had it relatively easy in *Brown* and its progeny. The blatant discrimination involved in segregation made it a simple matter for the Warren Justices to decide the cases that came before them. Moreover, in *Brown* and the later desegregation cases, judicial enforcement did not deprive anyone else of the right to an equal education. Yet such a result was precisely that with which the decision in the *Bakke* case (1978)[74] had to deal. The special admissions program in *Bakke* provided both for admitting a specified number of minority students and for excluding others who might have been admitted had there been no special program. During the oral argument, Justice Marshall put his finger on the case's dilemma in this respect, when he told Bakke's counsel, "You are arguing about keeping somebody out and the other side is arguing about getting somebody in."[75] A decision for the university would keep Bakke and others like him out of the Davis Medical School; a decision for Bakke would prevent minority applicants from "getting in" under the special program.

The decision in the *Bakke* case and the process by which it was reached well illustrates the manner in which the Burger Court operated. Instead of a flat decision upholding or striking down racially preferential programs, the Court handed down a fragmented decision that mirrored the internal atomization of the Court. *Bakke* points up the essential failure of Chief Justice Burger to assume an effective leadership role in molding his Court's jurisprudence. In *Bakke,* the lead in the decision process was assumed by other members of the Court—notably by Justices Brennan and Powell.

The *Bakke* decision was two decisions, each decided by a five-to-four vote. The first decision was that the Davis special admissions program was invalid and that Bakke should be admitted to the Davis Medical School. Justice Powell joined the Chief Justice and Justices Stevens, Stewart, and Rehnquist to make up the majority. The second decision was that race was a factor that might be considered in an admissions policy without violating the Constitution. Justice Powell joined Justices Brennan, White, Marshall, and Blackmun to make up the bare majority here.

The unifying factor behind both *Bakke* decisions was Justice Powell's vote, which—though it was fully endorsed by no other Justice—spoke for the bare majorities on each of the major issues. The result has been that commentators, courts, and admissions officers have treated Justice Powell's opinion as the authoritative opinion in the *Bakke* case. The Davis program was invalidated, but admissions officers are permitted to operate programs that grant racial preferences—provided that they do not do so as blatantly as was done under the sixteen-seat "quota" provided at Davis.

Justice Powell's crucial part in the *Bakke* decision process began with his circulation of a draft opinion.[76] It rejected the view that allotting places for preferred racial groups was the only effective means to secure diversity in the student body. But the draft went on to show how an admissions program that considered race among other factors might be "designed to achieve meaningful diversity in the broad sense of this term." Such a program permits race to be considered but, at the same time, "specifically eschews quotas."

The Powell draft ended with a categorical rejection of the Davis program as contrary to equal protection. At the *Bakke* conference on the merits, Justice Powell repeated the views expressed in his draft and said that he would vote to affirm the California decision in Bakke's favor. Justice Brennan interjected that he thought that, under the approach to the case he had stated, Powell should vote to affirm in part and reverse in part. The judgment of the lower court had required that Davis adopt a color-blind admissions system for the future. Justice Powell went along with the Brennan approach, telling the conference, "I agree that the judgment must be reversed insofar as it enjoins Davis from taking race into account."

The Powell concession was the key that led to the ultimate compromise resolution of the *Bakke* case. Under the Court's decision, the Davis program was ruled violative of equal protection; but the decision specifically held that race might be considered as a factor in admissions programs. This has permitted the use of race in a flexible admissions policy designed to produce diversity in a student body. Minority students may be admitted even though they may not fully measure up to the academic criteria.

Bakke has consequently meant anything but the end of programs providing for racial preferences. On the contrary, the later Burger Court decisions built upon *Bakke* in dealing with such programs. Unless there was proof of purposeful discrimination or a legislative or administrative finding to that effect, race as the sole determining factor in employment decisions was ruled invalid, as it was in *Bakke* itself. But properly tailored affirmative action programs were upheld. The *Bakke* decision that race may be considered as a factor has permitted the widespread use of affirmative action programs to be continued.

Other Equal Protection Cases

The Burger Court confirmed the trend toward applying judicial restraint in cases involving economic regulation. Thus Justice Marshall could refer to the "minimal standards of rationality that we use to test economic legislation that discriminates against business interests."[77] The Court did, however, expand the use of a stricter level of scrutiny in certain cases, such as the racial discrimination cases just discussed. All the Burger Court Justices agreed that racial classifications are suspect and must be subject to strict scrutiny under which the classification will be held to deny equal protection unless justified by a "compelling" governmental interest.

In particular, the Burger Court held that gender classifications were subject to stricter scrutiny than the rational-basis test governing economic classifications. While the Justices did not go as far as Justice Brennan urged and make sex a wholly suspect classification,[78] they did make it what has been called a "quasi-suspect" classification, subject to an intermediate review standard: "To withstand constitutional challenge, classifications by gender must serve important governmental objectives and must be substantially related to attainment of those objectives."[79]

Under this standard, the Court struck down discriminatory classifications in education,[80] dependent benefits,[81] clubs,[82] alimony,[83] estate administration,[84] and jury selection.[85] Indeed, it can be said that the removal of sexual disabilities in our constitutional law was almost entirely the handiwork of the Burger Court.

The same is true of the removal of disabilities for aliens and illegitimates. Classifications based upon alienage and illegitimacy were also treated as more suspect than economic classifications and hence also subject to stricter review. Laws restricting public employment or welfare benefits to citizens[86] or prohibiting illegitimate children from inheriting from their fathers[87] were ruled invalid under the stricter scrutiny standard.

The Court did, however, reject attempts to have classifications based upon wealth treated as suspect classifications—or at least as semisuspect. In *San Antonio Independent School District v. Rodriguez* (1973)[88] a challenge to the financing of public school education by the property tax claimed that equal protection was denied to children in poorer districts that had a low tax base. The lower court had held the rational-basis standard inappropriate because "lines drawn on wealth are suspect." The Supreme Court reversed; the fact that the law burdened poor persons in the allocation of educational benefits was held not enough to make the classification a suspect one.

In other cases, the Court refused to rule that classifications based upon age[89] or mental retardation[90] should be treated as suspect. The Court had, in effect, decided to draw the line and refused to create other classifications subject to broader review. There would be no further classifications to which heightened scrutiny would be applied.

The Burger Court also applied the strict-scrutiny standard in cases affecting fundamental rights. This meant broader review in cases involving the right to travel,[91] the right to vote,[92] the right of access to the judicial process,[93] and rights growing out of the marital relationship.[94] However, in the previously mentioned *San Antonio* case, the Court refused to hold that the right to education was a fundamental right. According to *San Antonio,* only rights protected by the Constitution are subject to strict-scrutiny review. Since no one has a constitutional right to education, infringements upon that right are reviewable only under "the usual standard for reviewing a State's usual social and economic legislation"[95]—that is, the rational-basis test.

First Amendment

During the Warren years, the preferred-position theory—that the Constitution gives a preferred status to personal, as opposed to property, rights—became accepted doctrine. Early in Chief Justice Burger's tenure, the Court stated "that the dichotomy between personal liberties and property rights is a false one. . . . In fact, a fundamental interdependence exists between the personal right to liberty and the personal right to property. Neither could have meaning without the other."[96]

But the Burger Court did not abandon the preferred-position approach; on the contrary, like its predecessor, it recognized that each generation must necessarily have its own scale of values. Both the Warren Court and the Burger Court were especially willing to recognize First Amendment rights as peculiarly suitable for inclusion in the preferred-position theory. Countless cases have recognized the special constitutional function of the First Amendment—a function both explicit and indispensable.[97] The free society itself is inconceivable without what Justice Holmes called "free trade in ideas."[98] In the Burger Court, as in the Warren Court, governmental power over First Amendment freedoms was narrower than that permitted over property rights.

However, the Burger Court did more than confirm its predecessor's general approach in First Amendment cases. It also extended constitutional protection to types of speech previously held beyond the Amendment's reach—in particular, commercial speech, which had previously been held beyond the scope of the First Amendment guaranty.[99] In *Virginia State Board of Pharmacy v. Virginia Consumer Council* (1976),[100] the Court held squarely that commercial speech came within the protection of the First Amendment. A state statute barring a pharmacist from advertising prescription drug prices was ruled invalid. Speech "which does 'no more than propose a commercial transaction'"[101] was ruled squarely within the protection of the First Amendment. The same approach was followed in cases involving the barring of advertising by attorneys and other professionals.[102] The states may not prohibit truthful advertisements concerning the availability and terms of routine legal and other professional services.

The Burger Court also broadened the protection given to political speech. It struck down limitations on direct expenditures by political candidates, treating such expenditures as a form of political expression protected by the First Amendment.[103] The notion of nonverbal speech had, however, been used to protect symbolic speech well before the Burger

Court. Over half a century ago, the Court ruled that display of a red flag as a symbol of opposition to organized government is covered by First Amendment protection.[104]

In *Spence v. Washington* (1974),[105] the Burger Court held that the same is true of a display of the flag upside down with a peace symbol affixed. The Court reversed a conviction under a state law forbidding the exhibition of the U.S. flag with figures, symbols, or other extraneous material attached.

The symbolic speech concept was also applied to a case involving a New Hampshire law that required motor vehicles to bear license plates embossed with the state motto, "Live Free or Die," and made it a misdemeanor to obscure the motto. Two Jehovah's Witnesses viewed the motto as repugnant to their moral, religious, and political beliefs and covered it up with tape. They were found guilty of violating the statute, but the Court reversed their convictions.[106]

Well before the Burger Court, the cases had developed the concept of the "public forum," giving a right of access to public places for First Amendment purposes. That concept was both applied and expanded during Chief Justice Burger's tenure. The earlier cases had upheld the right of individuals to use public places as forums for expression.[107] The Burger Court also recognized, in *Kleindienst v. Mandel* (1972),[108] a First Amendment right "to receive information and ideas," and that freedom of speech "'necessarily protects the right to receive.'"[109]

The public forum concept was applied to a municipal theater which refused to allow its facilities to be used for a controversial production.[110] The theater was ruled a public forum dedicated to expressive activities, in which content-based regulation was not permissible. For First Amendment purposes, the public auditorium was equated with streets, parks, and other public places that come within the public forum concept.

On the other hand, the public forum concept was ruled inapplicable to privately owned shopping centers. The Warren Court had held that such a center was the functional equivalent of the business district of a town.[111] The Burger Court rejected this view and categorically ruled that the First Amendment does not apply to privately owned shopping centers.[112]

Freedom of the Press

The Burger Court's most celebrated First Amendment decision was that in *New York Times Co. v. United States* (1971)[113]—usually known as the *Pentagon Papers Case*. The Court there struck down an attempt by the government, for the first time, to use the courts to censor news before it was printed. The decision was based upon the basic principle that, except in cases where "disclosure . . . will surely result in direct, immediate, and irreparable damage to the Nation or its people,"[114] there may be no prior restraint upon publication.

The *Pentagon Papers* decision marks the most dramatic assertion of the principle against prior restraint of the press in American jurisprudence. The rule against prior restraints led the Court to invalidate so-called gag orders—judicial attempts to censor the press by forbidding news media from publishing material that could impair the right to a fair trial.[115] The Court also ruled that the press had a right of access to the courtroom, reversing the order of a trial court that had closed a criminal proceeding to the public and the press.[116] However, it refused to decide that the press had a First Amendment right of access to news which gave a television station a right to conduct interviews in a prison.[117] The press was held to have only the same right of access as the general public. Nor did the Court accept the claim that the press right to gather news implies, in turn, a right to a confidential relationship between a reporter and his source.[118] Reporters had argued that to gather news, they had to protect their sources; otherwise, the sources would not furnish information, to the detriment of the free flow of information protected by the First Amendment. The argument was rejected: reporters have the same obligation as other citizens to respond to grand jury subpoenas. The press does not have any immunity from grand jury subpoenas beyond that possessed by other citizens.

The Burger Court displayed a more favorable attitude toward television in the courtroom than did its predecessor. The Warren Court had decided that televising a notorious trial over defendant's objections was inconsistent with the fair trial guaranteed by due process.[119] The Burger Court, on the contrary, upheld a Florida rule permitting television coverage of criminal trials, notwithstanding the objections of the accused.[120] The result has been a proliferation of televised trials that would have been impossible had the Warren Court jurisprudence remained unchanged.

Criminal Procedure

During the 1968 presidential election, Richard M. Nixon had run against Chief Justice Warren and his Court as much as he had run against his Democratic opponent, Hubert H. Humphrey. He accused the Court of "seriously weakening the peace forces and strengthening the criminal forces in our society."[121] Nixon pledged that his Court appointees would be different.

When Nixon's appointee, Warren E. Burger, took his place in the Court's center chair, it was widely expected that he would implement what Warren had called Nixon's "law and order" issue.[122] Many feared that the Burger Court would soon relegate the criminal-law landmarks of the Warren Court's constitutional-law "revolution" to legal limbo. But the anticipated reversals of the key Warren Court precedents did not materialize. Instead, the essentials of the Warren jurisprudential edifice were preserved. *Gideon, Mapp,* and *Miranda*[123]—the great Warren Court trilogy on the procedural rights of criminal-law defendants—all survived. To be sure, they were modified, even narrowed and blunted in some ways.

In many respects, indeed, the Burger Court expanded the rights of criminal defendants. The Warren Court's landmark *Gideon* decision had established the indigent defendant's right to an appointed attorney. But the *Gideon* Court, following the conference suggestion of Chief Justice Warren, limited the decision to the felony case at hand, without addressing the question of how far the new right to assigned counsel extended. That question was presented to the Burger Court in *Argersinger v. Hamlin* (1972).[124] The Court there held that the right extended to misdemeanors as well as felonies. In fact, *Argersinger* ruled "that absent a knowing and intelligent waiver, no person may be imprisoned for any offense, whether classified as petty, misdemeanor, or felony, unless he was represented by counsel at his trial."[125] As it was put in a letter by Justice Stewart, no person may be "sentenced to imprisonment unless he had a lawyer at his trial."[126]

In addition, the Burger Court extended the requirement of free transcripts to indigents seeking to appeal misdemeanor cases, even those involving only fines, not imprisonments.[127] The Court also expanded the *Gideon* requirement to include the assistance of a psychiatrist at state expense and even other expert witnesses where they were necessary for the defense.[128]

The Court also held that the Equal Protection Clause prohibits a prosecutor from using peremptory challenges to exclude blacks from the jury.[129] Racial equal protection claims in the jury context were treated in basically the same way as other kinds of racial discrimination claims.

But there were also decisions more restrictive than those of the Warren Court. Most important in this respect were the Burger Court decisions narrowing *Mapp* and *Miranda*—the Warren Court's paradigmatic criminal procedure decisions. A majority of the Burger Court was dissatisfied with what the Chief Justice termed "the monstrous price we pay for the exclusionary rule in which we seem to have imprisoned ourselves."[130] This led them to adopt a good-faith exception to the rule under which illegally seized evidence may be admitted where the police "officer conducting the search acted in objectively reasonable reliance on a warrant issued by a detached and neutral magistrate that subsequently is determined to be invalid."[131] Where officers had acted in good-faith reliance upon search warrants, the convictions were upheld.[132]

Miranda was treated the same way by the Burger Court. As the Chief Justice himself said, "I would neither overrule *Miranda*, disparage it, nor extend it at this late date."[133] *Miranda*, like *Mapp*, was narrowed but not overruled during the Burger years.

The Burger Court's blunting of *Miranda* included a decision holding that a statement obtained in violation of *Miranda* could be used to impeach petitioner's testimony at trial.[134] Thus at least some use could be made of statements inadmissible under *Miranda* as part of the state's direct case. In another case, the Court laid down a "public safety" exception to *Miranda*. According to the Court there, *Miranda* should not "be applied in all its rigor to a situation in which police officers ask questions reasonably prompted by a concern for the public safety."[135]

Oregon v. Elstad (1985)[136] was more far-reaching in its implications for *Miranda*. The Court there held that a voluntary confession obtained in violation of *Miranda* did not taint a later confession secured after proper *Miranda* warnings had been given. For the first time the Court indicated that the *Miranda* safeguards are not themselves rights guaranteed by the Fifth Amendment: "The *Miranda* exclusionary rule . . . sweeps more broadly than the Fifth Amendment itself. . . . [A] simple failure to administer Miranda warnings is not in itself a violation of the Fifth Amendment."[137]

Perhaps the most important thing, however, about *Mapp* and *Miranda* in Burger Court jurisprudence is not that they were narrowed, but that they were not overruled. Both *Mapp* and *Miranda* were among the most criticized cases decided by the Warren Court, and it was widely expected that they would be among the jurisprudential casualties of the Burger Court. Yet, while their rules were narrowed in significant respects by the Burger jurisprudence, they continued as foundations of the law of criminal procedure. Indeed, at the end of the Burger tenure, the Court declared that *Miranda* struck the proper balance between the competing interests of society and those accused of crime.[138]

"The Counter-Revolution that Wasn't"

In 1983, a book was published entitled *The Burger Court: The Counter-Revolution That Wasn't*.[139] The subtitle is a succinct summary of the Burger Court in operation.

There is no doubt that Chief Justice Burger came to the Court with an agenda that included some dismantling of the jurisprudential structure erected under his predecessor, particularly in the field of criminal justice. Similarly, Burger's strongest supporter, Justice Rehnquist, took his seat with a desire to correct some of the "excesses" of the Warren Court[140]—to, as he put it, see that the "Court has called a halt to a number of the sweeping rulings that were made in the days of the Warren Court."[141]

The Burger-Rehnquist agenda was not carried out in the Burger years. Instead, the intended counterrevolution served only as a confirmation of most of the Warren Court jurisprudence. It can, indeed, be said that no important Warren Court decision was overruled during the Burger tenure. Some of them were narrowed by Burger Court decisions; others, however, were not only fully applied but even expanded.

We have seen this even in the field of criminal justice, where Burger and Rehnquist were most eager to disown the Warren heritage—in Rehnquist's phrase, to make "the law dealing with the constitutional rights of accused criminal defendants . . . more even-handed now than it was when I came on the Court." It was, after all, to "the area of the constitutional rights of accused criminal defendants" that Rehnquist primarily referred when he referred to the "sweeping rulings of the Warren Court."[142]

Though the Warren criminal cases became a major issue of Richard Nixon's presidential campaign, the Justices appointed by Nixon did not tilt the Court to the point of repudiating them. In fact, we saw, one of the Warren criminal trilogy was even substantially expanded by the Burger Court. *Gideon v. Wainwright*[143] was extended to every case in which imprisonment may be imposed as a penalty, regardless of whether the crime involved is classified as a felony, misdemeanor, or even petty offense.[144] *Gideon* also upheld the right to counsel only at the criminal trial. The Burger Court extended the right to counsel to preliminary hearings[145]—that is, before any formal accusation and trial—and gave practical effect to the right of the accused to represent himself if he so chose and knowingly and intelligently waived the right to counsel.[146]

Other Warren Court landmarks also served as foundations of the Burger Court jurisprudence. *Griffin v. Illinois*[147]—in many ways the Warren watershed case in the Court's effort to ensure economic equality in the legal process—was extended to include the right of an indigent defendant to the psychiatric assistance needed for an effective defense[148] and even, outside the criminal law field, to invalidate court fees that prevented indigents from bringing divorce proceedings.[149]

A similar picture was presented in other areas. *Brown v. Board of Education*[150] and its Warren Court progeny were applied in the *Swann* case[151] and held to vest broad remedial power in the courts to ensure desegregation, including extensive busing. The *Brown* principle was expanded to uphold affirmative action programs to aid minorities.[152] No Justice, not even Justice Rehnquist, who may once have taken a different view,[153] questioned the antidiscrimination premise that underlay *Brown*. In fact, the most important thing about the Burger Court decisions in this area was that, as Justice O'Connor was to conclude in the last Burger Term, "[W]e have reached a common destination in sustaining affirmative action against constitutional attack."[154]

The same was true in the other areas of Warren Court jurisprudence, including the First Amendment, reapportionment, other aspects of equal protection, and judicial review in operation. In all these areas, the central premises of the Warren Court decisions were not really challenged by its successor. The core principles laid down in the Warren years remained rooted in our public law as they had been when Chief Justice Burger first took his seat.

We Are All Activists Now

According to one commentator, "The entire record of the Burger Court . . . is one of activism."[155] Yet the Justices now "decide without much self-conscious concern for whether this is a proper role for the Court.

We are all activists now."[156] At least so the history of the Burger Court tells us. One thing to be learned from the Burger years "is that the great conflict between judicial 'restraint' and 'activism' is history now."[157]

The statistics bear out the conclusion that the Burger Court record was, indeed, one of judicial activism. One measure was its willingness to strike down legislative acts. The Warren Court invalidated 21 federal and 150 state statutes; the Burger Court struck down 31 federal and 288 state laws. The laws invalidated were at least as significant as those ruled unconstitutional by the Warren Court. The federal statutes that failed to pass constitutional muster in the Burger years included laws governing election financing and judicial salaries,[158] granting eighteen-year-olds the vote in state elections,[159] establishing bankruptcy courts,[160] as well as laws based on gender classifications in the military[161] and in various social security programs.[162] In addition, the Burger Court struck down the legislative veto, a method used by Congress to control executive action in nearly 200 statutes,[163] and a law designed to deal with the endemic budget deficit.[164] "If deference to Congress be the acid test of judicial restraint, the litmus of the Burger Court comes out much the same color as that of its predecessor."[165]

But it is not numbers alone that mark the Burger Court as an activist one. More important than quantity was the quality of the decisions rendered. The Burger Court was as ready as its predecessor to resolve crucial constitutional issues and to do so in accordance with its own conceptions of what the law should be. *United States v. Nixon*[166] brought the Court into the center of the Watergate vortex and its decision led directly to the first resignation of a President. Nor was there any doubt among the Justices on the propriety of their exercise of power to resolve the crisis. *Nixon* demonstrates the willingness of the Justices to mold crucial constitutional principles to accord with their individual policy perceptions. The impact of judicial review was definitely broadened by *Nixon* and the Burger Court's other separation-of-powers decisions.

The Warren Court's activism was manifested in the number of new rights recognized by it, but the "rights explosion" was more than equaled by that under its successor. Few decisions were more far-reaching in their recognition of new rights than *Roe v. Wade,* to be discussed in the next chapter. The *Roe* decision may indeed be taken as the very paradigm of the activist decision: the decision was based not upon principles worked out in earlier cases but upon "policy judgments" made upon an ad hoc basis which led to recognition of a new right. Even here the Justices were influenced more by pragmatism than principle. "Too many wealthy women were flouting the law to get abortions from respected physicians. Too many poor women were being injured by inadequately trained mass purveyors of illegal abortions. Concerns of that sort, rather than issues of high principle, are what appeal to the centrist activists of the Burger Court."[167]

The hallmark of the activist Court is the *Roe*-type decision that creates a new right not previously recognized in law. The Burger Court recognized new rights for women and those dependent on public largess. During the Burger years the law on sexual classifications was completely changed. Though the Court did not go as far as some had wished, virtually all legal disabilities based upon sex were placed beyond the legal pale.[168]

It should, however, be stressed that there was a fundamental difference between the activism of the Burger Court and that of its predecessor. Chief Justice Warren and his supporters acted on the basis of overriding principles derived from their vision of the society the Constitution was intended to secure.

In particular the Warren Court acted on the basis of two broad principles: nationalism and egalitarianism. It preferred national solutions to what it deemed national problems and, to secure such solutions, was willing to countenance substantial growth in federal power. Even more important was the Warren Court's commitment to equality. If one great theme recurred in its jurisprudence, it was that of equality before the law—equality of races, of citizens, of rich and poor, of prosecutor and defendant. The result was what Justice Fortas termed "the most profound and pervasive revolution ever achieved by substantially peaceful means."[169] More than that, it was the rarest of all political animals: a judicially inspired and led revolution. Without the Warren Court decisions giving ever-wider effect to the right to equality before the law, most of the movements for equality that have permeated American society would never have gotten off the ground.

Yet the Warren Court was more than the judicial counterpart of Plato's philosopher-king. To Warren and his supporters, the Supreme Court was a modern Court of Chancery—a residual "fountain of justice" to rectify individual instances of injustice, particularly where the victims suffered from racial, economic, or similar disabilities. The Warren Justices saw themselves as present-day Chancellors, who secured fairness and equity in individual cases, fired above all by a vision of equal dignity, to be furthered by the Court's value-laden decisions.

No similar vision inspired the activism of the Burger Court. Instead of consciously using the law to change the society and its values, it rode the wave, letting itself be swept along by the consensus it perceived in the social arena—moving, for example, on gender discrimination when it became "fashionable" to be for women's rights. From this point of view, the Burger Court's activism has been well termed a "rootless activism," which dealt with cases on an essentially ad hoc basis, inspired less by moral vision than by pragmatic considerations.[170]

The rootless activism of the Burger Court was a direct consequence of the divisions between the Justices. Because of it, "[T]he hallmark of the Burger Court has been strength in the center and weakness on the wings."[171] The balance of power was held by the Justices "in the middle." In the last Burger terms, however, the center's grip started to weaken. As

Justice Blackmun put it just after Chief Justice Burger retired, "I think the center held generally . . . [but] it bled a lot, and it needs more troops. Where it's going to get them, I don't know."[172]

The shift toward the right did not occur until the end of the Burger years. Before that, the balance was with the pragmatic Justices who did not decide cases in accordance with a preconceived ethical philosophy. This was particularly true when Justices Stewart and Powell were the key swing votes. The Stewart reply, when he was asked whether he was a "liberal" or a "conservative," bears repeating. "I am a lawyer," Stewart answered. "I have some difficulty understanding what those terms mean even in the field of political life. . . . And I find it impossible to know what they mean when they are carried over to judicial work."[173]

Justices who felt this way had the lawyer's aversion to making fundamental value choices. Judicial policy-making was as frequent a feature during the Burger years as in the Warren years. But the policy choices were, in the main, made by Justices who, as relatively moderate pragmatists, were motivated by case-by-case judgments on how to make a workable judicial accommodation that would resolve a divisive public controversy. Inevitably, their decisions did not make for a logically consistent corpus such as that constructed by the Warren Court. In most areas of the law, the Burger Court decisions reflected less an overriding calculus of fundamental values than lawyerlike attempts to resolve the given controversy as a practical compromise between both sides of the issues involved.[174]

In the Warren Court, the leadership had come from the left; constitutional doctrine was, in the important cases, made by Chief Justice Warren and his liberal supporters, notably Justice Brennan. Under Warren's successor, Justice Brennan was shunted to one of the extremes that now more often played a lesser role. The Burger Court's activism was molded more by the moderate Justices "in the middle." As such, it was "inspired not by a commitment to fundamental constitutional principles or noble political ideals, but rather by the belief that modest injections of logic and compassion by disinterested, sensible judges can serve as a counterforce to some of the excesses and irrationalities of contemporary governmental decision-making."[175]

Thus judicial activism itself became a centrist philosophy, primarily practical in nature, without an agenda or overriding philosophy. Its essential approach was to adapt the answer of Diogenes, Solvitur gubernando,[176] and more or less on a case-by-case basis. Fundamental value choices were more often avoided than made. In its operation, "[T]he Burger Court has exhibited a notable determination to fashion tenuous doctrines that offer both sides of a social controversy something important."[177]

The Burger years appear to have marked a legal watershed. After the Warren Court's rewriting of so much of our public law, the Burger Court was bound to be primarily a Court of consolidation. Transforming innovation, in the law as elsewhere, can take place only for so long. In historical terms, indeed, the Burger Court's main significance was its consolidation and continuation of the Warren heritage. Its role in this respect seems all the more important now that the Burger Court has given way to the Rehnquist Court.

15

Watershed Cases:
Roe v. Wade, 1973

Roe v. Wade[1] was in many ways the paradigmatic Burger Court case. It was clearly the most controversial decision rendered by that Court, and no Court decision has been more bitterly attacked. "It is hard to think of any decision in the two hundred years of our history," declared Cardinal Krol, the President of the National Conference of Catholic Bishops, "which has had more disastrous implications for our stability as a civilized society."[2] Condemnatory letters were sent to the Justices in unprecedented volume, particularly to Justice Blackmun, the author of the opinion. Antiabortion pickets still show up at his speeches.

Griswold and Privacy

The right of privacy is the fundamental right upon which the *Roe* decision was based. It is most unlikely that the men who drew up the Constitution and the Bill of Rights intended to confer, as against government, a right of privacy in the sense in which that term is used today. But there are indications that the Founders did intend to include within the sphere of constitutional protection matters that the present-day observer would classify as coming under an overall right of privacy in the Fourth Amendment's prohibition against unreasonable searches and seizures.

The Fourth Amendment conception of privacy was, nevertheless, rudimentary compared to the right of privacy recognized in our law during the past quarter century. The Framers could think of a sphere peculiar to the individual in his home, his papers, his effects—a sphere that the Bill of Rights rendered immune from governmental encroachment. But the right to privacy in its present-day constitutional connotations is nothing less than the right of individuals to be protected from any wrongful intrusion into their private life.[3] "Liberty in the constitutional sense must mean more than freedom from unlawful governmental restraint; it must include privacy as well, if it is to be a repository of freedom. The right to be let alone is indeed the beginning of all freedoms."[4]

Griswold v. Connecticut,[5] decided in 1965, was the first case confirming the existence of a broad constitutional right of privacy "no less important than any other right . . . reserved to the people."[6] As Justice Black put it, *Griswold* elevated "a phrase which Warren and Brandeis used in discussing grounds for tort relief,[7] to the level of a constitutional rule which prevents state legislatures from passing any law deemed by this Court to interfere with 'privacy.'"[8]

A Connecticut law prohibited the use of contraceptive devices and the giving of medical advice on their use. Griswold and a doctor gave advice to married persons on preventing conception and prescribed contraceptive devices for the wives. They were convicted of violating the birth control law, and their conviction was affirmed by the state's highest court.

The *Griswold* conference found a seven-to-two majority in favor of striking down the Connecticut law, but the majority Justices did not articulate a clear theory on which to base the decision. Justice Douglas stated the simplest rationale. The law violated the defendant's First Amendment right of association, which was more than a right of assembly. Thus, he reasoned, the right to send a child to a religious school was "on the periphery" of the right of association.[9] He also used the analogy of the right to travel, which the Court had said "is in radiation of First Amendment and so is this right." There was nothing more personal than this right and it too was "on the periphery" and within First Amendment protection.

Justice Black was the strongest conference opponent of this Douglas approach. "The right of association is for me a right of assembly and the right of the husband and wife to assemble in bed is a new right of assembly for me."

Only Justice Stewart supported Justice Black's view. He stated that he could not "find anything [against this law] in the First, Second, Fourth, Fifth, Ninth or other Amendments. So I'd have to affirm." The others agreed that the law should be ruled unconstitutional, but differed in their reasoning.

Chief Justice Warren assigned the opinion to Justice Douglas, who had expressed the clearest theory upon which the Connecticut law might be invalidated. Douglas quickly prepared a draft opinion. The draft based the decision on the First Amendment, likening the husband-wife relationship to the forms of association given First Amendment protection. The draft did not mention, much less rely upon, a constitutional right of privacy, and had the draft come down as the *Griswold* opinion, the right of privacy might never have been recognized as a constitutional right by the Warren Court.

The Douglas draft was changed because of an April 24, 1965, letter to Justice Douglas from Justice Brennan. He wrote that the "association" of married couples had little to do with the advocacy protected by the First Amendment and urged that the *Griswold* decision be based upon the right of privacy. He suggested that the expansion of the First Amendment to include freedom of association be used as an analogy to justify a similar approach in the areas of privacy.

The final *Griswold* opinion of the Court stated that the "specific guarantees in the Bill of Rights have penumbras, formed by emanations from those guarantees that help give them life and substance." A constitutional right of privacy was included in these penumbras. The right of marital privacy—"older than our school system"[10]—was violated by the Connecticut law.

Griswold holds squarely that the Bill of Rights does establish a constitutionally protected zone of privacy. The *Griswold*-created right served as the foundation for some of the most controversial Burger Court decisions. By 1977, the Court could state, *"Griswold* may no longer be read as holding only that a State may not prohibit a married couple's use of contraceptives."[11] The right of privacy recognized in *Griswold* is not one that inheres only in the marital relationship. Instead, "If the right of privacy means anything, it is the right of the individual, married or single, to be free from unwarranted governmental intrusion into matters so fundamentally affecting a person."[12]

Conference and Assignment

During the *Griswold* conference Chief Justice Warren had stated that he could not say that the state had no legitimate interest, noting that that could apply to abortion laws—implying that he thought such laws were valid. Those who expected the Burger Court to be less activist than its predecessor relied upon false hopes. It was under Chief Justice Burger that the right of privacy was extended to include the right to an abortion. As Burger put it, in a "Memorandum to the Conference," "This is as sensitive and difficult an issue as any in this Court in my time."[13]

Roe v. Wade came before the Court together with a companion case, *Doe v. Bolton*. In both cases, pregnant women sought relief against state abortion laws, contending they were unconstitutional. At issue in *Roe* was a Texas statute that prohibited abortions except for the purpose of saving the mother's life. The statute in *Doe* was a Georgia law that proscribed an abortion except as performed by a physician who felt, in "his best clinical judgment," that continued pregnancy would endanger a woman's life or injure her health; the fetus would likely be born with a serious defect; or the pregnancy resulted from rape. In addition, the Georgia statutory scheme posed three procedural conditions: (1) that the abortion be performed in an accredited hospital; (2) that the procedure be approved by the hospital staff abortion committee; and (3) that the performing physician's judgment be confirmed by independent examinations by two other physicians.

The final *Roe v. Wade* opinion contrasted the Texas and Georgia statutes as follows: "The Texas statutes under attack here are typical of those that have been in effect in many States for approximately a century. The Georgia statutes, in contrast, have a modern cast and are a legislative product that, to an extent at least, obviously reflects the influences of recent attitudinal change, of advancing medical knowledge and techniques, and of new thinking about an old issue."[14]

Nevertheless, the lower court had held both laws invalid—in *Roe* because the statute infringed upon the plaintiff's "fundamental right . . . to choose whether to have children [which] is protected by the Ninth Amendment" and in *Doe* because the reasons listed in the statute improperly restricted plaintiff's right of privacy and of personal liberty.

Roe v. Wade and *Doe v. Bolton* were both discussed at the same post-argument conference in December 1971. The Chief Justice devoted much of his *Roe v. Wade* discussion to the question of standing. Referring to the lead plaintiff, he said, "Jane Roe is unmarried and pregnant. She doesn't claim health; just doesn't want the baby." In the Burger view, "The unmarried girl has standing. She didn't lose standing through mootness." This meant, the Chief Justice went on, that "she's entitled to an injunction if the statute is [un]constitutional." On the merits, Burger said, "The balance here is between the state's interest in protecting fetal life and the woman's interest in not having children." In weighing these interests, the Chief Justice concluded, "I can't find the Texas statute unconstitutional, although it's certainly archaic and obsolete."

Justice Douglas, who spoke next, declared categorically, "The abortion statute is unconstitutional. This is basically a medical and psychiatric problem" and not one to be dealt with by prohibitory legislation. Douglas also criticized the statute's failure to give "a licensed physician an immunity for good faith abortions." Justice Brennan, who followed, expressed a similar view, though he stressed more strongly than any of the others that the right to an abortion should be given a constitutional basis by the Court's decision.

Justice Stewart, next in order of seniority, also spoke in favor of standing. On the merits, Stewart stated, "I agree with Bill Douglas." Stewart did, however, indicate that there might be some state power in the matter. "The state," he said, "can legislate, to the extent of requiring a doctor and that, after a certain period of pregnancy, [she] can't have an abortion."

Justice White began his presentation, "I agree with Potter on all preliminaries, but on the merits I am on the other side. They want us to say that women have a choice under the Ninth Amendment." But White said that he refused to accept this "privacy argument."

Justice Marshall, on the other hand, declared, "I go with Bill Douglas, but the time problem concerns me." He thought that the state could not prevent abortions "in the early stage [of pregnancy]. But why can't the state prohibit after a certain stage?" In addition, Marshall said that he would use "'liberty' under the Fourteenth Amendment as the constitutional base."

Justice Blackmun, then the junior Justice, spoke last. He agreed that "the girl has standing." On the merits, Blackmun's presentation displayed an ambivalence that was to be reflected in his draft *Roe v. Wade* opinion. "Can a state properly outlaw all abortions?" he asked. "If we accept fetal life, there's a strong argument that it can. But there are opposing interests: the right of the mother to life and mental and physical health, the right of parents in case of rape, the right of the state in case of incest. I don't think there's an absolute right to do what you will with [your] body." Blackmun did, however, say flatly, "this statute is a poor statute that . . . impinges too far on her."

The discussion of *Doe v. Bolton* paralleled that in *Roe v. Wade*. The Chief Justice asserted, "I do not agree with this carving up of the statute by the three-judge court." As Burger saw it, "The state has a duty to protect fetal life at some stage, but we are not confronted with that question here . . . I would hold this statute constitutional."

The Georgia statute received a more favorable review than its Texas counterpart from Justice Douglas. "This is a much better statute than Texas," he declared. But Douglas had doubts on the statute's practical effects. "We don't know," he stated, "how this statute operates. Is it weighted on the side of only those who can afford this? What about the poor?" Douglas said that he was inclined "to remand to the district court to find out."

Justice Brennan had no doubts. He said that he would affirm the decision below "as far as it goes" but would also "go further to strike down the three-doctor thing as too restrictive." Justice Stewart agreed with the last point. But Justice White again spoke in favor of the state. As he saw it, "The state has power to protect the unborn child. This plaintiff didn't have trouble [taking]¹⁵ advantage of procedures. I think the state has struck the right balance here."

Once again, Justice Blackmun's position was ambivalent. "Medically," he pointed out, "this statute is perfectly workable." Blackmun emphasized the competing interests at stake. "I would like," he said, "to see an opinion that recognizes the opposing interests in fetal life and the mother's interest in health and happiness." Blackmun indicated interest in the approach stated by Justice Douglas earlier in the conference. "I would be perfectly willing," he stated, "to paint some standards and remand for findings as to how it operates: does it operate to deny equal protection by discriminating against the poor?"

The conference outcome was not entirely clear; the tally sheets of different Justices do not coincide on the votes.¹⁶ What was clear, however, was that a majority were in favor of invalidating the laws: in *Roe v. Wade,* five (Justices Douglas, Brennan, Stewart, Marshall, and Blackmun) to two (the Chief Justice and Justice White) in the tally sheet made available to me—but four to three (with Blackmun added to the dissenters) according to a Douglas "Dear Chief" letter to Burger.

Despite the fact that he was not part of the majority, the Chief Justice assigned the opinions in the two abortion cases to Justice Blackmun. On December 18, 1971, two days after the conference, Justice Douglas (whose tally sheet showed four votes for invalidating the laws, with himself as senior Justice in the majority) sent his "Dear Chief" missive: "As respects your assignment in this case, my notes show there were four votes to hold parts of the . . . Act unconstitutional. . . . There were three to sustain the law as written." Douglas concluded, "I would think, therefore, that to save future time and trouble, one of the four rather than one of the three, should write the opinion."

The Chief Justice replied with a December 20 "Dear Bill" letter. "At the close of discussion of this case, I remarked to the Conference that there were, literally, not enough columns to mark up an accurate reflection of the voting in either the Georgia or the Texas cases. I therefore marked down no votes and said this was a case that would have to stand or fall on the writing, when it was done."

According to the Burger letter, "That is still my view of how to handle these two . . . sensitive cases, which I might add, are quite probable candidates for reargument."

A few months later, the *Washington Post* reported the Burger–Douglas exchange. According to the *Post* story, Douglas sent his letter "asserting his prerogative to assign the case, but Burger held fast to his position."[17] Justices Douglas and Brennan, who had led the proabortion bloc at the conference, then decided to wait to see the Blackmun drafts before doing anything further in the matter. Though Douglas had, as seen, tallied Blackmun with the minority, others had noted his vote as one with the majority. This might well mean Blackmun opinions agreeable to Douglas and Brennan.

Douglas Draft

That Justices Douglas and Brennan did reach the conclusion just indicated is shown by a letter sent by Brennan to Douglas on December 30: "I gathered from our conversation yesterday that you too think we might better await Harry Blackmun's circulation in the *Texas* abortion case before circulating one." After all, when Justice Blackmun did circulate, it might make either a confrontation with the Chief Justice or a separate majority draft unnecessary.

The Brennan letter was called forth by an uncirculated Douglas draft in the Georgia case, which its author had sent to Justice Brennan soon after the assignment to Justice Blackmun. This draft is presumably what Justice

Douglas would have circulated as his draft opinion of the Court had he been able to assign the abortion opinions.

The Douglas draft, headed "Memorandum from MR. JUSTICE DOUGLAS,"[18] was a typical Douglas product. Written personally at break-neck speed—almost six months before the Blackmun first draft was circulated—it was finished before the others had even had a chance to reflect seriously on what had happened at the argument and conference. Unlike the situation with other Douglas drafts, however, this first effort was not to be the only or final Douglas product. The Justice spent considerable effort in refining his opinion; there would be seven drafts of the Douglas opinion (the last on May 22, 1972) before it was replaced by Justice Blackmun's draft opinions in the two abortion cases. Probably, up until the Blackmun drafts were circulated, Justice Douglas still hoped that he would be able to write the Court's abortion opinions—which would explain the unusual Douglas effort to improve his opinion through so many separate drafts. Ultimately, the Douglas seventh draft became the concurrence that the Justice issued in *Roe v. Wade*.

The Douglas draft was an opinion in the Georgia case, *Doe v. Bolton*. But its broad reasoning on the merits—grounding the right to an abortion on the constitutional right of privacy—was equally applicable to the Texas case, *Roe v. Wade*. "The right of privacy," Douglas declared "is a species of 'liberty' of the person as that word is used in the Fourteenth Amendment. It is a concept that acquires substance, not from the predilections of judges, but from the emanations of the various provisions of the Bill of Rights, including the Ninth Amendment."

The heart of the Douglas draft was its holding that the abortion right was protected by the right of privacy. That right, wrote Douglas, "covers a wide range" and is "broad enough to encompass the right of a woman to terminate an unwanted pregnancy in its early stages, by obtaining an abortion." This does not mean that "the 'liberty' of the mother, though rooted as it is in the Constitution, may [not] be qualified by the State." But where fundamental rights are involved, "this Court has required that the statute be narrowly and precisely drawn and that a 'compelling state interest' be shown in support of the limitation." This requirement is of cardinal significance. "Unless regulatory measures are so confined and are addressed to the specific areas of compelling legislative concern, the police-power would become the great leveller of constitutional rights and liberties."

As the Douglas draft saw it, the statute at issue failed the constitutional test. That was true because, as Justice Brennan summarized the draft in his December 30 letter, "The statute infringes the right of privacy by refusing abortions where the mother's mental, but not physical health is in jeopardy." The Douglas draft noted that "the vicissitudes of life produce pregnancies which may be unwanted or which may impair the 'health' in the broad . . . sense of the term or which may imperil the life of the mother or which in the full setting of the case may create such suffering, dislocations, misery, or tragedy as to make an early abortion the only civilized step to take. The suffering, dislocation, misery or tragedy just mentioned may be properly embraced in the 'health' factor of the mother as appraised by a person of insight." But the abortion statute did not embrace other than physical health and was therefore too narrow to meet constitutional requirements.

Justice Brennan sent his December 30 letter after reading the Douglas draft. The letter consisted of a ten-page analysis of the draft and suggestions for improvements. Brennan wrote, "I guess my most significant departure from your approach is in the development of the right-of-privacy argument." Brennan noted his agreement "that the right is a species of 'liberty' (although, as I mentioned yesterday, I think the Ninth Amendment . . . should be brought into this problem at greater length)."

The Brennan letter stressed that "[t]he decision whether to abort a pregnancy obviously fits directly within . . . the categories of fundamental freedoms . . . and, therefore, should be held to involve a basic individual right." This meant "that the crucial question is whether the State has a compelling interest in regulating abortion that is achieved without unnecessarily intruding upon the individual's right."

Justice Brennan wrote that, "although I would, of course, find a compelling State interest in requiring abortions to be performed by doctors, I would deny any such interest in the life of the fetus in the early stages of pregnancy." It follows "there is a right to an abortion in the early part of the term," and that the right of privacy in the matter of abortions means that the decision is that of the woman and her alone.

"In sum," the Brennan letter concluded, "I would affirm the district court's conclusion that the reason for an abortion may not be prescribed. I would further hold that the only restraint a State may constitutionally impose upon the women's individual decision is that the abortion must be performed by a licensed physician."

The later Douglas drafts as well as the ultimate Douglas concurrence adopted many of the Brennan suggestions. The Douglas draft also had an influence on Justice Blackmun's drafting process. There is a note in Justice Douglas's hand, dated March 6, 1972, indicating that a copy of the Douglas draft has been "sent . . . to HB several weeks ago."

Blackmun Drafts

In his December 30 letter, Justice Brennan had written, "I appreciate that some time may pass before we hear from Harry." Justice Blackmun was known as the slowest worker on the Court. The abortion cases were his first major assignment, and he worked at them during the next few months, mostly alone and unassisted in the Court library; he was still working on his draft as the Court term wore on. The opinions had been assigned in December 1971. However, it was mid-May before the Justice felt able to circulate anything.

Finally, on May 18, 1972, Justice Blackmun sent around his draft *Roe v. Wade* opinion.[19] "Herewith," began the covering memo, "is a first and tentative draft for this case." Blackmun wrote that "it may be somewhat difficult to obtain a consensus on all aspects. My notes indicate, however, that we were generally in agreement to affirm on the merits. That is where I come out on the theory that the Texas statute, despite its narrowness, is unconstitutionally vague."

The memo went on, "I think that this would be all that is necessary for disposition of the case, and that we need not get into the more complex Ninth Amendment issue. This may or may not appeal to you."

However, the Justice informed his colleagues, "I am still flexible as to results, and I shall do my best to arrive at something which would command a court."

As the covering memo explained, Blackmun's *Roe* draft avoided the broader constitutional issue and struck down the Texas statute on the ground of vagueness. The draft started by dealing with the issues of standing and mootness. It found that Roe had standing and that the termination of her pregnancy did not render the case moot.

On the merits, the Blackmun draft held the Texas statute unconstitutionally vague. The difficulty here was that, in *United States v. Vuitch*,[20] decided the year before, the Court had upheld a similar District of Columbia abortion law against a vagueness attack. The Blackmun draft distinguished *Vuitch* on the ground that the statute there prohibited abortion unless "necessary for the preservation of the mother's life or health," whereas the Texas statute permitted abortions only "for the purpose of saving the life of the mother." Consequently, the draft concluded, *Vuitch* "provides no answer to the constitutional challenge to the Texas statute."

In the Texas statute, "Saving the mother's life is the sole standard." According to the Blackmun draft, this standard was too vague to guide physicians' conduct in abortion cases. "Does it mean that he may procure an abortion only when, without it, the patient will surely die? Or only when the odds are greater than ever that she will die? Or when there is a mere possibility that she will not survive?"

In consequence, the draft declared, "We conclude that Art 1196, with its sole criterion for exemption as 'saving the life of the mother,' is insufficiently informative to the physician to whom it purports to afford a measure of professional protection but who must measure its indefinite meaning at the risk of his liberty, and that the statute cannot withstand constitutional challenge on vagueness grounds."

The Blackmun draft's vagueness analysis was extremely weak. If anything, the "life-saving" standard in the *Roe v. Wade* statute was more definite than the "health" standard upheld in *Vuitch*. But the draft's disposition of the case on the vagueness ground enabled the draft to avoid addressing what Justice Brennan, in a May 18 "Dear Harry" letter—after he had read the draft—called "the core constitutional question." As the Blackmun draft stated, "There is no need in Roe's case to pass upon her contention that under the Ninth Amendment a pregnant woman has an absolute right to an abortion, or even to consider the opposing rights of the embryo or fetus during the respective prenatal trimesters."

Indeed, so far as the draft contained intimations on the matter, they tended to support state substantive power over abortions. "Our holding today," the draft was careful to note, "does not imply that a State has no legitimate interest in the subject of abortions or that abortion procedures may not be subjected to control by the State." On the contrary, "We do not accept the argument of the appellants and of some of the *Amici* that a pregnant woman has an unlimited right to do with her body as she pleases. The long acceptance of statutes regulating the possession of certain drugs and other harmful substances, and making criminal indecent exposure in public, or an attempt at suicide, clearly indicate the contrary." This was, of course, completely different from the approach ultimately followed in the *Roe v. Wade* opinion of the Court.

In his covering memo transmitting his *Roe v. Wade* draft, Justice Blackmun also referred to the companion case, *Doe v. Bolton.* "The Georgia case, yet to come," he wrote, "is more complex. I am still tentatively of the view, as I have been all along, that the Georgia case merits reargument before a full bench. I shall try to produce something, however, so that we may look at it before any decision as to that is made."

On May 25, a week after he had sent around his *Roe v. Wade* draft, Justice Blackmun circulated his draft *Doe v. Bolton* opinion.[21] "Here, for your consideration," the covering memo began, "is a memorandum on the second abortion case." As summarized in the memo, his opinion "would accomplish . . . the striking of the Georgia statutory requirements as to (1) residence, (2) confirmation by two physicians, (3) advance approval by the hospital abortion committee, and (4) performance of the procedure only in [an] accredited hospital."

The *Roe v. Wade* draft did not deal at all with the right of privacy. It was, however, discussed in Justice Blackmun's *Doe* draft. Though it struck down the Georgia statute on the grounds noted in the covering memo, it also dealt with the claim that the law is an "invalid restriction of absolute fundamental right to personal and marital privacy." Here Justice Blackmun was substantially influenced by the treatment in the Douglas draft, particularly the later redrafts, which Justice Douglas noted in a May 19 letter, "I believe I gave you, some time back."

Blackmun's *Doe* draft dealt specifically with the claim that the law was an "invalid restriction of an absolute fundamental right to personal and marital privacy. . . . The Court, in varying contexts, has recognized a right of personal privacy and has rooted it in the Fourteenth Amendment, or in the Bill of Rights, or in the latter's penumbras." The draft flatly rejected the assertion "that the scope of this right of personal privacy includes, for a woman, the right to decide unilaterally to terminate an existing but *unwanted* pregnancy without any state interference or control whatsoever." As the draft put it, "Appellants' contention, however, that the woman's right to make the decision is absolute—that Georgia has either no valid interest in regulating it, or no interest strong enough to support any limitation upon the woman's sole determination—is unpersuasive."

The draft rejected as "unfair and illogical" the argument that "the State's present professed interest in the protection of embryonic and fetal 'life' is somehow to be downgraded. That argument condemns the State for past wrongs and also denies it the right to readjust its views and emphases in the light of the more advanced knowledge and techniques of today."

The *Doe* draft, utterly unlike the final Blackmun opinions, stressed the countervailing interest in fetal life. "The heart of the matter is that somewhere, either forthwith at conception, or at 'quickening,' or at birth, or at some other point in between, another being becomes involved and the privacy the woman possessed has become dual rather than sole. The woman's right of privacy must be measured accordingly." That being the case, "The woman's personal right . . . is not unlimited. It must be balanced against the State's interest." Hence "we cannot automatically strike down the remaining features of the Georgia statute simply because they restrict any right on the part of the woman to have an abortion at will."

The remainder of the *Doe* draft balanced "the impact of the statute upon the right, as it relates to the state interest being asserted." The balancing process led, as the covering memo summarized it, to invalidation of the residence, accreditation, approval, and confirmation requirements.

The Blackmun *Doe* draft was certainly an improvement over his first *Roe v. Wade* effort. Like the latter, however, it did not deal directly with Justice Brennan's "core constitutional question." Indeed, the implication here, too, was that substantial state power over abortion did exist. Under the *Doe* draft, as the Blackmun covering memo pointed out, the state may provide "that an abortion may be performed only if the attending physician deems it necessary 'based upon his best clinical judgment,' if his judgment is reduced to writing, and if the abortion is performed in a hospital licensed by the State through its Board of Health." This was, of course, wholly inconsistent with the Court's final decision in *Roe v. Wade.*

Justice Blackmun ended his *Doe* covering memo by again explaining his approach in the *Roe v. Wade* draft. "I should observe," he pointed out, "that, according to information contained in some of the briefs, knocking out the Texas statute in *Roe v. Wade* will invalidate the abortion laws in a *majority* of our States. Most States focus only on the preservation of the life of the mother. *Vuitch,* of course, is on the books, and I had assumed that the Conference, at this point, has no intention to overrule it. It is because of *Vuitch*'s vagueness emphasis and a hope, perhaps forlorn, that we might have a unanimous court in the Texas case, that I took the vagueness route."

Reargument and Second Conference

Had the Blackmun drafts in the abortion cases come down as the final decisions, the last twenty years in American life and politics might have been quite different. It soon became apparent, however, that the Blackmun drafts were not going to receive the majority imprimatur needed to transform them into Court opinions.

First came indications that the drafts were not satisfactory to the leaders of the conference majority. On May 18, soon after he had received the drafts, Justice Brennan sent a "Dear Harry" letter. "My recollection of the voting on this and the *Georgia* case," Brennan wrote, "was that a majority of us felt that the Constitution required the invalidation of abortion statutes save to the extent they required that an abortion be performed by a licensed physician within some limited time after conception. I think essentially this was the view shared by Bill, Potter, Thurgood and me. My notes also indicate that you might support this view at least in this *Texas* case."

This led Justice Brennan to urge a decision on the constitutional merits. "In the circumstances, I would prefer a disposition of the core constitutional question. Your circulation, however, invalidates the Texas statute only on the vagueness ground. . . . I think we should dispose of both cases on the ground supported by the majority."

The Brennan letter closed with an attempt to mollify Blackmun, who was, after all, the least firm of those willing to invalidate the laws. "This does not mean, however, that I disagree with your conclusion as to the vagueness of the Texas statute." But such deference did not indicate a Brennan inclination to allow the case to be decided narrowly. "I only feel that there is no point in delaying longer our confrontation with the core issue on which there appears to be a majority and which would make reaching the vagueness issue unnecessary."

The next day Justice Douglas wrote to Blackmun, "My notes confirm what Bill Brennan wrote yesterday in his memo to you—that abortion statutes were invalid save as they required that an abortion be performed by a licensed physician within a limited time after conception." That, according to Douglas, "was the clear view of the majority of the seven who heard the argument. . . . So I think we should meet what Bill Brennan calls the 'core issue.'" Justice Douglas also referred to the fact that, at the conference, "the chief had the opposed view, which made it puzzling as to why he made the assignment at all."

The Brennan and Douglas letters indicated opposition to the Blackmun attempt to avoid the "core [constitutional] issue" in *Roe v. Wade*. But now the conference minority sought to delay—and perhaps reverse—the abortion decision. *Roe* and *Doe* had come before a seven-Justice Court. The two vacancies were not filled until Justices Powell and Rehnquist took their seats in January 1972. After the Blackmun drafts were circulated in May, the Chief Justice directed his efforts to securing a reargument in the cases, arguing that the decisions in such important cases should be made by a full Court.

At this point Justice White sent around a brief draft dissent.[22] It effectively demonstrated the weakness of the Blackmun vagueness approach in striking down the Texas law. Referring to the *Vuitch* decision that a statute that permitted abortion on "health" grounds was not constitutionally vague,[23] the White draft declared, "If a standard which refers to the 'health' of the mother, a referent which necessarily entails the resolution of perplexing questions about the interrelationship of physical, emotional, and mental well-being, is not impermissibly vague, a statutory standard which focuses only on 'saving the life' of the mother would appear to be *a fortiori* acceptable. . . . [T]he relevant factors in the latter situation are less numerous and are primarily physiological."

On May 31, Chief Justice Burger sent around a Memorandum to the conference favoring reargument. The Chief Justice wrote, "[T]hese cases . . . are not as simple for me as they appear to be for others. The states have, I should think, as much concern in this area as in any within their province; federal power has only that which can be traced to a specific provision of the constitution." Moreover, the Burger memo went on, "This is as sensitive and difficult an issue as any in this Court in my time and I want to hear more and think more when I am not trying to sort out several dozen other difficult cases." Because of these factors, the memo concluded, "I vote to reargue early in the next Term."

The Burger move to secure reargument was opposed by the Justices in favor of striking down the abortion laws. They feared that, after the reargument, the two new Justices would vote for the laws. In addition the White draft dissent might lead another Justice to withdraw his support from the Blackmun *Roe* draft—maybe even Blackmun himself, whose position had been none too firm in the matter. At the least, the draft was subject to further erosion simply because it was based on the vulnerable vagueness argument.

On May 31, Justice Brennan wrote to Justice Blackmun, "I see no reason to put these cases over for reargument. I say that since, as I understand it, there are five of us (Bill Douglas, Potter, Thurgood, you and I) in substantial agreement with both opinions and in that circumstance I question that reargument would change things." Later that day, Justice Blackmun received a similar note from Justice Marshall: "Like Bill Brennan, I too, am opposed to reargument of these cases."

By now, however, Justice Blackmun was ready to break the five-man majority for immediate issuance of his opinions. He himself had become convinced that the cases should be reargued and circulated a "May 31 Memorandum to the Conference" to that effect: "Although it would prove costly to me personally, in the light of energy and hours expended, I have now concluded, somewhat reluctantly, that reargument in *both* cases at an early date in the next term, would perhaps be advisable." He gave two reasons for his position:

1. I believe on an issue so sensitive and so emotional as this one, the country deserves the conclusion of a nine-man, not a seven-man court, whatever the ultimate decision may be.

2. Although I have worked on these cases with some concentration, I am not yet certain about all the details. Should we make the *Georgia* case the primary opinion and recast Texas in its light? Should we refrain from emasculation of the Georgia statute and, instead, hold it unconstitutional in its entirety and let the state legislature reconstruct from the beginning? Should we spell out—although it would then necessarily be largely dictum—just what aspects are controllable by the State and to what extent? For example, it has been suggested that . . . Georgia's provision as to a licensed hospital should be held unconstitutional, and the Court should approve performance of an abortion in a "licensed medical facility." These are some of the suggestions that have been made and that prompt me to think about a summer's delay.

The Blackmun memo concluded with a vote supporting the Chief Justice: "I therefore conclude, and move, that both cases go over the Term."

Justice Douglas replied to Justice Blackmun with a letter that same day. "I feel quite strongly," Douglas wrote, "that they should not be reargued." He also had two reasons. "In the first place, these cases which were argued last October have been as thoroughly worked over and considered as any cases ever before the court in my time." The second reason was that reargument was not proper where an opinion was supported by a majority of the full Court. "I have a feeling," Douglas wrote, "that where the Court is split 4–4 or 4–2–1 or even in an important constitutional case 4–3, reargument may be desirable. But you have a firm 5 and the firm 5 will be behind you in these two opinions until they come down. It is a difficult field and a difficult subject. But where there is that solid agreement of the majority I think it is important to announce the cases, and let the result be known so that the legislatures can go to work and draft their new laws.

The Douglas letter concluded with a kudo to Justice Blackmun: "Again, congratulations on a fine job. I hope the 5 can agree to get the cases down this Term, so that we can spend our energies next Term on other matters."

The next day, June 1, an angry Douglas letter was sent to the Chief Justice:

Dear Chief:
I have your memo to the Conference dated May 31, 1972 re: *Abortion Cases*.
If the vote of the Conference is to reargue, then I will file a statement telling what is happening to us and the tragedy it entails.

The threatened Douglas statement was never issued, even though the Justices did vote to have the abortion cases reargued. The Douglas attempt to defeat reargument was doomed when the two new Justices voted in favor of reargument. On June 1, Justice Powell circulated a "Memorandum to the Conference" which began: "The question is whether the abortion cases should be reargued." Powell noted that he had not until then participated in the vote on reargument motions. This case, he wrote, was different. "I have been on the Court for more than half a term. It may be that I now have a duty to participate in this decision, although from a purely personal viewpoint I would be more than happy to leave this one to others."

The Powell memo went on: "I have concluded that it is appropriate for me to participate in the pending question. . . . I am persuaded to favor reargument primarily by the fact that Harry Blackmun, the author of the opinions, thinks the cases should be carried over and reargued next fall. His position, based on months of study, suggests enough doubt on an issue of large national importance to justify the few months delay."

Justice Rehnquist also sent around a June 1 memo voting in favor of reargument, as did Justice White on June 5. That gave the motion for reargument five votes. When, on June 29, the last day of the 1971 term, the Court issued its order setting the abortion cases for reargument, only Justice Douglas was listed as dissenting.[24]

In one of those tricks legal history sometimes plays, it was Justice Douglas and the others who favored invalidating the abortion laws who gained the most from the order for reargument. Had Douglas won his battle to prevent reargument, the original Blackmun drafts would have remained the final *Roe* and *Doe* opinions. They would have dealt narrowly with the issues before the Court and, as he predicted in his note to Blackmun, sent legislatures to work drafting new abortion laws, for these opinions were clearly not ringing affirmations of the right to abortion.

By moving for reargument, Chief Justice Burger hoped to secure the votes of the two new Justices and then persuade Justice Blackmun himself to switch to an opinion upholding the abortion laws. From his point of view, the Chief Justice would have been better off had the weak original *Roe* draft come down. As it turned out, he got a split vote from the new Justices and a vastly improved *Roe* opinion, with its broadside confirmation of the constitutional right to an abortion.

The abortion cases were reargued on October 11, 1972. At the conference following reargument, the Justices who had participated in the earlier conference took the same positions as before. The two new Justices took opposing positions. Justice Powell said that he was "basically in accord with Harry's position," whereas Justice Rehnquist stated, "I agree with Byron [White]"—who had declared "I'm not going to second guess state legislatures in striking the balance in favor of abortion laws."

Several Justices expressed dissatisfaction with the approach of the Blackmun *Roe* draft, agreeing with Justice Stewart's statement that "I can't join in holding that the Texas statute is vague." Stewart was for striking that law but urged a different approach. He said that he would "follow John Harlan's reasoning in the Connecticut case[25] and can't rest there on the Ninth Amendment. It's a Fourteenth Amendment right, as John Harlan said in *Griswold*."[26]

Second Draft and Final Opinion

The most significant part of Justice Blackmun's postreargument conference presentation was his announcement, "I am where I was last Spring." However, he made a much firmer statement this time in favor of invalidating the abortion laws. He also said, "I'd make Georgia the lead case," but he was opposed on this by several others, particularly Justice Powell, who felt that "Texas should be the lead case."

Most important of all, Justice Blackmun announced to the conference, "I've revised both the Texas and Georgia opinions of the last term." During the past summer, Blackmun had devoted his time to the abortion opinions and had completely rewritten them. On November 22, he circulated a completely revised draft of his *Roe v. Wade* opinion. "Herewith," began the covering memo, "is a memorandum (1972 fall edition) on the Texas abortion case."

Justice Blackmun's second *Roe* draft expressly abandoned the vagueness holding on which his first draft had turned. The holding on the constitutional merits, the new draft stated, "makes it unnecessary for us to consider the attack made on the Texas statute on grounds of vagueness." The covering memo explained, "I have attempted to preserve *Vuitch* in its entirety. You will recall that the attack on the Vuitch statute was restricted to the issue of vagueness. 420 U.S. at 73. I would dislike to have to undergo another assault on the District of Columbia statute based, this time, on privacy grounds."

The new Blackmun draft contained the essentials of the final *Roe v. Wade* opinion, including its lengthy historical analysis. Instead of the earlier "vagueness" approach, Justice Blackmun now grounded his decision upon *Griswold v. Connecticut*. According to Blackmun, "[T]he right of privacy, however based, is broad enough to cover the abortion decision." In addition, since the right at issue was a "fundamental" one, the law at issue was subject to strict-scrutiny review: the state regulation of the fundamental right of privacy "may be justified only by a 'compelling state interest.'"[27]

At the postargument conference, the Chief Justice had asked, "Is there a fetal life that's entitled to protection?" Justice Stewart said that the Court should deal specifically with this issue, saying, "[I]t seems essential that we deal with the claim that the fetus is not a person under the fourteenth Amendment." The *Roe v. Wade* opinion met this Stewart demand with a statement that the word "person" in the Fourteenth Amendment does not include a fetus[28]—a point that was specifically added at Justice Stewart's insistence.[29]

The second draft also adopted the time approach followed in the final opinion. However, it used the first trimester of pregnancy alone as the line between invalid and valid state power. "You will observe," Justice Blackmun explained in his covering memo, "that I have concluded that the end of the first trimester is critical. This is arbitrary, but perhaps any other selected point, such as quickening or viability, is equally arbitrary."

The draft stated that before the end of the first trimester, the state "must do no more than to leave the abortion decision to the best medical judgment of the pregnant woman's attending physician." However, "for the stage subsequent to the first trimester, the State may, if it chooses, determine a point beyond which it restricts legal abortions to stated reasonable therapeutic categories."

Later drafts refined this two-pronged time test to the tripartite approach followed in the final *Roe* opinion. In large part, this was in response to the suggestion in a December 12 letter from Justice Marshall: "I am inclined to agree that drawing the line at viability accommodates the interests at stake better than drawing it at the end of the first trimester. Given the difficulties which many women may have in believing that they are pregnant and in deciding to seek an abortion, I fear that the earlier date may not in practice serve the interests of those women, which your opinion does seek to serve."

The Marshall letter stated that his concern would be met "[i]f the opinion stated explicitly that, between the end of the first trimester and viability, state regulations directed at health and safety alone were permissible."

Marshall recognized "that at some point the State's interest in preserving the potential life of the unborn child overrides any individual interests of the woman." However, he concluded, "I would be disturbed if that point were set before viability, and I am afraid that the opinion's present focus on the end of the first trimester would lead states to prohibit abortions completely at any later date."

Justice Blackmun adopted the Marshall suggestion, even though Justice Douglas sent him a letter: "I favor the first trimester, rather than viability."

In addition, Justice Brennan sent a "Dear Harry" letter. "While as you know," the letter began, "I am in basic agreement with your opinions in these cases, I too welcome your giving second thoughts to the choice of the end of the first trimester as the point beyond which a state may appropriately regulate abortion practices." The Justice, however, questioned whether "viability" was the appropriate point.

Justice Brennan summarized the latest Blackmun drafts: "I read your proposed opinions as saying, and I agree, that a woman's right of personal privacy includes the abortion decision, subject only to limited regulation necessitated by the compelling state interests you identify. Moreover, I read the opinions to say that the state's initial interest (at least in point of time if not also in terms of importance) are in safeguarding the health of the woman and in maintaining medical standards."

The Brennan letter then asked, "[I]s the choice of 'viability' as the point where a state may begin to regulate abortions appropriate? For if we identify the state's initial interests as the health of the woman and the maintenance of medical standards, the selection of 'viability' (i.e., the point in time where the fetus is capable of living outside of the woman) as the point where a state may begin to regulate in consequence of these interests seems to me to be technically inconsistent."

As Justice Brennan saw it, "'Viability,' I have thought, is a concept that focuses upon the fetus rather than the woman." Brennan preferred an approach that he said corresponded more with the medical factors that give rise to the "cut-off" point. "For example," Brennan wrote, "rather than using a somewhat arbitrary point such as the end of the first trimester or a somewhat imprecise and technically inconsistent point such as 'viability,' could we not simply say that at that point in time where abortions become medically more complex, state regulation—reasonably calculated to protect the asserted state interests of safeguarding the health of the woman and of maintaining medical standards—becomes permissible[?]"

Despite the Douglas and Brennan letters, Justice Blackmun continued to use the "viability" approach, though he did modify it in later drafts to meet the Marshall suggestion. At this point, Justice Stewart delivered a more fundamental criticism of the Blackmun approach in a December 14 letter. "One of my concerns with your opinion as presently written is the specificity of its dictum—particularly in its fixing of the end of the first trimester as the critical point for valid state action. I appreciate the inevitability and indeed wisdom of dicta in the Court's opinion, but I wonder about the desirability of the dicta being quite so inflexibly 'legislative.'"

This is, of course, the common criticism that has since been directed at *Roe v. Wade*—that the Court was acting more like a legislature than a court; its drawing of lines at trimesters and viability was, in the Stewart letter's phrase, "to make policy judgements" that were more "legislative" than "judicial." Justice Stewart worked on a lengthy opinion giving voice to this criticism. In a December 27 letter, however, he informed Justice Blackmun, "I have now decided to discard the rather lengthy concurring opinion on which I have been working, and to file instead a brief monograph on substantive due process, joining your opinions."

In early December, Justice Blackmun had sent around a revised draft of his *Doe v. Bolton* opinion as well. This was substantially very close to the final opinion in the Georgia case. On December 21, the Justice circulated further drafts and then, on January 17, 1973, the final versions that came down as the Court opinions in the two cases on January 22.

Compelling Interest and Due Process

According to Justice Blackmun's November 22, 1972, covering memorandum transmitting his second *Roe v. Wade* draft, "As I stated in conference, the decision, however made, will probably result in the court's being severely criticized." Just before he circulated the final *Roe v. Wade* draft, Blackmun sent around a January 16, 1973, "Memorandum to the Conference," which began, "I anticipate the headlines . . . when the abortion decisions are announced." Because of this, the Justice was enclosing the announcement from the bench that he proposed to read when the two cases were made public. With this announcement, the memo expressed the hope that "there should be at least some reason for the press not going all the way off the deep end."

The Blackmun announcement did not, of course, have the calming effect for which its author hoped. If anything, indeed, the Justice had understated the outcry. The scare headlines and controversy were far greater than anything anticipated by the Court. Almost all the criticism was directed to the question of whether abortion should be permitted or prohibited. To one interested in Supreme Court jurisprudence, of even greater interest is the constitutional approach followed in striking down the state abortion laws.

The Blackmun *Roe* opinon was based upon two essential holdings: (1) "[T]he right of privacy, however based, is broad enough to cover the abortion decision." It follows from this that there is a "fundamental right" to an abortion; and (2) "Where certain 'fundamental rights' are involved, the court has held that regulation limiting these rights may be justified only by a 'compelling state interest.'"[30]

The state interest in protecting the health of the woman does not become "compelling" until the end of the first trimester of pregnancy. The interest in protecting "the potentiality of human life[31] becomes "compelling" only after viability. Hence state laws restricting the "fundamental right" to an abortion before that time are invalid.

Justice Rehnquist, in his dissent, points out what the Court had done in its *Roe* opinion. *Roe* was a due process, not an equal protection case. "The test traditionally applied in the area of social and economic legislation is whether or not a law such as that challenged has a rational relation to a valid state objective."[32] But that was not the review standard applied in *Roe*. The abortion laws were subjected to strict scrutiny under the compelling-interest test.

According to Justice Rehnquist, "The court eschews the history of the Fourteenth Amendment in its reliance on the 'compelling state interest' test."[33] The strict-scrutiny–compelling-interest approach had been developed to deal with equal protection claims. When the Court applied the test that had been used in suspect-classification cases to cases involving fundamental rights, those cases were also equal protection cases. Now, in *Roe,* the Court held that the compelling-interest test should be used when a statute that infringed upon fundamental rights was challenged on due process grounds. As Justice Rehnquist put it in *Roe,* "The court adds a new wrinkle to this test by transposing it from the legal considerations associated with the Equal Protection Clause of the Fourteenth Amendment to this case arising under the Due Process Clause of the Fourteenth Amendment."[34]

In Justice Rehnquist's view, "[T]he court's sweeping invalidation of any restriction on abortion during the first trimester is impossible to justify under that standard, and the conscious weighing of competing factors that the Court's opinion apparently substitutes for the established test is far more appropriate to a legislative judgment than to a judicial one."[35]

Certainly, there is danger that the importation of the compelling-interest standard into the Due Process Clause will lead to a revival of the substantive due process approach that prevailed in what Justice Stewart termed "the heyday of the Nine Old Men, who felt that the Constitution enabled them to invalidate almost any state laws they thought were unwise."[36]

From this point of view, there is validity to the Rehnquist charge that *Roe* marked a return to the substantive due process approach followed in cases such as *Lochner v. New York*,[37] "when courts used the Due Process Clause 'to strike down state laws . . . because they may be unwise, improvident, or out of harmony with a particular school of thought.'"[38] According to Justice Rehnquist, the *Roe* adoption of the compelling-interest standard in due process cases will inevitably require the Court once again to pass on the wisdom of legislative policies in deciding whether the particular interest put forward is or is not "compelling." As Rehnquist put it in a 1977 memorandum, "[T]he phrase 'compelling state interest' really asks the question rather than answers it, unless we are to revert simply to what Holmes called our own 'can't helps.'"[39] Just as important, the determination of what are and what are not "fundamental rights" is also left to the unfettered discretion of the individual Justice.[40]

It should, however, be noted that the Rehnquist animadversion is not as valid today as it may have been when *Roe* was decided. The Burger Court later drew the line at what rights may be considered "fundamental" for the purposes of strict scrutiny under the Due Process Clause.[41] This substantially narrowed the scope of its revival of substantive due process.

It should also be borne in mind that there are due process cases where review less flaccid than that under the rational-basis standard is plainly appropriate. The rights guaranteed by the First Amendment are fundamental rights, made applicable to the states by the Due Process Clause of the Fourteenth Amendment.[42] When a state burdens the exercise of First Amendment rights there is a violation of due process, and most observers would agree that the state law should be subjected to strict scrutiny under the compelling-interest test. The entire Burger Court expressly stated that that is the case.[43]

Presumably even Chief Justice Rehnquist would agree with this strict-scrutiny posture in cases involving the fundamental rights guaranteed under the First Amendment. His criticism of the *Roe v. Wade* approach was thus more a criticism of degree than one of kind and was presumably answered by the later cases which limit the applicability of the strict-scrutiny approach by restricting the rights deemed fundamental for purposes of the more exacting review standard. Those cases prevented the *Roe* approach from being extended much beyond the abortion cases themselves. For those favoring an even broader judicial role in protecting privacy, the *Roe* substantive due process revival was to become essentially a lost opportunity.

Nonenumerated Rights

Critics of *Roe* have also attacked the decision as one that departs drastically from the constitutional canon of "textual fidelity."[44] Where in the Constitution, they ask, is there any guaranty of the right to an abortion or of the right of privacy upon which it is based?

The view that the *Roe*-type protection of rights not specified in the Constitution is illegitimate has been stated forcefully by Judge Robert H. Bork. Referring specifically to the right of privacy, Bork points out that *Griswold* and *Roe* created "an overall right of privacy that applies even where no provision of the Bill of Rights does." Under this approach, Bork asserts, "the Bill of Rights was expanded beyond the known intentions of the Framers. Since there is no constitutional text or history to define the right, privacy becomes an unstructured source of judicial power."[45]

As Judge Bork sees it, the *Griswold-Roe* approach "requires the Court to say, without guidance from the Constitution, which liberties or gratifications may be infringed by majorities and which may not."[46] Since the judge is given no guide other than the constitutional text, which, of course, does not cover a case such as *Griswold,* the question of what nontextual rights should be protected depends completely on the judge's own discretion. Once we depart from the text of the Constitution," asks Justice Scalia, "just where . . . do we stop?"[47] Or as a much-quoted Bork aphorism has it, "The truth is that the judge who looks outside the historic Constitution always looks inside himself and nowhere else."[48]

It is, however, erroneous to assume that only those rights specifically mentioned are given constitutional protection. As Justice Brennan once put it, "[T]he protection of the Bill of Rights goes beyond the specific guarantees to protect from congressional abridgement those equally fundamental personal rights necessary to make the express guarantees fully meaningful."[49] The reason why was best stated in Justice Goldberg's *Griswold* concurrence: "To hold that a right so basic and fundamental and so deep-rooted in our society as the right of privacy in marriage may be infringed because that right is not guaranteed in so many words by the first eight amendments to the Constitution is to ignore the Ninth Amendment and to give it no effect whatsoever."[50] In Goldberg's words, "[T]he Ninth Amendment shows a belief of the Constitution's authors that fundamental rights exist that are not expressly enumerated in the first eight amendments and an intent that the list of rights included there not be deemed exhaustive."[51]

The Ninth Amendment shows that there are basic rights not specifically mentioned in the Bill of Rights that are nevertheless protected from governmental infringement. "In sum," Justice Goldberg concludes, "I believe that the right of privacy in the marital relation is fundamental and basic—a personal right 'retained by the people' within the meaning of the Ninth Amendment."[52]

Of course, the most controversial application of the *Griswold*-created right was in *Roe v. Wade,* where the Court decided that the state law proscribing most abortions was violative of the right to privacy. That right was held to include the right to terminate pregnancies. The right of privacy was ruled broad enough to encompass the abortion decision.

In his *Roe v. Wade* concurrence Justice Douglas stressed the role of the Ninth Amendment as a source of the right protected by the *Roe* decision. As Douglas saw it, this right is one of personal autonomy that "includes customary, traditional, and time-honored rights, amenities, privileges, and immunities that come within the sweep of 'the Blessings of Liberty' mentioned in the preamble to the Constitution." The Douglas concurrence then listed three groups of rights that "come within the meaning of the term 'liberty' as used in the Fourteenth Amendment."[53] The Douglas list was based upon a suggestion contained in the December 30, 1971, letter that had been sent to him by Justice Brennan after the latter had read the Douglas draft concurrence. The letter consisted of a ten-page analysis of the draft and suggestions for improvements. The Douglas draft had found the abortion law violative of the right of privacy. Brennan suggested a broader approach to the privacy concept.

The Brennan letter noted his agreement

> that the right [of privacy] is a species of "liberty" (although, as I mentioned yesterday, I think the Ninth Amendment . . . should be brought into this problem at greater length), but I would identify three groups of fundamental freedoms that "liberty" encompasses: first, freedom from bodily restraint or inspection, freedom to do with one's body as one likes, and freedom to care for one's health and person; second, freedom of choice in the basic decisions of life, such as marriage, divorce, procreation, contraception and the education and upbringing of children; and, third, autonomous control over the development and expression of one's intellect and personality.

The Brennan list, as adopted by Justice Douglas in his concurring opinion,[54] is the most comprehensive judicial statement of what is included in the right of personal autonomy protected, under *Griswold* and *Roe,* by the Constitution. To be sure, none of the liberties in the Brennan-Douglas list is mentioned in the Constitution. Yet all involve freedoms that most Americans rightly believe are protected against governmental intrusion.

The right of personal autonomy comes down to what was stated in classic language by Justice Brandeis: "The makers of our Constitution conferred, as against the Government, the right to be let alone—the most comprehensive right and the right most valued by civilized men. To protect that right, every unjustifiable intrusion by the Government upon the privacy of the individual, whatever the means employed, must be deemed a violation of the [Constitution]."[55]

If the individual is not to be overwhelmed by the state, equipped as it is with all the resources of modern science and technology, he or she must be permitted to retain the essential attributes of individuality—that intrinsic element of life that distinguishes not only our species from all others, but all of us within the human community from each other.[56] Above all, in a society where science offers governments a continually refined set of tools for intrusion and surveillance, it is essential that there remain an area of apartness in which the individual may live as Walt Whitman's "simple separate person"—one that is immune from intervention by the community itself. When man stands in constant danger of being overwhelmed by the machine, he can retain his individuality only if there is preserved for him "a privacy, an obscure nook"[57]—"A liberty of choice as to his manner of life, and neither an individual nor the public has a right to arbitrarily take away from him this liberty."[58]

Even though they are not specifically mentioned in the Constitution, few people would deny that the aspects of autonomy in the Brennan-Douglas list are entitled to constitutional protection. For example, there is what Justice Douglas calls the "freedom from bodily restraint or compulsion, freedom to walk, stroll, or loaf."[59] During a Court conference on a case involving freedom of religion,[60] Douglas stated, "I think we're entitled to our religious scruples, but I don't see how we can make everyone attune to them. I can't be required to goose-step because eight or ninety percent goose-step."

More recently, Justice John P. Stevens (at the time a circuit judge) gave the example of "a case in which the sovereign insists that every citizen must wear a brown shirt to demonstrate his patriotism."[61] Would any American judge, no matter how opposed in theory to nonenumerated rights, uphold a law requiring people to goose-step or wear a brown shirt?

A comparable illustration is given in Justice Goldberg's *Griswold* concurrence. "Surely," Goldberg writes, "the Government . . . could not decree that all husbands and wives must be sterilized after two children have been born to them."[62] Would even opponents of nontextual rights say that such a law "would not be subject to constitutional challenge because . . . no provision of the Constitution specifically prevents the Government from curtailing the marital right to bear children and raise a family?"[63] Clearly, as Justice Goldberg puts it, Americans would find it "shocking to believe that the personal liberty guaranteed by the Constitution does not include protection against such totalitarian limitation of family size, which is at complete variance with our constitutional concepts."[64]

In view of the recent attacks on the *Griswold-Roe* concept of nontextual rights, it is significant that the concept was vigorously supported by Justice Harlan, now considered the very paradigm of the true conservative judge. A few years before *Griswold*, Harlan anticipated the decision there. Speaking of the contraceptive ban in conference, Harlan asserted, "This is more offensive to the right to be let alone than anything possibly can be."[65]

In a 1961 dissent, Justice Harlan indicated why the right to be let alone was guaranteed even though it was not mentioned in the constitutional text. The Court, said Harlan, must approach "the text which is the only commission for our power not in any literalistic way, as if we had a tax statute before us, but as the basic charter of our society, setting out in spare but meaningful terms the principles of government." For Harlan, "[I]t is not the particular enumeration of rights . . . which spells out the reach of" constitutional protection. On the contrary, the "character of Constitutional provisions . . . must be discerned from a particular provision's larger context. And . . . this context is one not of words, but of history and purposes."66

Because of this, Justice Harlan went on,

> the full scope of the liberty guaranteed by the Due Process Clause cannot be found in or limited by the precise terms of the specific guarantees elsewhere provided in the Constitution. This "liberty" is not a series of isolated points pricked out in terms of the taking of property; the freedom of speech, press, and religion; the right to keep and bear arms; the freedom from unreasonable searches and seizures; and so on. It is a rational continuum which, broadly speaking, includes a freedom from all substantial arbitrary impositions and purposeless restraints . . . and which also recognizes . . . that certain interests require particularly careful scrutiny of the state needs asserted to justify their abridgment.67

It is true that the Harlan recognition of nonenumerated rights was based upon the Due Process Clause rather than the Ninth Amendment. Thus, in a 1970 case,68 Justice Harlan agreed that the reasonable-doubt standard in criminal trials was mandated by the Constitution. But his recognition of the nontextual right was explained on due process grounds: "I view the requirement of proof beyond a reasonable doubt in a criminal case as bottomed on a fundamental value determination of our society that it is far worse to convict an innocent man than to let a guilty man go free." Under this approach, "due process, as an expression of fundamental procedural fairness, requires a more stringent standard for criminal trials than for ordinary civil litigation."69

That Justice Harlan relied upon due process rather than the Ninth Amendment did not detract from his full acceptance of the concept of nonenumerated rights. As Harlan saw it, due process "is a discrete concept which subsists as an independent guaranty of liberty and procedural fairness, more general and inclusive than the specific prohibitions."70 Except for its terminology, the Harlan approach is essentially similar to the approach that relies on the Ninth Amendment. If the right is a basic right "which [is] fundamental; which belong[s] . . . to the citizens of all free governments,"71 it is one retained by the people under the Ninth Amendment or, in the Harlan view, included in the "liberty" protected by due process.

To be sure, the fact that Justice Harlan and other conservative judges, such as Chief Justice Burger himself, have supported the *Griswold-Roe* position that "fundamental rights even though not expressly guaranteed, have been recognized by the Court as indispensable to the enjoyment of rights explicitly defined,"[72] will scarcely change the bitter opposition to the Supreme Court jurisprudence on the matter—particularly that to *Roe v. Wade* itself. The controversy over that decision continues unabated to this day, to an extent far greater than anything anticipated by Justice Blackmun and his colleagues. All that would have been avoided had the original Blackmun *Roe v. Wade* draft come down as the final Court opinion. Instead of a *cause célèbre*, *Roe* might then have been a mere constitutional footnote used by law professors to illustrate how the Court can evade important legal issues.

16

Rehnquist Court, 1986–

The history of an institution such as the Supreme Court, like a tapestry, is made up of many strands that, interwoven, make a pattern; to separate a single one and look at it alone not only defaces the whole but gives the strand itself a false value.[1] All too many studies of the Supreme Court, or of its individual members, concentrate upon single strands of the Court's work, emphasizing, more often than not, those aspects that diverge most sharply from the overall pattern.

Such an approach is bound to give a distorted picture. Not infrequently, in truth, the reaction of commentators about the Court is akin to that of the blind man from Hindustan when first confronted with an elephant. The aspect of the Court's work emphasized by the particular observer tends to dominate his conception of the Court as a whole; yet he almost never comes to picture the Court as the institutional entity that it is. Small wonder, then, that the public, both legal and lay, has no clear picture of the working of our unique judicial organ and its proper place in a constitutional democracy.

Not long ago Chief Justice Rehnquist recalled his first visit to the Court after his appointment as a Justice: "We came over here, and it was kind of a grey afternoon. . . . And I just felt, literally, like I'd entered a monastery when I came over."[2] Neophytes on the high bench—even the strongest of them—are immediately aware of the overpowering institutional traditions. Such awareness continues throughout the Justices' tenure and, more than is generally realized, weaves into the Court's pattern all but the most eccentric of its members.

362

ㅤㅤIt has been said of one of the greatest Justices, Louis D. Brandeis, that he had an almost mystic reverence for the Court, whose tradition seemed to him not only to consecrate its own members, but to impress its sacred mission upon all who shared in any measure in its work.[3] Few members of the high tribunal may be capable of penetrating into its mystique with the perception of a Brandeis; still, says Chief Justice Rehnquist, "Everybody who comes here probably feels the constraints of the place."[4]

ㅤㅤThis history has been written upon the assumption that the pattern of the tapestry is more important than the single strands. Similarly, the Supreme Court as an institution is more significant than the individual Justices who make up its membership.

ㅤㅤTo be sure, to treat the Court as an institutional entity may seem outdated in an age when even the law has succumbed to our society's preoccupation with the behavioral sciences. Judges are only men, we were told[5]—which was, of course, an indisputable observation. All the same, it hardly follows from this that only study of the psychological makeup of the individual Justices is now worthwhile. The state of a man's mind is as much a fact as the state of his digestion, according to a nineteenth-century English judge. Now, however, we are told that the two are intimately related and that the state of a judge's mind can hardly be known without some knowledge of the state of his stomach.

ㅤㅤTo advocates of this sort of gastrological jurisprudence, all attempts to describe the Court as an institutional entity are fundamentally naive, and this is particularly true when the institutional ethos of the Court has seemed at a low ebb. At such a time, it is said, it is only the makeup of the individual Justice that is important if we are to understand the decisions of a fragmented Court.

ㅤㅤNo one not blind to the facts of legal life can deny that the Supreme Court has too often presented a far from edifying spectacle of internal atomization. But even that did not prevent that tribunal from functioning as an institutional entity. Even when the Court has been splintered, its work as a governmental organ had to go on. This is, in fact, a basic difference between an ultimate judicial tribunal and commentaries upon its work. The Court cannot adopt an either–or approach. It must decide the case before it, even though the decision requires it to choose between two conflicting truths. The theorist need wholly reject neither, where neither states an exclusive verity; the Court must choose between them. Yet it is a mistake to assume that because, in such cases, the individual members of the tribunal are sharply divided, the Court has ceased to function as an institution.

ㅤㅤOn the contrary, even amid a plethora of such cases, the institutional pattern continues to be woven. It may be harder to determine the boundary at which some Courts have balanced conflicting interests. Still, the Court is always engaged in drawing the line between conflicting interests. While we may not be able to determine it by a general formula, points in the line have been fixed by decisions that this or that concrete case fell on the nearer or farther side.

Mr. Right as Chief Justice

Drawing the line has, however, been easier since Chief Justice Burger retired in 1986 and was succeeded in the Court's center chair by Justice William H. Rehnquist. Nor can it be doubted that the line is being drawn farther to the right by the Rehnquist Court than it was by its immediate predecessors. The new Chief Justice was well characterized by a *Newsweek* article as "The Court's Mr. Right."[6] According to the *New York Times,* "William H. Rehnquist is a symbol. People who have trouble naming all nine Supreme Court Justices quickly identify him as its doctrinaire, right-wing anchor. . . . Justice Rehnquist is the Court's most predictable conservative member, using his considerable intelligence, energy and verbal facility to shape the law to his vision."[7]

The Rehnquist vision in this respect has always been a clear one. In a 1985 interview, he noted that he joined the Court with a desire to counteract the Warren Court decisions.[8] "I came to the court," Rehnquist said, "sensing . . . that there were some excesses in terms of constitutional adjudication during the era of the so-called Warren Court." Some of that Court's decisions, the Justice went on, "seemed to me hard to justify. . . . So I felt that at the time I came on the Court, the boat was kind of keeling over in one direction. Interpreting my oath as I saw it, I felt that my job was . . . to kind of lean the other way."[9]

By the time of his nomination as Chief Justice, Rehnquist had become the most influential member of the Burger Court, both because he stood out in a Court of generally bland personalities, which lacked a firm sense of direction, and because he was closely allied with the Chief Justice. During his early Court years, however, the majority remained largely unsympathetic to the Rehnquist entreaties from the right. It was then that he received a Lone Ranger doll as a gift from his law clerks, who called him the "lone dissenter" during that period. During his fourteen years as an Associate Justice, Rehnquist dissented alone fifty-four times—a Court record.

When asked about the origins of his conservatism, Rehnquist replied, "It may have something to do with my childhood."[10] He was raised in a modest house in a middle-class Milwaukee suburb. After World War II service in North Africa, he attended Stanford and its law school. He then served as a Supreme Court clerk to Justice Robert H. Jackson. Rehnquist wrote a memo on the *Brown* segregation case,[11] urging that the separate-but-equal doctrine, under which segregation had been upheld,[12] was "right and should be affirmed."[13]

Justice Rehnquist later stated that his views had changed and that he accepted *Brown* as the law of the land.[14] But his votes in cases involving racial issues clearly placed him at the opposite extreme from the Brennan wing of the Court. On most other issues, too, Rehnquist has reflected his conservative Republican background, with years of private practice and active involvement in the Goldwater wing of the party in Arizona and a position as Assistant Attorney General in the Nixon Department of Justice. Before his elevation to the Court's center chair, Rehnquist had anything but the appearance of a Justice; he still had the look of an overage college student—lumbering around the Court in his thick brown glasses, mismatched outfits, and Hush Puppies. As Chief Justice, however, he has trimmed his sideburns, begun to wear conservative suits and ties, and drives to work in a black limousine.

Despite his robust appearance and weekly tennis with his law clerks, Rehnquist has had health problems. In 1982, he was hospitalized because of his back and underwent withdrawal symptoms, with a period of mental confusion and slurred speech, when the heavy dosage of a powerful sedative, Placidyl, was reduced. In 1977, Rehnquist had written in reply to a letter comment on his draft opinion in a case: "It may be that my adverse reactions to your letter of March are partially induced by my doctor's insistence that I take valium four times a day. . . ."[15]

Time once called Justice Rehnquist the Court's "most self-consciously literate opinion writer."[16] He is arguably the best legal stylist and phrase-maker on the Court, though too much of his literary ability tends to be overshadowed by the extreme position which it supports. Rehnquist's literary talents are not limited to his opinion writing. A typical example is contained in a 1980 memo on a case involving a mandatory life sentence under a recidivist statute. The defendant had obtained $229 by fraud and forgery through his three criminal acts.[17] Rehnquist's draft opinion holding that the sentence did not constitute cruel and unusual punishment was disputed by a Powell draft dissent. Rehnquist took issue with the Powell view that, since the crimes were "properly related" and did not involve "violence," the sentence was unconstitutionally harsh. "The notions embodied in the dissent," Rehnquist wrote in his memo,

> that if the crime involved "violence," . . . a more severe penalty is warranted under objective standards simply will not wash, whether it be taken as a matter of morals, history, or law. Caesar's death at the hands of Brutus and his fellow conspirators was undoubtedly violent; the death of Hamlet's father at the hands of his brother, Claudius, by poison, was not. Yet there are few, if any states which do not punish just as severely murder by poison (or attempted murder by poison) as they do murder or attempted murder by stabbing. The highly placed executive who embezzles hugh sums from a state savings and loan association, causing many shareholders of limited means to lose substantial parts of their savings, has committed a crime very different from a man who takes a smaller amount of money from the same savings and loan at the point of a gun. Yet rational people could disagree as to which criminal merits harsher punishment. . . . In short, the "seriousness" of an offense or a pattern of offenses in modern society is not a line, but a plane.[18]

His extreme views have not prevented Rehnquist from being on good terms with the other Justices. Even his ideological opposites like Justice Brennan have commented on their cordial relations with the categorical conservative. To his colleagues, Rehnquist has been as well known for his good nature as for his rightist acumen. On a Court where, as Justice Blackmun once lamented, "[t]here is very little humor,"[19] Rehnquist stood out because of his impish irreverence and wit. When the Burger Court sat, one of Rehnquist's clerks would every now and then pass notes to the Justice. These were not legal memos but Trivial Pursuit-style questions. Rehnquist would answer them and then hand them to Justice Blackmun for that Justice to try his hand.[20]

Another illustration may also be seen in a 1973 Rehnquist memorandum: "In going over some material which had been stored for a long period of time in my present Chambers, I came across a manuscript poem entitled 'To a Law Clerk Dying Young,' written by someone named A. E. Schmaussman, or Schmousman[21] (the handwriting is not too good), who was apparently a law clerk here at one time." Rehnquist wrote, "I found the poem very moving and emotional, and thought that a public reading of it would be a suitable occasion for a gathering of present and retired members of the Court and their law clerks to toast a departing Term with sherry."

The Rehnquist memo concluded with what some considered a satiric allusion to Chief Justice Burger's constant emphasis on Court secrecy: "P.S. I debated circulating the actual text of the poem with this invitation, but decided that there was too great a chance that it mght be leaked to the newspapers before the party."[22]

The Rehnquist sense of humor sometimes degenerated into practical jokes. On April Fool's Day 1985, the Chief Justice was the Rehnquist victim. Rehnquist had a life-size photo cutout of Burger produced and sent a street photographer to a corner outside the Court with a sign: "Have your picture taken with the chief justice, $1." To make sure he wouldn't miss Burger's reaction, the Justice called him at home, saying he needed a ride to Court on April 1. Rehnquist "was laughing like crazy" when he drove past the scene that day with the overdignified Chief Justice.[23]

When all is said and done, however, the most important thing about Rehnquist is that he has proved to be a strong Chief Justice, more in the Warren than the Burger mold. It was said that Chief Justice Burger's discussion of cases at conference left the Justices with the feeling that he was "the least prepared member of the Court."[24] In a comment on Chief Justice Rehnquist, Justice Marshall said, "He has no problems, wishy-washy, back and forth. He knows exactly what he wants to do, and that's very important as a chief justice."[25] There is little doubt that Marshall was contrasting the Rehnquist performance with that of his predecessor.

Some years ago, the present Chief Justice compared the Burger and Warren Courts. Rehnquist stated that the impact of the Court had been diminished under Chief Justice Burger. "I don't think that the Burger Court has as wide a sense of mission. Perhaps it doesn't have any sense of mission at all."[26]

Certainly the Warren Court did have Rehnquist's "sense of mission" when it virtually rewrote the corpus of our constitutional law; concepts and principles that had appeared unduly radical became accepted rules of law. From this point of view, the Warren Court was the paradigm of the "result-oriented" Court, which used its power to secure the result it deemed right in the cases that came before it. Employing the authority of the ermine to the utmost, Warren and his colleagues never hesitated to do whatever they thought necessary to translate their own conceptions of fairness and justice into the law of the land. The same was plainly not true of the Burger Court; it did not have any "sense of mission" comparable to its predecessor.

Rehnquist himself has said, "I don't know that a court should really have a sense of mission."[27] Yet Rehnquist clearly is a judge who does have a sense of mission. From his first appointment to the Court Rehnquist has sought what he called "a halt to . . . the sweeping rules made in the days of the Warren Court"[28]—and not only a halt, but a rollback of much of the Warren jurisprudence. The Rehnquist conservative program has included enlargement of government authority over individuals, a check to the expansion of criminal defendants' rights, limitations on access to federal courts, and increased emphasis on protection of property rights. As Chief Justice, he has finally been in a position to advance his conservative agenda.

The Conservative Majority

Chief Justice Rehnquist has started to move the Justices in his desired direction because, for the first time in half a century, the Court had a conservative majority willing to support the shift to the right. This marked a definite change from prior Courts. Under Chief Justice Warren, the Court's liberal wing was dominant. In the Burger Court, the balance of power was held by the Justices who were, in Justice Blackmun's phrase, "in the middle."[29] In Chief Justice Burger's last terms, however, the center's grip started to weaken. As Justice Blackmun put it just after Burger retired, "I think the center held generally . . . [but] it bled a lot. And it needs more troops. Where it's going to get them, I don't know."[30]

As it turned out, it was not the center but the right which received the additional troops. The first of them, Antonin Scalia, was appointed in 1986 to fill the vacancy created by Chief Justice Rehnquist's elevation to the Court's center chair. Scalia had been a law professor (primarily at the University of Chicago), a government official, and a judge of the D.C. Circuit Court of Appeals. As a Justice, Scalia has been a doctrinaire conservative—even more extreme in his rightist views than the Chief Justice himself—though he has, as will be noted, also exhibited a libertarian streak that has led him to resist some governmental intrusions.

It was expected by many that Justice Scalia's intellectual brilliance would enable him to perform a leadership role on the Rehnquist Court. So far, it has not worked out that way. Instead, the Justice has persisted in extreme positions that have not been accepted even by a conservative majority that might be willing to accept a properly tempered Scalia posture.

Justice Scalia is the first noted law professor to be elevated to Olympus since Felix Frankfurter. Scalia has not, however, confined himself to the relatively restrained judicial role assumed by his predecessor. Instead, Scalia has been a judicial activist, not hesitating to import his own academic theories into our public law. But if Justice Scalia seems unduly rigid in his approach, at least he is always interesting in his opinions. This should make him the professoriate's favorite Justice. Both for those who agree and those who disagree with them, the Scalia opinions should provide grist for the academic mills for years to come.

Another vacancy occurred on the Rehnquist Court when Justice Powell retired in 1987. Anthony M. Kennedy, then a circuit judge, was chosen in Powell's place. After graduation from Harvard Law School, Kennedy was in private practice for fifteen years (six of those as a sole practitioner) and also taught part-time at the McGeorge School of Law in Sacramento. Justice Powell had been a leader in the Burger Court's centrist core. Justice Kennedy has, more often than not, been a vote for the Rehnquist Court's growing conservative majority. He has, however, been anything but a doctrinaire conservative in the Scalia sense. He has displayed a willingness to listen to opposing views and an openness to dialogue that contrast with Justice Scalia's often inflexible posture.

Perhaps the most notable thing about Justice Kennedy is the number of opinions he has written, particularly in critical cases. The unusually high number by a junior Justice indicates the Chief Justice's immediate confidence in the newest member of his Court. The confidence has been justified by Justice Kennedy's general adherence to the Rehnquist jurisprudence.

The Kennedy appointment meant that Chief Justice Rehnquist now had a five-Justice conservative core (the Chief Justice and Justices White, O'Connor, Scalia, and Kennedy). But the conservative majority was still a fragile one, which often saw defections by one or more members. That situation changed when President Bush was able to select replacements for Justices Brennan and Marshall, the last liberal holdovers from the Warren Court, who retired in 1990 and 1991.

To succeed Justice Brennan, the President nominated David H. Souter, a former Attorney General and Supreme Court Justice in New Hampshire, who had been appointed a federal appeals judge just before his elevation to the highest Court. The most striking thing about the Souter pre-Court record is its skeletonlike nature. For a man who had held eminent state positions, including seven years on his state's highest court, Justice Souter left practically no "paper trail." He was not informative about his views in professional writings or speeches and wrote no significant opinions. This record was continued during his early Court tenure; by June 1991, near the end of his first term, Justice Souter had written only three opinions and no dissents—the lowest number of any Justice appointed in the previous two decades. Still, he has been called a "polite and professional" Justice[31] whose questions have played a prominent part in oral argument—though he has been far less abrasive than Justice Scalia, the

Rehnquist Court's most active questioner. Recently, indeed, Justice Souter has taken a more active part in the Court's decision process; it was he who was most responsible for the eloquent 1992 opinion refusing to overrule *Roe v. Wade*.[32]

To replace Justice Marshall, the first black Justice, President Bush selected Clarence Thomas, another black, who had recently been appointed to the D.C. Circuit Court. Thomas was only forty-three when appointed—the youngest Justice since Justice Stewart had gone on the Court at the same age. Thomas also had less experience in the law than any Justice appointed in this century; his pre-Court career had been almost entirely in the field of civil-rights enforcement, primarily as Chairman of the Equal Employment Opportunity Commission—which, however important, is scarcely the type of legal work that provides the legal background that one desires in a Supreme Court appointee.

The Thomas confirmation was marred by charges of sexual harassment, which led to controversial televised hearings on the matter. More important was the Justice's refusal to give specific answers to questions on his jurisprudential posture, even when not framed in terms of specific cases. On the Court, however, Justice Thomas has already indicated that he is a doctrinaire conservative in the Scalia mold; he has, in fact, voted more with Justice Scalia than any other member of the Court.

Junior Supreme Court

Before we discuss the emerging jurisprudence of the Rehnquist Court, a word should be said about an important development during the present century that is changing the very nature of the Supreme Court—the expanding role of the law clerks.

In a congratulatory letter to Justice Rehnquist upon his court appointment, Justice Douglas wrote, "I realize that you were here before as a member of the so-called Junior Supreme Court."[33] Douglas was referring to Rehnquist's service as a law clerk to Justice Jackson. Once upon a time, the Douglas characterization of the clerk corps might have been taken as one made wholly in jest, but that was no longer the case.

Over half a century ago, Justice Brandeis stated, "The reason the public thinks so much of the Justices of the Supreme Court is that they are almost the only people in Washington who do their own work."[34] The legend that this remains true is still prevalent, and in his book on the Court, even Chief Justice Rehnquist tells us that "the individual justices still continue to do a great deal more of their 'own work' than do their counterparts in other branches of the federal government."[35]

The Rehnquist-type account has been accepted by both the press and the public. "Alone among Government agencies," Anthony Lewis wrote in the *New York Times*, "the court seems to have escaped Parkinson's Law. The work is still done by nine men, assisted by eighteen young law clerks. Nothing is delegated to committees or ghostwriters or task forces."[36] We saw in Chapter 9 how the Justices were provided with funds to pay secretaries or clerks in 1886, with provision for law clerks rather than stenographers in 1919. At that time, the law clerk would perform only the functions of an associate in a law firm, that is, research for senior members and assistance generally in the firm's work. It may be doubted that Justices such as Holmes or Brandeis used their clerks as more than research assistants. In fact, as we saw, Justice Hughes worried at the time that if the clerks were used too much, "it might be thought that they were writing our opinions."[37] That, indeed, is what happened. In recent years the Justices have given their clerks an ever-larger share of responsibility, including even the writing of opinions.

Complaints against the clerks' role have been common, including a noted 1957 article in *U.S. News & World Report* by William H. Rehnquist himself.[38] Rehnquist stated that the Justices were delegating substantial responsibility to their clerks, who "unconsciously" slanted materials to accord with their own views. The result was that the liberal point of view of the vast majority of the clerks had become the philosophy espoused by the Warren Court.

The situation has, if anything, gotten worse in recent years. "In the United States," notes a 1986 *London Times* article, "judges have 'clerks', i.e., assistants who prepare and frequently write judgments which their masters often merely adopt and which a qualified observer can easily recognize as the work of a beginner."[39]

An even harsher view of the clerk system was expressed by Professor Philip B. Kurland, a leading constitutional scholar, a year after Chief Justice Rehnquist was appointed. As he notes, the law clerks now exercise a major role in the two most important functions of the Justices: (1) the screening of cases to determine which the Court will hear and decide; and (2) the drafting of opinions. "I think Brandeis would be aghast."[40]

In a public lecture, Justice Stevens conceded that he did not read 80 percent of the certiorari petitions presented to the Court.[41] Instead his clerks prepare memoranda summarizing those cases and issues and recommending whether or not certiorari should be granted. The Justice reads only those where the granting of certiorari is recommended. The only member of the Burger and Rehnquist Courts who personally went over petitions for review was Justice Brennan, who customarily shared the work with his law clerks. In a letter to Brennan, who was temporarily away from the Court, his clerks stated, "We are all fascinated by the certs and shudder to think that when you get back you may take some of them away from us. But if you're very nice we won't fight too hard."[42]

In the 1972 Term, Justice Powell urged that the Justices combine their efforts in the screening process by having their clerks work together in one "cert pool."[43] The petitions would be divided equally among all the clerks in the pool, and the cert memos prepared by them would be circulated to each of the Justices participating. The Chief Justice and Justices White, Blackmun, Powell, and Rehnquist agreed to join in the cert pool. Justices

Douglas, Brennan, Stewart, and Marshall declined to participate. In the present Court, only Justice Stevens is not a member of the pool.

While the Justices make the final decision on what certiorari petitions to grant, *the* work on the petitions is done by the law clerks. In the vast majority of cases, the Justices' knowledge of the petitions and the issues they present is based on the clerks' cert memos, and they normally follow the recommendations in their memos. Sheer volume, if nothing else, has made this the prevailing practice.

The Justices themselves have expressed qualms about this delegation of the screening task. In declining to join the cert pool, Justice Douglas wrote to the Chief Justice: "The law clerks are fine. Most of them are sharp and able. But after all, they have never been confirmed by the Senate."[44]

An even more important delegation to the clerks involves the opinion-writing process. "As the years passed," says Justice Douglas in his *Autobiography*, "it became more evident that the law clerks were drafting opinions."[45] Almost all the Justices have made more extensive use of their clerks in the drafting process than outside observers have realized. In recent Courts, indeed, the routine procedure has been for the clerks to draft virtually all opinions.

Chief Justice Rehnquist has candidly described the opinion-writing process. "In my case," Rehnquist said, "the clerks do the first draft of almost all cases to which I have been assigned to write the Court's opinion." Only "when the case-load is heavy" does Rehnquist sometimes "help by doing the first draft of a case myself."[46] Rehnquist concedes that the "practice . . . may undoubtedly . . . cause raised eyebrows." Still, the Chief Justice asserts, "I think the practice is entirely proper: The Justice must retain for himself control not merely of the outcome of the case, but of the explanation of the outcome, and I do not believe this practice sacrifices either."[47]

It is, of course, true that the decisions are made by the Justices—though, even with regard to them, the weaker Justices have abdicated much of their authority to their clerks. In most chambers, the clerks are, to use a favorite expression of Chief Justice Warren, not "unguided missiles." The Justices normally outline the way they want opinions drafted. But the drafting clerk is left with a great deal of discretion. The Justices may "convey the broad outlines," but they "do not invariably settle exactly how the opinion will be reasoned through."[48] The details of the opinions are left to the clerk, in particular the specific reasoning and research supporting the decision. The technical minutiae and footnotes, so dear to the law professor, are left almost completely to the clerks. Thus footnote 11 of the *Brown* school segregation opinion[49]—perhaps the most famous footnote in Supreme Court history—was entirely the product of a Warren law clerk.

To be sure, the Justices themselves go over the drafts, and, said Chief Justice Rehnquist, "I may revise it in toto." But, he also admits, "I may leave it relatively unchanged."[50] Too many of the Justices circulate drafts that are almost wholly the work of their clerks.

The growing number of law clerks has naturally led to an increase in the length, though plainly not the quality, of opinions. What Justice Douglas once wrote about Court opinions has become increasingly true: "We have tended more and more to write a law-review-type of opinion. They plague the Bar and the Bench. They are so long they are meaningless. They are filled with trivia and nonessentials."[51]

Law clerks have a similar academic background and little other experience. For three years they have had drummed into them that the acme of literary style is the law review article. It is scarcely surprising that the standard opinion style has become that of the student-run reviews: colorless, prolix, platitudinous, always error on the side of inclusion, full of lengthy citations and footnotes—and above all dull.[52]

The individual flair that makes the opinions of a Holmes or a Cardozo literary as well as legal gems has become a thing of the past. There is all the difference in the world between writing one's own opinions and reviewing opinions written by someone else. It is hard to see how an editor can be a great judge. Can we really visualize a Holmes coordinating a team of law clerks and editing their drafts?[53]

According to a federal appellate judge, "We need to reduce our dependence on the system of judicial apprenticeships and on a mass production model that will soon swallow us up."[54] In the Supreme Court, as in most institutions, the balance of power has shifted increasingly to the bureaucrats and away from the nominal heads. The Justices have become the managers of a growing corps of law clerks, who increasingly write the opinions even in the most important cases. The swelling system of judicial apprenticeships threatens to repeat the story of the *Sorcerer's Apprentice*.

Rehnquist Jurisprudence

The outstanding point to bear in mind about the Rehnquist Court is that it is a Court that has reflected the general tilt toward the right that characterized American politics before the 1992 election. More than that, under the leadership of the conservative activist who now sits in the center chair, it has begun to shape a new constitutional case law that has already undone some of the work of its predecessors.

The decisions rendered by the Rehnquist Court through 1992 enable us to note a definite change in direction in its jurisprudence. The change has been manifested in the Court's decisions on civil rights and criminal law. In 1989, the Court struck down a Richmond affirmative action plan under which prime contractors awarded city contracts were required to subcontract at least 30 percent of each contract to minority business enterprises. Similar plans had been upheld by the Burger Court, but its successor ruled that, in the absence of proof of intentional discrimination by the city, the Richmond plan might not be upheld. The argument that the city was attempting to remedy societal discrimination, as shown by the disparity between contracts awarded in the past to minority businesses and the city's minority population, was rejected.[55]

Other decisions shifted the burden of proof in civil rights cases, holding that plaintiffs, not employers, had the burden of proving that a job requirement that was shown statistically to screen out minorities was not a "business necessity" and permitted employers to show by only a preponderance of the evidence rather than by clear and convincing evidence (a higher burden of proof) that their refusal to hire someone was based on legitimate and not discriminatory reasons.[56]

The Rehnquist Court also refused to invalidate a death sentence imposed upon a black defendant despite a detailed statistical study which showed that black defendants who killed white victims were far more likely to receive the death penalty than white defendants. The Court stressed that there was no proof "that the decisionmakers in *his* case acted with discriminatory purpose."[57]

Perhaps even more indicative of the changing emphasis of the Rehnquist Court are its decisions which may mark the beginning of a trend in favor of property rights. For the first time in years, the Court relied upon the constitutional prohibition against takings of property without compensation to invalidate governmental action that did not involve public acquisition of property.[58] Noteworthy in those cases was the Court's use of heightened scrutiny to review the merits of land-use regulations in order to decide whether a challenged regulation required judicial invalidation in the absence of compensation. Indeed, the Court implied that claims of unconstitutional takings (whether by acquisition or regulation) now fall into a particularly sensitive constitutional category comparable to that in which freedom of speech claims fall.

Its decisions on takings without compensation may signal a fundamental shift in Bill-of-Rights jurisprudence, with a tilt by the Rehnquist Court in favor of protection of property rights and away from the strong preference given to personal rights by the Warren and Burger Courts. Yet, though there has been a definite tilt to the right since Chief Justice Rehnquist was appointed, the change until now has not led to the overruling of any of the important decisions of the Warren and Burger Courts.

Most important in this respect was the Rehnquist Court's refusal to overrule *Roe v. Wade*—the Burger Court's landmark decision that the constitutionally protected right of privacy includes a woman's decision to have an abortion. Few Supreme Court decisions have been as controversial as that in *Roe v. Wade;* and the Rehnquist Court was strongly urged by the Bush Administration to overrule that decision. The Court, however, expressly declined to do so, though it did narrow its ruling to a limited extent.[59]

The significant Warren Court criminal procedure decisions have also thus far remained a part of the Rehnquist Court jurisprudence. The key Warren Court criminal trilogy—*Gideon, Mapp,* and *Miranda*[60]—are still followed by the Court, though some of their doctrines have been narrowed under Rehnquist's lead. When the Court struck down the legislative apportionment for New York City's Board of Estimate, it did so on the basis of the one-person, one-vote principle laid down in one of the Warren Court's most important decisions.[61]

It would be erroneous to assume, however, that the Rehnquist Court will follow the Burger Court in consolidating and continuing the essential Warren Court heritage. Chief Justice Rehnquist, as indicated, is a conservative activist who gives every indication of being a strong Chief Justice. Like Chief Justice Warren, he proceeded cautiously in his early years as head of the Court. After all, the Supreme Court is not normally the place for innovation. If the Warren Court was an exception to this, that was true primarily in Warren's later years after personnel changes gave him a strong majority willing to follow his lead in remolding so much of the corpus of our constitutional law.

The conservative combination in the Rehnquist Court's early years was a fragile one. Though usually enough to give the Chief Justice a scant majority, it could not be pushed too fast or too far. When the Chief Justice tried too hard to correct what he termed the "excesses" of the Warren Court, his coalition often would splinter and Rehnquist would wind up in disssent.

This has, however, changed with the appointments of Justices Scalia, Kennedy, Souter, and Thomas. They have reinforced the Court's conservative bloc—a change that has given Chief Justice Rehnquist the majority that may enable him to translate his constitutional vision into accepted jurisprudence. It may, indeed, be true that, as the *New York Times* stated in 1991, "[T]he Supreme Court is no longer in transition. It has become the Court it will most likely be for the next generation."[62] The 1991 and 1992 cases indicate that the conservative wing was substantially strengthened by the replacement of Justices Brennan and Marshall by Justices Souter and Thomas.

Particularly significant in this respect were 1991 and 1992 decisions limiting the use of habeas corpus by prisoners,[63] broadening the power of the police to search automobiles,[64] applying the harmless-error doctrine to a constitutional error committed at the trial,[65] and upholding regulations prohibiting abortion counseling, referrals, or advocacy by federally funded clinics.[66] The latter decision, one of the most controversial by the Rehnquist Court, ruled for the first time that the government spending power might be used to restrict First Amendment rights.

In addition, the Court has ruled that the First Amendment's Free Exercise Clause no longer requires that action that burdens a religious practice must be justified by a compelling governmental interest.[67] Burdens on religious practices are now not to be subject to the strict scrutiny that has governed judicial review in other First Amendment cases.

If Chief Justice Rehnquist continues to have a majority that will translate his conservative views into Supreme Court jurisprudence, the history of the Rehnquist Court may turn out to be the reverse image of the Warren Court. The law, like other institutions, has its epochs of ebb and flow.[68] In the face of the probable Rehnquist flood tide, critics of the Burger Court may come to look back upon its receding period with more than a little nostalgia.

A Moderate Core?

It would, nonetheless, be a mistake to think of the Rehnquist Court as a monolithic tribunal inexorably following the Chief Justice in the conservative cast of his jurisprudence. It should, in the first place, be noted that even the most conservative judge may resist governmental intrusions upon the right of privacy—what Justice Brandeis once called "the right to be let alone—the . . . right most valued by civilized men."[69] The Senate hearings on the nominations of Judge Robert Bork and Justice Clarence Thomas to the Supreme Court revealed a remarkable consensus among Americans that there is a constitutionally protected right of privacy, even though it is not enumerated in the rights safeguarded in the Bill of Rights. Justice Antonin Scalia, the most conservative member of the Rehnquist Court, has given indications (for example, in his dissent from a 1989 decision upholding a drug-testing program for customs employees)[70] that he still places the right to privacy in a preferred position.

It was also Justice Scalia and Justice Kennedy (both conservative Reagan appointees), who cast the decisive votes in the 1989 and 1990 decisions ruling that flag burning as a political protest was a protected form of individual expression guaranteed by the Constitution, as well as the 1992 decision striking down a law criminalizing "hate speech" as violative of the First Amendment.[71] It should not be forgotten that conservative thought encompasses a libertarian strain that resists intrusions upon the area of what William Faulkner termed "individual privacy lacking which [one] cannot be an individual and lacking which individuality [one] is not anything at all worth the having or keeping."[72]

More important perhaps is the fact that, during 1992, the conservative Justices in the Rehnquist Court itself divided into two blocs—one composed of the Chief Justice and Justices Scalia and Thomas and the other made up of Justices O'Connor, Kennedy, and Souter. The Justices in the first tend to follow faithfully the Rehnquist agenda of correcting the "excesses" of the Warren and Burger Courts, ultimately leading to overruling of the key decisions of those Courts—particularly that in *Roe v. Wade* and those protecting criminal defendants.

The other group of Justices is more moderate and seems to have taken as their model the second Justice Harlan, now usually seen as the very paradigm of the true conservative judge. Harlan had been the leading conservative in the last years of the Warren Court and he dissented from some of its most important decisions. Harlan's conservative philosophy did not, however, permit him to go along with those who urged the cavalier overruling of those decisions. "Respect for the Courts," Harlan once wrote to another Justice, "is not something that can be achieved by fiat."[73] He applied this principle to refusals to follow prior decisions with which he disagreed. The true conservative, Harlan believed, adhered to stare decisis, normally following even precedents against which he had originally voted.

The Harlan posture in this respect can be best seen in the 1970 case of *Coleman v. Alabama*.[74] At issue was the right to counsel at a preliminary hearing, where the defendants were bound over to the grand jury. At the conference there were seven votes to affirm the conviction, with only Harlan voting the other way. The Justice had dissented from the Warren Court's landmark *Miranda* decision.[75] Despite this, Harlan said at the *Coleman* conference that *Miranda* was "still on the books" and it should be followed here since the preliminary examination was as critical a stage as the custodial interrogation involved in *Miranda*. Ultimately, the Court agreed with Justice Harlan, holding that the conviction had to be reversed because the defendants had not been assigned counsel.

Coleman shows better than anything the Harlan conception of a conservative judge. Such a judge is not to use "judicial fiat" to disregard a precedent any more than he is to establish the precedent by fiat in the first place. While the precedent is "still on the books," it is to be followed in cases to which it logically applies.

This is precisely the approach followed by Justices O'Connor, Kennedy, and Souter in their 1992 refusal to follow the Rehnquist bloc in voting to overrule *Roe v. Wade*.[76] The three Justices wrote a joint opinion that might have been written by Justice Harlan himself. According to them, the rule of law itself requires a respect for precedent. As they see it, the only time stare decisis should not be followed is when the precedent's rule has been found unworkable or the facts have so changed "as to have robbed the old rule of significant application or justification."[77] Only two cases are said to have met these criteria during the past century: *Lochner v. New York*[78] and *Plessy v. Ferguson*.[79] The same was not true of *Roe v. Wade;* it meets none of the criteria and a decision to overrule it would be based only on "a present doctrinal disposition to come out differently from the Court of 1973."[80] To the more moderate conservative, that is not enough to justify failure to follow stare decisis: "[A] decision to overrule should rest on some special reason over and above the belief that a prior case was wrongly decided."[81] Justice Harlan could not have said it better than this quote from the 1992 opinion refusing to overrule *Roe v. Wade*.

It should, however, be recognized that the centrist bloc in the Rehnquist Court is still composed of Justices who are more conservative than their predecessors. The joint opinion may have been unwilling to overrule *Roe,* but it did interpret it more narrowly; indeed, according to the Chief Justice, under its interpretation "*Roe v. Wade* stands as a sort of judicial Potemkin Village . . . a mere facade to give the illusion of reality."[82] Justices O'Connor, Kennedy, and Souter have joined most of the Rehnquist Court decisions that have meant a change in direction in our constitutional law—particularly those involving civil rights, property, and criminal law. If the three Justices have been classified as moderates by Court commentators,[83] that is true only in comparison with the tendency to reweigh first principles of Chief Justice Rehnquist and his adherents.

In his separate opinon in the 1992 abortion case, Justice Blackmun concluded with a poignant observation: "I am 83 years old. I cannot remain on this Court forever, and when I do step down . . . the choice between the two worlds will be made."[84] The Justice was referring to the possible overruling of *Roe v. Wade*. But what he said is also true from a broader perspective. In this sense, the Supreme Court is always at a turning point. The direction it will take—the world it will choose—will depend most of all upon the new Justices who will assume their seats upon the bench. That will be as true of the Rehnquist Court and its successors as it has been of the prior Courts discussed in this volume.

Epilogue

Not long ago, the attitude of Americans to their constitutional system was that described by Burke: "We ought to understand it according to our measure; and to venerate where we are not able presently to comprehend."[1] As the nation begins its third century, Burke's attitude appears as quaint as the costume of his time. During the present century, veneration has too often given way to vituperation, as we have begun to doubt much that had always been taken for granted in the polity. At a minimum, we no longer assume that, with the contemporary constitutional system, the ultimate stage of organic evolution has been reached. We know that the system is continuing to evolve beyond the "perfection" that Americans at the turn of the century assumed it had attained.

American constitutional law in operation has directly reflected the needs of the nation. At the outset, the primary needs of establishing national power on a firm basis and vindicating property rights against excesses of state power were met in the now-classic decisions of the Marshall Court. A generation later, the needs of the society had changed. If the Taney Court was to translate the doctrines of Jacksonian Democracy, and particularly its emphasis on society's rights, into constitutional law, that was true because such doctrines were deemed necessary to the proper development of the polity. In addition, they furthered the growth of corporate enterprise and prevented its restriction by the deadening hand of established monopoly.

If in the latter part of the nineteenth century the Court was to elevate

378

the rights of property to the plane of constitutional immunity, its due process decisions were the necessary legal accompaniment of the industrial conquest of a continent. The excesses of a laissez-faire–stimulated industrialism should not lead us to overlook the vital part it played in American development. Nor should it be forgotten that the decisions exalting property rights may have been a necessary accompaniment of the post–Civil War economic expansion.

The picture has been completely altered during the present century. The Court came to recognize that property rights must be restricted to an extent never before permitted in American law. At the same time, unless the rights of the person are correlatively expanded, the individual will virtually be shorn of constitutional protection—hence the Court's shift in emphasis to the protection of personal rights. The Justices, like the rest of us, were disturbed by the growth of governmental authority and sought to preserve a sphere for individuality even in a society in which the individual stands dwarfed by power concentrations.

One must, however, concede that, despite the Court's efforts, the concentration of governmental power has continued unabated. The second half of the century has, if anything, seen an acceleration in the growth of such power. Indeed, the outstanding feature of the late twentieth century is the power concentrations that increasingly confront the individual. Even a more conservative Court may find it necessary to preserve a sphere for individuality in such a society.

Yeats tells us that "[a]ll states depend for their health upon a right balance between the One, the Few and the Many."[2] The maintenance of that balance is peculiarly the task of the Supreme Court since, following the famous Hughes aphorism, the Constitution is essentially what the judges say it is.[3] It is their unique function to serve as guardians of the organic ark. To enable them to do so effectively, they are armed with the awesome authority to nullify any governmental act deemed by them in conflict with any provisions of the basic document.

The historian who looks at the Supreme Court is struck with the generally successful way in which it has exercised this awesome authority. The Court's jurisprudence has illustrated the antinomy inherent in every system of law: the law must be stable and yet it cannot stand still.[4] The essential outlines of the constitutional system are still those laid down at the beginning in 1787; there is here a continuity in governmental structure that is all but unique in an ever-changing world. But the system still proves workable only because it has been continually reshaped to meet two centuries' changing needs.

There have been aberrations, but in the main the Supreme Court in operation has reflected the history of the nation: the main thrust has been to meet the "felt necessities"[5] of each period in the nation's history.

Now it is for the Rehnquist Court and its successors to construe the Constitution during the next stage of American development. Regardless of the Court's tilt, history gives confidence that, by and large at least, it will do the job so as to remain true to John Marshall's polestar—that we must never forget "it is a *constitution* we are expounding,"[6] a living instrument that must be construed to meet the practical necessities of the contemporary society.

Appendix: The Justices of the Supreme Court

Appointed by President Washington

JAY, JOHN*	1789†–1795	Resigned
Rutledge, John	1789–1791	Resigned
Cushing, William	1789–1810	Died
Wilson, James	1789–1798	Died
Blair, John	1789–1796	Resigned
Iredell, James	1790–1799	Died
Johnson, Thomas	1791–1793	Resigned
Paterson, William	1793–1806	Died
RUTLEDGE, JOHN	1795	Recess appointment; not confirmed
Chase, Samuel	1796–1811	Died
ELLSWORTH, OLIVER	1796–1800	Resigned

Appointed by President John Adams

Washington, Bushrod	1798–1829	Died
Moore, Alfred	1799–1804	Resigned
MARSHALL, JOHN	1801–1835	Died

---

*Block letters designate Chief Justices.

† Date of appointment.

Appointed by President Jefferson

Johnson, William	1804–1834	Died
Livingston, Brockholst	1806–1823	Died
Todd, Thomas	1807–1826	Died

Appointed by President Madison

Duval, Gabriel	1811–1835	Resigned
Story, Joseph	1811–1845	Died

Appointed by President Monroe

Thompson, Smith	1823–1843	Died

Appointed by President John Quincy Adams

Trimble, Robert	1826–1828	Died

Appointed by President Jackson

McLean, John	1829–1861	Died
Baldwin, Henry	1830–1844	Died
Wayne, James M.	1835–1867	Died
TANEY, ROGER B.	1835–1864	Died
Barbour, Philip P.	1835–1841	Died

Appointed by President Van Buren

Catron, John	1837–1865	Died
McKinley, John	1837–1852	Died
Daniel, Peter V.	1841–1860	Died

Appointed by President Tyler

Nelson, Samuel	1845–1872	Resigned

Appointed by President Polk

Woodbury, Levi	1845–1851	Died
Grier, Robert C.	1846–1870	Resigned

Appointed by President Fillmore

Curtis, Benjamin R.	1851–1857	Resigned

Appointed by President Pierce

Campbell, John A.	1853–1861	Resigned

Appointed by President Buchanan

Clifford, Nathan	1858–1881	Died

Appointed by President Lincoln

Swayne, Noah H.	1862–1881	Resigned
Miller, Samuel F.	1862–1890	Died
Davis, David	1862–1877	Resigned
Field, Stephen J.	1863–1897	Resigned
CHASE, SALMON P.	1864–1873	Died

Appointed by President Grant

Strong, William	1870–1880	Resigned
Bradley, Joseph P.	1870–1892	Died
Hunt, Ward	1872–1882	Resigned
WAITE, MORRISON R.	1874–1888	Died

Appointed by President Hayes

Harlan, John Marshall	1877–1911	Died
Woods, William B.	1880–1887	Died

Appointed by President Garfield

Matthews, Stanley	1881–1889	Died

Appointed by President Arthur

Gray, Horace	1881–1902	Died
Blatchford, Samuel	1882–1893	Died

Appointed by President Cleveland

Lamar, Lucius Q. C.	1887–1893	Died
FULLER, MELVILLE W.	1888–1910	Died

Appointed by President Harrison

Brewer, David J.	1889–1910	Died
Brown, Henry B.	1890–1906	Resigned
Shiras, George, Jr.	1892–1903	Resigned
Jackson, Howell E.	1893–1895	Died

Appointed by President Cleveland

White, Edward D.	1894–1910	Appointed Chief Justice
Peckham, Rufus W.	1895–1909	Died

Appointed by President McKinley

McKenna, Joseph	1897–1925	Resigned

Appointed by President Theodore Roosevelt

Holmes, Oliver Wendell	1902–1932	Resigned
Day, William R.	1903–1922	Resigned
Moody, William H.	1906–1910	Resigned

Appointed by President Taft

Lurton, Horace H.	1909–1914	Died
Hughes, Charles E.	1910–1916	Resigned
WHITE, EDWARD D.	1910–1921	Died
Van Devanter, Willis	1910–1937	Retired
Lamar, Joseph R.	1910–1916	Died
Pitney, Mahlon	1912–1922	Retired

Appointed by President Wilson

McReynolds, James C.	1914–1941	Retired
Brandeis, Louis D.	1916–1939	Retired
Clarke, John H.	1916–1922	Resigned

Appointed by President Harding

TAFT, WILLIAM H.	1921–1930	Resigned
Sutherland, George	1922–1938	Retired
Butler, Pierce	1922–1939	Died
Sanford, Edward T.	1923–1930	Died

Appointed by President Coolidge

Stone, Harlan F.	1925–1941	Appointed Chief Justice

Appointed by President Hoover

HUGHES, CHARLES E.	1930–1941	Retired
Roberts, Owen J.	1930–1945	Resigned
Cardozo, Benjamin N.	1932–1938	Died

Appointed by President Franklin D. Roosevelt

Black, Hugo, L.	1937–1971	Retired
Reed, Stanley F.	1938–1957	Retired
Frankfurter, Felix	1939–1962	Retired
Douglas, William O.	1939–1975	Retired
Murphy, Frank	1940–1949	Died
Byrnes, James F.	1941–1942	Resigned
STONE, HARLAN F.	1941–1946	Died

Jackson, Robert H.	1941–1954	Died
Rutledge, Wiley B.	1943–1949	Died

Appointed by President Truman

Burton, Harold H.	1945–1958	Retired
VINSON, FRED M.	1946–1953	Died
Clark, Tom C.	1949–1967	Retired
Minton, Sherman	1949–1956	Retired

Appointed by President Eisenhower

WARREN, EARL	1953–1969	Retired
Harlan, John Marshall	1955–1971	Retired
Brennan, William J., Jr.	1956–1990	Retired
Whittaker, Charles E.	1957–1962	Retired
Stewart, Potter	1958–1981	Retired

Appointed by President Kennedy

White, Byron R.	1962–1993	Retired
Goldberg, Arthur J.	1962–1965	Resigned

Appointed by President Lyndon B. Johnson

Fortas, Abe	1965–1969	Resigned
Marshall, Thurgood	1967–1991	Retired

Appointed by President Nixon

BURGER, WARREN E.	1969–1986	Retired
Blackmun, Harry A.	1970–	
Powell, Lewis F., Jr.	1971–1987	Retired
Rehnquist, William H.	1971–1986	Appointed Chief Justice

Appointed by President Ford

Stevens, John Paul	1975–

Appointed by President Reagan

O'Connor, Sandra Day	1981–
REHNQUIST, WILLIAM H.	1986–
Scalia, Antonin E.	1986–
Kennedy, Anthony M.	1987–

Appointed by President Bush

Souter, David H.	1990–
Thomas, Clarence	1991–

Notes

Introduction

1. *Bartlett's Familiar Quotations* 720 (15th ed. 1980).
2. *Marbury v. Madison,* 1 Cranch 137 (U.S. 1803).
3. Coke himself gives the date as November 10. *Prohibitions del Roy,* 12 Co. Rep. 63 (1608). The correct date, however, seems to have been November 14. See 5 Holdsworth, *A History of English Law* 430 (2d ed. 1937).
4. Holdsworth, op. cit. supra note 3, at 428.
5. *Prohibitions del Roy,* 12 Co. Rep. at 65.
6. Holdsworth, op. cit. supra note 3, at 431.
7. Ibid.
8. 8 Co. Rep. 113b (1610).
9. Id. at 118a.
10. Bowen, *The Lion and the Throne: The Life and Times of Sir Edward Coke* 514 (1956).
11. Schwartz, *The Law in America: A History* 14 (1974).
12. Bowen, op. cit. supra note 10, at 514.
13. Id. at 291.
14. 10 *Works of John Admas* 244–245 (C. F. Adams ed. 1856).
15. Quincy Reports 51 (Mass. 1761). The best account of the case is in 2 *Legal Papers of John Adams* 123 et seq. (Wroth and Zobel eds. 1965).
16. Id. at 141.
17. Qiuncy Reports, Appendix 520–521 (Mass.).
18. Adams, op. cit. supra note 15, at 127–128.
19. Bowen, *John Adams and the American Revolution* 217 (1950).
20. Loc. cit. supra note 17.

21. Bowen, op. cit. supra note 10, at 316.

22. Id. at 520.

23. 2 Farrand, *The Records of the Federal Convention of 1787,* 73 (1937).

24. Wood, *The Creation of the American Republic* 538 (1969).

25. The case was unreported and the best account of it is in Scott, *Holmes vs. Walton: The New Jersey Precedent: A Chapter in the History of Judicial Power and Unconstitutional Legislation,* 4 *American Historical Review* 456 (1899).

26. Id. at 459–460; Goebel, *History of the Supreme Court of the United States: Antecedents and Beginnings to 1801,* 124 (1971).

27. Scott, op. cit. supra note 25, at 459, 464.

28. *State v. Parkhurst,* 9 N.J.L. 427, 444 (1802).

29. 4 Call 5 (Va. 1782).

30. Id. at 20.

31. Id. at 8. Pendleton's notes were sketchy on Wythe's opinion, but they confirm the essentials of the Call account. 2 *The Letters and Papers of Edmund Pendleton* 426 (Mays ed. 1961). See Goebel, op. cit. supra note 26, at 127–128.

32. Pendleton, op. cit. supra note 31, at 422. See 2 Mays, *Edmund Pendleton 1721–1803: A Biography* 200 (1952).

33. Edmund Pendleton to James Madison, Nov. 8, 1782. Mays, op. cit. supra note 32, at 201.

34. Id. at 202.

35. So characterized in Goebel, op. cit. supra note 26, at 131.

36. The best account is in 1 *The Law Practice of Alexander Hamilton* 282–419 (Goebel ed. 1964).

37. Id. at 305.

38. Morris, *Witnesses at the Creation* 45 (1985).

39. Varnum, *The Case, Trevett against Weeden* (Providence 1787).

40. So characterized in Wood, op. cit. supra note 24, at 460.

41. Quoted ibid.

42. *Newport Mercury,* Oct. 6, 1786.

43. 1 N.C. 42 (1787).

44. Id. at 45.

45. 2 McRee, *Life and Correspondence of James Iredell* 169, 172–173 (1858).

46. Op. cit. supra note 36, at 312.

47. Id. at 314.

48. Goebel, op. cit. supra note 26, at 140.

49. Farrand, op. cit. supra note 23, at 28. Madison's account was not completely acccurate. See Goebel, op. cit. supra note 26, at 141.

50. 3 Farrand, op. cit. supra note 23, at 13.

51. *Penhallow v. Doane,* 3 Dall. 54, 80 (U.S. 1795).

52. 1 Farrand, op. cit. supra note 23, at 34.

53. Id. at 30.

54. Id. at 21.

55. Farrand, *The Framing of the Constitution* 154 (1913).

56. Id. at 80.

57. 1 Farrand, op. cit. supra note 23, at 21; 2 id at 298.

58. 1 id. at 97.

59. 2 id. at 73.

60. Infra note 65.

61. Farrand, op. cit. supra note 23, at 78.

62. Id. at 76. See also 1 id. at 109 (Rufus King); 2 id. at 28 (Gouverneur Morris); id. at 27 (Roger Sherman).

63. Id. at 299.

64. Wood, op. cit. supra note 24, at 304.

65. 2 Farrand, op. cit. supra note 23, at 93.

66. Goebel, op. cit. supra note 26, at 388.

67. Id. at 310.

68. 4 *Documentary History of the Supreme Court of the United States, 1789–1800,* 11 (1992).

69. Goebel, op. cit. supra note 26, at 311.

70. *The Federalist,* No. 78.

71. Ibid.

72. 1 Cranch 137 (U.S. 1803).

73. Op. cit. supra note 68, at 427.

74. Id. at 474.

75. Frankfurter and Landis, *The Business of the Supreme Court* 189 (1927).

76. 1 Warren, *The Supreme Court in United States History* 449 (1924).

77. Op. cit. supra note 68, at 417.

Chapter 1

1. Caplan, *The Tenth Justice* 162 (1987).

2. 1 Warren, *The Supreme Court in United States History* 48 (1924) [hereinafter cited as Warren].

3. 3 *Dictionary of American Biography* 635 (1930).

4. 1 Warren 48.

5. No. 78.

6. *Felix Frankfurter on the Supreme Court* 472 (Kurland ed. 1970).

7. 1 *The Documentary History of the Supreme Court of the United States, 1789–1800,* 688 (1985) [hereinafter cited as *Documentary*].

8. 10 *The Writings of George Washington* 34–36 (Sparks ed. 1847).

9. 1 *Documentary* 619.

10. Id. at 661.

11. Id. at 9.

12. Id. at 712.

13. Id. at 700–701.

14. Id. at 706.

15. Id. at 692.

16. Ibid.

17. Id. at 700.

18. Id. at 731.

19. 3 id. at 240.

20. 2 id. at 132.

21. 1 id. at 732.

22. 2 id. at 126.

23. Id. at 344.

24. Id. at 288.

25. Id. at 289–290.

26. Id. at 290.

27. Ibid.

28. Goebel, *History of the Supreme Court of the United States: Antecedents and Beginnings to 1801,* 567 (1971).

29. 2 *Documentary* 345.

30. 1 id. at 875.

31. 2 Dall. 402 (U.S. 1792).

32. Conway, *Omitted Chapters of History Disclosed in the Life and Papers of Edmund Randolph* 168 (1889).

33. 2 Dall. 419 (U.S. 1793).

34. Id. at 429.

35. Id. at 456.

36. Id. at 453, 457.

37. Id. at 466.

38. 1 Warren 496.

39. 1 Cranch 137 (U.S. 1803).

40. See Schwartz, *The Great Rights of Mankind: A History of the American Bill of Rights* 95–100 (expanded ed. 1992).

41. Infra.

42. 3 Dall. 199 (U.S. 1796).

43. 1 *Documentary* 754.

44. 3 Dall. at 201.

45. 2 Beveridge, *The Life of John Marshall* 187 (1916).

46. 3 Dall. at 236–237.

47. 3 Dall. 386 (U.S. 1798).

48. Id. at 399.

49. Id. at 387–388.

50. Id. at 392.

51. 3 Dall. 171 (U.S. 1796).

52. Quoted in Corwin, *The Constitution of the United States of America: Analysis and Interpretation* 318 (1953).

53. 3 Dall. at 181.

54. Id. at 171.

55. Id. at 175.

56. Ibid.

57. Id. at 172.

58. Compare Bickel, *The Least Dangerous Branch* 71 (1986).

59. Jackson, *The Supreme Court in the American System of Government* 11 (1955).

60. *Muskrat v. United States,* 219 U.S. 346, 361 (1911).

61. Hughes, *The Supreme Court of the United States* 30 (1936).

62. See *United States v. Congress of Industrial Organizations,* 335 U.S. 106, 124 (1948).

63. 3 *The Correspondence and Public Papers of John Jay* 487–489 (Johnston ed. 1891).

64. Hughes, loc. cit. supra note 61.

65. *Alabama Federation of Labor v. McAdory,* 325 U.S. 450, 451 (1945).

66. *Giles v. Harris,* 189 U.S. 475, 486 (1903).

67. *United States v. Congress of Industrial Organizations,* 335 U.S. at 125.

68. 2 Dall. 409 (U.S. 1792).

69. See id. at 409–410.

70. *United States v. Ferreira,* 13 How. 40, 50 (U.S. 1851).

71. 1 Warren 71.
72. 2 Dall. at 411.
73. Id. at 411–412.
74. 1 Warren 72. See Goebel, op. cit. supra note 28, at 561.
75. *United States v. Ferreira,* 13 How. at 53.
76. 1 Warren 71.
77. Id. at 72.
78. Id. at 73.
79. Goebel, op. cit. supra note 28, at 562.
80. Conway, op. cit. supra note 32, at 145.
81. 1 *Documentary* 913–914.
82. Id. at 759.
83. Id. at 816.
84. Id. at 834.
85. Id. at 840.
86. Id. at 847, 842.
87. Id. at 842.
88. Goebel, op. cit. supra note 28, at 777, 849.
89. Id. at 778.
90. 3 Dall. 321 (U.S. 1796).
91. Id. at 327.
92. Id. at 326.
93. *Barry v. Mercein,* 5 How. 103, 119 (U.S. 1847).
94. *Daniels v. Railroad Co.,* 3 Wall. 250, 254 (U.S. 1866).
95. 1 *Documentary* 857.
96. Id. at 895.
97. Id. at 894.
98. Id. at 900.
99. 1 Warren 208–209.
100. 1 Cranch 299 (U.S. 1803).
101. Id. at 309.

Chapter 2

1. Haskins and Johnson, *History of the Supreme Court of the United States: Foundations of Power: John Marshall, 1801–1815,* 75 (1981).
2. 2 Beveridge, *The Life of John Marshall* 121 (1919).
3. Haskins and Johnson, op. cit. supra note 1, at 79.
4. *Selected Writings of Benjamin Nathan Cardozo* 179 (Hall ed. 1947).
5. 9 *Memoirs of John Quincy Adams* 243 (C. F. Adams ed. 1876).
6. 1 *The Documentary History of the Supreme Court of the United States, 1789–1800,* 147 (1985).
7. John Marshall's Autobiographical Letter, 1827, reprinted in Schwartz, *A Basic History of the U.S. Supreme Court* 102 (1968).
8. Ibid.
9. Op. cit. supra note 6, at 903.
10. Loc. cit. supra note 7.
11. Op. cit. supra note 6, at 918.
12. Ibid.
13. Id. at 920–921.

14. Id. at 925. The letter has apparently been lost.

15. Schwartz, op. cit. supra note 7, at 103.

16. Op. cit. supra note 6, at 929–930.

17. Hooker, John Marshall on the Judiciary, the Republicans, and Jefferson, March 4, 1801, 53 *American Historical Review* 518, 519 (1948).

18. 1 Warren, *The Supreme Court in United States History* 178 (1924).

19. *The Mind and Faith of Justice Holmes* 385 (Lerner ed. 1943).

20. Warren, *A History of the American Bar* 402 (1913).

21. 1 Story, *Commentaries on the Constitution of the United States* v (1833).

22. Schwartz, op. cit. supra note 7, at 99.

23. Id. at 100.

24. 1 *The Papers of John Marshall* 41 (Johnson ed. 1974).

25. 1 Beveridge, op. cit. supra note 2, at 154.

26. Id. at 159–160.

27. *Felix Frankfurter on the Supreme Court* 536 (Kurland ed. 1970).

28. Frank, *Marble Palace: The Supreme Court in American Life* 62 (1958).

29. 1 Beveridge, op. cit. supra note 2, at 119.

30. *An Autobiographical Sketch by John Marshall* 9 (Adams ed. 1973).

31. Corwin, *John Marshall and the Constitution* (1919).

32. Haskins and Johnson, op. cit. supra note 1, at 82.

33. Warren, op. cit. supra note 18, at 185.

34. Haskins and Johnson, op. cit. supra note 1, at 80.

35. Id. at 82.

36. Ibid.

37. Id. at 104–105.

38. *The Miscellaneous Writings of Joseph Story* 692 (W. W. Story ed. 1852).

39. Quoted in Warren, op. cit. supra note 20, at 421.

40. *McCulloch v. Maryland,* 4 Wheat. 316, 407 (U.S. 1819).

41. *Felix Frankfurter Reminisces* 166 (Phillips ed. 1960) (emphasis omitted).

42. Op. cit. supra note 6, at 926.

43. Beveridge, op. cit. supra note 2, at 16.

44. Ibid.

45. Haskins and Johnson, op. cit. supra note 1, at 382.

46. Morgan, *Justice William Johnson: The First Dissenter* 182 (1954).

47. Thayer, Holmes, and Frankfurter, *John Marshall* 142 (1967).

48. 1 Cranch 137 (U.S. 1803).

49. Schwartz, *The Unpublished Opinions of the Burger Court* 219 (1988).

50. Id. at 279.

51. 1 Adams, *History of the United States of America during the First Administration of Thomas Jefferson* 275 (1903).

52. Warren, op. cit. supra note 18, at 232.

53. *Dred Scott v. Sanford,* 19 How. 393 (U.S. 1857). It is true that a section of the Judiciary Act was declared unconstitutional in *Hodgson v. Bowerbank,* 5 Cranch 303 (U.S. 1800). But the opinion there was essentially unreasoned, attracted no notice at the time, and has been virtually ignored by commentators.

54. See supra p. 6.

55. See Gipson, *The Coming of the Revolution* 53–54 (1954).

56. See supra pp. 7–10.

57. 1 Cranch at 180.

58. Supra p. 23.

59. Beveridge, op. cit. supra note 2, at 118.
60. Corwin, *John Marshall and the Constitution* 70, 67 (1919).
61. 1 Cranch at 177–178.
62. 3 *Howell's State Trials* 45 (1627).
63. Holmes, *Collected Legal Papers* 295–296 (1920).
64. 6 Cranch 87 (U.S. 1810).
65. Id. at 136.
66. 1 Wheat. 304 (U.S. 1816).
67. 4 Beveridge, op. cit. supra note 2, at 164.
68. 6 Wheat. 264 (U.S. 1821).
69. Id. at 415.
70. 4 Beveridge, op. cit. supra note 2, at 343.
71. Corwin, op. cit. supra note 60, at 225.
72. 4 Wheat. 316 (U.S. 1819).
73. 8 *The Papers of Alexander Hamilton* 102 (Syrett ed. 1965).
74. 4 Wheat. at 421.
75. *John Marshall's Defense of* McCulloch v. Maryland 93, 99 (Gunther ed. 1969).
76. Warren, *The Story–Marshall Correspondence (1810–1831)* 3 (1942).
77. 4 Wheat. at 436, 433.
78. Id. at 436.
79. 9 Wheat. 1 (U.S. 1824).
80. Id. at 210–211.
81. Article I, section 8.
82. 9 Wheat. at 189–190.
83. Id. at 196.
84. White, *History of the Supreme Court of the United States: The Marshall Court and Cultural Change, 1815–35,* 578–579 (1988).
85. Douglas, *We the Judges* 192 (1956).
86. Wickard v. Filburn, 317 U.S. 111, 120 (1941).
87. 39 *Annals of Congress* 1833 (1822). There had been a similar veto by President Madison in 1817. 8 *The Writings of James Madison* 386 (Hunt ed. 1908).
88. 2 Parrington, *Main Currents in American Thought: The Romantic Revolution in America* 22 (1954).
89. See White, op. cit. supra note 84, at 597.
90. *Ogden v. Saunders,* 12 Wheat. 213, 346 (U.S. 1827).
91. 4 Beveridge, op. cit. supra note 2, at 479.
92. *Coster v. Lorillard,* 14 Wend. 265, 374–375 (N.Y. 1835).
93. See 4 Kent, *Commentaries on American Law* 3 (1830).
94. Dodd, *American Business Corporations until 1860,* 13 (1954).
95. *Barrow Steamship Co. v. Kane,* 170 U.S. 103, 106 (1898).
96. *Sutton's Hospital Case,* 10 Co. Rep. 1a, 23a, 32b (1612).
97. *Dartmouth College v. Woodward,* 4 Wheat. 518 (U.S. 1819).
98. Id. at 636.
99. Maine, *Popular Government* 248 (1886).
100. 1 *Life and Letters of Joseph Story* 331 (W. W. Story ed. 1851).
101. According to Wright, *Economic History of the United States* 388 (1941), the first frequent use of the corporation in this country came in the 1820s and 1830s—i.e., after the Dartmouth College decision.

102. White, op. cit. supra note 84, at 828.

103. 12 Wheat. 64 (U.S. 1827).

104. 2 Warren, op. cit. supra note 18, at 156–157.

105. Marshall to Bushrod Washington, July 12, 1823, Marshall Papers, Library of Congress. Marshall's circuit court decision is not reported. See 2 *The Papers of John Marshall: A Descriptive Calendar* (Rhodes ed. 1969).

106. 2 Warren, op. cit. supra note 18, at 157.

107. Warren, op. cit. supra note 76, at 20.

108. White, op. cit. supra note 84, at 791.

109. Id. at 751.

110. 2 Tocqueville, *Democracy in America* 166 (Bradley ed. 1954).

111. Compare White, op. cit. supra note 84, at 751.

112. 2 Wheat. 66 (U.S. 1817).

113. Id. at 72, 75, 76.

114. *Hopkirk v. Page,* 2 Brockenburgh 20, 41 (Cir. Ct. 1822).

115. 2 Wheat. at 74.

116. Loc. cit. supra note 114.

117. Compare White, op. cit. supra note 84, at 798.

118. *Swift v. Tyson,* 16 Pet. 1, 20 (U.S. 1842).

119. White, op. cit. supra note 84, at 813.

120. Id. at 828.

121. 10 *The Writings of Thomas Jefferson* 140 (Ford ed. 1899).

122. Beveridge, op. cit. supra note 2, at 78.

123. Supra note 75, at 1.

124. Id. at 16.

125. The quotes from the Roane and Marshall essays are taken from op. cit. supra note 75.

126. Supra p. 38.

127. Ibid.

128. Beveridge, op. cit. supra note 2, at 144.

129. Id. at 157 (italics omitted).

130. Warren, op. cit. supra note 18, at 294–295.

131. Elsmere, *Justice Samuel Chase* 225 (1980); 2 Adams, op. cit. supra note 51, at 227.

132. Beveridge, op. cit. supra note 51, at 227.

133. *N.Y. Times,* Dec. 16, 1991, at B11.

134. Warren, op. cit. supra note 18, at 250.

135. 4 Beveridge, op. cit. supra note 2, at 60.

136. *Jacobellis v. Ohio,* 378 U.S. 184, 197 (1964).

137. Holmes, *The Common Law* 1 (1881).

138. Morgan, op. cit. supra note 46, at 289.

139. Supra note 79.

140. 4 Beveridge, op. cit. supra note 2, at 443.

141. Morgan, op. cit. supra note 46, at 288.

142. Op. cit. supra note 100, at 84.

143. Quoted in Dunne, *Justice Joseph Story and the Rise of the Supreme Court* 77 (1970).

144. Quoted id. at 91.

145. Supra note 66.

146. Supra note 103.

147. Id. at 70.

148. Quoted in 2 Warren, op. cit. supra note 18, at 157.

149. Story, *Commentaries on the Constitution* § 148.

150. *Van Ness v. Pacard,* 2 Pet. 137, 144 (U.S. 1829).

151. 2 Pet. 137 (U.S. 1829).

152. Dunne, op. cit. supra note 143, at 283.

153. Fixtures, 10 *The American Jurist and Law Magazine* 53 (1833).

154. 2 Pet. at 145.

155. 2 Story, op. cit. supra note 142, at 318.

156. Morgan, op. cit. supra note 46, at 182.

157. Warren, op. cit. supra note 18, at 464.

158. Haskins and Johnson, op. cit. supra note 1, at 99.

159. 6 Fed. Cas. 546 (C.C.E.D. Pa. 1823).

160. Particularly in the *Slaughter-House Cases,* 16 Wall. 36 (U.S. 1873).

161. Haskins and Johnson, op. cit. supra note 1, at 101.

162. Morgan, op. cit. supra note 46.

163. Op. cit. supra note 27, at 540.

164. Morgan, op. cit. supra note 46, at 178, 189.

165. Id. at 181–182.

166. Id. at 183.

167. Id. at 185.

168. Id. at 190.

169. Id. at 157.

170. Supra note 72.

171. Morgan, op. cit. supra note 46, Chapter VII.

172. *Anderson v. Dunn,* 6 Wheat. 204 (U.S. 1821).

173. Morgan, op. cit. supra note 46, at 123–124.

174. Id. at 125.

175. 4 Beveridge, op. cit. supra note 2, at 359.

176. Ibid.

177. White, op. cit. supra note 84, at 327.

178. Op. cit. supra note 100, at 499.

179. See White, op. cit. supra note 84, at 316.

180. Op. cit. supra note 4, at 342–343.

181. Corwin, op. cit. supra note 60, at 124.

182. Baker, *John Marshall: A Life in Law* 353 (1974).

183. Id. at 65.

184. Marshall, letter headed "Washington, November 27, 1800," no addressee, Marshall Papers, Library of Congress.

185. Warren, op. cit. supra note 18, at 15.

186. Marshall to Samuel Chase, Jan. 23, 1804, Marshall Papers, Library of Congress.

187. 2 Beveridge, op. cit. supra note 2, at 437.

188. Compare Corwin, op. cit. supra ntoe 60, at 123–124.

189. Holmes, *The Common Law* 1.

190. Compare ibid.

191. Marshall to Richard Peters, July 21, 1815, Marshall Papers, Library of Congress.

Chapter 3

1. White, *History of the Supreme Court of the United States: The Marshall Court and Cultural Change, 1815–35,* 157 (1988).

2. 1 Warren, *The Supreme Court in United States History* 460 (1924).

3. 2 *American Law Register* 750, 706 (1954).

4. 2 Warren, op. cit. supra note 2, at 252.

5. Id. at 461.

6. Butler, *A Century at the Bar of the Supreme Court of the United States* 29 (1942).

7. Id. at 29–30.

8. Quoted in Lewis, *Without Fear or Favor: A Biography of Chief Justice Roger Brooke Taney* 250 (1965).

9. Actually, Taney presided over the Court on August 1, 1836, but no other Justice was present and the Court was adjourned until the January Term. See Frankfurter, *The Commerce Clause under Marshall, Taney and Waite* 46 (1964).

10. 2 *Life and Letters of Joseph Story* 227, 226 (W. W. Story ed. 1851).

11. 9 *Memoirs of John Quincy Adams* 243–244 (C. F. Adams ed. 1876).

12. Quoted in Schwartz, *The American Heritage History of the Law in America* 111 (1974).

13. Op. cit. supra note 10, at 277.

14. Id. at 173.

15. Letter from J. Q. Adams to General P. Porter, Jan. 11, 1831, Parke-Bernet Galleries, Sale No. 3103, Item 3 (1970).

16. 2 Richardson, *A Compilation of the Messages and Papers of the Presidents 1789–1897,* 576–591 (1896).

17. Letter from Joseph Story to Richard Peters, July 24, 1835, op. cit. supra note 10, at 202.

18. Richardson, op. cit. supra note 16, at 582.

19. Quoted in Warren, op. cit. supra note 2 at 284.

20. Id. at 290.

21. Letter from John Tyler to Mrs. Mary Jones, Jan. 20, 1836, Charles Hamilton, Auction No. 12, Item 190 (1966).

22. Letter from Francis Scott Key to Mrs. R. B. Taney, Mar. 15, 1836, Charles Hamilton, Auction No. 24, Item 183 (1968). Amos Kendall was nominated as Postmaster General and Phillip P. Barbour as a Supreme Court Justice.

23. See 2 Poore *Perley's Reminiscences of Sixty Years in the National Metropolis* 85 (1886).

24. See Swisher, *Roger B. Taney* 359 (1935).

25. Smith, *Roger B. Taney: Jacksonian Jurist* 3 (1936).

26. Orestes A. Brownson, quoted in Schlesinger, *The Age of Jackson* 312 (1945).

27. Letter from Roger B. Taney to Andrew Jackson, Mar. 17, 1836, in 5 Bassett, *Correspondence of Andrew Jackson* 390 (1933). The reference at the end of the letter is to Martin Van Buren, whom the Senate had refused to confirm as Minister to England in 1831.

28. Taney states that he was "mortified" by one of these early defeats. Lewis, supra note 8, at 38.

29. His autobiography is contained in Tyler, *Memoir of Roger Brooke Taney, LL.D.* 17–95 (1872).

30. Id. at 78, 79.

31. Letter from Roger B. Taney to James M. Campbell, Dec. 21, 1845, Parke-Bernet Galleries, Sale No. 2310, Item 125 (1964).

32. Letter from Roger B. Taney to James M. Campbell, Jan. 18, 1841, Parke-Bernet Galleries, Sale No. 2310, Item 125 (1964).

33. Compare Adler, *The Great Ideas: A Syntopicon, 2 Great Books of the Western World* 305, 221 (1952).

34. Richardson, op. cit. supra note 16, at 590.

35. Compare Harris, *The Quest for Equality* 17 (1960).

36. Frankfurter, op. cit. supra note 9, at 4.

37. See Story, J., dissenting, in *New York v. Miln,* 11 Pet. 102, 161 (U.S. 1837); *Briscoe v. Bank of Kentucky,* 11 Pet. 257, 328, 350 (U.S. 1837). It is probable that Marshall, after the first argument, had opted in favor of constitutionality in the *Charles River Bridge* case, the third of the key 1837 decisions. Kutler, *Privilege and Creative Destruction: The Charles River Bridge Case* 172–179 (1971); Dunne, *Justice Joseph Story and the Rise of the Supreme Court* 364 (1970).

38. *Charles River Bridge v. Warren Bridge,* 11 Pet. 420, 548 (U.S. 1837).

39. 11 Pet. 420 (U.S. 1837).

40. 2 Warren, op. cit. supra note 2, at 295–296.

41. Op. cit. supra note 11, at 267.

42. 11 Pet. at 546.

43. Id. at 547, 548.

44. Letter from Joseph Story to Mrs. Joseph Story, Feb. 14, 1837, op. cit. supra note 10, at 268.

45. 11 Pet. at 608.

46. *Dartmouth College v. Woodward,* 4 Wheat. 518 (U.S. 1819).

47. 11 Pet. at 552–553, per Taney, C. J.

48. Id. at 598.

49. *Felix Frankfurter on the Supreme Court* 121 (Kurland ed. 1970).

50. Letter from Joseph Story to Harriet Martineau, Apr. 7, 1837, op. cit. supra note 10, at 277.

51. Letter from Daniel Webster to Joseph Story, n.d., id. at 269.

52. Campbell, J., dissenting, in *Piqua Branch v. Knoop,* 16 How. 369, 409 (U.S. 1853).

53. 11 Pet. 257 (U.S. 1837); 11 Pet. 102 (U.S. 1837).

54. *Briscoe v. Bank of Kentucky,* 11 Pet. at 328, 350.

55. Compare Swisher, op. cit. supra note 24, at 375.

56. Jackson, Farewell Address, 3 Richardson, op. cit. supra note 16, at 305.

57. Compare Frankfurter, op. cit. supra note 9, at 69.

58. 8 Adams, op. cit. supra note 11, at 315–316.

59. 11 Pet. at 139, 141.

60. Id. at 142.

61. The cases are cited in *Edwards v. California,* 314 U.S. 160, 176–177 (1941).

62. *Matter of Chirillo,* 283 N.Y. 417, 436 (1940) (dissenting opinion).

63. *Edwards v. California,* 314 U.S. 160, 177 (1941). For a more recent case, see *Shapiro v. Thompson,* 394 U.S. 618 (1969).

64. Swisher, op. cit. supra note 24, at 309.

65. 11 Pet. at 552.

66. 5 How. 504, 583 (U.S. 1847).

67. See *Passenger Cases,* 7 How. 283, 424 (U.S. 1849).

68. *West River Bridge v. Dix,* 6 How. 507, 532 (U.S. 1848).

69. 7 Cush. 53 (Mass. 1851).

70. *State v. Searcy,* 20 Mo. 489, 490 (1855).

71. Corwin, The Doctrine of Due Process of Law before the Civil War, 24 *Harvard Law Review* 460, 461 (1911).

72. *Charles River Bridge v. Warren Bridge,* 11 Pet. at 547.

73. *License Cases,* 5 How. at 579.

74. Supra note 53.

75. 5 How. 504 (U.S. 1847).

76. 7 How. 283 (U.S. 1849).

77. Rutledge, *A Declaration of Legal Faith* 33 (1947).

78. Id. at 45.

79. Compare *Gibbons v. Ogden,* 9 Wheat. 1, 226 (U.S. 1824).

80. 11 Pet. at 158.

81. 2 Pet. 245 (U.S. 1829).

82. Id. at 251.

83. *License Cases,* 5 How. at 579.

84. Ibid.

85. 9 Wheat. 1 (U.S. 1824); supra p. 47.

86. Id. at 181.

87. Id. at 178.

88. Frankfurter, op. cit. supra note 9, at 24.

89. 13 How. 299 (U.S. 1852).

90. Letter from Benjamin R. Curtis to Mr. Ticknor, Feb. 29, 1852, 1 *Memoir of Benjamin Robbins Curtis, LL.D.* 168 (1879).

91. *Dred Scott v. Sandford,* 19 How. 393 (U.S. 1857).

92. Blaustein and Mersky, Rating Supreme Court Justices, 58 *American Bar Association Journal* 1183, 1185 (1972).

93. Quoted in 2 Friedman and Israel, *The Justices of the United States Supreme Court* 905 (1969).

94. He did not graduate, having left halfway through his course to work in a law office.

95. For the Curtis argument, see The Case of the Slave Med, 2 op. cit. supra note 90, at 69.

96. Quoted in Swisher, *History of the Supreme Court of the United States: The Taney Period 1836–64,* 239 (1974).

97. Frankfurter, op. cit. supra note 9, at 57.

98. Warren, op. cit. supra note 2, at 429.

99. Swisher, *American Constitutional Development* 205 (1943).

100. Op. cit. supra note 90, at 168.

101. Compare Powell, *Vagaries and Varieties in Constitutional Interpretation* 152–153 (1956).

102. 9 Wheat. at 14.

103. 12 How. at 318.

104. Id. at 319.

105. Ibid.

106. Ibid.

107. *Crandall v. Nevada,* 6 Wall. 35, 42 (U.S. 1868).

108. Op. cit. supra note 90, at 168.

109. The phrase of Holmes, J., in *Le Roy Fibre Co. v. Chicago, Mil. & St. P. Ry.*, 232 U.S. 340, 354 (1914).

110. *California v. Zook*, 336 U.S. 725, 728 (1949).

111. *Southern Pacific Co. v. Arizona*, 325 U.S. 761, 767 (1945).

112. *United States V. South-Eastern Underwriters Assn'n.*, 322 U.S. 533, 548 (1944).

113. *Parker v. Brown*, 317 U.S. 341, 362 (1943).

114. Gillette, in op. cit. supra note 93, at 901.

115. Frankfurter, op. cit. supra note 9, at 164.

116. Jackson Bank Veto Message, Richardson, op. cit. supra note 16, at 590.

117. William Leggett, quoted in Commager, *The Era of Reform*, 1830–1860, 94 (1960).

118. 3 Richardson, op. cit. supra note 16, at 305–306.

119. Speech by Taney, Aug. 6, quoted in Swisher, op. cit. supra note 24, at 297.

120. Quoted in Smith, op. cit. supra note 25, at 67.

121. Corporations, 4 *American Jurist* 298 (1830), in *The Golden Age of American Law* (Haar ed. 1965).

122. Letter from Peter V. Daniel to Martin Van Buren, Dec. 16, 1841, quoted in Frank, *Justice Daniel Dissenting: A Biography of Peter V. Daniel, 1784–1860*, 164 (1964).

123. 13 Pet. 519 (U.S. 1839).

124. Warren, op. cit. supra note 2, at 324.

125. Ibid.

126. Ibid.

127. 13 Pet. at 567.

128. Id. at 592.

129. Compare Lewis, op. cit. supra note 8, at 292.

130. Compare Frankfurter, op. cit. supra note 9, at 64–65.

131. Quoted in Tyler, op. cit. supra note 29, at 288.

132. Quoted in Warren, op. cit. supra note 2, at 332.

133. *Barrow Steamship Co. v. Kane*, 170 U.S. 100, 106 (1898).

134. Letter from John J. Crittenden to J. Meredith, 1842, Charles Hamilton, Auction No. 19, Item 340 (1967).

135. Concurring, in *Brown v. Allen*, 344 U.S. 443, 540 (1953).

136. Taney had been a leader of the Federalist party in Maryland before the War of 1812.

137. 4 *The Diary of James K. Polk during His Presidency* 137 (Quaife ed. 1910).

138. Id. at 138.

139. Letter from Roger B. Taney to Martin Van Buren, May 8, 1860, quoted in Lewis, op. cit. supra note 8, at 164.

140. Ibid.

141. Charge to the grand jury, Circuit Court of the U.S., April Term, 1836, Taney's Circuit Court Reports 615, 616.

142. Letter from Roger B. Taney to the Secretary of the Treasury, Feb. 16, 1863, quoted in Tyler, op. cit. supra note 29, at 433.

143. 21 How. 506 (U.S. 1859).

144. Id. at 514, 525.

145. Id. at 515, 517.

146. Id. at 518, 522–523.

147. Letter from J. H. Eaton to Andrew Jackson, Apr. 13, 1829, Charles Hamilton, Auction No. 25, Item 141 (1968).

148. See Richardson, op. cit. supra note 16, at 458.

149. *Worcester v. Georgia,* 6 Pet. 515 (U.S. 1832).

150. 8 op. cit. supra note 11, at 262–263.

151. Compare Warren, op. cit. supra note 2, at 219, with James, *The Life of Andrew Jackson* 603–604 (1938).

152. See Freehling, *Prelude to Civil War: The Nullification Controversy in South Carolina, 1816–1836,* 233 (1966).

153. Letter from Andrew Jackson to John Coffee, Apr. 7, 1832, 4 Bassett, op. cit. supra note 27, at 430.

154. Compare Frankfurter, op. cit. supra note 9, at 71.

155. Acheson, Roger Brooke Taney: Notes upon Judicial Self Restraint, 31 *Illinois Law Review* 705 (1937).

156. The famous term of Frankfurter, J., in *Colegrove v. Green,* 328 U.S. 549, 556 (1946).

157. Dissenting, in *Pennsylvania v. Wheeling & B. Bridge Co.,* 13 How. 518 (U.S. 1852).

158. 13 How. 518 (U.S. 1852).

159. Quoted in Warren, op. cit. supra note 2, at 509.

160. Quoted in Frank, op. cit. supra note 122, at 198.

161. The Court had original jurisdiction over such an action brought by a state.

162. Compare Frank, loc. cit. supra note 160.

163. 10 Stat. 112 (1852).

164. *Pennsylvania v. Wheeling & B. Bridge Co.,* 18 How. 421 (U.S. 1856).

165. *Decatur v. Paulding,* 14 Pet. 497, 516 (U.S. 1840).

166. *Kentucky v. Dennison,* 24 How. 66, 109–110 (U.S. 1861).

167. 7 How. 1 (U.S. 1849).

168. Id. at 42.

169. Id. at 43.

170. Now 10 U.S.C. § 331.

171. 7 How. at 43.

172. See Acheson, op. cit. supra note 155, at 714.

173. Swisher, op. cit. supra note 96, at 12.

174. 1 Hoar, *Autobiography of Seventy Years* 141 (1905).

175. Though Justices McLean and Wayne had served briefly under Marshall, they played no significant role in his Court.

176. Op. cit. supra note 10, at 277.

177. Swisher, op. cit. supra note 96, at 45.

178. 8 op. cit. supra note 11, at 304.

179. 3 Warren, op. cit. supra note 2, at 122.

180. 2 id. at 543.

181. Ibid.

182. Ibid.

183. Swisher, op. cit. supra note 96, at 47.

184. Warren, op. cit. supra note 2, at 644.

185. Weisenberger, *The Life of John McLean* 140 (1937).

186. Id. at 141.

187. White, op. cit. supra note 1, at 297.
188. Op. cit. supra note 90, at 168.
189. 1 Friedman and Israel, op. cit. supra note 93, at 609.
190. White, op. cit. supra note 1, at 294.
191. Ibid.
192. Swisher, op. cit. supra note 96, at 55.
193. *Autobiography of Martin Van Buren* 578 (1973 reprint).
194. White, op. cit. supra note 1, at 298.
195. Swisher, op. cit. supra note 96, at 55.
196. 2 Warren, op. cit. supra note 2, at 293.
197. 8 op. cit. supra note 11, at 315–316.
198. Op. cit. supra note 10, at 277.
199. 1 Bryce, *The American Commonwealth* 73 (1889).
200. 20 Wall. ix.
201. Swisher, op. cit. supra note 24, at 427.
202. Swisher, op. cit. supra note 96, at 717–718.
203. 3 Warren, op. cit. supra note 2, at 84.
204. Frankfurter, op. cit. supra note 9, at 48.
205. Compare McLaughlin, *A Constitutional History of the United States* 456 (1935).
206. Frankfurter, *Of Law and Men: Papers and Addresses, 1939–1956,* 133 (Elman ed. 1956).
207. Jackson, Farewell Address, 3 Richardson, op. cit. supra note 16, at 306.
208. Wright, *Economic History of the United States* 388 (1941).
209. Evans, *Business Incorporations in the United States* 13 (1948).
210. Faulkner, *American Economic History* 243 (Spriggs ed. 8th ed. 1960).
211. Hughes, Roger Brooke Taney, 17 *American Bar Association Journal* 787 (1931).
212. 12 How. 443 (U.S. 1852).
213. *The Thomas Jefferson,* 10 Wheat. 428 (U.S. 1825).
214. 12 How. at 457.
215. The term used in 2 Warren, op. cit. supra note 2, at 513.
216. 12 How. at 465.
217. Quoted in Steiner, *Life of Roger Brooke Taney* 292 (1922).
218. Op. cit. supra note 49, at 462.
219. Frankfurter, op. cit. supra note 9, at 71.
220. Quoted in 2 Warren, op. cit. supra note 2, at 290.
221. Lewis, op. cit. supra note 8, at 477.
222. Frankfurter, op. cit. supra note 9, at 72–73.

Chapter 4

1. Dissenting, in *Northern Securities Co. v. United States,* 193 U.S. 197, 401 (1904).
2. *Dred Scott v. Sandford,* 19 How. 393 (U.S. 1857).
3. *Cong. Globe,* 38th Cong., 2d Sess. 1013.
4. Lewis, *Without Fear or Favor: A Biography of Chief Justice Roger Brooke Taney* 470, 471 (1965).
5. Swisher, *Roger B. Taney* 578 (1961 ed.).
6. Jackson, *The Struggle for Judicial Supremacy* 327 (1941).

7. *Felix Frankfurter on the Supreme Court* 554 (Kurland ed. 1970).

8. But see Ehrlich, *They Have No Rights: Dred Scott's Struggle for Freedom* 173 (1979): "Bernard Schwartz was too kind in calling it merely a 'mistake' on the Court's part 'to imagine that a flaming political issue could be quenched by calling it a "legal" question.' It was not a mistake; it was a tragedy."

9. 1 Stat. 50 (1789).

10. 1 Stat. 106 (1790).

11. 3 Stat. 545 (1820).

12. 4 *Memoirs of John Quincy Adams* 502 (C. F. Adams ed. 1876).

13. Quoted in Coit, *John C. Calhoun* 146 (1950).

14. 2 *The Diary of James K. Polk during His Presidency* 75 (Quaife ed. 1910).

15. 3 id. at 366.

16. Quoted in Van Deusen, *The Jacksonian Era: 1828 to 1848,* 245 (1959).

17. Compare Dumond, *Antislavery Origins of the Civil War in the United States* 76–77 (1939).

18. Op. cit. supra note 12, at 530.

19. 5 id. at 5.

20. 4 *The Works of John C. Calhoun* 343 (Cralle ed. 1854).

21. Id. at 347.

22. Id. at 344–345.

23. Spain, *The Political Theory of John C. Calhoun* 24 (1951).

24. Op. cit. supra note 20, at 348.

25. Jefferson Davis, Notes for W.T.W. (n.d.), Charles Hamilton, Auction No. 31, Item 83 (1968).

26. Op. cit. supra note 20, at 348.

27. Mendelson, in S. I. Kutler, *The Dred Scott Decision: Law or Politics* 153 (1967).

28. 4 op. cit. supra note 14, at 297–298.

29. Compare Morison, *The Oxford History of the American People* 567 (1965).

30. 4 op. cit. supra note 14, at 20.

31. *Cong. Globe,* 30th Cong., 1st Sess. 950.

32. Id. at 1002; 4 op. cit. supra note 14, at 21.

33. *Cong. Globe,* 30th Cong., 1st Sess. 950.

34. 4 op. cit. supra note 14, at 207.

35. 4 Richardson, *A Compilation of the Messages and Papers of the Presidents, 1789–1897,* 642 (1896).

36. 5 id. at 431.

37. Jefferson Davis, Notes for W.T.W. (n.d.), Charles Hamilton, Auction No. 31, Item 83 (1968).

38. *Cong. Globe Appendix,* 31st Cong., 1st Sess. 95. See also id. at 154.

39. 9 Stat. 446, 449, 453, 454 (1850).

40. 9 Stat. 446, 450, 453, 456 (1850).

41. McLaughlin, *Constitutional History of the United States* 534 (1935).

42. *Cong. Globe,* 31st Cong. 1st Sess. 1155.

43. See Kutler, op. cit. supra note 27, at 156.

44. *Cong. Globe Appendix,* 33rd Cong., 1st Sess. 232. See also *Cong. Globe,* 35th Cong., 2d Sess. 1258.

45. *Cong. Globe Appendix,* 34th Cong., 1st Sess. 797.

46. To be sure, *Dred Scott* itself did not reach the Supreme Court by the procedural route provided in the 1850 Compromise or the Kansas-Nebraska Act;

yet it did dispose of the substantive issue contemplated by those laws. Kutler, op. cit. supra note 27, at 160.

47. 2 *The Collected Works of Abraham Lincoln* 355 (Basler ed. 1953). But see 4 id. at 67.

48. *Washington Union,* quoted in Lewis, op. cit. supra note 3, at 420.

49. Defendant's name was misspelled in the official report, so it is as *Dred Scott v. Sandford* that the case is still known.

50. The quote is from Emerson's "The Snow-Storm."

51. Letter from Roswell Field to Montgomery Blair, Dec. 24, 1855, quoted in Marke, *Vignettes of Legal History* 85 (1965).

52. Quoted in Hopkins, *Dred Scott's Case* 38 (1967).

53. Aristotle, *Politics* bk. 3, 5.

54. 1 *Memoir of Benjamin Robbins Curtis, LL.D.* 179 (1879).

55. Swisher, op. cit. supra note 5, at 493.

56. Klein, *President James Buchanan* 269 (1962).

57. The Catron letters are summarized in Swisher, op. cit. supra note 5, at 496.

58. Lawrence, *James Moore Wayne: Southern Unionist* 143 (1943).

59. Op. cit. supra note 54, at 235.

60. Lawrence, op. cit. supra note 58, at 155.

61. 2 Curtis, *Constitutional History of the United States* 275 (1896). Other commentators have asserted that dissents on the merits planned by Justices Curtis and McLean led to Justice Wayne's motion and its adoption—a view I formerly held. See Schwartz, *From Confederation to Nation: The American Constitution 1835–1877,* 118–119 (1973).

62. Curtis, op. cit. supra note 61, at 274–275.

63. Op. cit. supra note 54, at 206.

64. Quoted in Swisher, *History of the Supreme Court of the United States: The Taney Period, 1836–64,* 591 (1974).

65. Hopkins, op. cit. supra note 52, at 56.

66. Ibid.

67. 3 Warren, *The Supreme Court in United States History* 17 (1924).

68. Op. cit. supra note 47, at 466.

69. The actual order of reading was somewhat different, since the dissenters read their opinions first on March 7, followed by Justices Daniel, Grier, Campbell, and Wayne.

70. Warren, op. cit. supra note 67, at 27.

71. Justice Nelson alone stuck to that ground; he filed as a separate concurrence the opinion written by him as the original opinion of the Court.

72. According to the monumental study on the law of slavery, 5 Catterall, *Judicial Cases Concerning American Slavery and the Negro* 121 (1968), the decision on the third point was "unquestionably correct."

73. Notably *Strader v. Graham,* 10 How. 82 (U.S. 1850).

74. 19 How. at 455.

75. Lewis, op. cit. supra note 3, at 420–421.

76. Op. cit. supra note 7, at 212.

77. Justices Wayne, Grier, Daniel, Campbell, and Catron.

78. *Cong. Globe,* 35th Cong., 1st Sess. 617.

79. 19 How. at 450.

80. Benton, *Historical and Legal Examination of That Part of the Decision of the*

Supreme Court of the United States in the Dred Scott Case Which Declares the Unconstitutionality of the Missouri Compromise Act 11–12 (1857).

81. Taney, C.J., 19 How. at 449.

82. *De Lima v. Bidwell,* 182 U.S. 1 (1901); *Dooley v. United States,* 182 U.S. 222 (1901); *Downes v. Bidwell,* 182 U.S. 244 (1901), infra p. 186.

83. *Reid v. Covert,* 354 U.S. 1, 6 (1957).

84. 13 N.Y. 378 (1856).

85. See 3 Schwartz, *A Commentary on the Constitution of the United States: The Rights of Property* 31 (1965).

86. See, e.g., *Annals of Cong.,* 16th Cong., 1st Sess. 1262, 1521.

87. Baldwin, J., in *Groves v. Slaughter,* 15 Pet. 449, 515 (U.S. 1841).

88. *Greene v. Briggs,* 1 Curtis 311 (C.C.R.I. 1852). See also *Greene v. James,* 2 Curtis 187 (C.C.R.I. 1854). The Curtis approach here was not exactly that of substantive due process, but it went far in that direction.

89. Op. cit. supra note 54, at 173.

90. See Hopkins, op. cit. supra note 52, at 137.

91. *Cong. Globe,* 42nd Cong., 1st Sess. 576.

92. Kutler, op. cit. supra note 27, at 49.

93. 19 How. at 403.

94. Id. at 405, 406.

95. Id. at 407.

96. Ibid.

97. Id. at 416, 427.

98. Id. at 572, 575.

99. Id. at 614.

100. Id. at 614, 616.

101. 1 Stat. 50 (1789).

102. 19 How. at 623, 621.

103. Id. at 572.

104. North Carolina, Massachusetts, New Hampshire, New York, and New Jersey. Id. at 572–574.

105. Id. at 575.

106. Id. at 420.

107. Opinions of the Justices, 44 Maine 505, 573 (1857).

108. Ibid.

109. 1 Op. Att'y Gen. 506 (1821); 7 Op. Att'y Gen. 746, 753 (1856). See also an unpublished 1832 Taney opinion to such effect quoted in Swisher, op. cit. supra note 5, at 154 (1961).

110. E.g., *Amy v. Smith,* 1 Littell 326 (Ky. 1822); *Crandall v. State,* 10 Conn. 339 (Conn. 1834); *Hobbs v. Fogg,* 6 Watts 553 (Pa. 1837).

111. 4 Deveraux and Battle 144 (N.C. 1839).

112. 5 Iredell 203 (N.C. 1840).

113. Id. at 206–207.

114. See Hopkins, op. cit. supra note 52, at 99.

115. Op. cit. supra note 47, at 500.

116. Id. at 407.

117. Loc. cit. supra note 26.

118. 19 How. at 420.

119. Taney's term, id. at 451.

120. *Downes v. Bidwell,* 182 U.S. 244, 274 (1901).

121. Id. at 275.

122. Fehrenbacher, *Slavery, Law, and Politics: The Dred Scott Case in Historical Perspective* 304 (1981).

123. *Remarks of the Hon. Stephen A. Douglas on Kansas, Utah, and the Dred Scot Decision* 6 (1857).

124. Op. cit. supra note 47, at 453.

125. Id. at 467.

126. *Barron v. Mayor of Baltimore,* 7 Pet. 243 (U.S. 1833). See also 3 op. cit. supra note 27, at 100–101, where Lincoln assumed that the Fifth Amendment was applicable to the states.

127. Jefferson Davis, Notes for W.T.W., Charles Hamilton, Auction No. 31, Item 83 (1968). This apparently unpublished manuscript is not dated, but appears, from its condition and context, to have been written shortly after the Civil War.

128. 12 Stat. 432 (1862).

129. Loc. cit. supra note 127.

Chapter 5

1. Political Causes of the American Revolution, in Acton, *Essays on Freedom and Power* 196 (Himmelfarb ed. 1948).

2. *Texas v. White,* 7 Wall. 700, 725 (U.S. 1869).

3. *The Spirit of Liberty: Papers and Addresses of Learned Hand* 189–190 (3d ed. 1960).

4. *The Great Rights* 90 (Cahn ed. 1963).

5. Harper, *Lincoln and the Press* 109 (1951).

6. *Ex parte Merryman,* 17 Fed. Cas. 144, 153 (D. Md. 1861).

7. 17 Fed. Cas. 144 (D. Md. 1861).

8. Id. at 153.

9. Ibid.

10. 5 *The Collected Works of Abraham Lincoln* 436–437 (Basler ed. 1953).

11. See Randall, *Constitutional Problems under Lincoln* 161 (rev. ed. 1951); Sprague, *Freedom under Union* 43–44 (1965).

12. Jackson, *The Supreme Court in the American System of Government* 76 (1955).

13. Mikell, in 4 *Great American Lawyers* 188–189 (W. D. Lewis ed. 1908).

14. Dunning, *Essays on the Civil War and Reconstruction* 20–21 (1910).

15. Randall, op. cit. supra note 11, at 1.

16. Id. at 2.

17. 4 Lincoln op. cit. supra note 10, at 281.

18. 7 id. at 100.

19. Id. at 281.

20. 5 id. at 394.

21. Id. at 372.

22. 7 id. at 488.

23. 5 id. at 343.

24. Ibid.

25. Id. at 261, 265.

26. Id. at 264.

27. *Ex Parte Milligan,* 4 Wall. 2 (U.S. 1866).

28. 1 Wall. 243 (U.S. 1864).

29. 2 Black 635 (U.S. 1862).

30. Sandburg, *Abraham Lincoln* 275 (1966).

31. 7 Moore, *A Digest of International Law* 190 (1906).

32. *The Protector,* 12 Wall. 700, 702 (U.S. 1872).

33. *Prize Cases,* 17 L. Ed. 459, 465 (U.S. 1862). The Evarts argument is not given in the official report of the case.

34. 2 Black at 669.

35. Rossiter, *The Supreme Court and the Commander-in-Chief* 75 (1951).

36. *Williams v. Bruffy,* 96 U.S. 176, 189 (1878).

37. *Prize Cases,* 2 Black at 669–670.

38. Id. at 666.

39. This was the view of the dissent, id. at 690.

40. Id. at 668.

41. Compare Jackson, J. dissenting, in *Terminiello v. Chicago,* 337 U.S. 1, 37 (1949).

42. A. Lincoln to W. H. Herndon, Feb. 15, 1848, 1 op. cit. supra note 10, at 451.

43. Letter from John Quincy Adams to Gerrit Smith, July 31, 1839, Charles Hamilton, Auction No. 10, Item 1a (1965).

44. Ibid.

45. 4 op. cit. supra note 10, at 432–433.

46. Id. at 435.

47. Oliver Ellsworth, in 3 Farrand, *The Records of the Federal Convention of 1787,* at 241 (1911).

48. 4 Churchill, *A History of the English Speaking Peoples* 176 (1958).

49. Dabbs, *The Southern Heritage* 116 (1958).

50. 7 Wall. 700 (U.S. 1869).

51. Id. at 726.

52. Cooley, *Constitutional History of the United States* 49 (1889).

53. 7 Wall. at 725.

54. Bradley, J., concurring, in *Legal Tender Cases,* 12 Wall. 457, 554–555 (U.S. 1871).

55. 6 *The Works of John C. Calhoun* 309–311 (Cralle ed. 1857).

56. Id. at 311.

57. Maine, *Popular Government* 245 (1886).

58. Woodward, *Reunion and Reaction: The Compromise of 1877 and the End of Reconstruction* 14 (1956).

59. Warren, C.J., in op. cit. supra note 4, at 106.

60. *Ex parte Milligan,* 4 Wall. 2 (U.S. 1866).

61. *Cross v. Harrison,* 16 How. 164 (U.S. 1853).

62. *Texas v. White,* 7 Wall. 700, 725 (U.S. 1869).

63. Wright, The Programme of Peace, by a Democrat of the Old School 18 (1862), quoted in Hyman, *New Frontiers of the American Reconstruction* 25 (1966).

64. McKittrick, *Andrew Johnson and Reconstruction* 102 (1960).

65. House of Representatives, June 13, 1866, quoted in 1 Schwartz, *Statutory History of the United States: Civil Rights* 282 (1970).

66. Stampp, *The Era of Reconstruction: 1865 to 1877,* 87 (1966).

67. *Cong. Globe,* 39th Cong., 1st Sess. 142.

68. John Jay to Charles Sumner, Feb. 12, 1862, in McPherson, *The Struggle for Equality: Abolitionists and the Negro in the Civil War and Reconstruction* 238 (1964).

69. 7 Wall. 700 (U.S. 1869).

70. Id. at 726.

71. Id. at 727.

72. Id. at 730.

73. See *Dooley v. United States,* 182 U.S. 222 (1901); *Madsen v. Kinsella,* 343 U.S. 341 (1952).

74. See *Dooley v. United States,* 182 U.S. at 234.

75. 7 Wall. at 727–728.

76. 7 How. 1 (U.S. 1849).

77. 7 Wall. at 730.

78. *Georgia v. Stanton,* 6 Wall. 50 (U.S. 1868).

79. *Baker v. Carr,* 369 U.S. 186, 225 (1962).

80. Belz, *Reconstructing the Union* 207 (1969).

81. Woodward, in Hyman, op. cit. supra note 63, at 131.

82. 14 Stat. 428 (1867).

83. Though no death sentence could be executed without presidential approval.

84. Burgess, *Reconstruction and the Constitution* 113, 111 (1902).

85. 4 Wall. 2 (U.S. 1866).

86. Op. cit. supra note 4, at 100.

87. *Cong. Globe,* 39th Cong., 2d Sess. 251.

88. 4 Wall. at 121.

89. Fairman, *History of the Supreme Court of the United States: Reconstruction and Reunion, 1864–88, Part One* 232 (1971).

90. 7 Wall. 506 (U.S. 1869).

91. Fairman, op. cit. supra note 89, at 437.

92. 14 Stat. 385, 386 (1867).

93. *Ex parte McCardle,* 6 Wall. 318 (U.S. 1868).

94. 3 *The Diary of Gideon Welles* 314 (Beale ed. 1960).

95. *Cong. Globe,* 40th Cong. 2d Sess. 489.

96. Op. cit, supra note 94, at 258.

97. *Cong. Globe,* 40th Cong. 2d Sess. 1204, 1428.

98. 15 Stat. 44 (1868).

99. 7 Wall. at 514, 515.

100. See Fairman, op. cit. supra note 89, at 494; Hughes, Salmon P. Chase: Chief Justice, 18 *Vanderbilt Law Review* 586, 591 (1965).

101. Op. cit. supra note 94, at 320.

102. 3 Warren, *The Supreme Court in the United States History* 199 (1924).

103. Id. at 205.

104. Corwin, *The Constitution of the United States of America: Analysis and Interpretations* 615 (1953).

105. Roberts, Now Is the Time: Fortifying the Court's Independence, 35 *American Bar Association Journal* 1, 4 (1949).

106. *Ex parte Yerger,* 8 Wall. 85, 104 (U.S. 1869).

107. *United States v. Bitty,* 208 U.S. 393, 400 (1908).

108. Ratner, Congressional Power over the Appellate Jurisdiction of the Supreme Court, 109 *University of Pennsylvania Law Review* 157, 181 (1960).

109. 7 Wall. at 515.

110. Under the 1867 statute previously referred to.

111. Under § 14 of the first Judiciary Act, 1 Stat. 73, 81 (1789). Under it, a person confined under color of federal authority and denied release by a circuit court could petition the Supreme Court for habeas corpus, as well as for the common-law writ of certiorari to bring up the record below. The Supreme Court, in issuing habeas corpus in such a case, after its denial by a circuit court, has been held to be acting in the exercise of its appellate jurisdiction. *Ex parte Yerger*, 8 Wall. at 102.

112. 8 Wall. 85 (U.S. 1869).

113. Id. at 96.

114. 8 Wall. at 102–103.

115. Id. at 103.

116. Compare Hart and Wechsler, *The Federal Courts and the Federal Judicial System* 312 (1953).

117. *The Diary of Edward Bates* 187 (1933).

118. 1 Stat. 23 (1789).

119. 12 Stat. 326 (1861).

120. See Hyman, *Era of the Oath* 2 (1954).

121. 12 Stat. 502 (1862).

122. Letter from Judah P. Benjamin to Mr. Mason, Oct. 25, 1866, Charles Hamilton, Auction No. 31, Item 78 (1968).

123. 15 Stat. 2 (1867).

124. § 9, 15 Stat. at 16.

125. 15 Stat. 344 (1869).

126. See 13 Op. Att'y Gen. 390 (1871).

127. 12 Stat. 610 (1862).

128. Ibid.

129. See Hyman, op. cit, supra note 120, at 20–21.

130. 13 Stat. 424 (1865).

131. 12 Stat. 424 (1865).

132. Clemenceau, *American Reconstruction, 1865–1870,* 84–85 (1928).

133. Lincoln, op. cit. supra note 10, at 284.

134. 4 Wall. 333 (U.S. 1867).

135. 4 Wall. 277 (U.S. 1867).

136. *United States v. Brown*, 381 U.S. 437, 447 (1965). In addition, the law at issue in *Garland* was held an invalid infringement upon the presidential power of pardon.

137. *Cong. Globe,* 41st Cong., 3d Sess. 886.

138. 23 Stat. 21 (1884).

Chapter 6

1. 2 *The Collected Works of Abraham Lincoln* 405–407 (Basler ed. 1953).

2. T. Paine, *Rights of Man* 155 (Heritage Press ed. 1961) (italics omitted).

3. Frederick Douglass, quoted in Lynd, *Class Conflict, Slavery, and the United States Constitution* 155 (1967).

4. Jefferson Davis, Notes for W.T.W., Charles Hamilton, Auction No. 31, Item 83 (1968).

5. Quoted in McPherson, *The Struggle for Equality: Abolitionists and the Negro in the Civil War and Reconstruction* 100 (1964).

6. Lincoln, op. cit. supra note 1, at 323.

7. 2 Boutwell, *Reminiscences of Sixty Years in Public Affairs* 29 (1902).

8. 3 Warren, *The Supreme Court in United States History* 135 (1924).

9. Belden and Belden, *So Fell the Angels* 138 (1956).

10. Wambaugh, in 5 *Great American Lawyers* 344 (Lewis ed. 1908).

11. Hughes, Salmon P. Chase, Chief Justice, 18 *Vanderbilt Law Review* 568, 572 (1965).

12. Circular, "S. P. Chase, Attorney, Solicitor and Counsellor of Cincinnati, Ohio," attached to letter to John H. James, Dec. 14, 1839 (in author's possession).

13. 2 *The Diary of Gideon Welles* 187 (Beale ed. 1960).

14. Warren, op. cit. supra note 8, at 122.

15. Adams, *The Education of Henry Adams* 250 (1931 ed).

16. Magrath, *Morrison R. Waite: The Triumph of Character* 281 (1963).

17. Fairman, *History of the Supreme Court of the United States: Reconstruction and Reunion 1864–88, Part One* 26–27 (1971).

18. 16 Wall. 36 (U.S. 1873), infra p. 158.

19. *In re Neagle*, 135 U.S. 1 (1890).

20. Frankfurter, Mr. Justice Holmes and the Constitution, 41 *Harvard Law Review* 121, 141 (1927).

21. Fairman, op. cit. supra note 17, at 63, 66.

22. Haskins and Johnson, *History of the Supreme Court of the United States: Foundations of Power: John Marshall 1801–15,* 384–386 (1981).

23. Fairman, op. cit. supra note 17, at 66.

24. Ibid.

25. Schwartz, *The American Heritage History of the Law in America* 87 (1974).

26. Swisher, *History of the Supreme Court of the United States: The Taney Period 1836–64,* 250 (1974).

27. Id. at 258.

28. Fairman, op. cit. supra note 17, at 4.

29. Id. at 541–545.

30. Bowers, *The Tragic Era: The Revolution after Lincoln* v (1929).

31. Kutler, *Judicial Power and Reconstruction Politics* 6 (1968).

32. See id. at 114. Kutler cites as the two previous decisions invalidating congressional acts *Marbury v. Madison,* 1 Cranch 137 (U.S. 1803) and the *Dred Scott* case. Like most commentators he overlooks *Hodgson v. Bowerbank,* 5 Cranch 303 (U.S. 1809), where a section of the Judiciary Act was declared unconstitutional.

33. *Gordon v. United States,* 2 Wall. 561 (U.S. 1865); *Reichart v. Felps,* 6 Wall. 160 (U.S. 1868); *The Alicia,* 7 Wall. 571 (U.S. 1869); *United States v. Dewitt,* 9 Wall. 41 (U.S. 1870); *The Justices v. Murray,* 9 Wall. 274 (U.S. 1870); *Collector v. Day,* 11 Wall. 113 (U.S. 1871); *United States v. Railroad Company,* 17 Wall. 322 (U.S. 1873). Of these, only *Collector v. Day* was of great potential importance, but it had no immediate effect because the federal income tax at issue expired a few months after the decision. It may also be doubted that *United States v. Railroad Company* belongs among such decisions, since it turned on statutory interpretation, though the constitutional issue was dealt with by way of obiter.

34. 8 Wall. 603 (U.S. 1870).

35. *Legal Tender Cases,* 12 Wall. 457 (U.S. 1871).

36. Bradley, J. concurring, id. at 570.

37. 13 Wall. 128 (U.S. 1872).

38. See Schwartz, *Constitutional Law: A Textbook* 20 (2d ed. 1979).

39. *Hart v. United States* 118 U.S. 62 (1886) (Congress need not appropriate funds to satisfy claims of pardoned persons).

40. 4 Wall. 333 (U.S. 1867).

41. 4 Wall. 277 (U.S. 1867).

42. Galbraith, *Money: Whence It Came, Where It Went* 91 (1975).

43. 8 Wall. 603 (U.S. 1870).

44. Adams, op. cit. supra note 15, at 250.

45. 7 *American Law Review* 146 (1872).

46. *Legal Tender Cases,* 12 Wall. at 652–653, per Field, J., dissenting.

47. Id. at 653.

48. Id. at 583–584, per Chase, C.J., dissenting.

49. Johnson, *A Dictionary of the English Language* (1755).

50. 2 Farrand, *The Records of the Federal Convention of 1787,* 168 (1911).

51. The debate extracts and Madison notes are in id. at 309–310.

52. 3 id. at 172.

53. *Legal Tender Cases,* 12 Wall. at 529.

54. Warren, op. cit. supra note 8, at 236.

55. 14 Stat. 209 (1866).

56. 16 Stat. 44 (1869).

57. 12 Wall. 457 (U.S. 1871).

58. See Fairman, op. cit. supra note 17, at 1395.

59. Warren, op. cit. supra note 8, at 247.

60. Hughes, *The Supreme Court of the United States* 52 (1936).

61. 16 Wall. 36 (U.S. 1873).

62. Warren, op. cit. supra note 8, at 261.

63. Fairman, op. cit. supra note 17, at 1349.

64. 16 Wall. at 79.

65. Corwin, *The Constitution of the United States of America: Analysis and Interpretation* 965 (1953).

66. 16 Wall. at 81.

67. Id. at 105.

68. Id. at 116.

69. Id. at 122.

70. Supra p. 117.

71. Compare Frankfurter, Mr. Justice Holmes and the Constitution, 41 *Harvard Law Review* 141 (1927).

72. Referring to *Walker v. Sauvinet,* 92 U.S. 90, 93 (1876).

73. Hough, Due Process of Law—Today, 32 *Harvard Law Review* 226 (1919).

74. Samuel Shellabarger, in *In Memoriam—Morrison Remick Waite,* 126 U.S. 585, 599–600 (1888).

75. Warren, op. cit. supra note 8, at 283.

76. Frankfurter, *The Commerce Clause under Marshall, Taney and Waite* 76–77 (1964).

77. Quoted in Fairman, *Mr. Justice Miller and the Supreme Court* 373 (1939).

78. *Felix Frankfurter on the Supreme Court* 476 (Kurland ed. 1970).

79. Id. at 250.

80. Id. at 505.

81. Magrath, op. cit. supra note 16, at 299.

82. Ibid.

83. Id. at 182.

84. Id. at 271.

85. Fairman, op. cit. supra note 77, at 384.

86. Fairman, *History of the Supreme Court of the United States: Reconstruction and Reunion 1864–88, Part Two* 529 (1987).

87. Op. cit. supra note 78, at 476.

88. Schwartz, *Super Chief: Earl Warren and His Supreme Court* 259 (1983).

89. Magrath, op. cit. supra note 16, at 151.

90. Op. cit. supra note 78, at 477.

91. See Magrath, op. cit. supra note 16, at 300.

92. Ibid.

93. Id. at 301.

94. *Munn v. Illinois,* 94 U.S. 113 (1877) and its companion cases.

95. Frankfurter, op. cit. supra note 76, at 83.

96. *Munn v. Illinois,* 94 U.S. at 126.

97. See Fairman, The So-Called Granger Cases, 5 *Stanford Law Review* 592 (1953).

98. Of private right.

99. Sir Matthew Hale, De Portibus Maris, in Hargrave, *Collection of Tracts Relative to the Law of England* 77–78 (1787).

100. Note, 25 *American Law Register* 545 (1877).

101. Compare Hamilton, Affectation with a Public Interest, 39 *Yale Law Journal* 1092, 1097 (1930).

102. Frankfurter, supra note 76, at 86.

103. See *Wolff Packing Co. v. Industrial Court,* 262 U.S. 522, 536 (1923).

104. Brewer, J., dissenting, in *Budd v. New York,* 143 U.S. 517, 549 (1892).

105. Supra note 103.

106. *Nebbia v. New York,* 291 U.S. 502 (1934).

107. 18 Stat. 433 (1871).

108. 1 Schwartz, *Statutory History of the United States: Civil Rights* 727 (1970).

109. Pierce, *Memoir and Letters of Charles Sumner* 598 (1893).

110. 109 U.S. 3 (1883).

111. Id. at 24, 11.

112. Id. at 48.

113. Magrath, op. cit. supra note 16, at 145.

114. *Shelley v. Kramer,* 334 U.S. 1, 13 (1948).

115. Notably Senators Allen G. Thurman and James K. Kelly. See Schwartz, op. cit. supra note 108, at 671–676, 703–707.

116. Id. at 764.

117. 78 Stat. 241 (1964).

118. Schwartz, op. cit. supra note 108, at 764.

119. *Heart of Atlanta Motel v. United States,* 379 U.S. 241 (1964).

120. 347 U.S. 483 (1954).

121. See Black, J., dissenting, in *Connecticut General Life Ins. Co. v. Johnson,*

303 U.S. 77, 87 (1938). The argument in question occurred in *San Mateo County v. Southern Pac. R.R. Co.*, 116 U.S. 138 (1885).

122. Quoted in Graham, "The Conspiracy Theory" of the Fourteenth Amendment, 47 *Yale Law Journal* 371 (1938). The Conkling argument to such effect is not contained in the law reports.

123. Quoted in id. at 378.

124. The best-reasoned rejection of the Conkling thesis is to be found in the Graham article, referred to in the two prior notes. See also Corwin, *Liberty against Government: The Rise, Flowering, and Decline of a Famous Juridical Concept* 191–193 (1948). It should, however, be noted that earlier writers, following the lead of Beard and Beard, *Rise of American Civilization* 111–113 (1927), tended to follow the Conkling view.

125. Black, J., dissenting, in *Connecticut General Life Ins. Co. v. Johnson*, 303 U.S. 77, 87 (1938).

126. Compare Harris, *The Quest for Equality: The Constitution, Congress and the Supreme Court* 40 (1960).

127. Black, J., dissenting, in *Adamson v. California*, 332 U.S. 46, 74 (1947).

128. Schwartz, op. cit. supra note 108, at 305–306.

129. Which, in the Fifth Amendment, has always been construed to include corporations.

130. See Faulkner, *Economic History of the United States* ch. 9 (1938).

131. Especially *Bank of Augusta v. Earle*, 13 Pet. 519 (U.S. 1839), supra p. 89.

132. Supra note 94.

133. 118 U.S. 394 (1886).

134. Id. at 396.

135. Douglas, J., dissenting, in *Wheeling Steel Corp. v. Gander*, 337 U.S. 562, 576–577 (1949).

136. Id. at 576.

137. Id. at 574, per Jackson, J.

138. Freund, in Fairman, *History of the Supreme Court of the United States: Five Justices and the Electoral Commission of 1877*, xiii (1988).

139. Haworth, *The Hayes–Tilden Disputed Presidential Election of 1876*, 168 (1966 ed.).

140. Ibid.

141. Quoted in Woodward, *Reunion and Reaction: The Compromise of 1877 and the End of Reconstruction* 120 (1956).

142. Quoted in Magrath, op. cit. supra note 16, at 289.

143. Jackson, *The Struggle for Judicial Supremacy* xi (1941).

144. Woodward, op. cit. supra note 141, at 33.

145. Magrath, op. cit. supra note 16, at 291.

146. 19 Stat. 227 (1877).

147. McLaughlin. *A Constitutional History of the United States* 707 (1935).

148. 24 Stat. 373 (1887).

149. There was also an Oregon statute permitting the electors to fill any vacancy in their ranks.

150. Magrath, op. cit. supra note 16, at 293.

151. Id. at 294.

152. Adams, op. cit. supra note 15, at 277.

153. Id. at 280–281.

154. Rep. Charles Foster, 5 *Cong. Rec.* 1708.

Chapter 7

1. Holmes, *Collected Legal Papers* 26 (1920).
2. Holmes, *The Common Law* 1 (1881).
3. Dissenting, in *Lochner v. New York,* 198 U.S. 45, 75 (1905).
4. *Planned Parenthood v. Casey,* 112 S.Ct. 2791, 2812 (1992).
5. King, *Melville Weston Fuller* 109–110 (1950) [hereinafter cited as King].
6. King 114.
7. Schiffman, in 2 Friedman and Israel, *The Justices of the United States Supreme Court* 1479 (1969).
8. *Felix Frankfurter on the Supreme Court* 476 (Kurland ed. 1970).
9. King 125, 127.
10. Op. cit. supra note 8, at 476, 477.
11. King 138; op. cit. supra note 8, at 477–478.
12. Friedman and Israel, op. cit. supra note 7, at 1480.
13. Op. cit. supra note 8, at 477.
14. Ibid.
15. King 134.
16. King 123.
17. 1 *Holmes–Laski Letters* 579 (Howe ed. 1963).
18. King 158, 149.
19. Strong, Relief for the Supreme Court, *North American Review* 567, 570 (1881).
20. Fairman, *Mr. Justice Miller and the Supreme Court* 404 (1939).
21. King 150.
22. Supra p. 164.
23. *Budd v. New York,* 143 U.S. 517, 548, 551 (1892).
24. Friedman and Israel, op. cit. supra note 7, at 1521.
25. Id. at 1577.
26. Id. at 1603.
27. 1 *Holmes–Laski Letters* 356.
28. 198 U.S. 45 (1905).
29. Bickel, *The Unpublished Opinions of Justice Brandeis* 164 (1967).
30. Acheson, *Morning and Noon* 65 (1965).
31. Bickel, loc. cit. supra note 29.
32. Holmes, op. cit. supra note 1, at 184.
33. *Allgeyer v. Louisiana,* 165 U.S. 578 (1897), infra note 45.
34. *Lochner v. New York,* 198 U.S. 45 (1905).
35. *Mr. Justice Holmes* 78 (Frankfurter ed. 1931).
36. *Adkins v. Children's Hospital,* 261 U.S. 525, 568 (1923).
37. *American Federation of Labor v. American Sash Co.,* 335 U.S. 538, 543 (1949).
38. Compare *Hume v. Moore-McCormack Lines,* 121 F.2d 336, 339 (2d Cir. 1941).
39. Loc. cit. supra note 37.
40. Ibid.
41. Except for recognized physical and mental disabilities. See *State v. F. C. Coal & Coke Co.,* 33 W.Va. 188, 190 (1889).
42. *People v. Gillson,* 109 N.Y. 389, 398 (1888).
43. *People v. Budd,* 117 N.Y. 1, 48 (1889).

44. Venable, Growth or Evolution of Law, 23 *American Bar Association Reports* 278, 298 (1900).

45. 165 U.S. 578 (1897).

46. Id. at 590–591.

47. Id. at 589.

48. *People v. Gillson,* 109 N.Y. at 405.

49. *People v. Budd,* 117 N.Y. at 69.

50. 16 Wall. 36 (U.S. 1873), supra p. 158.

51. *Munn v. Illinois,* 94 U.S. 113 (1877) and its companion cases, supra p. 164.

52. 94 U.S. at 134.

53. *Railroad Commission Cases,* 116 U.S. 307, 331 (1886).

54. *Chicago & C. Railway Co. v. Minnesota,* 134 U.S. 418, 458 (1890).

55. See *Smyth v. Ames,* 169 U.S. 466, 526 (1898).

56. *Matter of Application of Jacobs,* 98 N.Y. 98, 105 (1885).

57. Ibid.

58. Hough, Due Process of Law—Today, 32 *Harvard Law Review* 218, 228 (1919).

59. Op. cit. supra note 8, at 31.

60. Frankfurter, Mr. Justice Holmes and the Constitution, 41 *Harvard Law Review* 121, 142 (1927).

61. Hough, supra note 58, at 228.

62. Loc. cit. supra note 37.

63. Jackson, *The Struggle for Judicial Supremacy* 48 (1941).

64. Dissenting, in *Poe. v. Ullman,* 367 U.S. 497, 517 (1961).

65. Supra p. 47.

66. 128 U.S. 1 (1888).

67. 156 U.S. 1 (1895).

68. Id. at 16.

69. Dissenting, in *Lochner v. New York,* 198 U.S. 45, 75 (1905).

70. Ibid.

71. King 193.

72. 157 U.S. 429, 158 U.S. 601 (1895).

73. 39 L. Ed. at 786.

74. 157 U.S. at 532.

75. Id. at 533.

76. 39 L. Ed. at 807.

77. 158 U.S. at 622.

78. 157 U.S. at 569.

79. See Corwin, *Court over Constitution* 196 (1937); King 220.

80. Corwin, op. cit. supra note 79, at 194.

81. 158 U.S. at 674, 684, 685.

82. Latham, *The Great Dissenter: John Marshall Harlan* 114 (1970).

83. King 216.

84. 157 U.S. at 607.

85. *DeLima v. Bidwell,* 182 U.S. 1 (1901); *Dooley v. United States,* 182 U.S. 222 (1901): *Downes v. Bidwell,* 182 U.S. 244 (1902).

86. Bander, *Mr. Dooley on the Choice of Law* 52 (1963).

87. Id. at 47.

88. 182 U.S. at 249.

89. Id. at 341–342.

90. Op. cit. supra note 8, at 484.

91. King 265.

92. 182 U.S. at 380.

93. 195 U.S. 138 (1904).

94. Latham, op. cit. supra note 82, at 141.

95. 163 U.S. 537 (1896). Nor is it mentioned in 3 Warren, *The Supreme Court in United States History* (1924).

96. *Planned Parenthood v. Casey,* 112 S.Ct. 2791, 2813 (1992).

97. 163 U.S. at 551, 544.

98. *To Secure These Rights, Report of the President's Committee on Civil Rights* 81 (1947).

99. 163 U.S. at 562.

100. Id. at 559.

Chapter 8

1. 198 U.S. 45 (1905).

2. Supra p. 35.

3. Bowen, *Yankee from Olympus: Justice Holmes and His Family* (1944).

4. Bartlett, *Familiar Quotations* 519 (15th ed. 1980).

5. Bowen, op. cit. supra note 3, at 201.

6. Kaplan, Encounters with O. W. Holmes, Jr., 96 *Harvard Law Review* 1828, 1829 (1983).

7. See Bowen, op. cit. supra note 3, at 285.

8. Ibid.

9. Holmes, The Path of the Law, in *Collected Legal Papers* 167, 180 (1920).

10. Holmes, *The Common Law* 1 (1881).

11. Ibid.

12. Biddle, *Mr. Justice Holmes* 61 (1986 ed.).

13. Supra p. 178.

14. Ibid.

15. 117 N.Y. 1 (1889).

16. *Munn v. Illinois,* 94 U.S. 113 (1877) and its companion cases, supra p. 164.

17. 117 N.Y. at 46, 45.

18. Id. at 46, 47.

19. Id. at 69.

20. Id. at 68, 71.

21. 3 Friedman and Israel, *The Justices of the United States Supreme Court* 1693 (1969).

22. Brief for plaintiff in error 18.

23. Butler, *A Century at the Bar of the Supreme Court of the United States* 172 (1942).

24. Notably, in *Holden v. Hardy,* 169 U.S. 366 (1898).

25. King, *Melville Weston Fuller* 298 (1950).

26. Novick, *Honorable Justice: The Life of Oliver Wendell Holmes* 463 (1989).

27. Supra p. 180.

28. *Holden v. Hardy,* 169 U.S. 366 (1898).

29. 198 U.S. at 53.

30. Id. at 54, 57, 56.

31. Id. at 57.

32. Ibid.

33. Id. at 58, 59.

34. Id. at 61, 62.

35. Id. at 61.

36. Id. at 63, 64.

37. *Selected Writings of Benjamin Nathan Cardozo* 81 (Hall ed. 1947).

38. Id. at 134.

39. 198 U.S. at 76, 75.

40. Compare Posner, *Law and Literature* 283 (1988).

41. 198 U.S. at 75.

42. "Every man has freedom to do all that he will, provided he infringes not the equal freedom of any other man." Spencer, *Social Statics* 55 (abridged and revised ed. 1892).

43. 198 U.S. at 75.

44. Ibid.

45. *Holden v. Hardy,* 169 U.S. 366 (1898); *Otis v. Parker,* 187 U.S. 606 (1903); *Jacobson v. Massachusetts,* 197 U.S. 11 (1905).

46. 198 U.S. at 75–76.

47. Id. at 76.

48. Ibid.

49. Posner, op. cit. supra note 40, at 285.

50. Compare Posner, *The Federal Courts* 106–107 (1985).

51. Posner, op. cit. supra note 40, at 285 (emphasis added).

52. Supra p. 179.

53. *Home Life Ins. Co. v. Fisher,* 188 U.S. 726 (1903).

54. King, op. cit. supra note 25, at 291.

55. 198 U.S. at 56.

56. Posner, op. cit. supra note 40, at 285.

57. 198 U.S. at 76.

58. Loc. cit. supra note 56.

59. 1 *Holmes–Laski Letters* 51 (Howe ed. 1953).

60. *Adkins v. Children's Hospital,* 261 U.S. 525, 570 (1923).

61. *Hebe Co. v. Shaw,* 248 U.S. 297, 303 (1919).

62. See Holmes, J., dissenting, in *Meyer v. Nebraska,* 262 U.S. 390, 412 (1923).

63. Frankfurter, J., dissenting, in *West Virginia Board of Education v. Barnette,* 319 U.S. 624, 647 (1943).

64. Pollock, Note, 21 *Law Quarterly Review* 211, 212 (1905).

65. *Ferguson v. Skrupa,* 372 U.S. 726, 731–732 (1963).

66. Id. at 732.

67. King, op. cit. supra note 25, at 109.

68. *Matter of Jacobs,* 98 N.Y. 98, 104–105, 115 (1885).

69. Supra note 42.

70. Supra p. 178.

71. *Budd v. New York,* 143 U.S. 517, 551 (1892).

72. *Poe v. Ullman,* 367 U.S. 497, 517 (1961).

73. *People v. Budd,* 117 N.Y. at 47.

74. Acheson, *Morning and Noon* 211 (1965).

75. Stevens, Judicial Restraint, 22 *San Diego Law Review* 437, 448 (1985).

76. Siegan, Rehabilitating *Lochner,* 22 *San Diego Law Review* 453 (1985).

77. Posner, *Economic Analysis of the Law* 409 (3d ed. 1986).

78. Id. at 593.

79. Fox and Sullivan, Retrospective and Perspective: Where Are We Coming From? Where Are We Going? 62 *New York University Law Review* 936, 957 (1987).

80. Posner, The Constitution as an Economic Document, 56 *George Washington Law Review* 4, 20, and note 25 (1987).

81. Id. at 20.

82. See *Hume v. Moore-McCormack Lines*, 121 F.2d 336, 339–340 (2d Cir. 1941).

83. *American Federation of Labor v. American Sash Co.*, 335 U.S. 538, 543 (1949).

84. Jackson, *The Struggle for Judicial Supremacy* 48 (1941).

85. *Hume v. Moore-McCormack Lines*, 121 F.2d at 340.

86. Holmes, op. cit. supra note 9, at 184.

87. *Baldwin v. Missouri*, 281 U.S. 586, 595 (1930).

88. *Otis v. Parker*, 187 U.S. 606, 609 (1903).

89. 1 *Holmes–Pollock Letters* 167 (Howe ed. 1961).

90. Holmes, op. cit. supra note 9, at 295.

91. *Commonwealth v. Perry*, 155 Mass. 117, 123 (1891).

Chapter 9

1. *Felix Frankfurter Reminisces* 86 (1960).

2. Quoted in Jackson, *The Supreme Court in the American System of Government* 54 (1955).

3. Ibid.

4. Rodell, *Nine Men* 179 (1955).

5. 1 Pringle, *The Life and Times of William Howard Taft* 529 (1964).

6. Id. at 530.

7. Id. at 531.

8. Id. at 532, 533.

9. 1 Pusey, *Charles Evans Hughes* 281 (1963).

10. *Felix Frankfurter on the Supreme Court* 481 (Kurland ed. 1970).

11. Id. at 480.

12. 241 U.S. xvii.

13. Umbreit, *Our Eleven Chief Justices* 372 (1938).

14. *Guinn v. United States*, 238 U.S. 347 (1915); *Buchanan v. Warley*, 245 U.S. 60 (1917).

15. Pringle, op. cit. supra note 5, at 536.

16. 2 id. at 971.

17. Op. cit. supra note 10, at 487.

18. Bickel and Schmidt, *History of the Supreme Court of the United States: The Judiciary and Responsible Government 1910–21*, 54–55 (1984).

19. Id. at 42.

20. Frank, *Marble Palace* 44 (1958).

21. Bickel and Schmidt, op. cit. supra note 18, at 334.

22. 2 *Holmes–Pollock Letters* 113 (Howe ed. 1961).

23. Pringle, op. cit. supra note 5, at 536.

24. Bickel and Schmidt, op. cit. supra note 18, at 9.

25. 2 Pringle, op. cit. supra note 5, at 854.

26. 1 *Holmes–Pollock Letters* 190.

27. Bickel and Schmidt, op. cit. supra note 18, at 80.

28. 1 *Holmes–Laski Letters* 579 (Howe ed. 1953).

29. Loc. cit. supra note 27.

30. Op. cit. supra note 1, at 101.

31. Bickel and Schmidt, op. cit. supra note 18, at 81.

32. Pusey, op. cit. supra note 9, at 275.

33. Frankfurter, *Of Law and Men* 112 (Elman ed. 1956).

34. Bickel and Schmidt, op. cit. supra note 18, at 82.

35. Id. at 83.

36. Id. at 84.

37. Id. at 85.

38. Id. at 97.

39. *Standard Oil Co. v. United States,* 221 U.S. 1 (1911); *United States v. American Tobacco Co.,* 221 U.S. 106 (1911).

40. 3 Friedman and Israel, *The Justices of the United States Supreme Court* 1650 (1969).

41. 221 U.S. at 66.

42. Acheson, *Morning and Noon* 62 (1965).

43. Bickel and Schmidt, op. cit. supra note 18, at 109.

44. Ibid. The quotes are from the oral dissent delivered by Harlan. They do not appear in the printed version.

45. 221 U.S. at 63.

46. Loc. cit. supra note 42.

47. Frankfurter, *Law and Politics* 4 (1962 ed.).

48. Bickel and Schmidt, op. cit. supra note 18, at 201.

49. 13 *Columbia Law Review* 294 (1913).

50. To show this, Warren cited, among others, *Minnesota Iron Co. v. Kline,* 199 U.S. 593 (1905); *Chicago, B. & Q. R.R. v. McGuire,* 219 U.S. 549 (1911); *Atkin v. Kansas,* 191 U.S. 207 (1903): *Muller v. Oregon,* 208 U.S. 412 (1908).

51. *Second Employers' Liability Cases,* 223 U.S. 1 (1912).

52. Bickel and Schmidt, op. cit. supra note 18, at 207.

53. *New York Central R.R. v. Winfield,* 244 U.S. 147, 165 (1917).

54. Smith, Sequel to Workmen's Compensation Acts, 27 *Harvard Law Review* 235, 251 (1914).

55. *Arizona Employer's Liability Cases,* 250 U.S. 400, 433 (1919).

56. 250 U.S. 400 (1919).

57. Id. at 452.

58. Acheson, op. cit. supra note 42, at 67.

59. Bickel and Schmidt, op. cit. supra note 18, at 588.

60. 2 *Holmes–Pollock Letters* 15.

61. *Lochner v. New York,* 198 U.S. at 76.

62. 250 U.S. at 431–432.

63. Id. at 419.

64. McKenna, J., dissenting, id. at 436.

65. *Wilson v. New,* 243 U.S. 332 (1917).

66. Bickel and Schmidt, op. cit. supra note 18, at 473.

67. 1 *Holmes–Laski Letters* 69.

68. 243 U.S. at 426.

69. Op. cit. supra note 1, at 102.

70. 208 U.S. 412 (1908).

71. Friedman and Israel, op. cit. supra note 40, at 1732.

72. Supra p. 194.

73. 243 U.S. at 435, 438.

74.. Id. at 438.

75. *Coppage v. Kansas,* 236 U.S. 1 (1915); *Hitchman Coal & Coke Co. v. Mitchell,* 245 U.S. 229 (1917); *Duplex Printing Press Co. v. Deering,* 254 U.S. 443 (1921).

76. 247 U.S. 251 (1918).

77. Supra p. 183.

78. 1 *Holmes–Pollock Letters* 267.

79. Ibid.

80. 247 U.S. at 280.

81. Pringle, op. cit. supra note 5, at 535.

82. 2 id. at 957.

83. Id. at 967.

84. Op. cit. supra note 1, at 85.

85. 2 Pringle, op. cit. supra note 5, at 961.

86. Op. cit. supra note 10, at 488.

87. Mason, *William Howard Taft: Chief Justice* 233 (1965).

88. 1 *Holmes–Pollock Letters* 555; 2 id. at 113–114; 1 *Holmes–Laski Letters* 290.

89. 1 *Holmes–Laski Letters* 79.

90. Mason, op. cit. supra note 87, at 199.

91. Bindler, *The Conservative Court 1910–1930,* 205 (1986).

92. Mason, op. cit. supra note 87, at 165.

93. Schwartz, *Super Chief: Earl Warren and His Supreme Court* 279 (1983).

94. Friedman and Israel, op. cit. supra note 40, at 2024.

95. Mason, op. cit. supra note 87, at 216–217.

96. Id. at 215.

97. 1 *Holmes–Laski Letters* 493.

98. Strum, *Louis D. Brandeis: Justice for the People* 293 (1984).

99. Mason, op. cit. supra note 87, at 74.

100. Loc. cit. supra note 70.

101. Brandeis, Brief for Defendant in Error, *Muller v. Oregon* 10.

102. Frankfurter, Hours of Labor and Realism in Constitution Law, 29 *Harvard Law Review* 353, 365 (1916).

103. Compare Holmes, *Collected Legal Papers* 187 (1920).

104. Acheson, op. cit. supra note 42, at 82.

105. 1 *Holmes–Laski Letters* 209.

106. *American Federation of Labor v. American Sash Co.,* 335 U.S. 538, 543 (1949).

107. Mason, *Brandeis: A Free Man's Life* 581 (1946).

108. *The Words of Justice Brandeis* 154 (Goldman ed. 1953).

109. Mason, op. cit. supra note 87, at 228.

110. 2 Pringle, op. cit. supra note 5, at 960.

111. Bindler, op. cit. supra note 91, at 197.

112. Op. cit. supra note 10, at 488.

113. 2 Pringle, op. cit. supra note 5, at 1000.

114. Id. at 998.

115. Op. cit. supra note 10, at 488.

116. Frankfurter and Landis, *The Business of the Supreme Court: A Study in the Federal Judicial System* 280 (1927).

117. 2 Pringle, op. cit. supra note 5, at 967.

118. 257 U.S. 312 (1921).

119. 2 Pringle, op. cit. supra note 5, at 967.

120. 257 U.S. at 342.

121. *Bailey v. Drexel Furniture Co.*, 259 U.S. 20 (1922).

122. *Hammer v. Dagenhart,* supra note 76.

123. 259 U.S. at 37.

124. 2 Pringle, op. cit. supra note 5, at 1013–1014.

125. Mason, op. cit. supra note 87, at 247–248.

126. Quoted id. at 292–293. The decisions referred to were *Adkins v. Children's Hospital*, infra note 127; *Burns Baking Co. v. Bryan*, 264 U.S. 504 (1924); *Tyson Brothers v. Banton*, 273 U.S. 418 (1927); *Ribnik v. McBride*, 277 U.S. 350 (1928); *Baldwin v. Missouri*, 281 U.S. 586 (1930).

127. 261 U.S. 525 (1923).

128. *Planned Parenthood v. Casey,* 112 S.Ct. 2791, 2812 (1992).

129. 261 U.S. at 554–555, 557–558.

130. Id. at 546.

131. Op. cit. supra note 1, at 103.

132. Ibid.; id. at 104.

133. 1 *Holmes–Laski Letters* 495.

134. 2 *Holmes–Pollock Letters* 117.

135. Schwartz, op. cit. supra note 93, at 46.

136. Frankfurter, The Early Writings of O. W. Holmes, Jr., 44 *Harvard Law Review* 717, 724 (1931).

137. *Truax v. Corrigan,* 257 U.S. at 343–344.

138. 1 *Holmes–Pollock Letters* 167.

139. *Baldwin v. Missouri,* 281 U.S. at 595.

140. *Missouri, Kansas and Texas Ry. Co. v. May,* 194 U.S. 267, 270 (1904).

141. See Jackson, *The Struggle for Judicial Supremacy* 323 (1941).

142. 2 *Holmes–Laski Letters* 1209.

143. Compare *Burns Baking Co. v. Bryan,* 264 U.S. at 534.

144. See *North Dakota Board of Pharmacy v. Snyder's Drug Stores*, 414 U.S. 156, 167 (1973).

145. See *Duke Power Co. v. Carolina Environmental Study Group*, 438 U.S. 59, 84 (1978).

146. *New Orleans v. Dukes,* 427 U.S. 297, 303 (1976).

147. Compare Cardozo, *The Nature of the Judicial Process* 79 (1921).

148. Loc. cit. supra note 106.

149. Frankfurter, *Mr. Justice Holmes and the Supreme Court* 50 (1938).

150. *United States v. Schwimmer,* 279 U.S. 644, 654–655 (1928).

151. Frankfurter, op. cit. supra note 149, at 51.

152. 250 U.S. 616 (1919).

153. Lerner, *The Mind and Faith of Justice Holmes* 306 (1989 ed.)

154. Compare id. at 290.

155. 250 U.S. at 630.

156. 1 *Holmes–Laski Letters* 153.

157. Frankfurter, op. cit. supra note 149, at 51.

158. 250 U.S. at 630.
159. Id. at 628, 629.
160. Id. at 630–631.
161. *Schenck v. United States,* 249 U.S. 47, 52 (1919).
162. *Schaefer v. United States,* 251 U.S. 466, 482 (1920).
163. Supra note 161.
164. Compare Chafee, *Thirty Five Years with Freedom of Speech* 7 (1952).
165. Hughes, *The Supreme Court of the United States* 68 (1936).
166. *2 Holmes–Laski Letters* 1219.
167. Holmes, *The Common Law* 1, 35 (1881).

Chapter 10

1. Mason, *Harlan Fiske Stone: Pillar of the Law* 406 (1956).
2. 2 Pringle, *The Life and Times of William Howard Taft* 1076 (1964).
3. Mason, *William Howard Taft: Chief Justice* 137 (1965).
4. Id. at 136.
5. 2 Pusey, *Charles Evans Hughes* 651 (1963).
6. 3 Friedman and Israel, *The Justices of the United States Supreme Court* 1903 (1969).
7. *Felix Frankfurter on the Supreme Court* 491 (Kurland ed. 1970).
8. Frankfurter, *Of Law and Men* 149 (Elman ed. 1956).
9. Id. at 141.
10. Op. cit. supra note 7, at 492.
11. Freund, Book Review, 65 *Harvard Law Review* 370 (1951).
12. Mason, op. cit. supra note 1, at 789.
13. Holmes, *Collected Legal Papers* 292 (1920).
14. The quote is from Emerson, The Snow-Storm.
15. Supra p. 218.
16. Dissenting, in *Baldwin v. Missouri,* 281 U.S. 586, 595 (1930).
17. Op. cit. supra note 7, at 517.
18. Id. at 528.
19. Schwartz, *The American Heritage History of the Law in America* 200 (1974).
20. Frankfurter, op. cit. supra note 8, at 202.
21. Posner, *Cardozo: A Study in Reputation* 32 (1990).
22. Cardozo, *The Nature of the Judicial Process* 28, 102 (1921).
23. Posner, op. cit. supra note 21, at 126.
24. Frankfurter, op. cit. supra note 8, at 148.
25. Pringle, op. cit. supra note 2, at 1044.
26. Ibid.
27. 290 U.S. 398 (1934).
28. Article I, section 10.
29. Mason, op. cit. supra note 1, at 362.
30. 291 U.S. 502 (1934).
31. Op. cit. supra note 7, at 523.
32. *Railroad Retirement Board v. Alton,* 295 U.S. 330 (1935).
33. *The Public Papers and Addresses of Franklin D. Roosevelt* 133 (1941).
34. Id. at 14, 1933 vol. (1938).

35. *Schechter Poultry Corp. v. United States,* 295 U.S. 495 (1935); *United States v. Butler,* 297 U.S. 1 (1936).

36. *Carter v. Carter Coal Co.,* 298 U.S. 238 (1936); *Ashton v. Cameron County Dist.,* 298 U.S. 513 (1936); *Louisville Joint Stock Land Bank v. Radford,* 295 U.S. 555 (1935).

37. Hughes, *The Supreme Court of the United States* 95 (1928).

38. See, e.g., Jackson, *The Struggle for Judicial Supremacy* 174 (1941).

39. Id. at 175.

40. *Morehead v. New York ex rel. Tipaldo,* 298 U.S. 587 (1936).

41. Jackson, op. cit. supra note 38, at 175.

42. Op. cit. supra note 33, at 126.

43. Dissenting, in *Lochner v. New York,* 198 U.S. 45, 75 (1905).

44. *Mandeville Island Farms v. American C.S. Co.,* 334 U.S. 219, 230 (1948).

45. Jackson, op. cit. supra note 38, at v.

46. Corwin, *Constitutional Revolution, Ltd.* (1941).

47. Op. cit. supra note 33, at lxix.

48. Mason, op. cit. supra note 1, at 363.

49. Roberts, *The Court and the Constitution* 61 (1951).

50. *Planned Parenthood v. Casey,* 112 S.Ct. 2791, 2812 (1992).

51. Roberts, op. cit. supra note 49, at 62.

52. *West Coast Hotel Co. v. Parrish,* 300 U.S. 379 (1937).

53. Pusey, op. cit. supra note 5, at 757.

54. Powell, quoted in Mason, *The Supreme Court: Vehicle of Revealed Truth or Power Group* 39 (1953).

55. *West Coast Hotel Co. v. Parrish,* 300 U.S. 379 (1937).

56. Supra supra note 40.

57. Jackson, op. cit. supra note 38, at 207–208.

58. *Wright v. Vinton Branch,* 300 U.S. 440 (1937); *Virginian Ry. v. Federation,* 300 U.S. 515 (1937); *Sonzinsky v. United States,* 300 U.S. 506 (1937).

59. Jackson, op. cit. supra note 38, at 213.

60. Corwin, op. cit. supra note 46, at 65.

61. 301 U.S. 1 (1937).

62. Jackson, op. cit. supra note 38, at 214.

63. 301 U.S. at 41.

64. *Helvering v. Davis,* 301 U.S. 619 (1937).

65. *Steward Machine Co. v. Davis,* 301 U.S. 548 (1937); *Carmichael v. Southern Coal Co.,* 301 U.S. 495 (1937).

66. *The Supreme Court under Earl Warren* 135 (Levy ed. 1972).

67. Douglas, *Go East Young Man: The Early Years* 450 (1974).

68. Gerhart, *America's Advocate: Robert H. Jackson* 274 (1958).

69. *The Fourteenth Amendment Centennial Volume* 34 (Schwartz ed. 1970).

70. Holmes, The Path of the Law, 10 *Harvard Law Review* 457, 469 (1897).

71. Dunne, *Hugo Black and the Judicial Revolution* 414 (1977).

72. 5 Burke, *Works* 67 (rev. ed. 1865).

73. Hugo Black, Jr., *My Father: A Remembrance* 239 (1975).

74. Hugo Black to Alan Washburn, Dec. 17, 1958, Black Papers, Library of Congress.

75. Frankfurter, op. cit. supra note 8, at 133.

76. Simon, *The Antagonists: Hugo Black, Felix Frankfurter and Civil Liberties in Modern America* 61 (1989).

77. Id. at 64.

78. Stanley Reed to Felix Frankfurter, n.d., Frankfurter Papers, Library of Congress.

79. Schwartz, *Super Chief: Earl Warren and His Supreme Court* 268 (1983).

80. 310 U.S. 381 (1940).

81. 307 U.S. 38 (1939).

82. *Waialua Agricultural Co. v. Maneja*, 216 F.2d 466, 476 (9th Cir. 1954).

83. *Maneja v. Waialua Agricultural Co.*, 349 U.S. 254, 259 (1955).

84. 301 U.S. at 99.

85. 312 U.S. 100 (1941).

86. Supra p. 212.

87. 312 U.S. at 117.

88. Id. at 113.

89. Compare Holmes, J., dissenting, in *Lochner v. New York*, 198 U.S. 45, 76 (1905).

90. *American Federation of Labor v. American Sash Co.*, 335 U.S. 538, 543 (1949).

91. 198 U.S. at 75.

92. *West Coast Hotel Co. v. Parrish*, 300 U.S. at 391.

93. *Olsen v. Nebraska*, 313 U.S. 236 (1941), overruling *Ribnik v. McBride*, 277 U.S. 350 (1928).

94. *Williamson v. Lee Optical Co.*, 348 U.S. 483, 488 (1955).

95. *Ferguson v. Skrupa*, 372 U.S. 726, 730 (1963).

96. Dissenting, in *Adkins v. Children's Hospital*, 261 U.S. at 570.

97. Jackson, op. cit. supra note 38, at xv–xvi.

Chapter 11

1. Quoted in Schwartz, *A Basic History of the U.S. Supreme Court* 172 (1968).

2. *Felix Frankfurter on the Supreme Court* 491 (Kurland ed. 1970).

3. 2 Pusey, *Charles Evans Hughes* 788 (1951).

4. Mason, *Harlan Fiske Stone: Pillar of the Law* 569 (1956).

5. Id. at 790.

6. Op. cit. supra note 2, at 492–493.

7. Mason, op. cit. supra note 4, at 791.

8. Frank, *Marble Palace* 81 (1958).

9. Mason, op. cit. supra note 4, at 792.

10. Ibid.

11. Lash, *From the Diaries of Felix Frankfurter* 152 (1974).

12. Frank, loc. cit. supra note 8.

13. Simon, *The Antagonists: Hugo Black, Felix Frankfurter and Civil Liberties in Modern America* (1989).

14. Harold H. Burton Diary, Library of Congess.

15. With Phil Kurland, n.d., Frankfurter Papers, Library of Congress.

16. *Brown v. Allen*, 344 U.S. 443, 540 (1953).

17. *Ex parte Quirin*, 317 U.S. 1, 35 (1942).

18. *Hirabayashi v. United States*, 320 U.S. 81, 100 (1943).

19. 323 U.S. 214 (1944).

20. 323 U.S. 283 (1944).

21. Id. at 304.

22. Id. at 310.

23. Jackson, *The Supreme Court in the American System of Government* 60 (1955).

24. *Selective Draft Law Cases,* 245 U.S. 366 (1918).

25. *Lichter v. United States,* 334 U.S. 742, 756 (1948).

26. Id. at 755.

27. *United States v. Bethlehem Steel Corp.,* 315 U.S. 289, 305 (1942).

28. 321 U.S. 414 (1944).

29. *Steuart & Bro. v. Bowles,* 322 U.S. 398 (1944).

30. *Felix Frankfurter Reminisces* 62 (1960).

31. Lash, op. cit. supra note 11, at 270.

32. Id. at 274.

33. Philip Elman to Frankfurter, Sept. 3, Frankfurter Papers, Library of Congress.

34. Lash, op. cit. supra note 11, at 274, 283.

35. Rodell, *Nine Men* 311 (1955).

36. Kluger, *Simple Justice: The History of Brown v. Board of Education and Black America's Struggle for Equality* 585 (1975).

37. Gerhart, *America's Advocate: Robert H. Jackson* 260 (1958).

38. Frankfurter to C. C. Burlingham, Oct. 27, 1954, Frankfurter Papers, Library of Congress.

39. Jackson to Frankfurter, June 19, 1946, Frankfurter Papers, Harvard Law School.

40. Jackson to Frankfurter, Sept. 20, 195[?], Frankfurter Papers, Library of Congress.

41. Black to Jerome Cooper, 1946, *"Sincerely your friend": Letters of Mr. Justice Hugo L. Black to Jerome A. Cooper* (n.d.).

42. Rodell, op. cit. supra note 35, at 310.

43. Berry, *Stability, Security, and Continuity: Mr. Justice Burton and Decision-Making in the Supreme Court 1945–1948,* vii (1978).

44. Id. at 89.

45. Burton to Frankfurter, Apr. 29, 1964, Frankfurter Papers, Harvard Law School.

46. Clark to Warren, Oct. 24, 1955, Clark Papers, University of Texas.

47. Harlan to Clark, Postcard, July 29, 1959, Clark Papers, University of Texas.

48. Miller, *Plain Speaking* 225–226 (1974).

49. Frankfurter to C. C. Burlingham, Sept. 8, 1956, Frankfurter Papers, Library of Congress.

50. 341 U.S. 494 (1951).

51. Id. at 510.

52. Id. at 589.

53. *Harisiades v. Slaughnessy,* 342 U.S. 580 (1952).

54. *Garner v. Los Angeles Board,* 314 U.S. 716, 724 (1951).

55. *Bailey v. Richardson,* 341 U.S. 918 (1951).

56. Dissenting, in *West Virginia Board of Education v. Barnette,* 319 U.S. 624, 665 (1943).

57. *Youngstown Sheet and Tube Co. v. Sawyer,* 343 U.S. 579 (1952).

58. Id. at 587, 589, 587.

59. Id. at 597.

60. 310 U.S. 586 (1940).

61. Frankfurter to Stone, May 27, 1940, Frankfurter Papers, Harvard Law School.

62. 319 U.S. 624 (1943).

63. Id. at 641.

64. Loc. cit. supra note 61.

65. Freedman, *Roosevelt and Frankfurter: Their Correspondence 1928–1945,* 701 (1967).

66. 319 U.S. at 639.

67. *Kovacs v. Cooper,* 336 U.S. 77, 95 (1949).

68. Ibid.

69. Mason, op. cit. supra note 4, at 626.

70. *United States v. Carolene Products Co.,* 304 U.S. 144, 152, n.4 (1938).

71. The most famous was that in *Brown v. Board of Education,* 347 U.S. 483, 494, n.11 (1954).

72. Loc. cit. supra note 70.

73. 323 U.S. at 216.

74. *Smith v. Allwright,* 321 U.S. 649 (1944).

75. *Shelley v. Kraemer,* 334 U.S. 1 (1948).

76. 163 U.S. 537 (1896), supra p. 188.

77. *Sipuel v. Board of Regents,* 332 U.S. 631 (1948).

78. *Sweatt v. Painter,* 339 U.S. 629, 633 (1950).

79. The term used by Harlan, J., dissenting, in *Plessy v. Ferguson,* 163 U.S. 537, 562 (1896).

Chapter 12

1. Frankfurter to Fred Vinson, n.d., Frankfurter Papers, Library of Congress.

2. 1 Bryce, *The American Commonwealth* 274 (1917).

3. Frankfurter to Reed, Apr. 13, 1939, Frankfurter Papers, Library of Congress.

4. Yeats, Easter 1916.

5. *Brown v. Board of Education,* 347 U.S. 483 (1954), infra Chapter 13.

6. *Reynolds v. Sims,* 377 U.S. 533 (1964).

7. Weaver, *Warren: The Man, The Court, The Era* 44 (1967).

8. Pollack, *Earl Warren: The Judge Who Changed America* 193 (1979).

9. Dunne, *Justice Joseph Story and the Rise of the Supreme Court* 91 (1970).

10. Id. at 307–308.

11. *N.Y. Times,* July 10, 1974, p. 24.

12. Lash, *From the Diaries of Felix Frankfurter* 152 (1975).

13. June 15, 1983, at A16.

14. Supra note 5.

15. 163 U.S. 537 (1896).

16. Schwartz, *Super Chief: Earl Warren and His Supreme Court* 4 (1983).

17. Woodward and Armstrong, *The Brethren: Inside the Supreme Court* (1979).

18. Douglas, *The Court Years 1939–1975,* 173 (1980).

19. 347 U.S. 483, 494, n.11 (1954).

20. Dunne, *Hugo Black and the Judicial Revolution* (1977).

21. Hutchinson, Hail to the Chief: Earl Warren and the Supreme Court, 81 *Michigan Law Review* 922, 923 (1983).

22. Dear Chief, June 23, 1969, Black Papers, Library of Congress.

23. Douglas, op. cit. supra note 18, at 240.

24. June 15, 1983, at A16.

25. Rodell, *Nine Men* 284 (1955).

26. Frankfurter, *Of Law and Men* 133 (1956).

27. Supra note 15.

28. Frankfurter to C. C. Burlingham, Apr. 24, 1955, Frankfurter Papers, Library of Congress.

29. Clark to Harlan, Sept. 12, Clark Papers, University of Texas.

30. Clayton, *The Making of Justice: The Supreme Court in Action* 217 (1964).

31. Douglas, op. cit. supra note 18, at 250.

32. *Jacobellis v. Ohio,* 378 U.S. 184, 197 (1964).

33. *N.Y. Times,* March 18, 1986, at 7.

34. Clayton, op. cit. supra note 30, at 43.

35. Schlesinger, *Robert Kennedy and His Times* 379 (1978).

36. Clayton, op. cit. supra note 30, at 159.

37. Minton to Frankfurter, Oct. 22, 1964, Frankfurter Papers, Library of Congress.

38. Clayton, op. cit. supra note 30, at 293.

39. Pound, *Interpretations of Legal History* 1 (1923).

40. Holmes, *The Common Law* (1881).

41. Id. at 1, 35–36.

42. Compare Cardozo, J., in *Jacob and Youngs v. Kent,* 230 N.Y. 239, 242–243 (1921).

43. Compare Cardozo, *Paradoxes of Legal Science* 57 (1927).

44. Lewis, *Portrait of a Decade: The Second American Revolution* 139 (1964).

45. 384 U.S. 436 (1966).

46. 1 *Holmes–Pollock Letters* 167 (Howe ed. 1961).

47. *U.S. News & World Report,* July 15, 1968, at 64.

48. *Reynolds v. Sims,* 377 U.S. 533 (1964).

49. *Johnson v. Virginia,* 373 U.S. 61, 62 (1963).

50. *Brown v. Board of Education,* 349 U.S. 294 (1955).

51. 377 U.S. 218 (1964).

52. 391 U.S. 430 (1968).

53. E.g., *Bell v. Maryland,* 378 U.S. 226 (1964); *Brown v. Louisiana,* 383 U.S. 131 (1966).

54. E.g., *Cox v. Louisiana,* 379 U.S. 536, 559 (1965); *Amalgamated Food Employees Union v. Logan Valley Plaza,* 391 U.S. 308 (1968).

55. 109 U.S. 3 (1883); supra p. 166.

56. 379 U.S. 241 (1964).

57. *South Carolina v. Katzenbach,* 383 U.S. 301 (1966).

58. 369 U.S. 186 (1962).

59. *Colegrove v. Green,* 328 U.S. 549, 556 (1946).

60. 377 U.S. 533 (1964).

61. Id. at 562, 568.

62. Pollock, op. cit. supra note 8, at 209.

63. *Fourteenth Amendment Centennial Volume* 34 (Schwartz ed. 1970).

64. 351 U.S. 12 (1956).

65. Id. at 19.

66. 372 U.S. 335 (1963).

67. Id. at 344.

68. 384 U.S. 436 (1966).

69. 367 U.S. 643 (1961).

70. Warren, The Law and the Future, *Fortune* 106, 126 (Nov. 1955).

71. 304 U.S. 144 (1938); supra p. 261.

72. Letter of Stone, J., Apr. 12, 1941, quoted in Mason, The Core of Free Government, 1938–40: Mr. Justice Stone and "Preferred Freedoms," 65 *Yale Law Journal* 597, 626 (1956).

73. Compare *Kovacs v. Cooper,* 336 U.S. 77, 95 (1949).

74. *Barron v. Mayor of Baltimore,* 2 Pet. 243 (U.S. 1833).

75. Black, J., dissenting, in *Adamson v. California,* 332 U.S. 46, 71–72 (1947).

76. *Benton v. Maryland,* 395 U.S. 784 (1969).

77. *Malloy v. Hogan,* 378 U.S. 1 (1964).

78. *Duncan v. Louisiana,* 391 U.S. 145 (1968).

79. *Klopfer v. North Carolina,* 386 U.S. 213 (1967).

80. *Pointer v. Texas,* 380 U.S. 400 (1965).

81. These are listed in Schwartz, *The Law in America: A History* 174 (1974).

82. See *New York Times Co. v. United States,* 403 U.S. 713, 761 (1971).

83. *Edwards v. South Carolina,* 372 U.S. 229 (1963); *Cox v. Louisiana,* 379 U.S. 536 (1965).

84. *New York Times Co. v. Sullivan,* 376 U.S. 254 (1964).

85. See 4 Schwartz, *A Commentary on the Constitution of the United States,* pt. 3, *Rights of the Person* 374 et seq. (1968).

86. Id. at 313 et seq.

87. *Poe v. Ullman,* 367 U.S. 497, 521 (1961).

88. 381 U.S. 479 (1965).

89. *N.Y. Times,* July 7, 1966, at 22.

90. Compare Hand, Mr. Justice Cardozo, 52 *Harvard Law Review* 361 (1939).

91. 4 Friedman and Israel, *The Justices of the United States Supreme Court* 2727 (1969).

Chapter 13

1. *Brown v. Board of Education,* 347 U.S. 483 (1954).

2. May 20, 1954, Frankfurter Papers, Harvard Law School.

3. Kluger, *Simple Justice: The History of Brown v. Board of Education and Black America's Struggle for Equality* 709 (1975).

4. Id. at 614.

5. 163 U.S. 537 (1896).

6. The conference quotes are from notes taken by Justice Burton, in his papers in the Library of Congress, and Justice Douglas, contained in Tushnet, What Really Happened in *Brown v. Board of Eduation,* 91 *Columbia Law Review* 1867, 1902–1907 (1992).

7. Id. at 1908.

8. Kluger, op. cit. supra note 3, at 601.

9. Burton Diary, Dec. 12, 1952, Burton Papers, Library of Congress.

10. Kluger, op. cit. supra note 3, at 614.

11. Ibid.

12. *Brown v. Board of Education,* 345 U.S. 972 (1953).

13. The memo is reprinted in Nomination of Justice William Hubbs Rehnquist, Hearings before the Committee on the Judiciary, United States Senate, 99th Cong., 2d Sess. 324 (1986). The original is in the *Brown* file, Robert H. Jackson Papers, Library of Congress.

14. Supra Chapter 8.

15. This, of course, is from the dissent by Justice Holmes in *Lochner v. New York,* 198 U.S. 45, 75 (1905).

16. Myrdal, *An American Dilemma* (1944).

17. *Washington Post* July 28, 1986 (national weekly edition), at 8.

18. See, e.g., supra note 13, at 322, 328–333.

19. Kluger, op. cit. supra note 3, at 609.

20. Id. at 606–609.

21. Memorandum by Mr. Justice Jackson, Mar. 15, 1954, Brown file, Jackson Papers, Library of Congress.

22. Frankfurter Memorandum, Sept. 29, 1959, Frankfurter Papers, Harvard Law School.

23. Frankfurter, Memorandum for the Conference, n.d., Frankfurter Papers, Harvard Law School.

24. 347 U.S. at 489.

25. Burton Diary, Dec. 17, 1953.

26. Ulmer, Earl Warren and the Brown Decision, 33 *Journal of Politics* 689, 698 (1971).

27. Frankfurter, Memorandum, Jan. 15, 1954, Frankfurter Papers, Library of Congress.

28. Frankfurter Papers, Harvard Law School. These notes have not been used by other writers on the *Brown* case.

29. Warren, *Memoirs* 285 (1977).

30. Kluger, op. cit. supra note 3, at 692.

31. Id. at 698.

32. Ibid. The clerk in question confirms this statement.

33. Reed to Frankfurter, May 21, 1954, Frankfurter Papers, Harvard Law School.

34. Frankfurter, "Dear Brethren," Jan. 15, 1954, Burton Papers, Library of Congress.

35. Loc. cit. supra note 29.

36. Warren, To the Members of the Court, May 7, 1954, Tom Clark Papers, University of Texas. For the text of Warren's draft, see Schwartz, *The Unpublished Opinions of the Warren Court* 451 (1985).

37. 5/5 54-II. The first draft of this memo is also in Warren's writing.

38. A comparison with Earl Pollock's bench memo in *Brown* also shows that the Warren draft was not in any way based on that memo.

39. 347 U.S. 497 (1954).

40. Burton Diary.

41. These are discussed in Schwartz, *Super Chief: Earl Warren and His Supreme Court* 97–98 (1983).

42. May 18, 1954, quoted in Kluger, op. cit. supra note 3, at 713.

43. 347 U.S. at 494, note 11.

44. *Argument: The Oral Argument before the Supreme Court in Brown v. Board of Education of Topeka, 1952–55,* xxxvii (Friedman ed. 1969).

45. Kluger, op. cit. supra note 3, at 706.

46. Ibid.

47. Douglas to Warren, May 11, 1954.

48. Burton Diary.

49. Warren, op. cit. supra note 29, at 3.

50. Id. at 2.

51. See Kluger, op. cit. supra note 3, at 701.

52. The quotes from the Brown opinion are in 347 U.S. at 492–495.

53. Warren, op. cit. supra note 29, at 3.

54. *N.Y. Times,* May 18, 1954, quoted in Kluger, op. cit. supra note 3, at 113.

55. Scudder interview, Earl Warren Oral History Project, Bancroft Library, University of California, Berkeley.

56. Kluger, op. cit. supra note 3, at 709.

57. Frankfurter to Warren, May 17, 1954, Warren Papers, Library of Congress.

58. Frankfurter to Warren, May 13, 1954, Warren Papers, Library of Congress.

59. 347 U.S. at 500.

60. *Dawson v. Mayor,* 220 F.2d 386 (4th Cir. 1955).

61. *Mayor v. Dawson,* 350 U.S. 877 (1955).

62. *Holmes v. Atlanta,* 350 U.S. 879 (1955).

63. *Gayle v. Browder,* 352 U.S. 903 (1956).

64. *Johnson v. Virginia,* 373 U.S. 61 (1963).

65. *New Orleans Park Improvement Assn. v. Detiege,* 358 U.S. 54 (1959).

66. Loc. cit. supra note 64.

67. *Watson v. Memphis,* 373 U.S. 526 (1963).

68. *Burton v. Wilmington Parking Authority,* 365 U.S. 715 (1961).

69. *Johnson v. Virginia,* 373 U.S. 61, 62 (1963).

70. Burton Papers, Library of Congress.

71. *McCulloch v. Maryland,* 4 Wheat. 315, 407 (U.S. 1819).

72. Goldberg, Equality and Governmental Action, 39 *New York University Law Review* 205 (1964).

Chapter 14

1. Justice Scalia, *N.Y. Times,* Feb. 2, 1988, at A16.

2. Rehnquist, *The Supreme Court: How It Was, How It Is* 290 (1987).

3. *N.Y. Times,* Mar. 3, 1985, Magazine, at 35.

4. *Gonzalez v. Freeman,* 534 F.2d 570 (D.C. Cir. 1964).

5. Harlan to Burger, May 21, 1970, Harlan Papers, Princeton.

6. Savage, *Turning Right: The Making of the Rehnquist Supreme Court* 4 (1992).

7. *N.Y. Times,* Feb. 20, 1983, Magazine, at 20.

8. Drury, *Decision* 61 (1983).

9. *Washington Post,* July 7, 1986 (national weekly edition), at 8.

10. *N.Y. Times,* Feb. 22, 1988, at A16. This statement was made in 1986, while Burger was still Chief Justice.

11. WHR, Re: *Goldwater v. Carter,* Memorandum to the Conference, Dec. 10, 1979.

12. *Mapp v. Ohio,* 367 U.S. 643 (1961); *Miranda v. Arizona,* 384 U.S. 436 (1966).

13. Infra p. 322.

14. Harlan to Burger, June 9, 1990, Harlan Papers, Princeton.

15. Harlan to Frankfurter, Aug. 21, [1963], Frankfurter Papers, Library of Congress.

16. *N.Y. Times,* Mar. 8, 1986, at 7.

17. Infra p. 322.

18. Schwartz, *Swann's Way: The School Busing Case and the Supreme Court* 178 (1986).

19. Loc. cit. supra note 16.

20. On *Communist Party v. Whitcomb,* 414 U.S. 441 (1974).

21. Schwartz, *The Unpublished Opinions of the Burger Court* 412 (1988).

22. Schwartz, *Behind Bakke: Affirmative Action and the Supreme Court* 53 (1988).

23. Infra Chapter 15.

24. *N.Y. Times,* Feb. 20, 1983, Magazine, at 20.

25. Infra p. 324.

26. Supra note 11.

27. *Guardians Association v. Civil Service Commission,* 463 U.S. 582 (1983).

28. *Lau v. Nichols,* 414 U.S. 563 (1974).

29. July 23, 1979, p. 68.

30. *N.Y. Times,* July 23, 1984, at 8.

31. *The Burger Court: The Counter-Revolution That Wasn't* 252 (Blasi ed. 1983).

32. *N.Y. Times,* Jan. 17, 1986, at A14.

33. Rehnquist, op. cit. supra note 2, at 301.

34. Oct. 5, 1981, p. 22.

35. Apr. 19, 1982, p. 49.

36. *N.Y. Times,* Mar. 8, 1986, at 7.

37. *Moran v. Burbine,* 475 U.S. 412 (1986); *Oregon v. Elstad,* 470 U.S. 298 (1985).

38. *Garcia v. San Antonio Metropolitan Transit Authority,* 469 U.S. 528, 580 (1985) (dissent).

39. *Metropolitan Life Insurance Co. v. Ward,* 470 U.S. 869, 884 (1985) (dissent); *Akron v. Akron Center for Reproductive Health,* 462 U.S. 416, 453 (1983) (dissent).

40. *Mississippi University for Women v. Hogan,* 458 U.S. 718 (1982).

41. *Wygant v. Jackson Board of Education,* 476 U.S. 267, 284 (1986).

42. *Minnesota Star Co. v. Commissioner of Revenue,* 460 U.S. 575 (1983).

43. *McKaskle v. Wiggins,* 465 U.S. 168, 187–188 (1984).

44. Supra p. 203.

45. *Washington Post,* Oct. 1, 1984 (national weekly edition), at 33.

46. Ibid.

47. 1 Tocqueville, *Democracy in America* 290 (Bradley ed. 1954).

48. *Economist,* July 12, 1983, p. 16.

49. *United States v. Nixon,* 418 U.S. 683 (1974).

50. Infra Chapter 15.

51. *Dred Scott v. Sandford,* 19 How. 393 (U.S. 1857).

52. *Pollock v. Farmers' Loan & Trust Co.,* 157 U.S. 429 (1895).

53. Supra Chapter 13.

54. *Powell v. McCormack,* 395 U.S. 486 (1969).

55. *Frontiero v. Richardson,* 411 U.S. 677 (1973).

56. *Roe v. Wade,* infra Chapter 15.

57. *O'Connor v. Donaldson,* 422 U.S. 563 (1975).

58. *Houchins v. KQED,* 438 U.S. 1 (1978).

59. *Goldberg v. Kelly,* 397 U.S. 254 (1970).

60. *Richmond Newspapers v. Virginia,* 448 U.S. 555 (1980).

61. Supra Chapter 13.

62. *Brown v. Board of Education,* 349 U.S. 294, 301 (1955).

63. 396 U.S. 19 (1969).

64. Schwartz, op. cit. supra note 18, at 85–86.

65. 396 U.S. at 20.

66. 402 U.S. 1 (1971).

67. Reprinted in Schwartz, op. cit. supra note 18, at 208.

68. 402 U.S. at 15, 30.

69. *United States v. Montgomery County Board of Education,* 395 U.S. 225, 231 (1969).

70. *Keyes v. School District No. 1,* 413 U.S. 189 (1973).

71. 418 U.S. 717 (1974).

72. Schwartz, *The Ascent of Pragmatism: The Burger Court in Action* 266 (1990).

73. 426 U.S. 229 (1976).

74. *Regents of the University of California v. Bakke,* 438 U.S. 265 (1978).

75. Schwartz, op. cit. supra note 22, at 54.

76. Reprinted id. at 198.

77. *Massachusetts Board of Retirement v. Murgia,* 427 U.S. 307, 321 (1976).

78. *Frontiero v. Richardson,* 411 U.S. 677 (1973).

79. *Craig v. Boren,* 429 U.S. 190, 197 (1976).

80. *Mississippi University for Women v. Hogan,* 458 U.S. 718 (1982).

81. *Weinberger v. Weisenfeld,* 420 U.S. 636 (1975).

82. *Roberts v. United States Jaycees,* 468 U.S. 609 (1984).

83. *Orr v. Orr,* 440 U.S. 268 (1979).

84. *Reed v. Reed,* 404 U.S. 71 (1971).

85. *Taylor v. Louisiana,* 419 U.S. 522 (1975).

86. *Sugarman v. Dougall,* 413 U.S. 634 (1973); *Graham v. Richardson,* 403 U.S. 365 (1971).

87. *Trimble v. Gordon,* 430 U.S. 762 (1977).

88. 411 U.S. 1 (1973).

89. *Massachusetts Board of Retirement v. Murgia,* 427 U.S. 307 (1976).

90. *Cleburne v. Cleburne Living Center,* 473 U.S. 432 (1985).

91. *Memorial Hospital v. Maricopa County,* 415 U.S. 250 (1974).

92. *Dunn v. Blumstein,* 405 U.S. 330 (1972).

93. *Boddie v. Connecticut,* 401 U.S. 371 (1971).

94. *Zablocki v. Redhail,* 434 U.S. 374 (1978).

95. 411 U.S. at 35.

96. *Lynch v. Household Finance Corp.,* 405 U.S. 538, 552 (1972).

97. *Greer v. Spock,* 424 U.S. 828, 852 (1976).

98. Dissenting, in *Abrams v. United States,* 250 U.S. 616, 630 (1919).

99. *Valentine v. Chrestenson,* 316 U.S. 52 (1942).

100. 425 U.S. 748 (1976).

101. Id. at 762.

102. *Bates v. State Bar,* 433 U.S. 350 (1977).

103. *Buckley v. Valeo,* 424 U.S. 1 (1976).

104. *Stromberg v. California,* 283 U.S. 359 (1931).

105. 418 U.S. 405 (1974).

106. *Wooley v. Maynard,* 430 U.S. 705 (1977).

107. Supra p. 283.

108. 408 U.S. 753 (1972).

109. Id. at 762–763.

110. *Southeastern Productions v. Conrad,* 420 U.S. 546 (1975).

111. *Amalgamated Food Employees Union v. Logan Valley Plaza,* 391 U.S. 308 (1968).

112. *Hudgens v. National Labor Relations Board,* 424 U.S. 507 (1976).

113. 403 U.S. 713 (1971).

114. Id. at 730.

115. *Nebraska Press Association v. Stuart,* 427 U.S. 539 (1976).

116. *Richmond Newspapers v. Virginia,* 448 U.S. 555 (1980).

117. *Houchins v. KQED,* 438 U.S. 1 (1978).

118. *Branzburg v. Hayes,* 408 U.S. 665 (1972).

119. *Estes v. Texas,* 381 U.S. 532 (1965).

120. *Chandler v. Florida,* 449 U.S. 560 (1981).

121. Schwartz, *Super Chief: Earl Warren and His Supreme Court* 763 (1983).

122. Ibid.

123. *Gideon v. Wainwright,* 372 U.S. 335 (1963); *Mapp v. Ohio,* 367 U.S. 643 (1961); *Miranda v. Arizona,* 384 U.S. 436 (1966).

124. 407 U.S. 25 (1972).

125. Id. at 37.

126. Stewart to Douglas, Apr. 12, 1972.

127. *Mayer v. Chicago,* 404 U.S. 189 (1971).

128. *Ake v. Oklahoma,* 470 U.S. 68 (1985).

129. *Batson v. Kentucky,* 476 U.S. 79 (1986).

130. *Coolidge v. New Hampshire,* 403 U.S. 443, 492 (1971).

131. *Massachusetts v. Sheppard,* 468 U.S. 981, 987–988 (1984).

132. Ibid.; *United States v. Leon,* 468 U.S. 897 (1984).

133. *Rhode Island v. Innis,* 446 U.S. 291, 304 (1980).

134. *Harris v. New York,* 401 U.S. 222 (1970).

135. *New York v. Quarles,* 467 U.S. 649, 656 (1984).

136. 470 U.S. 298 (1985).

137. Id. at 306.

138. *Moran v. Burbine,* 475 U.S. 412, 424 (1986).

139. *The Burger Court: The Counter-Revolution That Wasn't* (Blasi ed. 1983).

140. *N.Y. Times,* Feb. 28, 1988, section 4, at 1.

141. *N.Y. Times,* Mar. 3, 1985, Magazine, at 35.

142. Id. at 34, 35.

143. Supra note 123.

144. *Argersinger v. Hamlin,* supra note 124.

145. *Coleman v. Alabama,* 399 U.S. 1 (1970).

146. *Faretta v. California,* 422 U.S. 806 (1975).

147. 351 U.S. 12 (1956).

148. *Ake v. Oklahoma,* 470 U.S. 68 (1985).

149. *Boddie v. Connecticut,* 401 U.S. 371 (1971).

150. Supra Chapter 13.

151. Supra note 66.

152. Supra p. 324.

153. Supra p. 289.

154. *Wygant v. Jackson Board of Education,* 476 U.S. 267 (1986).

155. *The Burger Years: Rights and Wrongs in the Supreme Court 1969–1986,* xx (H. Schwartz ed. 1987).

156. Op. cit. supra note 139, at ix.

157. Ibid.

158. *United States v. Will,* 449 U.S. 200 (1980); *Buckley v. Valeo,* 424 U.S. 1 (1976).

159. *Oregon v. Mitchell,* 400 U.S. 112 (1970).

160. *Northern Pipeline Construction Co. v. Marathon Pipe Line Co.,* 458 U.S. 50 (1982).

161. *Frontiero v. Richardson,* 411 U.S. 677 (1973).

162. The cases are listed in op. cit. supra note 139 at 306 n. 14.

163. *Immigration & Naturalization Service v. Chadha,* 462 U.S. 919 (1983).

164. *Bowsher v. Synar,* 478 U.S. 714 (1986).

165. Op. cit. supra note 139 at 200.

166. 418 U.S. 683 (1974).

167. Op. cit. supra note 139, at 212–213.

168. Supra p. 333.

169. Fortas, in *The Fourteenth Amendment Centennial Volume* 34 (Schwartz ed. 1970).

170. Op. cit. supra note 139, at 198.

171. Id. at 211.

172. *N.Y. Times,* Sept. 25, 1986, at B10.

173. Supra p. 272.

174. Compare op. cit. supra note 139, at 216.

175. Id. at 211.

176. Compare Pound, *Administrative Law* 56 (1942).

177. Op. cit. supra note 139, at 216.

Chapter 15

1. 410 U.S. 113 (1973).

2. Woodward and Armstrong, *The Brethren: Inside the Supreme Court* 238 (1979).

3. *McGovern v. Van Riper,* 43 A.2d 514, 518 (N.J. 1945).

4. Douglas, J., dissenting, in *Public Utilities Commission v. Pollak,* 343 U.S. 451, 467 (1952).

5. 381 U.S. 479 (1965).

6. *Mapp. v. Ohio,* 367 U.S. 643, 656 (1961).

7. In their seminal article, Warren and Brandeis, The Right to Privacy, 4 *Harvard Law Review* 193 (1890).

8. Dissenting, 381 U.S. at 510, n.1.

9. Citing *Pierce v. Society of Sisters,* 281 U.S. 370 (1925).

10. 381 U.S. at 484.

11. *Carey v. Population Services International,* 431 U.S. 678, 687 (1977).

12. *Eisenstadt v. Baird,* 405 U.S. 438, 453 (1972).

13. Re: *Abortion Cases,* Memorandum to the Conference, May 31, 1972.

14. 410 U.S. at 116.

15. The conference notes I used read "getting" but this seems an error.

16. See Woodward and Armstrong, op. cit. supra note 2, at 170.

17. *Washington Post,* July 4, 1972, at 1.

18. For its text, see Schwartz, *The Unpublished Opinions of the Burger Court* 93 (1988).

19. The draft is reprinted in id. at 103.

20. 402 U.S. 62 (1971).

21. For its text, see Schwartz, op. cit. supra note 18, at 120.

22. Reprinted id. at 141.

23. Supra note 20.

24. 408 U.S. 919 (1972).

25. *Poe v. Ullman,* 367 U.S. 497 (1961).

26. *Griswold v. Connecticut,* supra note 5.

27. 410 U.S. at 155.

28. Ibid.

29. Woodward and Armstrong, op. cit. supra note 2, at 233.

30. 410 U.S. at 155.

31. Id. at 162.

32. Id. at 173.

33. Ibid.

34. Ibid.

35. Ibid.

36. Stewart to Lewis F. Powell, Feb. 8, 1973.

37. 198 U.S. 45 (1905), supra Chapter 8.

38. *Ferguson v. Skrupa,* 372 U.S. 726, 731 (1963).

39. Re: No. 76-811 *Regents of the University of California v. Allen Bakke,* Memorandum to the Conference, Nov. 11, 1977.

40. Compare Rehnquist, J., dissenting, in *Weber v. Aetna Casualty & Surety Co.,* 406 U.S. 164, 179 (1972).

41. Particularly in *San Antonio School District v. Rodriguez,* 411 U.S. 1 (1973).

42. *Gitlow v. New York,* 268 U.S. 652 (1925).

43. *Buckley v. Valeo,* 424 U.S. 1, 64 (1976).

44. Kozinski, It Is a Constitution We Are Expounding: A Debate, 1987 *Utah Law Review* 977, 980.

45. Bork, in *The Great Debate: Interpreting Our Written Constitution* 48, 49 (1986).

46. Id. at 11.

47. Dissenting, in *Morrison v. Olson,* 487 U.S. 654, 711 (1988).

48. Bork, *The Tempting of America* 242 (1989).

49. *Lamont v. Postmaster General,* 381 U.S. 301, 308 (1965).

50. 381 U.S. at 491.

51. Id. at 492.

52. Id. at 499.

53. 410 U.S. at 210–211.

54. Ibid.

55. Dissenting, in *Olmstead v. United States,* 277 U.S. 438, 478 (1928).

56. Strunsky, The Invasion of Privacy: The Modern Case of Mistaken Identity, 28 *The American Scholar* 219 (1959).

57. The phrase is from Browning's "Paracelsus."

58. *Pavesich v. New England Life Ins. Co.,* 50 S.E. 68, 70 (Ga. 1905).

59. 410 U.S. at 213 (italics omitted).

60. *McGowan v. Maryland,* 366 U.S. 420 (1961).

61. *Miller v. School District No. 167,* 495 F.2d 658, 664, n.25 (7th Cir. 1974).

62. 381 U.S. at 496.

63. Id. at 496.

64. Ibid.

65. The conference was on *Poe v. Ullman,* 367 U.S. 497 (1961).

66. Id. at 540, 541, 542–543.

67. Ibid.

68. *In re Winship,* 397 U.S. 358 (1970).

69. Id. at 372.

70. *Poe v. Ullman,* 367 U.S. at 542.

71. Id. at 541.

72. *Richmond Newspapers v. Virginia,* 448 U.S. 555, 580 (1980) (opinion per Burger, C.J.).

Chapter 16

1. Compare Judge Learned Hand, in 317 U.S. xi (1942).

2. *N.Y. Times,* Mar. 3, 1985, Magazine, at 100.

3. L. Hand, supra note 1.

4. *N.Y. Times,* supra note 2.

5. 1 Bryce, *The American Commonwealth* 274 (1917).

6. *Newsweek,* July 23, 1979, at 68.

7. *N.Y. Times,* July 12, 1981, § 4, at 22.

8. *N.Y. Times,* Feb. 28, 1988, § 4, at 1.

9. *N.Y. Times,* Mar. 3, 1985, Magazine, at 33.

10. Id. at 31.

11. *Brown v. Board of Education,* 347 U.S. 483 (1954).

12. *Plessy v. Ferguson,* 163 U.S. 537 (1896).

13. The Rehnquist memo is reprinted in Nomination of Justice William Hubbs Rehnquist, Hearings before the Senate Judiciary Committee, 99th Cong., 2d Sess. 324 (1986).

14. *N.Y. Times,* supra note 9, at 32.

15. WHR to WJB, Re: No. 75-1064 *Kremens v. Bartley,* Mar. 8, 1977.

16. *Time,* Oct. 8, 1984, at 28.

17. *Rummell v. Estelle,* 445 U.S. 263 (1980).

18. WHR, Memorandum to the Conference, Feb. 21, 1980.

19. *Time,* loc. cit. supra note 16.

20. *N.Y. Times,* Oct. 8, 1986, at A32.

21. This was, of course, a takeoff on A. E. Housman, "To An Athlete Dying Young."

22. WHR, Memorandum to the Chambers of All Active and Retired Justices, June 12, 1973, Earl Warren Papers, Library of Congress.

23. Savage, *Turning Right: The Making of the Rehnquist Supreme Court* 16 (1992).

24. O'Brien, *Storm Center: The Supreme Court in American Politics* 189 (1986).

25. *N.Y. Times,* Dec. 13, 1987, at 37.

26. *N.Y. Times,* Mar. 3, 1985, Magazine, at 35.

27. Ibid.

28. *N.Y. Times,* supra note 26.

29. *N.Y. Times,* Mar. 8, 1986, at 7.

30. *N.Y. Times,* Sept. 25, 1986, at B10.

31. 2 *Almanac of the Federal Judiciary* 7 (1992).

32. *Planned Parenthood v. Casey,* 112 S.Ct. 2791 (1992).

33. *The Douglas Letters* 146 (Urofsky ed. 1987).

34. Wyzanski, *Whereas—A Judge's Premises: Essays in Judgment, Ethics, and the Law* 61 (1985).

35. Rehnquist, *The Supreme Court: How It Was, How It Is* 261 (1987).

36. *N.Y. Times,* Apr. 21, 1962, at 17.

37. Supra p. 208.

38. Rehnquist, Who Writes Decisions of the Supreme Court? *U.S. News & World Report,* Dec. 13, 1957, at 74.

39. *The Times (London),* July 11, 1986.

40. *N.Y. Times,* Sept. 20, 1987, Book Review, at 40.

41. Stevens, Madison Lecture at New York University Law School, Oct. 27, 1982.

42. Miky, Mike, & Ricki to WJB, Monday, Nov. 23, [19   ].

43. Rehnquist, op. cit. supra note 35, at 263.

44. Op. cit. supra note 33, at 141.

45. Douglas, *The Court Years 1939–1975,* 173 (1980).

46. *Harvard Law School Bulletin,* Winter 1986, p. 28.

47. Rehnquist, op. cit. supra note 35, at 299–300.

48. Id. at 300.

49. *Brown v. Board of Education,* 347 U.S. 483, 494 n.11 (1954).

50. Supra note 46.

51. William O. Douglas, Memorandum to the Conference, Oct. 23, 1961, Hugo Black Papers, Library of Congress.

52. Compare Posner, *The Federal Courts: Crisis and Reform* 106–107 (1985).

53. Compare id. at 111.

54. *A Dialogue about Legal Education as It Approaches the 21st Century* 29 (J. Kelso ed. 1987).

55. *Richmond v. J. A. Croson Co.,* 488 U.S. 469 (1989).

56. *Wards Cove Packing Co. v. Antonio,* 490 U.S. 642 (1989).

57. *McCloskey v. Kemp,* 481 U.S. 279 (1987).

58. *Nollan v. California Coastal Commission,* 483 U.S. 825 (1987).

59. *Planned Parenthood v. Casey,* 112 S.Ct. 2791 (1992).

60. *Gideon v. Wainwright,* 372 U.S. 335 (1963); *Mapp v. Ohio,* 367 U.S. 643 (1961); *Miranda v. Arizona,* 384 U.S. 436 (1967); supra pp. 280–281.

61. *Board of Estimate v. Morris,* 489 U.S. 688 (1989).

62. *N.Y. Times,* May 26, 1991, § 4, at 1.

63. *McCleskey v. Zant,* 111 S.Ct. 1454 (1991).

64. *California v. Acevedo,* 111 S.Ct. 1982 (1991).

65. *Yates v. Evatt,* 111 S.Ct. 1884 (1991).

66. *Rust v. Sullivan,* 111 S.Ct. 1759 (1991).

67. *Employment Division v. Smith,* 494 U.S. 872 (1990).

68. Cardozo, A Ministry of Justice, 35 *Harvard Law Review* 113, 126 (1921).

69. *Olmstead v. United States,* 277 U.S. 438, 478 (1928).

70. *National Treasury Employees Union v. Von Raab,* 489 U.S. 656 (1989).

71. *Texas v. Johnson,* 491 U.S. 397 (1989); *United States v. Eichmann,* 110 S.Ct. 2404 (1990); *R.A.V. v. St. Paul,* 112 S.Ct. 2538 (1992).

72. Faulkner, Privacy—The American Dream: What Happened to It? *Harper's,* July 1955, at 33, 36.

73. Harlan to Felix Frankfurter, Aug. 21, [1963], Frankfurter Papers, Library of Congress.

74. 399 U.S. 1 (1970).

75. *Miranda v. Arizona,* 384 U.S. 436 (1966).

76. *Planned Parenthood v. Casey,* 112 S.Ct. 2791 (1992).

77. Id. at 2809. In addition, the overruling must not lead to "serious inequity to those who have relied upon it or significant damage to the stability of the society governed by the rule." Ibid.

78. 198 U.S. 45 (1905), supra Chapter 8.

79. 163 U.S. 537 (1896), supra p. 188.

80. 112 S.Ct. at 2813–2814.

81. Id. at 2814.

82. Id. at 2866.

83. E.g., *N.Y. Times,* July 5, 1992, § 4, at 1.

84. *Planned Parenthood v. Casey,* 112 S.Ct. at 2854–2855.

Epilogue

1. Quoted in Dicey, *Law of the Constitution* 1 (9th ed. 1939).

2. Yeats, *Explorations* 351 (1962).

3. Hughes, *Addresses of Charles Evans Hughes* 185 (2d ed. 1916).

4. Pound, *Interpretations of Legal History* 1 (1967).

5. Holmes, *The Common Law* 1 (1881).

6. *McCulloch v. Maryland,* 4 Wheat. 316, 407 (U.S. 1819).

Bibliography

Abraham, Henry J. *Freedom and the Court: Civil Rights and Liberties in the U.S.* 5th ed. New York: Oxford University Press, 1988.

Abraham, Henry J. *Justices and Presidents: A Political History of Appointments to the Supreme Court.* 3rd ed. New York: Oxford University Press, 1991.

Baker, Leonard. *Brandeis and Frankfurter: A Dual Biography.* New York: Harper & Row, 1984.

Baker, Leonard. *John Marshall: A Life in Law.* New York: Macmillan, 1974.

Baker, Liva. *Felix Frankfurter.* New York: Coward-McCann, 1969.

Baker, Liva. *The Justice from Beacon Hill: The Life and Times of Oliver Wendell Holmes.* New York: Harper Collins, 1991.

Ball, Howard, and Phillip J. Cooper. *Of Power and Right: Hugo Black, William O. Douglas, and America's Constitutional Revolution.* New York: Oxford University Press, 1992.

Bates, Ernest S. *The Story of the Supreme Court.* South Hackensack: Fred B. Rothman & Co., 1982.

Baum, Lawrence. *The Supreme Court.* 3rd ed. Washington: Congressional Quarterly, 1988.

Belden, Thomas Graham, and Marva Robins Belden. *So Fell the Angels.* Boston: Little, Brown, 1956.

Berry, Mary. *Stability, Security and Continuity: Mr. Justice Burton and Decision-Making in the Supreme Court 1945–1958.* Westport: Greenwood Press, 1978.

Beth, Loren P. *John Marshall Harlan: The Last Whig Justice.* Lexington: University Press of Kentucky, 1992.

Beveridge, Albert Jeremiah. *The Life of John Marshall.* 4 vols. Boston: Houghton Mifflin, 1916–1919.

Bickel, Alexander M. *Politics and the Warren Court*. New York: Harper & Row, 1965.

Bickel, Alexander M. *The Least Dangerous Branch: The Supreme Court at the Bar of Politics*. Indianapolis: Bobbs-Merrill, 1962, 1986.

Bickel, Alexander M. *The Supreme Court and the Idea of Progress*. New York: Harper & Row, 1970.

Bickel, Alexander M. (ed.). *The Unpublished Opinions of Mr. Justice Brandeis: The Supreme Court at Work*. Chicago: University of Chicago Press, 1967.

Biddle, Francis. *Mr. Justice Holmes*. New York: Scribner, 1942.

Black, Hugo, Jr. *My Father: A Remembrance*. New York: Random House, 1975.

Blasi, Vincent (ed.). *The Burger Court: The Counter-Revolution That Wasn't*. New Haven: Yale University Press, 1986.

Blue, Frederick J. *Salmon P. Chase: A Life in Politics*. Kent: Kent State University Press, 1987.

Boles, Donald E. *Mr. Justice Rehnquist, Judicial Activist: The Early Years*. Ames: Iowa State University Press, 1987.

Boudin, Louis Boudianoff. *Government by Judiciary*. New York: W. Godwin, 1932.

Bowen, Catherine Drinker. *Yankee from Olympus: Justice Holmes and His Family*. Boston: Little, Brown, 1944.

Brown, William Garrott. *The Life of Oliver Ellsworth*. New York: Da Capo Press, 1970.

Bryce, James. *The American Commonwealth*. 2 vols. London and New York: Macmillan, 1889, 1917.

Cahn, Edmond (ed.). *Supreme Court and Supreme Law*. Bloomington: Indiana University Press, 1954.

Cardozo, Benjamin N. *The Nature of the Judicial Process*. New Haven: Yale University Press, 1921.

Cate, Wirt Armistead. *Lucius Q. C. Lamar, Secession and Reunion*. Chapel Hill: University of North Carolina Press, 1935.

Choper, Jesse H. *Judicial Review and the National Political Process: A Functional Reconsideration of the Role of the Supreme Court*. Chicago: University of Chicago Press, 1980.

Christman, Henry (ed.). *The Public Papers of Chief Justice Earl Warren*. New York: Simon & Schuster, 1959.

Clayton, James M. *The Making of Justice: The Supreme Court in Action*. New York: Dutton, 1964.

Clinton, Robert L. *Marbury v. Madison and Judicial Review*. Lawrence: University Press of Kansas, 1989.

Corwin, Edward Samuel. *Court over Constitution: A Study of Judicial Review as an Instrument of Popular Government*. Princeton: Princeton University Press, 1938.

Corwin, Edward Samuel. *John Marshall and the Constitution: A Chronicle of the Supreme Court*. New Haven: Yale University Press, 1919.

Corwin, Edward Samuel. *The Twilight of the Supreme Court: A History of Our Constitutional Theory*. New Haven: Yale University Press, 1934.

Cox, Archibald. *The Court and the Constitution*. Boston: Houghton Mifflin, 1988.

Cox, Archibald. *The Role of the Supreme Court in American Government*. New York: Oxford University Press, 1976.

Currie, David P. *The Constitution in the Supreme Court: The First Hundred Years, 1789–1888*. Chicago: University of Chicago Press, 1986.

Currie, David P. *The Constitution in the Supreme Court: The Second Century, 1888–1986*. Chicago: University of Chicago Press, 1990.

Curtis, Benjamin R., Jr. (ed.). *A Memoir of Benjamin Robbins Curtis*. New York: Da Capo Press, 1970.

Davis, Sue. *Justice Rehnquist and the Constitution*. Princeton: Princeton University Press, 1989.

Documentary History of the Supreme Court of the United States, 1789–1800. 4 vols. New York: Columbia University Press, 1985–1992.

Douglas, William O. *Go East Young Man: The Early Years*. New York: Random House, 1974.

Douglas, William O. *The Court Years 1939–1975*. New York: Random House, 1980.

Dunham, Allison, and Philip B. Kurland (eds.). *Mr. Justice*. Chicago: University of Chicago Press, 1956.

Dunne, Gerald T. *Hugo Black and the Judicial Revolution*. New York: Simon & Schuster, 1977.

Dunne, Gerald T. *Justice Joseph Story and the Rise of the Supreme Court*. New York: Simon & Schuster, 1970.

Elsmere, Jane Shaffer. *Justice Samuel Chase*. Muncie: Janevar Publ. Co., 1980.

Estreicher, Samuel, and John Sexton. *Redefining the Supreme Court's Role: A Theory of Managing the Federal Judicial Process*. New Haven: Yale University Press, 1987.

Fairman, Charles. *Mr. Justice Miller and the Supreme Court, 1862–1890*. Cambridge: Harvard University Press, 1939.

Fine, Sidney. *Frank Murphy*. 3 vols. Ann Arbor: University of Michigan Press; Chicago: University of Chicago Press, 1975–1984.

Flanders, Henry. *The Lives and Times of the Chief Justices of the Supreme Court of the United States*. Buffalo: William S. Hein, [Reprint of the 1881 edition] n.d.

Frank, John Paul. *Justice Daniel Dissenting: A Biography of Peter V. Daniel, 1784–1860*. Cambridge: Harvard University Press, 1964.

Frank, John Paul. *Marble Palace: The Supreme Court in American Life*. New York: Knopf, 1958.

Frank, John Paul. *Mr. Justice Black: The Man and His Opinions*. New York: Knopf, 1949.

Frank, John Paul. *The Warren Court*. New York: Macmillan, 1964.

Frankfurter, Felix. *Felix Frankfurter Reminisces*. New York: Doubleday, 1962.

Frankfurter, Felix. *Mr. Justice Holmes and the Constitution*. Cambridge: Dunster House Bookshop, 1927.

Frankfurter, Felix. *Mr. Justice Holmes and the Supreme Court*. 2d ed. Cambridge: Harvard University Press, 1961.

Frankfurter, Felix. *Of Law and Men: Papers and Addresses of Felix Frankfurter*. New York: Harcourt Brace, 1956.

Frankfurter, Felix. *The Commerce Clause under Marshall, Taney and Waite*. Chapel Hill: University of North Carolina Press, 1937, 1960.

Frankfurter, Felix, and James M. Landis. *The Business of the Supreme Court: A Study in the Federal Judicial System*. New York: Macmillan, 1927.

Freund, Paul A. *On Understanding the Supreme Court*. Boston: Little, Brown, 1949; New York: Greenwood, 1977.

Friedman, Leon (ed.). *The Justices of the United States Supreme Court: Their Lives and Major Opinions.* Vol. 5, *The Burger Court, 1969–1978.* New York: Chelsea House, 1979.

Friedman, Leon, and Fred Israel (eds.). *The Justices of the United States Supreme Court 1789–1969: Their Lives and Major Opinions.* 4 vols. New York: Chelsea House, 1969.

Gerhart, Eugene C. *America's Advocate: Robert H. Jackson.* Indianapolis: Bobbs-Merrill, 1958.

Goldman, Roger. *Thurgood Marshall: Justice for All.* New York: Carroll & Graf, 1992.

Goldstein, Joseph. *The Intelligible Constitution: The Supreme Court's Obligation to Maintain the Constitution as Something We the People Can Understand.* New York: Oxford University Press, 1992.

Haines, Charles Grove. *The Role of the Supreme Court in American Government and Politics, 1789–1835.* New York: Russell & Russell, 1960.

Harper, Fowler Vincent. *Justice Rutledge and the Bright Constellation.* Indianapolis: Bobbs-Merrill, 1965.

Hart, Albert Busnell. *Salmon Portland Chase.* Boston; New York: Houghton Mifflin, 1899.

Hellman, George Sidney. *Benjamin N. Cardozo, American Judge.* New York: McGraw-Hill, 1940.

Hendel, Samuel. *Charles Evans Hughes and the Supreme Court.* New York: King's Crown Press, 1951.

Hirsch, H. N. *The Enigma of Felix Frankfurter.* New York: Basic Books, 1981.

Howard, J. Woodford. *Mr. Justice Murphy: A Political Biography.* Princeton: Princeton University Press, 1968.

Howe, Mark De Wolfe. *Justice Oliver Wendell Holmes.* 2 vols. Cambridge: Harvard University Press, 1957.

Hughes, Charles E. *The Supreme Court of the United States.* New York: Columbia University Press, 1936, 1966.

Huston, Luther. *Pathway to Judgment: A Study of Earl Warren.* Philadelphia: Chilton Books, 1966.

Jackson, Robert Houghwout. *The Struggle for Judicial Supremacy: A Study of a Crisis in American Power Politics.* New York: Knopf, 1941.

Jackson, Robert Houghwout. *The Supreme Court in the American System of Government.* Cambridge: Harvard University Press, 1955.

Kalman, Laura. *Abe Fortas: A Biography.* New Haven: Yale University Press, 1990.

Katcher, Leo. *Earl Warren: A Political Biography.* New York: McGraw-Hill, 1967.

King, Willard Leroy. *Lincoln's Manager, David Davis.* Cambridge: Harvard University Press, 1960.

King, Willard Leroy. *Melville Weston Fuller, Chief Justice of the United States, 1888–1910.* New York: Macmillan, 1950.

Kluger, Richard. *Simple Justice: The History of Brown v. Board of Education and Black America's Struggle for Equality.* New York: Knopf, 1975.

Konefsky, Samuel Joseph. *Chief Justice Stone and the Supreme Court.* New York: Macmillan, 1945.

Konefsky, Samuel Joseph. *The Legacy of Holmes and Brandeis: A Study in the Influence of Ideas.* New York: Macmillan, 1956.

Kurland, Philip B. *Mr. Justice Frankfurter and the Constitution.* Chicago: University of Chicago Press, 1971.

Kurland, Philip B. (ed.). *Felix Frankfurter on the Supreme Court: Extrajudicial*

Essays on the Court and the Constitution. Cambridge: Harvard University Press, 1970.

Kurland, Phillip B., and Gerhard Casper (eds.). *Landmark Briefs and Arguments of the Supreme Court of the United States: Constitutional Law.* Arlington: University Publishers of America, 1977–19[ ].

Lamb, Charles M., and Stephen C. Halpern (eds.). *Burger Court: Political and Judicial Profiles.* Champaign: University of Illinois Press, 1990.

Latham, Frank Brown. *The Great Dissenter: John Marshall Harlan, 1833–1911.* New York: Cowles Book Co., 1970.

Lawrence, Alexander A. *James Moore Wayne: Southern Unionist.* Chapel Hill: University of North Carolina Press, 1943; Westport: Greenwood Press, 1970.

Levy, Leonard W. (ed.). *The Supreme Court under Earl Warren.* New York: Quadrangle Books, 1972.

Lewis, Anthony. *Gideon's Trumpet.* New York: Random House, 1964.

Lewis, Anthony. *Portrait of a Decade: The Second American Revolution.* New York: Random House, 1964.

Lewis, Walker. *Without Fear or Favor: A Biography of Chief Justice Roger Brooke Taney.* Boston: Houghton Mifflin, 1965.

MacKenzie, John P. *The Appearance of Justice.* New York: Scribner's, 1974.

Magrath, C. Peter. *Morrison R. Waite: The Triumph of Character.* New York: Macmillan, 1963.

Mason, Alpheus Thomas. *Harlan Fiske Stone: Pillar of the Law.* New York: Viking Press, 1956.

Mason, Alpheus Thomas. *The Supreme Court from Taft to Burger.* Baton Rouge: Louisiana State University Press, 1979.

Mason, Alpheus Thomas. *William Howard Taft: Chief Justice.* New York: Simon & Schuster, 1965.

Mason, Alpheus Thomas, and William M. Beaney. *The Supreme Court in a Free Society.* Englewood Cliffs: Prentice-Hall, 1959.

McClellan, James. *Joseph Story and the American Constitution.* Norman: University of Oklahoma Press, 1971.

McCloskey, Robert Green. *The Modern Supreme Court.* Cambridge: Harvard University Press, 1972.

McDevitt, Matthew. *Joseph McKenna.* New York: Da Capo Press, 1974.

McLean, Joseph Erigina. *William Rufus Day, Supreme Court Justice from Ohio.* Baltimore: Johns Hopkins Press, 1946.

Mendelson, Wallace. *Justices Black and Frankfurter: Conflict in the Court.* Chicago: University of Chicago Press, 1961.

Mendelson, Wallace. *The Supreme Court: Law and Discretion.* Indianapolis: Bobbs-Merrill, 1967.

Morgan, Donald Grant. *Justice William Johnson: The First Dissenter.* Columbia: University of South Carolina Press, 1954.

Morris, Richard Brandon. *John Jay, the Nation and the Court.* Boston: Boston University Press, 1967.

Murphy, Bruce Allen. *Fortas: The Rise and Ruin of a Supreme Court Justice.* New York: Morrow, 1988.

Murphy, Bruce Allen. *The Brandeis/Frankfurter Connection.* New York: Oxford University Press, 1982.

Murphy, James B. *L.Q.C. Lamar: Pragmatic Patriot.* Baton Rouge: Louisiana State University Press, 1973.

Newmyer, R. Kent. *Supreme Court Justice Joseph Story: Statesman of the Old Republic*. Chapel Hill: University of North Carolina Press, 1985.

Novick, Sheldon M. *Honorable Justice: The Life of Oliver Wendell Holmes*. Boston: Little, Brown, 1989.

O'Brien, David M. *Storm Center: The Supreme Court in American Politics*. 2d ed. New York: W. W. Norton, 1990.

Palmer, Jan. *The Vinson Court Era: The Supreme Court's Conference Votes, Data and Analysis*. New York: AMS Press, 1990.

Paper, Lewis J. *Brandeis*. Englewood Cliffs: Prentice-Hall, 1983.

Parrish, Michael E. *Felix Frankfurter and His Times: The Reform Years*. New York: Free Press, 1982.

Paschal, Joel Francis. *Mr. Justice Sutherland: A Man Against the State*. Princeton: Princeton University Press, 1951.

Pellew, George. *John Jay*. Boston; New York: Houghton Mifflin, 1890.

Pohlman, H. L. *Justice Oliver Wendell Holmes, Free Speech and the Living Constitution*. New York: New York University Press, 1991.

Pollack, Jack Harrison. *Earl Warren: The Judge Who Changed America*. Englewood Cliffs: Prentice-Hall, 1979.

Pollard, Joseph Percival. *Mr. Justice Cardozo: A Liberal Mind in Action*. New York: Yorktown Press, 1935.

Posner, Richard A. *Cardozo: A Study in Reputation*. Chicago: University of Chicago Press, 1990.

Posner, Richard A. (ed.). *The Essential Holmes: Selections from the Letters, Speeches, Judicial Opinions and Other Writings of Oliver Wendell Holmes, Jr*. Chicago: University of Chicago Press, 1992.

Pringle, Henry F. *The Life and Times of William Howard Taft: A Biography*. 2 vols. New York; Toronto: Farrar & Rinehart, 1939, 1964.

Pritchett, Charles Herman. *The Roosevelt Court: A Study in Judicial Politics and Values, 1937–1947*. New York: Macmillan, 1948.

Pusey, Merlo J. *Charles Evans Hughes*. 2 vols. New York: Macmillan, 1951, 1963.

Rehnquist, William H. *The Supreme Court: How It Was, How It Is*. New York: Morrow, 1987.

Rodell, Fred. *Nine Men: A Political History of the Supreme Court from 1790 to 1955*. New York: Random House, 1955.

Rostow, Eugene V. *The Sovereign Prerogative: The Supreme Court and the Quest for Law*. New Haven: Yale University Press, 1962; New York: Greenwood, 1974.

Savage, David G. *Turning Right: The Making of the Rehnquist Supreme Court*. New York: Wiley, 1992.

Schubert, Blendon A. *Constitutional Politics: The Behavior of the Supreme Court Justices and the Constitutional Policies That They Make*. New York: Holt, Rinehart and Winston, 1960.

Schubert, Glendon A. *The Judicial Mind: The Attitudes and Ideologies of the Supreme Court Justices, 1946–1963*. Evanston: Northwestern University Press, 1965.

Schwartz, Bernard. *Behind Bakke: Affirmative Action and the Supreme Court*. New York: New York University Press, 1988.

Schwartz, Bernard. *Super Chief: Earl Warren and His Supreme Court—A Judicial Biography*. New York: New York University Press, 1983.

Schwartz, Bernard. *The Ascent of Pragmatism: The Burger Court in Action*. Reading: Addison-Wesley, 1990.

Schwartz, Bernard. *The Law in America: A History*. New York: McGraw-Hill, 1974.

Schwartz, Bernard. *The Reins of Power: A Constitutional History of the United States*. New York: Hill and Wang, 1963.

Schwartz, Bernard. *The Supreme Court: Constitutional Revolution in Retrospect*. New York: Ronald Press, 1957.

Schwartz, Bernard. *The Unpublished Opinions of the Burger Court*. New York: Oxford University Press, 1988.

Schwartz, Bernard. *The Unpublished Opinions of the Warren Court*. New York: Oxford University Press, 1985.

Shogan, Robert. *A Question of Judgment: The Fortas Case and the Struggle for the Supreme Court*. Indianapolis: Bobbs-Merrill, 1972.

Sickels, Robert J. *John Paul Stevens and the Constitution: The Search and the Balance*. University Park: Pennsylvania State University Press, 1988.

Siegan, Bernard H. *The Supreme Court's Constitution*. New Brunswick: Transaction, 1987.

Simon, James F. *Independent Journey: The Life of William O. Douglas*. New York: Harper & Row, 1980.

Simon, James F. *The Antagonists: Hugo Black, Felix Frankfurter and Civil Liberties in Modern America*. New York: Simon & Schuster, 1989.

Smith, Charles William. *Roger B. Taney: Jacksonian Jurist*. Chapel Hill: University of North Carolina Press, 1936.

Smith, Page. *James Wilson, Founding Father, 1742–1798*. Chapel Hill: University of North Carolina Press, 1956.

Story, William Wetmore. *Life and Letters of Joseph Story, Associate Justice of the Supreme Court of the United States, and Dane Professor of Law at Harvard University*. 2 vols. Boston: Little and J. Brown, 1851.

Streamer, Robert J. *Chief Justice: Leadership and the Supreme Court*. Columbia: University of South Carolina Press, 1986.

Strum, Phillippa. *Louis D. Brandeis: Justice for the People*. Cambridge: Harvard University Press, 1984.

Swindler, William Finley. *Court and Constitution in the Twentieth Century*. 2 vols. Indianapolis: Bobbs-Merrill, 1969, 1974.

Swisher, Carl Brent. *Roger B. Taney*. New York: Macmillan, 1935.

Swisher, Carl Brent. *Stephen J. Field: Craftsman of the Law*. Hamden: Archon Books, 1963.

Swisher, Carl Brent. *The Supreme Court in Modern Role*. Rev. ed. New York: New York University Press, 1965.

Thomas, Helen Shirley. *Felix Frankfurter: Scholar on the Bench*. Baltimore: Johns Hopkins Press, 1960.

Tocqueville, Alexis de. *Democracy in America*. 2 vols. New York: Vintage Books, 1954.

Tribe, Laurence H. *God Save This Honorable Court: How the Choice of the Supreme Court Justices Shapes Our History*. New York: Random House, 1985.

Twiss, Benjamin. *Lawyers and the Constitution: How Laissez Faire Came to the Supreme Court*. Princeton: Princeton University Press, 1942; New York: Greenwood, 1974.

Umbreit, Kenneth Bernard. *Our Eleven Chief Justices*. New York: Harper & Brothers, 1938.

Urofsky, Melvin I., and Philip E. Urofsky (eds.). *The Douglas Letters: Selections from the Private Papers of Justice William O. Douglas*. Bethesda: Adler & Adler, 1987.

Warner, Hoyt Landon. *The Life of Mr. Justice Clarke: A Testament to the Power of Liberal Dissent in America*. Cleveland: Case Western Reserve University Press, 1959.

Warren, Charles. *The Supreme Court in United States History*. 3 vols. Boston: Little, Brown, 1922, 1924.

Warren, Earl. *The Memoirs of Earl Warren*. New York: Doubleday, 1977.

Wasby, Stephen L. (ed.). *He Shall Not Pass This Way Again: The Legacy of Justice William O. Douglas*. Pittsburgh: University of Pittsburgh Press, 1990.

Weaver, John D. *Warren: The Man, The Court, The Era*. Boston: Little, Brown, 1967.

White, G. Edmund. *Earl Warren: A Public Life*. New York: Oxford University Press, 1982.

Wiecek, William M. *Liberty under Law: The Supreme Court in American Life*. Baltimore: Johns Hopkins University Press, 1988.

Witt, Elder. *Congressional Quarterly's Guide to the U.S. Supreme Court*. 2d ed. Washington: Congressional Quarterly, 1990.

Woodward, Bob, and Scott Armstrong. *The Brethren: Inside the Supreme Court*. New York: Simon & Schuster, 1979.

Wright, Benjamin Fletcher. *The Growth of American Constitutional Law*. New York: H. Holt, 1942.

Oliver Wendell Holmes Devise

History of the Supreme Court of the United States. New York: Macmillan, 1971–1988.

Volume I: *Antecedents and Beginnings to 1801,* by Julius Goebel, Jr. (1971).

Volume II: *Foundations of Power: John Marshall, 1801–15,* by George L. Haskins and Herbert A. Johnson (1981).

Volumes III–IV: *The Marshall Court and Cultural Change, 1815–35,* by G. Edward White (1988).

Volume V: *The Taney Period, 1836–64,* by Carl B. Swisher (1974).

Volume VI: *Reconstruction and Reunion, 1864–88, Part One,* by Charles Fairman (1971).

Volume VII: *Reconstruction and Reunion, 1864–88, Part Two,* by Charles Fairman (1987).

Supplement to Volume VII: *Five Justices and the Electoral Commission of 1877,* by Charles Fairman (1988).

Volume IX: *The Judiciary and Responsible Government, 1910–21,* by Alexander M. Bickel and Benno C. Schmidt, Jr. (1984).

List of Cases

Index